VITAMINS AND HORMONES

VOLUME 49

VITAMINS AND HORMONES
ADVANCES IN RESEARCH AND APPLICATIONS

STEROIDS

Editor-in-Chief

GERALD LITWACK

Department of Pharmacology
Jefferson Cancer Institute
Thomas Jefferson University Medical College
Philadelphia, Pennsylvania

Volume 49

ACADEMIC PRESS

San Diego   New York   Boston
London   Sydney   Tokyo   Toronto

Academic Press, Inc.
A Division of Harcourt Brace & Company
525 B Street, Suite 1900, San Diego, California 92101-4495

United Kingdom Edition published by
Academic Press Limited
24-28 Oval Road, London NW1 7DX

International Standard Serial Number: 0083-6729

International Standard Book Number: 0-12-709849-6

PRINTED IN THE UNITED STATES OF AMERICA
94   95   96   97   98   99   BB   9   8   7   6   5   4   3   2   1

Contents

The Steroid/Nuclear Receptors: From Three-Dimensional Structure to Complex Function

BEN F. LUISI, JOHN W. R. SCHWABE, AND LEONARD P. FREEDMAN

Function/Activity of Specific Amino Acids in Glucocorticoid Receptors

S. STONEY SIMONS, JR.

Genetic Diseases of Steroid Metabolism

PERRIN C. WHITE

Structure, Function, and Regulation of Androgen-Binding Protein/Sex Hormone-Binding Globulin

DAVID R. JOSEPH

Molecular Biology of Vitamin D Action

TROY K. ROSS, HISHAM M. DARWISH, AND HECTOR F. DELUCA

Nuclear Retinoid Receptors and Their Mechanism of Action

MAGNUS PFAHL, RAINER APFEL, IGOR BENDIK, ANDREA FANJUL, GERHART GRAUPNER, MI-OCK LEE, NATHALIE LA-VISTA, XIAN-PING LU, JAVIER PIEDRAFITA, MARIA ANTONIA ORTIZ, GILLES SALBERT, AND XIAO-KUN ZHANG

Molecular Mechanisms of Androgen Action

Jonathan Lindzey, M. Vijay Kumar, Mike Grossman, Charles Young, and Donald J. Tindall

Role of Androgens in Prostatic Cancer

John T. Isaacs

Preface

The publisher has welcomed the opportunity to bring forth a thematic volume as part of the serial *Vitamins and Hormones*. My suggestion was to emphasize the areas where the most dramatic advances have been made. Consequently, I chose to review the subject of steroid hormones where there recently have been large gains in our understanding of the mechanism by which DNA binding domains of receptors contact DNA and also of the activities displayed by specific amino acids in receptor sequences. Large steps toward greater understanding have been taken regarding the mode of action of steroid hormones and in studying steroid hormone-related diseases of genetic origin. More recently, there has been increased emphasis on cancer and the need for translational research. Consequently, this volume is entitled "Steroids." The reviews in this volume focus on these and related areas and provide up-to-date information both for reference and possibly as a basis for formal course presentation.

The volume is arranged so that the first chapters deal with advances in the structure–function of steroid receptors. This area is represented by contributions from Ben Luisi, John Schwabe, and Leonard Freedman and then from S. Stoney Simons. Next, Perrin White covers the topic of genetic diseases of steroid metabolism and David Joseph approaches the subject of sex hormone binding proteins. This is followed by contributions detailing the mechanisms of actions of various hormones. Hector DeLuca's laboratory describes the molecular biology of vitamin D action, Magnus Pfahl's laboratory describes nuclear retinoid receptors and their mechanism of action, and Donald Tindall's laboratory defines the molecular mechanisms of androgen action. Finally, John Isaacs discourses on the role of androgens in prostatic cancer.

This volume is filled with details of modern research in this field in general and also provides a much-needed update on the subject of steroids and their actions.

As always, the publisher continues to be cooperative and indulgent. The revamped Editorial Board will be relied on heavily for ideas for future publications and especially for other thematic volumes.

GERALD LITWACK

The Steroid/Nuclear Receptors: From Three-Dimensional Structure to Complex Function

BEN F. LUISI,* JOHN W. R. SCHWABE,† AND LEONARD P. FREEDMAN‡

*Medical Research Council, Virology Unit
Glasgow G11 5JR, United Kingdom

†Medical Research Council
Laboratory of Molecular Biology
Cambridge CB2 2QH, United Kingdom

‡Cell Biology and Genetics
Memorial Sloan–Kettering Cancer Center
New York, New York 10021

I. INTRODUCTION

Since their discovery, lipophilic hormones such as the steroids and thyroids have provided a time-honored paradigm for the mechanism of long-distance intercellular communication. The diffusable hormone, representing the "signal," originates in one tissue but subsequently affects the growth or activity of target cells in a second tissue at some distance. In contrast to these hormones, the chemically related vitamins have traditionally been thought to play more static roles, being metabolites or ingested factors required throughout the body for routine metabolism—or so it seemed. Despite these descriptive differences, it has emerged that both groups of chemical messengers exert their physiological effects by activating specialized receptor proteins. In contrast to the membrane-bound receptors, exemplified by the growth hormone receptor, the vitamin and hormone receptors are found within either the cell cytoplasm or nucleus. The thyroid and vitamin receptors appear to be primarily nuclear, whereas the steroid receptors are cytoplasmic, but on binding the hormone translocate to the nucleus. Once there, the steroid receptors, like the ligand-activated thyroid or vitamin receptors, bind avidly to specific DNA sequences and modulate the expression of target genes. In some cases, the ligand-induced activation of the receptor brings about an enhancement of transcription; in other cases, repression results. To date, some 30 distinct receptors have been identified from organisms as diverse as arthropods, such as the fruit fly *Drosophila,* to mammals. This group includes a number of "orphan receptors" whose ligands and functions are presently not known. All of these proteins are clearly related functionally and may be classified into a collective group, which we refer to as the steroid/nuclear receptor family.

The receptor proteins are evolutionarily related, as evidenced by their high sequence homology (Evans, 1989). They share a characteristic modular organization whereby separable functions are encoded by discrete functional domains (Fig. 1) that encompass ligand-binding, DNA-binding, nuclear localization, and transcriptional modulation. The most strongly conserved region is the DNA-binding domain, a segment of approximately 90 residues (Fig. 2). Molecular genetic analyses of a number of receptors show that, despite the small size, this domain is sufficient to direct the recognition of DNA targets. A compilation of sequences from the DNA-binding domains of representative receptors is shown in Fig. 2B. This region behaves as a true domain in that it folds stably in isolation and retains full DNA-binding activity. The folded structure is stabilized by two zinc ions, each of which is coordinated by four cysteines (Freedman *et al.,* 1988a). The spacing of the zinc ligands is reminiscent of the zinc-finger motif (Rhodes and Klug, 1993), but three-dimensional structural analyses have shown that these metal-binding motifs are structurally distinct (Hard *et al.,* 1990a; Schwabe *et al.,* 1990, 1993a; Luisi *et al.,* 1991; Knegtel *et al.,* 1993; Lee *et al.,* 1993).

In all steroid/nuclear receptors, the DNA-binding domain is followed by a region of extensive sequence variability, which in turn is followed by the comparatively well conserved ligand-binding domain. The transcriptional activation function maps to the ligand-binding domain. The amino termini of the receptors, preceding the DNA-binding domain, are hypervariable in length and sequence composition. In the steroid receptors, this region encodes a second transcriptional activation domain. Their sequence homology and phylogenetic diversity suggest that the receptors diverged some 500 million years ago from a primordial ancestor, and have subsequently evolved to serve many

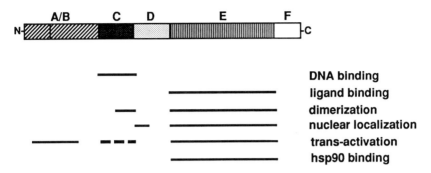

FIG. 1. The functional domains of the steroid/nuclear receptors. The capitalized letters indicate a convention for identification of the various domains.

important regulatory roles in most multicellular eukaryotes (Amero *et al.*, 1992; Laudet *et al.*, 1992).

The steroid/nuclear receptors are involved in both the control of development and the maintenance of homeostasis. For example, specialized nuclear receptors regulate *Drosophila* development in response to ecdysone and related molting hormones (Seagraves, 1991). The *Drosophila* FTZ-F1 protein (which also has a mouse homologue) plays an active role in early embryogenesis (Tsukiyama *et al.*, 1992). The glucocorticoid and mineralocorticoid receptors regulate the homeostatic processes of gluconeogenesis, whereas the sexual steroid receptors affect sex gland development and activity in all vertebrates. The regulatory roles of vitamins is a topic of expanding interest. The retinoids, a group of vitamin A derivatives, have been found to play central roles in development, growth, reproduction, vision, and general homeostasis of numerous tissues (Lohnes *et al.*, 1992a,b; Lufkin *et al.*, 1993). Vitamin D_3 mediates control of intestinal calcium and phosphorus absorption, bone remodeling, and conservation of minerals in the kidney, but also appears to play an important role in regulating cells of immune system origin (Reichel *et al.*, 1992).

In light of their key control of cellular functions, it is unsurprising that mutant forms of several receptors have been implicated in oncogenic processes. The oncogenic *v-erbA* gene product is related to the thyroid hormone receptor. A truncated version of the estrogen receptor (missing the ligand-binding domain and therefore constitutively active) has been implicated in certain breast cancers. The fusion of a putative transcription factor called PML and the retinoic acid receptor is associated with acute promyelocytic leukemias (de-Thé *et al.*, 1991;

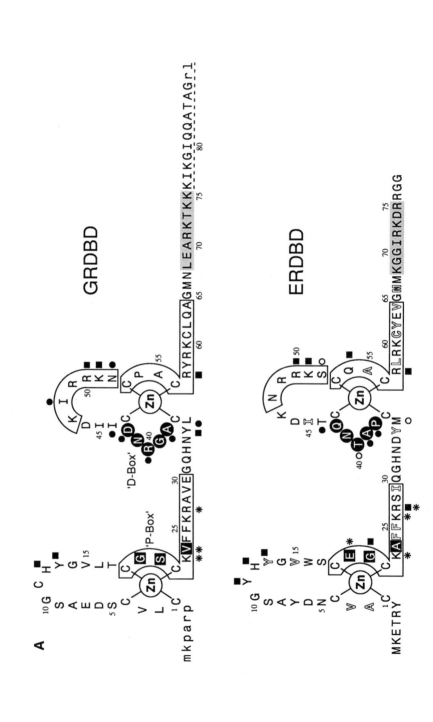

B

```
Steroid subgroup:
                                          +    *  *  *         +   *  *  *
Human glucocorticoid          CLVCSDEASGCHYGVLTCGSCKVFFKRAVEGQHN--Y-LCAGRNDCIIDKIRRKNCPACRYRKCLQAGMNLEARKTKKKIKGIQQATTG
Human androgen                CLICGDEASGCHYGALTCGSCKVFFKRAAEGKQK--Y-LCASRNDCTIDKFRRKNCPSCRLRKCYEAGMTLGARKLKKLGNLKLQEEGE
Human mineralocorticoid       CLVCGDEASGCHYGVVTCGSCKVFFKRAVEGQHN--Y-LCAGRNDCIIDKIRRKNCPACRLQKCLQAGMNLGARKSKKLGKLKGIHEEQ
Human estrogen                CAVCNDYASGYHYGVWSCEGCKAFFKRSIQGHND--Y-MCPATNQCTIDKNRRKSCQACRLRKCYEVGMMKGGIRKDRRGGRMLKHKRQ

Nuclear subgroup:
                                                   *         **      *
Human vitamin D3              CGVCGDRATGEHFNAMTCEGCKGFFRRSMKRKA-L-F-TCPENGDCRITKDNRRHCQACRLKRCVDIGMKKEEILTDEEVQRKREMILK
Human retinoic acid α         CFVCQDKSSGYHYGVSACEGCKGFFRRSIQKNM-V-Y-TCHRDKMCIINKVTRNRCQYCRLQKCFEVGMSKESVRNDRNKKKKDVPKPE
Chicken COUP-TF               CVVCGDKSSGKHYGQFTCEGCKSFFKRSVRRNL--TY-TCRANRMCPIDQHHRNQCQYCRLKKCLKVGMRREAVQRGRMPPTQPNPGQY
Human thyroid hormone α       CVVCGDKATGYHYRCITCEGCKGFFRRTIQKNLHPTY-SCKYDSCCVIDKITRNCQLCRFKKCIAVGMAMDLVLDDSKRVAKRLIEQ
Dros. ultraspiracle (USP 77)  CSICGDRASGKHYGVYSCEGCKGFFKRTVRKDL--TY-ACRENRMCIIDKRQRNRCQYCRYQKCLTGMKREAVQEERQRGARNAAGRL
Mouse NGFI-B (nur 77)         CAVCGDNASCQHYGVRTCEGCKGFFKRTVQKSAK--Y-ICLANKDCPVDKRRRNRCQFCRFQKCLAVGMVKEVVRTDSLKGRRGRLPSK
C. elegans nuclear receptor-1 (CEB-1)  CAVCNDRAVCLHYGARTCEGCKGFFKRTVQKNSK--Y-TCAGNKTCPIDKRYRSRCQYCRYQKCLEVGMKEIVRHGSLSGRRGRLSSK
Xen. peroxisome proliferator-activated-β  CKICGDRASGFHYGVHACECGKGFFRRTIRMRLQ--YEHC--DRNCKIQKKNRNKCQYCRFNKCLSLGMSHNAIRFGRMPESEKRKIVQ
Mouse peroxisome proliferator-activated  CRICGDKASGYHYGVHACEGCKGFFRRTIRLKL--V-YDKC--DRSCKIQKKNRNKCQYCRFHKCLSVGMSHNAIRFGRMPRSEKAKLKA
Mouse RXR-β                   CAICGDRSSGKHYGVYSCEGCKGFFKRTIRKDL--TY-SCRDNKDCTVDKRQRNRCQYCRYQKCLATGMKREAVQEERQRGKDKDGDGD
                             <       Module 1       >  <        Module 2       >
```

FIG. 2. (A) Primary and secondary structure of the GR and ER DNA-binding domains as determined by X-ray crystallography (Luisi et al., 1991; Schwabe et al., 1993a,b). The helical elements are boxed. Outlined residues in the ERdbd participate in the hydrophobic core (see Fig. 5). Asterisks indicate residues that interact with base pairs (only for the specific monomer in the GRDBD), solid boxes those making phosphate backbone contacts, and solid circles those participating in dimerization. The shaded region in both the ER and GR DNA-binding domains corresponds to a helical element so far only observed in the RXRdbd. Residues that form the "P-box" and "D-box" (see text) are highlighted by square and circles, respectively. (B) Sequence compilation of the DNA-binding domain of representative steroid/nuclear receptors. Metal coordinating residues are in bold and elements of the D-box are italicized. Asterisks at the top of the alignment indicate residues of the conserved hydrophobic core, and the plus sign indicates a conserved Asp and Arg, which may always form a stabilizing salt bridge. Insertions have been introduced to optimally align sequences based on knowledge of tertiary structure of estrogen, glucocorticoid, retinoic acid, and RXR receptor DNA-binding domains.

Kastner *et al.*, 1992, Kakizuka *et al.*, 1991, and references therein). The administration of retinoic acid induces these leukemia cell lines to differentiate and causes remission of the disease in patients. Interestingly, the PML–RAR fusion protein appears to disrupt a novel nuclear organelle which contains the native PML protein, and retinoic acid reverses this disruption (Dyck *et al.*, 1994; Weis *et al.*, 1994; Koken *et al.*, 1994). Evidence suggests that the vitamin D_3 receptor can also suppress activation of T lymphocytes and induce the differentiation of promyelocytic leukemia cell lines (Reichel *et al.*, 1992).

It is recently been established that certain receptors modulate each other's function by forming heterodimeric complexes. The receptors for retinoic acid, vitamin D_3, and thyroid form heterodimers with the receptor for 9-*cis*-retinoic acid, also known as RXR. In analogy with the regulatory heterodimerization found in other classes of transcription factors, such as max/myc, the resulting interplay between receptors results in an additional level of physiological control and hierarchical complexity (Segars *et al.*, 1993).

Despite the distinct physiological effects mediated by the different hormones, the upstream regulatory elements to which their receptors bind share remarkably similar consensus sequences. In particular, the hormone response elements (HREs) of the steroid receptors are very closely related (Fig. 3). The near palindromic nature of their response element indicates that the steroid receptors bind their HREs as homodimers. Mutagenesis has pinpointed three residues in the DNA-binding domain that are largely responsible for target discrimination of estrogen and glucocorticoid response elements (Danielson *et al.*, 1989; Mader *et al.*, 1989; Umesono and Evans, 1989). These residues are close together in a region called the "proximal" or P-box (Umesono and Evans, 1989). Surprisingly, most steroid/nuclear receptors have P-boxes resembling that of either the estrogen or glucocorticoid receptors. In nearly all nuclear receptors, the P-boxes resemble that of the estrogen receptor; correspondingly, these receptors recognize HREs that contain half-sites from the estrogen response element, but these sites can be arranged in either direct or inverted repeats with various spacings. DNA target discrimination thus appears to be achieved in part by recognizing the arrangement of half-sites. The question as to how the functional diversity of the steroid/nuclear receptors is generated from a limited repertoire of amino acid residues presents an intriguing structure–function problem.

In this article, we focus principally on the current understanding of the DNA-binding domain. Three-dimensional structures of the DNA-binding domains from several members of the steroid/nuclear receptor

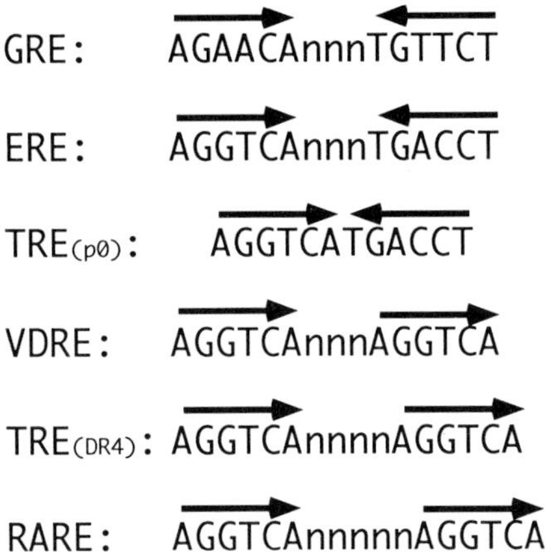

GRE: AGAACAnnnTGTTCT

ERE: AGGTCAnnnTGACCT

TRE(p0): AGGTCATGACCT

VDRE: AGGTCAnnnAGGTCA

TRE(DR4): AGGTCAnnnnAGGTCA

RARE: AGGTCAnnnnnAGGTCA

FIG. 3. The hormone response elements. Arrows have been drawn over the hexameric consensus sequences to emphasize the orientation of the repeating units. A DBD monomer is presumed to bind to each of the indicated hexamers in each case. The GRE and ERE represent idealized consensus (Klock et al., 1987). In nature, deviations from exact palindrome are the rule rather than the exception; however, the half-site spacing is always conserved at an invariant three bases. The directly repeating elements listed here were found to function in vivo in transient transfection assays. Natural response elements of the nonsteroid receptors tend to be more complex and deviate from these ideal sequences. For instance, TR can activate from both an inverted repeating element with zero spacing between half-sites (p0) as well as the direct repeat (DR4). The function of the various nonsteroid response elements may be complex and depend on the context of the element, including the requirement for the receptor to form a heterodimer with RXR.

It is interesting to note that the GRE can function virtually as effectively as response elements for progesterone, androgen, and mineralocorticoid receptors, although the hormones result in distinct physiological effects. The distinguishing effects of these hormones may arise from tissue-specific expression of the corresponding receptors (Strahle et al., 1989).

superfamily have been determined both in the absence of DNA using nuclear magnetic resonance spectroscopy (NMR) (Hard et al., 1990a; Schwabe et al., 1990; Knegtel et al., 1993; Lee et al., 1993) and also in complex with DNA using X-ray crystallography (Luisi et al., 1991; Schwabe et al., 1993a,b). These structures have yielded an understanding of the stereochemical basis of how receptors recognize the half-site sequence, spacing, and orientation, and thus are able to discriminate between different response elements. Through these three elements of recognition, the receptors are able to regulate diverse sets of genes from a seemingly limited set of core response elements. Finally, we show that these structures can explain how known mutations in the steroid/nuclear receptors result in clinical disorders.

II. THE DNA-BINDING DOMAIN DIRECTS DNA RECOGNITION

To facilitate a comparison of the DNA-binding domains from different receptors we have employed a common numbering scheme such that the first metal-binding cysteine is residue 1. Residues preceding this are numbered negatively.

A. A STRUCTURALLY CONSERVED DNA-BINDING MOTIF

To date, the DNA-binding domain is the best structurally characterized region of the steroid/nuclear receptors. Nuclear magnetic resonance spectroscopy has been used to determine the structures of the DNA-binding domains from the receptors for estrogen (Schwabe et al., 1990), retinoic acid (β isoform; Knegtel et al., 1993), 9-cis-retinoic acid (RXR; Lee et al., 1993), and glucocorticoid (Hard et al., 1990a). The structures of both the glucocorticoid and estrogen receptor DNA-binding domains have also been solved in complex with DNA using X-ray crystallography (Luisi et al., 1992; Schwabe et al., 1993a). These structures all reveal a similar three-dimensional protein fold indicative of their high sequence conservation. This three-dimensional homology is illustrated in Fig. 4, which shows a spatial superposition of the DNA-binding domains (DBDs) of the estrogen and glucocorticoid receptors (from the crystal structures). Comparison of the structures in the crystalline complexes with DNA with those in the absence of DNA in solution reveals a small difference in one region of secondary structure that appears to become well ordered only on binding to DNA.

One of the striking characteristic features of the DNA-binding domain is that it is composed of two metal-binding motifs that have some structural similarities, but on closer examination have clearly discernible features. As had been previously revealed by spectroscopic techniques (Freedman et al., 1988a), each motif binds a zinc ion that is tetrahedrally coordinated by four conserved cysteine residues. The zinc ions appear to serve a structural role with each zinc capping the N terminus of an amphipathic α helix and supporting a peptide loop (Figs. 2A and 4). The role of zinc as a structural buttress within small DNA-binding domains has been seen in four other protein classes. In particular (and as will be seen later) the zinc appears to stabilize and orient an element of α helix for interaction in the major groove of the DNA double helix. Such motifs were first observed in the classical "zinc-finger" proteins (reviewed in Rhodes and Klug, 1993) found in many eukaryotic transcription factors, for example, Xfin (Lee et al.,

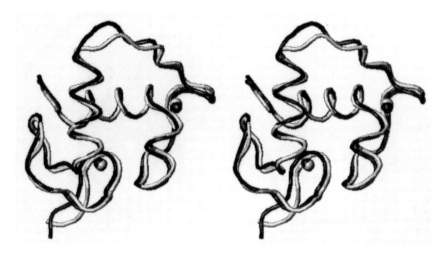

FIG. 4. Comparison of superimposed crystal structures (solved in complex with DNA) of the DNA-binding domains from the estrogen (dark) and glucocorticoid (light) receptors. The zinc ions are shown as spheres. The stereoscopic view shows the DNA-binding surface of the DBD with the N termini at the bottom of the figure.

1989), ADR1 (Klevitt *et al.*, 1990), SWI5 (Neuhaus *et al.*, 1992), Zif268 (Pavletich and Pabo, 1991), and Tramtrack (Fairall *et al.*, 1993), in which the metal is coordinated by two histidine and two cysteine residues. The retroviral nucleocapsid proteins coordinate a zinc with three cysteines and one histidine (Summers *et al.*, 1991; Morellet *et al.*, 1992), and the family of yeast genetic regulatory proteins, represented by GAL4 (Baleja *et al.*, 1992; Kraulis *et al.*, 1992; Marmorstein *et al.*, 1992), coordinates two metal ions with shared cysteine ligands, in a manner analogous to the metal-binding sites of metallothionein. In the transcriptional elongation factor, a zinc-binding site is formed by two loops of an antiparallel β sheet (Qian *et al.*, 1993). Despite the common role of the metal ion, the zinc-binding motifs in the receptors are structurally distinct from those in other proteins and to distinguish them, we refer to the nuclear receptor metal-binding motifs as "Zn modules." Residues involved in coordinating the zinc are conserved throughout the superfamily, suggesting that the supporting role of the zinc ions is an absolute requirement among the receptors (see sequence alignment in Fig. 2B).

The two amphipathic helices of the nuclear receptors pack together to form a well-conserved hydrophobic core that ultimately links the two zinc-binding pockets together. Figure 5A illustrates the hydro-

A

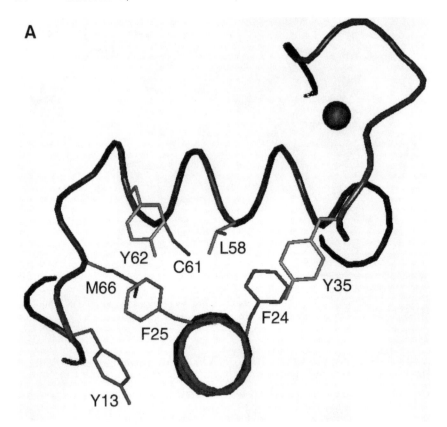

FIG. 5. The hydrophobic core of the estrogen (A) and glucocorticoid (B) receptor DBDs. The core is illustrated by a view down the axis of the α helix of Module 1. The amphipathic helix of Module 2 runs right to left. The Zn ion in the second module is indicated by a sphere (top right). Residues comprising this core are conserved or semiconserved (the positions of L58 and Y62 in the ER are reversed in the GR) among all members of the superfamily (see Fig. 2B).

phobic cores of the estrogen and glucocorticoid receptors' DNA-binding domains. They are essentially identical except that Leu58 and Tyr62 of the estrogen receptor exchange positions in the glucocorticoid receptor. Two absolutely conserved phenylalanines lie at the center of the core, and the bulky aromatic side chains pack tightly with other conserved hydrophobic residues (Fig. 2B). NMR analyses show that the structure of the core is conserved in the 9-*cis*-retinoic acid and retinoic acid receptors (Lee *et al.*, 1993; Knegtel *et al.*, 1993). The importance of this

B

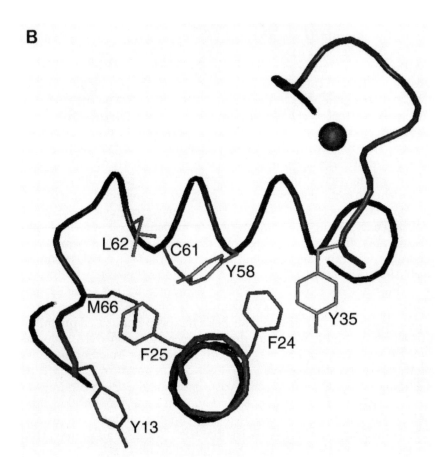

core is illustrated by the observation that nonconservative amino acid substitutions in the core of the glucocorticoid receptor result in loss of function (Schena *et al.*, 1991a). Also, substitutions occurring in the core of the human androgen receptor are associated with androgen insensitivity syndrome. These and other mutations are described in Section V.

Another well-conserved feature in the DNA-binding domain is a salt bridge that links the two modules (Fig. 2B). In the glucocorticoid receptor, the side chains of Asp6 and Arg50 form a hydrogen bond, and this interaction orients the Arg NH1 to interact with the phosphate backbone of DNA. Asp and Arg are conserved at the corresponding positions throughout the superfamily, suggesting that the hydrogen bond is conserved and is structurally important. The Arg is substituted by Gln in the vitamin D receptor of a patient suffering from hereditary, vitamin D-resistant rickets (Hughes *et al.*, 1988; see Section V).

Although they have some structural similarities, the two Zn modules play different functional roles. The amino-terminal module (Module 1) exposes an α helix to the major groove. Residues on the surface of this helix contact the bases of the target site. Residues within the carboxy-terminal module (Module 2) contribute to a dimerization interface. The nature of this interface appears to differ between the steroid and nuclear receptors. In the steroid receptors, Module 2 forms a dyad-symmetrical dimerization interface that enforces the protein to match the spacing between half-sites of the palindromic target; it is therefore an important feature of DNA target recognition. Molecular genetic analyses suggest that, in certain nuclear receptors, the second module may direct an asymmetrical dimerization that is required for binding to directly repeating half-sites (Towers *et al.*, 1993; Perlmann *et al.*, 1993; Kurokawa *et al.*, 1993). NMR studies (Lee *et al.*, 1993) show that the RXR receptor DNA-binding domain contains an additional helical region at the carboxy terminus of Module 2 (the position of which is indicated in Fig. 2A). This helix has been implicated in binding site selection, and it may play a role in recognition of DNA sequence and the symmetry of the target. Both modes of steroid receptor and nuclear receptor dimerization are described in detail in Section II, C.

Although the structural similarity between the DNA-binding domains was clear from the sequence conservation alone, proteins that share little or no apparent sequence homology can often exhibit remarkable congruence over part or the entirety of their three-dimensional structures (Chothia, 1992). Data suggest that Module 1 of the steroid/hormone nuclear receptor DNA-binding domain is related to the zinc-binding motif found in a family of transcription factors represented by chicken GATA-1 (Omichinski *et al.*, 1993). The structural similarity is such that the recognition α helix of the GATA-1 zinc finger makes DNA contacts much like the corresponding helix of steroid/nuclear receptors. Interestingly, the two zinc-mediated subdomains diverge structurally at the end of the α helix, and this point coincides with an intron–exon junction in the hormone/nuclear receptors. Perhaps this common module of the two protein groups originated from a primordial shuffling exon, or "shuffle-on."

Detailed structural analyses reveals another evolutionary curiosity. The linear representation of Modules 1 and 2 as shown in Fig. 2A suggests a direct structural repeat of Zn loop/amphipathic helix, and it is tempting to speculate that they may have arisen from duplication of a shuffle-on. Closer examination of the three-dimensional structure reveals an interesting "twist" to this proposal. The metal-binding sites can be defined as having a chirality if one uses the numbers of the coordinating residues to assign the priorities. With this scheme, Module 1 has the S configuration, like the TFIIIA and retroviral finger classes. On the other hand, Module 2 has the R configuration. Consequently, the Zn-binding sites are topologically distinct and cannot be equivalently superimposed. Furthermore, it is clear that the hydrophobic and hydrophilic surfaces of the helices are reversed in the two modules. The implication is that, if the modules did arise from a common origin, one must have dramatically adopted a new fold in the course of evolution. The other possibility is that the two modules arose by a remarkable evolutionary convergence. As we will describe in greater detail, the distinctive R configuration of Module 2 exposes a peptide loop to make a dimerization interface. Nature is a creative topologist!

B. An Overview of the ERdbd/ERE and GRdbd/GRE Complexes

The global features of the complexes of both the glucocorticoid and estrogen receptors with their DNA targets sites are shown in Fig. 6. The details of their interaction with phosphates and base pairs are shown in Figs. 7 and 8. Reflecting the palindromic nature of their response elements, the DBDs bind to their targets as symmetrical dimers. The conserved spacing of three bases between the hexameric half-sites separates their centers by nine base pairs, which is nearly a structural repeat of the DNA helix (10.5 base pairs per turn for random sequence). Consequently, two monomers lie on the same surface of the DNA duplex. The relative orientation of the two DBDs in the complex is determined not only by the target site but also by extensive protein–protein interactions between the two monomers (Fig. 9). As we will discuss in detail (Section II, C, 2), these contacts are critical for recognition of spacing and orientation of the hexameric half-sites.

In the crystallographic studies of the GRdbd interactions with DNA, one of the complexes included DNA with a modification of the GRE half-site spacing. Here, the separation of the two GRE half-sites was increased by one base, from the natural three to an artificial four (GRE$_{s4}$). The GRdbd binds as a dimer to DNA harboring this sequence. As a consequence of the dimerization, one subunit loses alignment with its half-site in the GRE$_{s4}$ complex and faces a sequence having no relationship to the preferred GRE half-site. The latter subunit can only form a few interactions with the DNA, and we propose that these

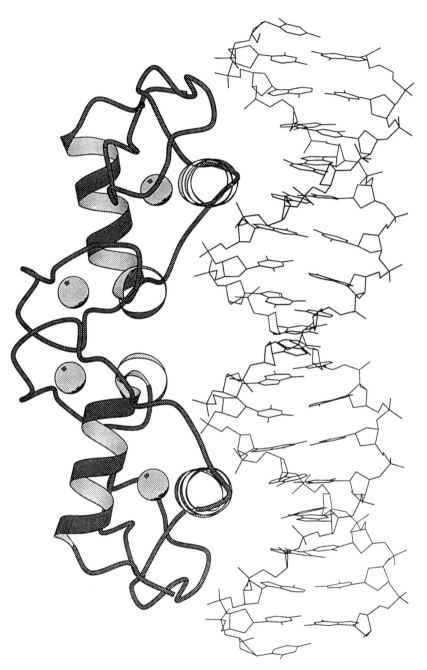

FIG. 6. Two ERdbd monomers bound to the ERE (Schwabe *et al.*, 1993a). The Zn ions are indicated by spheres. The recognition helix lies in successive major grooves of the DNA. The figure was drawn with MOLSCRIPT (Kraulis, 1991). Using this type of illustration, the complex of two GRdbd monomers bound to DNA is indistinguishable.

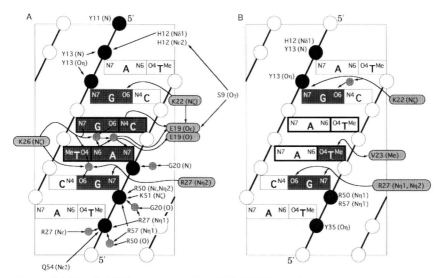

FIG. 7. (A) Protein DNA contacts in the ERdbd/ERE complex. (B) Protein DNA contacts of the GRdbd/GRE complex. In the case of polar interactions, arrows indicate the direction of the hydrogen bond (donor to acceptor); arrows at both ends indicate a van der Waals interaction. Bases outlined in bold differ between the two consensus half-sites. Phosphates shown in black are in contact with protein. Ordered water molecules are shown as small gray circles. Figure adapted from Schwabe *et al.* (1993a).

are nonspecific (Fig. 10). The structure of the GRdbd/GRE$_{s4}$ complex has been studied at 2.9 Å resolution, which is sufficiently accurate to examine details of the protein–base contacts. The crystal structure has also been solved for the complex of GRdbd with a GRE having the natural spacing of three bases. Because the crystals were of lower quality, the structure of the GRdbd/GRE$_{s3}$ complex was studied at 4 Å resolution, which is inadequate to satisfactorily resolve side chain–base contacts. However, it is sufficient to examine the global structural differences between the GRE$_{s3}$ and GRE$_{s4}$ complexes. The nature of the dimerization interface can be inferred to be the same between the two complexes. These observations illustrate the strength of subunit dimerization, which will assure that the three-base-pair spacing will be recognized as a feature of an optimal target site. Higher-resolution models of both complexes (at 2.5 and 3.1 Å, respectively) are now being refined and corroborate the features of the present models (Weixin Xu, B. Luisi, and P. Sigler, unpublished results).

The ERdbd/ERE complex has been determined at 2.4 Å, giving an accurate picture of the stereochemistry of the protein–protein and protein–DNA interfaces (Schwabe *et al.,* 1993a). The target site has

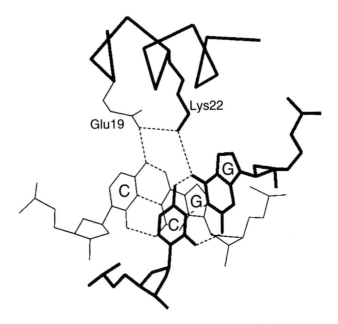

FIG. 8. An important sequence discriminating contact in the ERdbd/ERE complex (Schwabe *et al.*, 1993a). The Glu19 accepts a hydrogen bond from the cytosine and from the nearby Lys22, which in turn donates a hydrogen bond to the carbonyl group of a guanine in the next base pair. In the GRE, a thymine occurs at the position corresponding to the cytosine. Water molecules are not shown for clarity.

the native three-base-pair spacing and the two DBDs bind as a perfectly symmetrical dimer.

The complexes of ERdbd and GRdbd dimers with DNA show that each DBD straddles the phosphate backbone and exposes an α helix to the bases in the major groove of the DNA. The binding constants for the GRdbd and ERdbd are much smaller than those of the prokaryotic gene regulatory proteins, by two or three orders of magnitude and is reflected by the comparatively fewer contacts. The relatively weak DNA binding is typical of eukaryotic transcription factors. In the GRdbd/DNA complex three side chains make base contacts at a half-site, whereas in the ERdbd/DNA complex four side chains make similar contacts, although the network of interactions is enhanced through a number of ordered water molecules sandwiched between protein and DNA (summarized in Fig. 7). Details of the base contacts will be described in Section II, C.

As found in other protein/DNA complexes, the recognition surfaces of the GRdbd and ERdbd are supported in the major groove by a con-

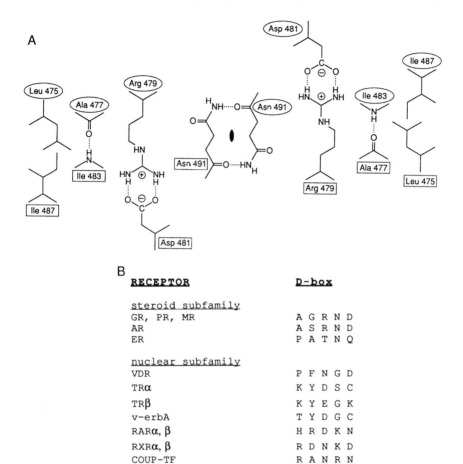

FIG. 9. (A) The details of the dimer interface interactions in the GRdbd/DNA complex. (B) Sequence alignment of the D-box elements of the steroid receptors: androgen, progesterone, glucocorticoid, mineralocorticoid, and estrogen. Also shown are the D-box sequences of selected nuclear receptors: the vitamin D, thyroid hormone α and β, v-erbA, retinoic acid, 9-*cis*-retinoic acid, chicken COUP, and mouse orphan NGFI-B. The D-boxes of the latter group may not support symmetrical dimerization.

stellation of hydrogen-bonding interactions with the phosphate backbone. Each monomer of GRdbd and ERdbd contacts the sugar–phosphate backbone on either side of the major groove it faces. The phosphate strands farthest from and closest to the central dyad are contacted by Modules 1 and 2, respectively (Fig. 6 and 7). Nonconservative substitutions of the residues contacting the phosphates in the

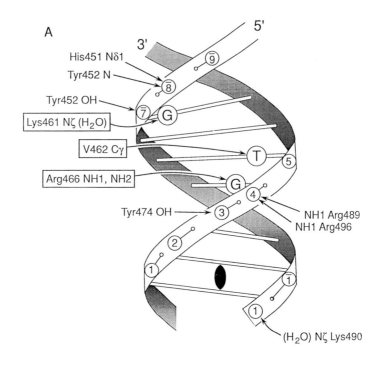

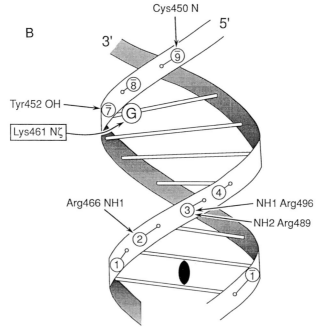

FIG. 10. Comparison of specific/nonspecific complexes from the perspective of the GRdbd recognition helix.

glucocorticoid receptor result in loss of function *in vivo* and reduced DNA-binding activity *in vitro* (Schena *et al.,* 1989).

Finally, the response elements for a number of other steroid receptors have similar palindromic symmetry. It seems likely that the nature of the interactions between these receptors and their targets sites will be directly analogous to that seen in the GRdbd/GRE and ERdbd/ERE complexes.

C. DNA TARGET RECOGNITION AND DISCRIMINATION

The consensus DNA targets of the steroid receptors differ by only two bases within a hexanucleotide sequence element (called a half-site; see Fig. 3). The targets of the nuclear subgroup of receptors, represented by the thyroid, retinoic acids and vitamin D receptors, generally share a hexanucleotide sequence element similar to that found in the estrogen response element; however, the targets differ in how pairs of these small elements are orientated and spaced (Fig. 3). The key to understanding the discrimination of targets within the family is to explain how receptors recognize the hexameric element, its orientation, and pair spacings. In the next two sections, we will use the crystal structure of the GRdbd and ERdbd DNA complexes to explain each of these facets of recognition.

1. *Reading the Sequence*

Studies using *in vitro* transactivation assays have identified three amino acids in the DNA-binding domain of both the glucocorticoid and estrogen receptors that direct this discrimination (Mader *et al.,* 1989; Danielson *et al.,* 1989; Umesono and Evans, 1989). These residues are located in the amino terminus of the α-helical region of the first Zn module, in a region referred to as the P-box (Fig. 2). If the three amino acids of the P-box are changed, from **Glu-Gly**-cys-lys-**Ala,** as in the estrogen receptor, to **Gly-Ser**-cys-lys-**Val,** as in the glucocorticoid receptor, the specificity of the receptor is completely changed so that this mutant ER transactivates from a GRE-driven reporter and not at all from an ERE. Substitution of two of these amino acids in the GR to the corresponding ER residues (**Gly-Ser** → **Glu-Gly**) partially switches the receptor's specificity from a GRE to an ERE; the third substitution, Val→Ala, is required for the full switch (Danielson *et al.,* 1989; Umesono and Evans, 1989). The binding affinities of these mutants in the context of the DNA-binding domain have been studied *in vitro* and correlate with the *in vivo* effects (Alroy and Freedman, 1992). The discrimination corresponds to an affinity difference of roughly 50-fold in both cases, which might appear to be quite small considering the stringent discrimination of targets *in vivo.*

The crystal structure of the GRdbd/DNA complex shows that only the Val of the discriminatory triplet (Gly-Ser-cys-lys-**Val**) makes a base contact. The hydrophobic interaction of the γ-methyl group of Val23 with the 5-methyl group of the T of the GRE (tg**T**tct) appears to be an attractive contact. Substituting the Val with Ala (i.e., removing the methyl group) or the thymine with uracil (i.e., removing the 5-methyl group from the T) diminishes binding affinity by a factor of 10- and 6-fold, respectively (Alroy and Freedman, 1992; Truss *et al.,* 1990). Interestingly, if the 5-methyl group is replaced by the more bulky bromine (i.e., 5-bromo-uracil substituted for thymine), the affinity is actually slightly increased (Truss *et al.,* 1990). Perhaps the substitution enhances the van der Waals interaction with the Val methyl group. In the ERE, an A occurs at the corresponding position, and this base offers no hydrophobic surface for the Val. The substitution of Val for Ala in the ER has only a small effect on its affinity for the ERE (Mader *et al.,* 1989).

The specificity of the ERdbd for the ERE appears to be achieved through an intricate network of direct and water-mediated base contacts. Glu of the ER P-box triplet (**Glu**-Gly-cys-lys-Ala) accepts a hydrogen bond from the N4 of C of tga**C**ct (Fig. 8) and from a water molecule (Fig. 7A), which in turn accepts and donates bonds to the A of tg**A**cct. The complementary T of the same A (agg**T**ca) accepts a hydrogen bond from the side chain of Lys26, which also contacts G (a**G**tc) both directly and through a water molecule, and G of t**G**acct through a second water molecule. In the GRdbd, the corresponding lysine makes a salt bridge with a glutamic acid at the end of the recognition helix: the same interaction cannot occur in the ERdbd because a glutamine is found instead at the corresponding position of the glutamate. The interwoven, water-mediated network specifies the central elements of the ERE half-site (t**GGAC**t). The same network would not be conserved if the ER were to bind a GRE, nor if the GR were to bind an ERE.

In summary, ERE/GRE discrimination appears to be mediated by surprisingly few residues. In the ER/ERE complex, it is principally achieved through a lysine in the recognition helix, a glutamine in the P-box, and an ordered water network. In the GR/GRE complex, it is mostly through a discriminatory van der Waals contact by one residue in the P-box. The remaining two residues of the "recognition triplet" may not be affecting DNA recognition in either case per se. For instance, Gly in the ER P-box makes water-mediated contacts to the phosphate backbone. However, the Cα makes a van der Waals contact (within 3.6 Å of the phosphate backbone) and so may be recognizing the local conformation of the DNA. If this were changed to a serine (as in the GRdbd) it would almost certainly perturb the orientation of the recognition helix. Finally, the alanine in the ER P-box (that is equivalent to the valine in the GRdbd) makes no contact with the DNA. If it were charged to a valine, it would almost certainly perturb the polar nature of the interface.

Two of the base contacts are nearly equivalent in the GR/GRE and ER/ERE complexes. These may also occur in the DNA complexes of most other steroid/nuclear receptors, because the bases tend to be common to all hormone response elements, and amino acids contacting them are conserved throughout the superfamily. For instance, Lys22 is conserved between the ERdbd and GRdbd, and it interacts with the corresponding G in both response elements, although details differ. The Lys of the estrogen receptor also interacts with the side chain of Glu from the discriminatory P-box. Probably the Lys–Glu interaction is conserved among the nuclear receptors with the same P-box. Another conserved residue, Arg27, interacts with a G conserved in the ERE and GRE, but in the ERdbd complex, a different orientation of this side chain allows it to contact the phosphate backbone of the DNA at the same time.

The differences in these two conserved side chain–base contacts may reflect a more subtle aspect of half-site recognition. A superposition of the GRdbd and ERdbd complexes on their recognition helices illustrates that the DNA conformation differs markedly and also that the recognition helices assume subtly different orientations in the major groove. Given the precise complementarity of the protein–DNA interfaces, it would seem that these differences may be an important component of half-site discrimination.

Using the P-box to group steroid/nuclear receptors, two principal subgroups emerge, the members of which presumably recognize the same response element half-site. Receptors carrying the GS-V motif in the P-box all recognize a GRE with high affinity (e.g., the glucocorticoid, progesterone, mineralocorticoid, and androgen receptors). Receptors harboring the EG-A or EG-G motif (e.g., estrogen, thyroid hormone, retinoic acid, ecdysone, and vitamin D_3 receptors) all appear to bind the ERE half-site (5'-AGGTCA-3'). Probably in all of these latter cases, the interaction is similar to that found in the ERdbd/ERE half-site, with the Glu of the P-Box accepting a hydrogen bond from the C of the complementary strand, tgaCct and making water-mediated contacts to the A of tgAct, and Lys26 contacting both the G and T of ag**GT**ca.

This classification clearly serves only as a coarse approximation, with residues outside the conventional P-box also playing a role in discrimination. This is illustrated by the DNA-binding properties of purified human vitamin D_3 receptor, which binds well to a sequence derived from the mouse secreted phosphoprotein 1 gene (Spp-1; also known as osteopontin). The site is composed of a direct repeat with a three-base separation: 5'-GGTTCAcgaGGTTCA-3'. Although hVDR and ER share a similar P-box, the purified hVDR binds the foregoing target better than a direct repeat of the estrogen response element half site: 5'-AGGTCAnnnAGGTCA-3' (Freedman and Towers, 1991; Freedman *et al.*, 1994; Towers *et al.*, 1993). The P-box of VDR contains the Glu that usually indicates a preference for a C in the ERE-type half-site (tgaCct). In contrast, the foregoing VDRE has an A at the corresponding position (tgaAcc). Glu can make a hydrogen-bonding interaction with the N6 of the A, just as it does with the N4 of the C in the ER/ERE complex. Freedman *et al.* (1994) tested this hypothesis by changing the A–T base pair to an I:5–methyl C base pair. The effect of the substitution is to exchange the amino group of the pyrimidine for a carbonyl, which cannot interact favorably with the carboxylate of the Glu at physiological pH (Fig. 11). As anticipated, the substitution weakened hVDR binding. A second-site revertant was designed by changing the Glu to Gln. It was reasoned that the side chain of the Gln, which can either donate or accept a hydrogen bond, could make a compensating interaction with the carbonyl group in the I:5–methyl C

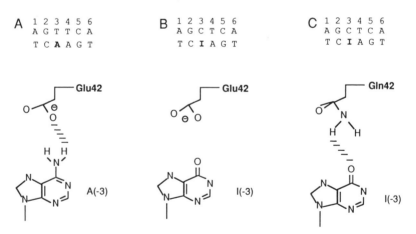

FIG. 11. Details of the putative Glu-A and Gln-I interactions of the vitamin D receptor. Substitution of A for I in each half-site results in diminished DNA binding, but the effect can be reversed by substituting the Glu of the VDR for a Gln.

base pair. The activity was indeed restored by this "second-site" substitution, supporting the role of the Glu–A interaction in recognizing the target DNA (Fig. 11). Additional interactions with the bases probably are directed with the side chain of Arg26 (which is a lysine in the ERdbd recognition helix; see Fig. 7). It is interesting to note that an Arg occurs at the same position in TR and RAR (Fig. 2B).

2. *How the Steroid Receptors Recognize the Spacing of Half-sites*

The DNA-binding domains of the estrogen and glucocorticoid receptors are monomeric in solution (Freedman *et al.*, 1988b; Hard *et al.*, 1990b; Schwabe *et al.*, 1990), but in both cases two DBDs bind cooperatively to their response element (Dahlman-Wright *et al.*, 1990; Alroy and Freedman, 1992; Schwabe *et al.*, 1993b), that is, binding of the first monomer increases the affinity of the second. The increase in affinity may be as great as two orders of magnitude (Hard *et al.*, 1990b). When the second monomer binds, the first monomer helps to orient it on its half-site through the formation of favorable protein–protein contacts. The enthalpy gained by these contacts, as well as any entropy change due to the orienting effect of the first monomer, must outweigh the entropy penalty of protein association (Perutz, 1990). Interestingly, there is some evidence that several contacts to the phosphate backbone of the DNA are only formed as the protein dimerizes on the response element (Schwabe *et al.*, 1993b). This might explain in part the highly cooperative binding. A short segment of α-helix within the second module may be stabilized by DNA binding and could also contribute to the cooperative effect. An analogous effect, whereby secondary structure formation is coupled with DNA association, plays a role in the binding mechanism of the basic zipper proteins to their DNA targets, which is associated with a random coil-to-helix transition of the sequence-recognition helix.

If the spacing between half-sites in a GRE or ERE is increased or decreased by a single base, two monomers will still bind the element, but in this case the binding affinity of the second DBD is lower than that of the first, that is, the cooperativity is apparently lost or even reversed (Freedman and Towers, 1991; Dahlman-Wright *et al.*, 1991; Schwabe *et al.*, 1993b). Correspondingly, the GREs with modified spacing cannot direct transactivation *in vivo* (Dahlman-Wright *et al.*, 1990; Nordeen *et al.*, 1992). There are two implicit and subtle effects at play here that account for the loss of cooperative free energy. First, if two monomers were to each bind a half-site where the spacing is 2 or 4, they would probably clash sterically, resulting in "virtual" negative cooperativity. Indeed, this is supported by experimental observations, as the affinity for the second monomer binding to such sites is less than that when two distant sites are bound independently (Schwabe *et al.*, 1993b). The negative cooperativity is "virtual" because the second monomer does not bind to the second bona fide half-site, but instead binds such that it is shifted with respect to the bona fide half-site and thus faces noncognate sequence. However, this allows the second DBD to form a native dimer interface with the monomer already bound to the specific site. The lowest energy configuration is therefore one in which a dimer binds such that one subunit faces a nontarget sequence (Fig. 10). This gives rise to the second effect: the loss of free energy of cooperativity arises from the difference in binding to a nonspecific versus a specific site. In the crystal structure of GRdbd bound to an element with a spacing of four bases, only a dimer is found and not two monomers bound independently to the two half-sites. The effects described here account for the palindromic symmetry of the targets of the steroid receptors and strict adherence to a three-base-pair half-site spacing.

The preceding principles have been found to be applicable to target recognition by other DNA-binding proteins. Like the GRdbd and ERdbd, the *E. coli* DNA-binding protein, LexA, is also monomeric, but binds its target DNA as a dimer with positive cooperativity (Kim and Little, 1992). Analogously, the cooperativity probably also arises from favorable protein–protein interactions. The DNA-binding domain of the GAL4 family of proteins also behaves in a similar fashion.

The crystal structures of the GRdbd/GRE and ERdbd/ERE complexes show that the dimer interface is self-complementary and is stabilized through salt bridges, hydrophobic interactions, and hydrogen bonds between main and side chains (Fig. 9A). The contacts are made by residues in the carboxy-terminal module, principally by residues in a segment between the first and second coordinating cysteines. This region, the "D-box," was identified genetically as a dimerization region (Umesono and Evans, 1989). The distinctive R chirality of the second zinc-binding site imparts a specific conformation to the D-box that is important for its dimerization role. Dimerization contacts are also made by residues in the short segment of α helix within the loop of the module and a residue in the extended peptide segment preceding the module itself.

The dimerization domain appears to be highly conserved among the steroid receptors. Figure 9B shows a sequence comparison of the dimerization domains of the receptors for glucocorticoid, mineralocorticoid, estrogen, progesterone, and androgen. Many of the substitutions in this region are conservative such that they would maintain the

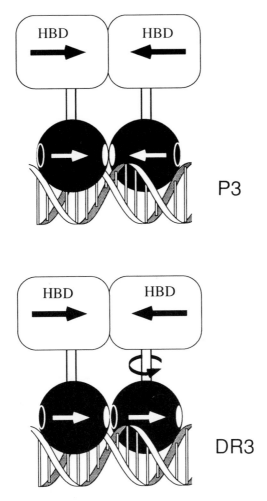

Fig. 12. Mode of binding of steroid receptor to inverted repeats (P3) versus the binding of the vitamin D receptor to direct repeats (DR3). The orientation of the recognition element in the major groove is implied by the arrows of the DBD. The symmetrical dimerization region of the steroid receptor DBDs is indicated by white ovals, whereas the heterologous dimerization region of the vitamin D is indicated by dark and white ovals. Like the vitamin D receptor, other nuclear receptors that bind to direct repeats may form analogous heterologous associations. The hormone-binding domain, which has a stronger dimerization function, may form a symmetrical dimerization interface. The two domains may be linked by a flexible tether.

features of the common interface. In the context of the superfamily sequences, the D-box region does vary (Fig. 2B). Its role in dimerization may also differ between different receptor subgroups. For instance, residues stabilizing the interface in the steroid receptors are not conservatively substituted in the nuclear receptors. Interestingly, Freedman and Towers (1991) found that substitution of the D-box residues of the GRdbd with those of the vitamin D receptor abolishes cooperative DNA binding. The substitution would disrupt a number of interactions that would be mediated by the GRdbd D-box and thereby weaken protein–protein association (Fig. 12). The vitamin D receptor targets appear to be direct repeats, in contrast to the inverted repeats—or palindromes—characteristic of the steroid receptors. The inability of the DNA-binding domain to stabilize symmetrical homodimers may therefore have a functional implication for target selection by the vitamin D receptor.

Although the composition varies, the length of the D-box is usually conserved throughout the superfamily at five residues, but in a number of characterized receptors for peroxisomal activators, only three residues are found (see Fig. 2; Issemann and Green, 1990; Dreyer *et al.*, 1992). Probably these proteins cannot interact symmetrically through this segment of their DNA-binding domains.

3. *How the Nuclear Receptors Recognize the Spacing Between Half-sites*

In contrast to the palindromic (inverted repeat) response elements characteristic of the steroid receptors, the targets of the nuclear receptors usually consist of a direct repeat ((Naar *et al.*, 1992; Umesono *et al.*, 1991; and Marks *et al.*, 1992). The repeating element is usually the half-site of the consensus ERE: 5'-AGGTCA (Fig. 3). The nuclear receptors that recognize these elements have been shown to sense the relative orientation and spacing of the two half-sites. Again as in the case of the steroid receptors described earlier, the recognition of spacing almost certainly occurs through cooperative protein–protein interactions (Freedman and Luisi, 1993; Towers *et al.*, 1993; Kurokawa *et al.*, 1993; Perlmann *et al.*, 1993; Zechel *et al.*, 1994), but this time between DBDs arranged as nonsymmetrical head-to-tail dimers.

The receptors that recognize directly repeated half-sites include the thyroid hormone, retinoic acid, and vitamin D_3 receptors, as well as certain orphan receptors such as COUP-TF. These receptors have been shown to form heterodimers *in vitro* with the receptor for 9-*cis*-retinoic acid, or RXR. The heterodimer formation affects their affinity for DNA and has been proposed to be involved in response element selectivity. The vitamin D_3 receptor can bind and activate from a target composed of ERE half-sites arranged as a direct repeat with a spacing of three bases (Umesono *et al.*, 1991). TR can activate genes from elements with two ERE half-sites separated by four spaces or arranged in an inverted repeat with no spacing (Fig. 3). In the latter case, the center-to-center separation is approximately half a helical turn, which implies that two DBDs lie on opposite faces of the DNA and thus could not make protein–protein interactions in a manner analogous to the steroid receptors. Indeed, this mode of binding has been corroborated by crystallographic studies of a chimeric GR/TR DBD (D. Gewirth and P. Sigler, personal communication). The retinoic acid receptor preferentially activates reporter genes under the control of an element composed of ERE half-sites arranged as direct repeats with a spacing of five bases. The preference of VDR, TR, and RAR to activate genes containing response elements that are direct repeats of ERE half-sites with a spacing of three, four, and five bases, respectively, has been called the "3-4-5" rule (Umesono *et al.*, 1991). It is now clear that TR and RAR require RXR for these spacing preferences (Perlmann *et al.*, 1993; Kurokawa *et al.*, 1993), although it is not required for spacing recognition by VDR (Towers *et al.*, 1993).

As suggested by the organization of these elements containing direct repeats, two receptors must bind in tandem as asymmetrical homodimers or, with RXR, as heterodimers (i.e., in a head-to-tail fashion). This is illustrated with a model of two DBDs bound to a direct repeat (Figs. 12 and 13). The model suggests that a hVDR homodimer or hVDR/RXR heterodimer could form protein–protein contacts through the DNA-binding domain when bound to a direct repeat with a three-base spacer (DR+3). Supporting this hypothesis, the hVDR DBD cooperatively binds a DR+3 target (Towers *et al.*, 1993). Specific contacts may be made between residues in the carboxy terminus of the DNA-binding domain with residues in Module 2 of a neighboring receptor (Fig. 13). Mutagenesis studies show that changing residues in these regions weakens cooperative binding to the DR+3 element (Towers *et al.*, 1993). Cooperative protein–protein interactions could in principal also occur with binding to direct repeats with spacings of four or five bases, but because of the different relative orientations of the monomers, different protein–protein interfaces would have to be formed in each case (Zechel *et al.*, 1994a,b).

The region in the carboxy terminus of the hVDR DBD that has been proposed to be involved in heterologous dimer contacts corresponds to the "T-box" of the mouse orphan receptor NGFB-I, which has been implicated by genetic studies as directing protein–protein contacts in

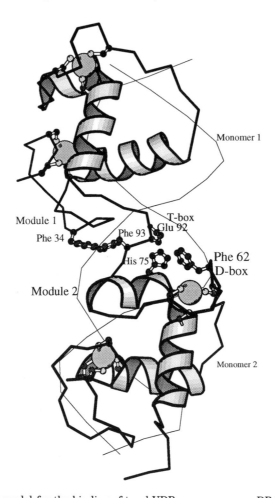

Fig. 13. A model for the binding of two hVDR monomers on a DR+3 element.

the NGFB-I dimer (Wilson *et al.*, 1992). Interestingly, the NMR structure of the RXRα DNA-binding domain reveals an α helix in this same
region. The orphan receptor NGFB-I appears to recognize sequences
flanking the consensus hexameric element, and this occurs through
residues in a region adjacent to the T-box referred to as the "A-box."
Residues of the A-box may make additional stabilizing contacts with
the DNA outside the hexameric half-site. For receptors that are
thought to bind as monomers, such as *Drosophila* FTZ-F1 or mouse
ELP protein (Tsukiyama *et al.*, 1992), these additional contacts may
play an important role in contributing to their DNA affinity.

4. *In Vivo Recognition of Nonconsensus Response Elements*

The foregoing discussion has been concerned entirely with "consensus" response elements. As the name implies, these sequences have been inferred from the common features of response elements found *in vivo* for the hormone in question. It is perhaps surprising that the response elements for most hormones contain half-sites whose sequences differ significantly from the ideal consensus (e.g., Payvar *et al.*, 1983; reviewed in Beato, 1989). Why would this deviation from ideality, which presumably results in complexes with suboptimal binding affinity, have evolved? Perhaps such variation in binding affinity is itself an important part of the control of gene expression. For our purposes here we need to ask how the DNA-binding domain might cope with such nonideality. The GRdbd/GRE$_{s4}$ complex probably represents an example of such a complex with a less than optimal binding site. Figure 10 compared the specific and nonspecific sites of the GRE$_{s4}$ complex, which have been superimposed on the GRdbd recognition helix. The perspective is therefore from the recognition surface of the exposed helix. The major groove of the specific site is more open than that at the nonspecific site, and the recognition helix is more deeply buried within it. LaBaer (1989) and Hard *et al.* (1990b) have shown that the glucocorticoid receptor DNA-binding domain association constant for a specific site, relative to a nonspecific site, differs by only two orders of magnitude (10^8 M^{-1} versus 10^6 M^{-1}, respectively). This difference presumably arises from the smaller buried surface area and the smaller number of base–protein contacts made at the nonspecific site.

As the GRdbd/GREs4 complex suggests, the receptor could still form a dimer on the surface of the most extreme GRE deviation, but one subunit would be forced to form a nonspecific complex. The affinity of such a complex would clearly be weaker than that of a complex with a consensus response element. Because naturally occurring GREs and EREs tend not to be ideal, we must assume that dimerization allows one of the monomers to bind to an otherwise poorly selected site. Thus, dimerization may serve a means of modulating target affinities and, possibly, transcriptional responsiveness.

III. DIMERIZATION THROUGH THE LIGAND-BINDING DOMAIN

Molecular genetics studies have identified a dimerization region in the ligand-binding domain of the estrogen receptor that forms a stronger dimer interface than that made by the receptor's DNA-binding domain (Kumar and Chambon, 1988). The dimerization domain has been localized to a segment (Fawell *et al.*, 1990) that is indicated in Fig. 14B. Other steroid/nuclear receptors have been isolated from cells as homodimers, whereas certain nuclear receptors are mostly found as heterodimers with the RXR receptor. Like the estrogen receptor, these receptors have a strong dimerization function that appears to be primarily located in the ligand-binding domain. The preferential heterodimerization of RAR and RXR has been shown to be mediated through their ligand-binding regions *in vivo* (Nagpal *et al.*, 1993).

In addition to the dimerization signal, the ligand-binding domain may be dissected into three other principal subdomains: at the carboxy and amino termini are regions that appear to contact the ligand, and these flank the transcriptional activation (Ti) and dimerization segments (Fig. 14A). The sequences of the Ti and dimerization region have been aligned partially on the basis of consistent patterns of secondary structure prediction (Fig. 14B). The sequence conservation of the Ti ligand-binding domain is apparent, but homology is much weaker for the dimerization subdomain. Nonetheless, it is clear that the dimerization region is rich in leucine and other small hydrophobic residues. These residues generally occur at every third or fourth posi-

FIG. 14. (A) Schematic illustrating the subdomains of the ligand-binding region. The amino end of the domain is to the left. (B) Sequence alignment of the Ti and dimerization regions of the ligand-binding domain. The figure is patterned in part on the alignment of Laudet *et al.* (1992) and Fawell *et al.* (1990). The approximate boundaries of the Ti and dimerization regions are indicated at the top of the list. The individual sequences were analyzed by secondary structural prediction, and the consistent consensuses are indicated at the bottom of the list for helices (corkscrew), beta strands (arrow), and loops. These were used in part for preparing the alignment. The two indicated loop regions were predicted by Hugh Savage (York University) using the sequence of the human thyroid hormone receptor. Fawell *et al.* (1990) have noted a sequence similarity between rabbit uteroglobin and the mouse estrogen and human progesterone receptor dimerization domains. The latter has been structurally characterized by X-ray crystallography, and on the basis of the sequence similarity, Fawell *et al.* also predict two helices and a loop in roughly the same region as the last three putative elements (at the C terminus) of the alignment provided here. All three of the indicated putative helical regions could in principle form amphipathic α helices. Residues that are cross-linked to the glucocorticoid and estrogen receptors by hormone analogues or affect ligand affinity and discrimination when mutated are indicated at the top of the list by asterisks (Danielian *et al.*, 1993; Benhamou *et al.*, 1992; Harlow *et al.*, 1989; Ratajczak *et al.*, 1992) and map to the dimerization domain. Conserved residues are shown in bold and underlined residues have similar properties. All sequences are from human proteins. Abbreviations: AR, androgen receptor; GR, glucocorticoid; ER, estrogen; PR, progesterone; TR, thyroid hormone; VDR, vitamin D.

A

| ligand 1 | Ti | dimerization | ligand 2 |

B

<pre>
           <   Ti   > <
           *   *
AR ELGERQLVHVVKWAKALPGFRNLHVDD QMAVIQYSWMG LMVF AMGWRSFTNVNSRM LYFA

GR MLGGRQVIAAVKWAKAIPGFRNLHLDD QMTLLQYSWMF LMAF ALGWRSYRQSSANL LCFA

ER NLADRELVHMINWAKRVPGFVDLTLHD QVHLLECAWLEILMI   GLVWRSMEHPVK   LLFA

PR QLGERQLLSVVKWSKSLPGFRNLHIDD QITLIQYSWMS LMVF GLGWRSYKHVSGQM LYFA

TR KIITPAITRVVDFAKKLPMFSELPCED QIILLKGCCME IMSL RAAVRYDPESD   TLTLS

VDR DLVSYSIQKVIGFAKMIPGFRDLTSED QIVLLKSSAIEVIML RSNESFTMDD   MSWTC
</pre>

dimerization

<pre>
           *
AR PDLVFNEYRMHKS RMYSQC   VRMRHLSQEFGWLQITPQEFLCMKALLLFSI IPVDGLKN

GR PDLIINEQRMTLP CMYDQ   CKHMLYVSSELHRLQVSYEEYLCMKTLLLLSS VPKDGLKS

ER PNLLLDRNQGKCVEGMVEI   FDMLLATSSRFRMMNLQGEEFVCLKSIILLNSGVYTFLSSTLKSL

PR PDLILNEQRMKESS FYSLCL   TMWQIPQEFVKLQVSQEEFLCMKVLLLLNT IPLEGLRS

TR GEMAVKREQLKNGG LGVVSD   AIFELGKSLSAFNLDDTEVALLQAVLLMST DRSGLLCV

VDR GNQDYKYRVSDVTKAGHSLFL   IEPLIKFQVGLKKLNLHEEEHVLLMAICIVSP DRPGVQDAA
</pre>

loop

<pre>
                                          ** *   *
AR QKFFDELRMNYIKELDRIIACK RKNPTSCS   RRFYQLTKLLDSVQPIARELHQ   FT

GR QELFDEIRMTYIKELGKAIVKR EGNSSQNW   QRFYQLTKLLDSMHEVVENLLN   YC

ER EEKDHIHRVLDKITDTLIHLMAK AGLTLQQQ   HQRLAQLLLILSHIRHMSNKGM   EHL

PR QTQFEEMRSSYIRELIKAIGL   RQKGVVSSS   QRFYQLTKLLDNLHDLVKQLH   LYC

TR DKIEKSQEAYLLAFEHYVN   HRKHNIP   HFWPKLLMKVTDLRMIGACHA   SRF

VDR LIEAIQDRLSNTLQTYIRCRH   PPPG   SHLLYAKMIQKLADLRSLNEEH   SKQY
</pre>

loop

tion, with a seven-residue periodicity (Forman *et al.*, 1989). This pattern is characteristic of the coiled-coil helix assumed by the dimerization motif of the "leucine zipper" transcription factors (Landschultz *et al.*, 1988). Forman and Samuels (1990) have proposed that this region forms a dimerization domain, or "regulatory zipper," that permits homo- and heterodimerization. Secondary structure predictions suggest that amphipathic α helices may occur in this region.

In analogy to the dimerization specificity of certain zipper proteins, such as fos and jun, the ligand-binding domain motif may make discriminatory interfaces. For instance, the five steroid receptors form homodimers, and there is no evidence that they cross-dimerize. It seems unlikely that this discrimination is mediated by the weaker dimerization interface within the DNA-binding domain since residues here are conserved or have functionally conservative substitutions that would probably maintain the same dimerization pattern (Fig. 9B). Instead, the steroid receptors perhaps recognize and distinguish each other through the ligand-binding domain. Residues within the putative coiled-coil region might serve to distinguish a given steroid receptor's "self" partner from the other steroid receptors. In this regard, it is interesting to note that the ninth heptad repeat in the TR and RAR ligand-binding domains appears to direct heterodimer formation with the RXR receptor (Au-Fliegner *et al.*, 1993).

A number of bulky aromatic residues are conserved in the Ti subdomain (Fig. 14B). The conservation suggests a central role in stabilizing a hydrophobic core that supports the fold of the domain. Hydrophobic residues elsewhere are conserved or have been conservatively changed, and perhaps these contribute to the interior. Some of the conserved hydrophobic residues of the Ti and dimerization regions appear to be contacted by the ligand (Fig. 14B). It is quite likely that these domains share a common fold, perhaps with a few key residue changes yielding functional diversity.

Several studies have implicated specific residues of the conserved region in ligand binding and discrimination. Benhamou *et al.* (1992) have identified a single amino acid that determines the sensitivity of the human progesterone receptor to the abortifacient RU486. The importance of this residue was first suggested by sequence comparison of the human receptor, which binds RU486, with the sequences of the chicken and hamster receptors, which do not. The role of this residue has been corroborated by mutagenesis. Substitution of the corresponding residue in the glucocorticoid receptor ligand-binding domain results in loss of binding of both dexamethasone and RU486. These residues have been indicated by asterisks in the sequence alignment of the ligand-binding domains shown in Fig. 14B. Studies with a photo-affinity label in the glucocorticoid steroid indicate that Met 610, Cys 644 (shown in sequence alignment; Fig. 14B), and Cys 772 (in ligand-2 subdomain) are in proximity of the ligand. Residues in the estrogen receptor that confer differential sensitivity to estrogen and the anti-estrogen 4-hydroxytamoxifen (Danielian *et al.*, 1993) map to the di-merization domain (Fig. 14B). Affinity labeling experiments suggest that Cys 530 (ligand-2 subdomain) in human estrogen receptor (or 534 in the mouse receptor) is in proximity to the ligand (Harlow *et al.*, 1989; Ratajczak *et al.*, 1992). Mutations of Lys 529 and 531 (ligand-2 subdomain) in estrogen receptor to Gln reduce estradiol binding 5- to 10-fold (Pakdel and Katzenellenbogen, 1993). The pattern accumulat-ing from these data clusters into groups of residues that are likely to define the ligand-binding pocket of the hormone-binding domain. Ste-reochemical details await structural evaluation.

As we discussed in Sections II,C,2 and II,C,3, recognition of both inverted and directly repeating targets may be mediated by the DNA-binding domain itself. In contrast to the apparently variable symmetry of these protein–protein interactions, the ligand-binding domain ap-pears to direct formation of symmetrical homodimers or imperfectly symmetrical heterodimers. If the dimerization interface made by the DNA-binding domain can be either heterologous (i.e., as found in di-rect repeats) or homologous (found in inverted repeats), but the latter is always homologous, how are the two domains linked? It seems most likely that the variable region linking the two domains must be intrin-sically flexible to allow the DNA-binding domain to rotate with respect to the fixed ligand-binding domain dimerization interface (Fig. 12).

The ligand-binding domain appears to be multifaceted. Not only does it provide dimerization and ligand-binding functions, but it may mediate interactions with other proteins, such as the heat-shock pro-tein 90 (hsp90), which forms a complex with the steroid receptors in the cytoplasm (Pratt *et al.*, 1988) but is dissociated upon the recep-tor's ligand binding (Howard and Distelhorst 1988; Cadepond *et al.*, 1991).

In vivo, hormone is required to elicit activation of reporter genes by steroid/nuclear receptors; however, it is curious that ligands are not required for DNA binding *in vitro* for any receptor studied to date (i.e., see Towers *et al.*, 1993; Perlmann *et al.*, 1993). Indeed, Schmitt and Stunnenberg (1993) have demonstrated that ligand is not required for transcriptional activation by the glucocorticoid receptor *in vitro*. Perhaps the role of ligand binding is to unmask the DNA-binding domain from complex with auxiliary proteins (Picard *et al.,* 1990). Alternatively, it may also unmask the ligand-dependent activation function found in the LBD (AF-2) and identified genetically in the estrogen, glucocorticoid, and progesterone receptors (Meyer *et al.,* 1990). In estrogen and glucocorticoid receptors, activity of the AF-2 region requires a 14-residue segment that is conserved in many steroid/nuclear receptors (Danielian *et al.,* 1992). Mutations here have little effect on ligand binding, although they do diminish transcriptional activation. Presumably ligand binding does not affect the autonomous activation function (AF-1) found in N-terminal A/B regions of estrogen, progesterone (Meyer *et al.,* 1990), glucocorticoid receptors (Godowski *et al.,* 1987), or RAR α, β, and γ and RXR α and γ (Nagpal *et al.,* 1993).

IV. INTERACTIONS OF RECEPTORS WITH OTHER PROTEINS

A number of hormone/nuclear receptors have been found to retain function when expressed in yeast, including hormone-dependent activation of target genes (Schena and Yamamoto, 1988; Schena *et al.,* 1989; Hall *et al.,* 1993). The rat glucocorticoid receptor will also function in insect (Freedman *et al.,* 1989; Yoshinaga and Yamamoto, 1992) and even plant cells (Schena *et al.,* 1991), indicating that even these diverse organisms share some similar genetic regulatory mechanisms. The activity of the receptor is found to be optimal in the presence of hsp90 (Picard *et al.,* 1990). Perhaps hsp90 serves to stabilize the receptor against degradation or other catabolic processes and sequesters it until triggered into action by hormone. Proteins that are required for transcriptional regulation of yeast genes have been found to be important for glucocorticoid receptor- or estrogen receptor-mediated gene expression in yeast (Yoshinaga *et al.,* 1992). The receptor failed to activate in strains harboring mutants of the yeast proteins SWI1, SWI2, or SWI3, and appears to directly interact with SWI3.

The glucocorticoid receptor and AP-1, which is the heterodimer of the oncogenic proteins fos and jun, appear to interact (Kerppola and Curran, 1993; reviewed in Miner and Yamamoto, 1992). Since this discovery, other hormone/nuclear receptors have also been noted to interact with AP-1, such as the receptors for estrogen, thyroid hormones, retinoic acid, and vitamin D_3 (Desbois, 1991). The interaction may occur through direct competition for overlapping binding sites in the DNA target, but there is also evidence that AP-1 and the hormone/nuclear receptors may directly interact (Schule and Evans, 1992). There may be interactions of other transcriptional regulatory proteins of importance, for instance, the glucocorticoid receptor may interact with CREB (Imai *et al.*, 1993), retinoic acid receptor with E1A/TFIID (Berkenstam *et al.*, 1992), and thyroid hormone and progesterone receptors with TFIIB (Ing *et al.*, 1992; Fondell *et al.*, 1993). Phosphorylation of certain steroid receptors may provide another regulatory mechanism.

V. RECEPTOR MUTATIONS AND DISEASE

Many clinically diagnosed diseases arise from mutations in the genes encoding the steroid hormone or nuclear receptors. These have been characterized primarily in the vitamin D, thyroid, and some sex hormone receptors, where it appears they are generally not lethal but are readily identified from characteristic clinical aberrations. Mutations in other receptors (e.g., the estrogen and retinoic acid receptors) appear to be less common, which may be explained by their playing central roles in development and regulation, such that any loss of function is lethal. Where clinical aberrations can be studied it has emerged that the mutations occur principally in the DNA-binding or ligand-binding domains of the receptors. Many mutations have now been identified from DNA sequencing, and a data base is now developing (Fig. 15).

The effects of the mutations in the DNA-binding domain can be understood with reference to the crystal structures of the GRdbd/GRE and ERdbd/ERE complexes. Given the structural similarity of the DBDs of receptors for glucocorticoid, retinoic acid β, estrogen, and 9-*cis*-retinoic acid, this modeling is likely to be sufficiently accurate to explain the mutational effects. Many of the substitutions in the vitamin D and androgen receptor DBDs involve residues that are conserved throughout the superfamily and, based on the GRdbd/GRE and ERdbd/ERE structures, are contacting bases or the phosphate backbone of the nucleic acid. The explanations for these mutations are summarized in Tables I and II, and we discuss a few in detail.

Arg27, Arg50, and Arg57 are conserved at the corresponding positions throughout the superfamily, and all three play important roles in the GRdbd/GRE and the ERdbd/ERE complexes. Each of these residues independently occurs as Gln in the vitamin D_3 receptor of three patients with hereditary vitamin D-resistant rickets (Hughes *et al.*, 1988; Sone *et al.*, 1990, 1991). In all cases, the substitution cannot support the same interactions of the arginines. Arg27 donates hydrogen bonds to the bases of the conserved sequence core of the hormone

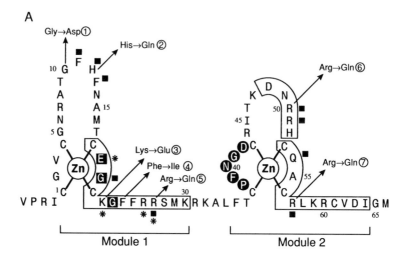

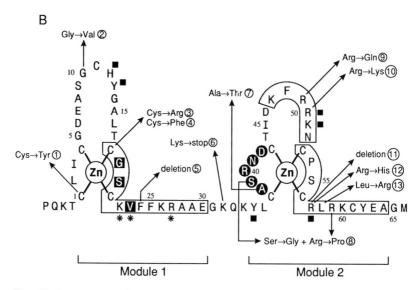

FIG. 15. A summary of mutations in the vitamin D and androgen receptor DNA-binding domains that are associated with the congenital diseases hereditary rickets (mutation in hVDR) (A) and androgen insensitivity syndrome (B). See Tables I and II.

TABLE I
Point Mutations in hVDR DBD Associated with Hereditary Rickets

Number	Substitution	Structural consequence	Reference
1	Gly → Asp	Electrostatic repulsion of DNA and destabilization of Module 1	Hughes et al. (1988)
2	His → Gln	Perturbation of conserved phosphate contact	Yagi et al. (1993)
3	Lys → Glu	Loss of conserved base contact	Rut et al. (1994)
4	Phe → Ile	Destabilization of conserved hydrophobic core	Rut et al. (1994)
5	Arg → Gln	Loss of conserved base contact	Hughes et al. (1988)
6	Arg → Gln	Loss of phosphate contact and conserved salt bridge	Sone et al. (1990)
7	Arg → Gln	Loss of conserved phosphate contact	Saijo et al. (1991)

TABLE II
Point Mutations in hAR DBD Associated with Androgen Insensitivity Syndrome

Number	Substitution	Structural consequence	Reference
1	Cys → Tyr	Abolish Zn binding and hence destabilization of Module 1	Zoppi et al. (1992)
2	Gly → Val	Destabilize Module 1	Chang et al. (1991)
3	Cys → Arg	Abolish Zn binding and destabilization of Module 1	Zoppi et al. (1992)
4	Cys → Phe	Abolish Zn binding and destabilization of Module 1	Chang et al. (1991)
5	Phe deletion	Destabilization of conserved hydrophobic core	Pinsky et al. (1992)
6	Lys → Stop	Termination of DBD and LBD	Marcelli et al. (1991); see also Quigley et al. (1992)
7	Ala → Thr	Disrupt dimerization ?	Klocker et al. (1992)
8	Ser → Gly Arg → Pro	Ser → Gly probably little effect Arg → Pro disrupts hydrophobic core	Zoppi et al. (1992)
9	Arg → Gln	Electrostatic effect ?	Wooster et al. (1992)
10	Arg → Lys	Loss of phosphate contact or conserved salt bridge	Saunders et al. (1992)
11	Arg deletion	Loss of conserved phosphate contact and disruption of hydrophobic core	Pinsky et al. (1992)
12	Arg → His	Loss of conserved phosphate contact and disruption of hydrophobic core	Mowszowicz et al. (1993)
13	Leu → Arg	Disruption of hydrophobic core	de Bellis et al. (1992)

response element. Gln at this position could not reach the base. Arg50 interacts with a DNA phosphate and is buttressed in this interaction by a hydrogen bond with Asp6, a residue that is absolutely conserved in the corresponding position throughout the superfamily. Substitution of the Arg with Gln would allow one but not both of these interactions to be made. Arg57 forms a phosphate backbone contact, and a Gln side chain here would not be sufficiently long to duplicate this interaction.

It seems curious that Arg has a propensity to be the site of substitutions in the receptors. The substitution involve C → T transitions at CG doublets, which are potential methylation sites. This type of mutation occurs in roughly half of human mutations associated with diseases and probably results from deamination of C at sites that either are or can be methylated (Bird, 1992; Perutz, 1990).

Gly is strongly conserved in the nuclear receptor superfamily at the position corresponding to Gly10 of GR, and it is substituted by Asp in the VDR from a patient suffering from the rickets syndrome (Hughes *et al.*, 1988) and by Val in the androgen receptor of a case of androgen insensitivity syndrome (Chang *et al.*, 1991). The Gly is probably required for two purposes. First, because it has no substituent, Gly permits the peptide backbone to assume a special conformation that orients His12 such that it can donate a hydrogen bond to the DNA phosphate and for Tyr13 to pack against the hydrophobic interior of the domain. The Val substitution in the AR probably has serious structural consequences for the protein, because the forked side chain of Val has restricted rotational freedom and would collide with the peptide backbone. Asp at the same position of VDR might not seriously affect the peptide fold, but would electrostatically repel the phosphate backbone and prevent a close approach to the DNA. It is interesting to note in this regard that in the recently reported sequences for the mouse RXR-β receptor and the *Caenorhabditis elegans* nuclear receptor-like gene, the Gly occurs as cysteine (Fig. 2), yet these receptors bind DNA (Wilson *et al.*, 1992). Probably in these cases, the module's loop region has a different backbone conformation from that found in GRdbd and ERdbd, and it is possible that compensating substitutions may have occurred elsewhere in the protein to permit the formation of a favorable complex.

A final mutation in the vitamin D receptor is, at first sight, not so easy to explain (Yagi *et al.*, 1993). In this case His12 is substituted by a glutamine. As discussed earlier, His12 contacts the phosphate backbone in both the ERdbd and GRdbd complexes. One might have imagined that glutamine could make a similar phosphate contact; however, in both the ERdbd and the GRdbd the histidine is involved in a more extensive hydrogen bonding network. It is itself buttressed by the conserved Ser9. In the ERdbd this is in turn hydrogen-bonded to Glu19 (which contacts a cytosine base, see foregoing discussion). In the GRdbd the serine is hydrogen-bonded to the peptide backbone. This network is clearly important for reinforcing the phosphate anchor such that any substitution of the conserved His12 would perturb the complex.

VI. Conclusions

The crystal structures of the GR and ER DBD/DNA complexes have provided stereochemical details of protein–DNA and protein–protein interactions, and the importance of these contacts have been corroborated by mutational analyses. The marriage of structural and functional methods has permitted a better understanding of not only GR and ER DNA binding, but also binding by its highly homologous relatives comprising the nuclear receptor superfamily. The combined methodologies have provided a framework for understanding hereditary disorders associated with defects in the androgen and vitamin D_3 receptors mapping to the DNA-binding domain. The structural and functional data from GR and ER, which allowed us to explain how a few amino acids direct discrimination between closely related GREs and EREs, have been consolidated and extrapolated to describe the DNA-binding modes of other nuclear receptors, such as the vitamin D_3, thyroid hormone, and retinoic acid receptors. These latter receptors appear to bind DNA specifically through discrimination of the spacing and orientation of homologous half-sites.

The results described here represent an early stage of an expanding investigation of an intriguing and important superfamily. In our work, only one aspect of nuclear receptor action has been addressed, leaving unanswered many questions concerning other functions. For example, how are ligands recognized and discriminated? How does ligand binding activate nuclear localization and/or transcriptional activation? How do interactions with other proteins, such as hsp90 and AP-1, affect function, and where is the site of interaction? What converts a nuclear receptor from a transcriptional activator to a repressor? These and other questions are fundamental to a detailed understanding of nuclear receptor action, and may be answered by further studies combining structural and functional approaches.

Acknowledgments

We are grateful to Dan Gewirth (Yale University) for permission to refer to unpublished work and Hugh Savage (York University) for his program to predict the turn regions in proteins. We also thank Mark Patterson (Addenbrooke's Hospital, Cambridge, UK) for providing the compilation of androgen receptor mutants and Andrew Rut (Middlesex Hospital, London) for permission to refer to unpublished results on vitamin D receptor mutants.

REFERENCES

Alroy, I., and Freedman, L. P. (1992). DNA binding analysis of glucocorticoid specificity mutants. *Nucleic Acids Res.* **20,** 1045–1052.

Amero, S. A., Kretsinger, R. H., Moncrief, N. D., Yamamoto, K. R., and Pearson, W. R. (1992). The origin of nuclear receptor proteins: A single precursor distinct from other transcription factors. *Mol. Endocrinol.* **6,** 3–7.

Au-Fliegner, M., Helmer, E., Casanova, J., Raaka, B. M., and Samuels, H. H. (1993). The conserved ninth c-terminal heptad in thyroid hormone and retinoic acid receptors mediates diverse responses by affecting heterodimer but not homodimer formation. *Mol. Cell. Biol.* **13,** 5725–5737.

Baleja, J. D., Marmorstein, R., Harrison, S. C., and Wagner, G. (1992). Solution structure of the DNA-binding domain of Cd2-GAL4 from *S. cerevisiae. Nature (London)* **356,** 450–453.

Beato, M. (1989). Gene regulation by steroid hormones. *Cell (Cambridge, Mass.)* **56,** 335–344.

Benhamou, B., Garcia, T., Lerouge, T., Vergezac, A., Gofflo, D., Bigogne, C., Chambon, P. and Gronemeyer, H. (1992). A single amino acid that determines the sensitivity of progesterone receptors to RU486. *Science* **255,** 206–209.

Berkenstam, A., del Mar Vivianci Ruiz, M., Barettino, D., Horikoshi, M., and Stunnenberg, H. G. (1992). Cooperativity in transactivation between retinoic acid receptor and TFIID requires an activity analogous to E1A. *Cell (Cambridge, Mass.)* **69,** 401–412.

Bird, A. (1992). The essentials of DNA methylation. *Cell (Cambridge, Mass.)* **70,** 5–8.

Cadepond, F., Schweizer-Groyer, G., Segard, I., Jibard, N., Hollenberg, S., Giguère, V., Evans, R. M., and Baulieu, E. E. (1991). Heat shock protein 90 as a critical factor in maintaining glucocorticoid receptor in a nonfunctional state. *J. Biol. Chem.* **266,** 5834–5841.

Chang, Y. T., Higeon, C. J., and Brown, T. R. (1991). Human androgen insensitivity syndrome due to androgen receptor gene point mutations in subjects with normal androgen receptor levels but impaired biological activity. Abstract 28, 73rd Meeting of the American Endocrine Society, Washington, DC.

Cheskis, B., and Freedman, L. P. (1994). Ligand modulates the conversion of DNA-bound vitamin D_3 receptor (VDR) homodimers into VDR-retinoid X receptor heterodimers. *Moll Cell. Biol.,* **14,** 3329–3338.

Chothia, C. (1992). One thousand protein families for the molecular biologist. *Nature (London)* **357,** 543–544.

Dahlman-Wright, K., Siltala-Roos, H., Carlstedt-Duke, J., and Gustafsson, J.-A. (1990). Protein–protein interactions facilitate DNA binding by the glucocorticoid receptor DNA-binding domain. *J. Biol. Chem.* **265,** 14030–14035.

Dahlman-Wright, K., Wright, A., Gustafsson, J.-A., and Carstedt-Duke, J. (1991). Interaction of the glucocorticoid receptor DNA-binding domain with DNA as a dimer is mediated by a short segment of five amino acids. *J. Biol. Chem.* **266,** 3107–3112.

Danielian, P. S., White, R., Hoare, S. A., Fawell, S. E., and Parker, M. G. (1993). Identification of residues in the estrogen receptor which confer differential sensitivity to estrogen and hydroxytamoxifen. *Mol. Endocrinol.* **7,** 232–240.

Danielian, P. S., White, R., Lees, J. A., and Parker, M. G. (1992). Identification of a conserved region required for hormone dependent transcriptional activation by steroid hormone receptors. *EMBO J.* **11**, 1025–1033.

Danielson, M., Hinck, L., and Ringold, G. M. (1989). Two amino acids within the knuckle of the first zinc finger specify response element activation by the glucocorticoid receptor. *Cell* **57**, 1131–1138.

Desbois, C. (1991). Retinoid receptors and their role in cellular proliferation and differentiation. *Nucleic Acids Mol. Biol.* **7**, 148–157.

de Thé, H., Lavau, C., Marchilo, A., Chommienne, C., Degos, L., and Dejean, A. (1991). The PML–RARα fusion mRNA generated by the t(15;17) translocation in acute promyelocytic leukemia encodes a functionally altered RAR. *Cell (Cambridge, Mass.)* **66**, 675–684.

de Bellis, A., Quigley, C. A., Cariello, N. F., Elawady, M. K., Sar, M., Lane, M. V., Wilson, E. M., and French, F. S. (1992). Single base mutations in the human androgen receptor gene causing complete androgen insensitivity: Rapid detection by a modified denaturing gradient gel-electrophoresis technique. *Mol. Endocrinol.* **6**, 1909–1920.

Dreyer, C., Krey, G., Keller, H., Givel, F., Helftenbein, G., and Wahli, W. (1992). Control of the peroxisomal β-oxidation pathway by a novel family of nuclear hormone receptors. *Cell (Cambridge, Mass.)* **68**, 879–887.

Dyck, J. A., Maul, G. G., Miller, W. H., Chen, J. D., Kakizuka, A., and Evans, R. M. (1994). A novel macromolecular structure is a target of the promyleocyte-retinoic acid receptor oncoprotein. *Cell* **76**, 333–343.

Evans, R. M. (1989). The steroid and thyroid hormone receptor superfamily. *Science* **240**, 889–895.

Fairall, L., Schwabe, J. W. R., Chapman, L., Finch, J. T., and Rhodes, D. (1993). The crystal structure of two zinc fingers from tramtrack bound to DNA. *Nature (London)* **366**, 483–487.

Fawell, S. E., Lees, J. A., White, R., and Parker, M. G. (1990). Characterization and localization of steroid binding and dimerization activities in the mouse oestrogen receptor. *Cell (Cambridge, Mass.)* **60**, 953–962.

Fondell, J. D., Roy, A. L., and Roeder, R. G. (1993). Unliganded thyroid hormone receptor inhibits formation of a functional preinitiation complex: Implications for active repression. *Genes Dev.* **7**, 1400-1410.

Forman, B. M., and Samuels, H. H. (1990). Interactions among a subfamily of nuclear hormone receptors: The regulatory zipper model. *Mol. Endocrinol.* **4**, 1293–1300.

Forman, B. M., Yang, C.-R., Au, M., Casanova, J., Ghysdael, J., and Samuels, H. H. (1989). A domain containing leucine zipper like motifs mediate novel *in vivo* interactions between the thyroid hormone and retinoic acid receptors. *Mol. Endocrinol.* **3**, 1610–1626.

Freedman, L. P., and Luisi, B. F. (1993). On the mechanism of DNA binding by nuclear hormone receptors: A structural and functional perspective. *J. Cell. Biochem.* **51**, 140–150.

Freedman, L. P., and Towers, T. L. (1991). DNA binding properties of the vitamin D3 receptor zinc finger region. *Mol. Endocrinol.* **5**, 1815–1826.

Freedman, L. P., Luisi, B. F., Korszun, Z. R., Basavappa, R., Sigler, P. B., and Yamamoto, K. R. (1988a). The function and structure of the metal coordination sites within the glucocorticoid receptor DNA binding domain. *Nature (London)* **334**, 543–546.

Freedman, L. P., Yamamoto, K. R., Luisi, B. F., and Sigler, P. J. (1988b). More fingers in hand. *Cell (Cambridge, Mass.)* **54**, 444.

Freedman, L. P., Yoshinaga, S. K., Vanderbilt, J., and Yamamoto, K. R. (1989). *In vitro* transcriptional enhancement by purified glucocorticoid receptor derivatives. *Science* **245**, 298–301.

Freedman, L. P., Arce, V., and Perez-Fernandez, R. (1994). DNA sequences that act as high affinity targets for the vitamin D_3 receptor in the absence of RXR. *Mol. Endocrinol.* **8**, 265–273.

Godowski, P. J., Rusconi, S., Miesfeld, R., and Yamamoto, K. R. (1987). Glucocorticoid receptor mutants that are constitutive activators of transcriptional enhancement. *Nature (London)* **325**, 365–368.

Hall, B. L., Smit-McBride, Z., and Privalsky, M. L. (1993). Reconstitution of retinoid X receptor function and combinatorial regulation of other nuclear hormone receptors in the yeast *Saccharomyces cerevisiae. Proc. Natl. Acad. Sci. U.S.A.* **90**, 6929–6933.

Hard, T., Kellenbach, E., Boelens, R., Maler, B. A., Dahlman, K., Freedman, L. P., Carlstedt-Duke, J., Yamamoto, K. R., Gustafsson, J.-Å., and Kaptein, R. (1990a). Solution structure of the glucocorticoid receptor DNA-binding domain. *Science* **249**, 157–160.

Hard, T., Dahlman, K., Carstedt-Duke, J., Gustafsson, J.-A., and Rigler, R. (1990b). Cooperativity and specificity in the interactions between DNA and the glucocorticoid receptor DBA-binding domain. *Biochemistry* **29**, 5358–5364.

Harlow, K. W., Smith, D. N., Katzenellenbogen, Greene, G. L., and Katzenellenbogen, B. S. (1989). Identification of cysteine 530 as the covalent attachment site of an affinity-labelling estrogen (Ketononestrol aziridine) and antiestrogen (tamoxifen aziridine) in the human estrogen receptor. *J. Biol. Chem.* **264**, 17476–17485.

Howard, K. J., and Distelhorst, C. W. (1988). Evidence for intracellular association of the glucocorticoid receptor with the 90-kDa heat shock protein. *J. Biol. Chem.* **263**, 3474–3481.

Hughes, M. R., Malloy, P. J., Kieback, D. G., Kesterson, R. A., Pike, J. W., Feldman, D., and O'Malley, B. W. (1988). Point mutations in the human vitamin D receptor gene associated with hypocalcemic rickets. *Science* **242**, 1702–1705.

Imai, E., Miner, J. N., Yamamoto, K. R., and Granner, D. K. (1993). Glucocorticoid receptor–cAMP response element-binding protein interaction and the response of the phosphoenolpyruvate carboxykinase gene to glucocorticoids. *J. Biol. Chem.* **268**, 5853–5856.

Ing, N. H., Beekman, J. M., Tsai, S. Y., Tsai, M.-J., and O'Malley, B. W. (1992). Members of the steroid hormone receptor superfamily interact with TFIIB (S300–II). *J. Biol. Chem.* **267**, 17617–17623.

Issemann, I., and Green, S. (1990). Activation of a member of the steroid hormone receptor superfamily by peroxisome proliferators. *Nature (London)* **347**, 645–650.

Kakizuka, A., Miller, W. H., Umesono, K., Warrell, R. P., Frankel, S. R., Murty, V. V. V. S., Dmitrovsky, E., and Evans, R. M. (1991). Chromosomal translocation t(15;17) in human acute promyelocytic leukemia fuses RARα with a novel putative transcription factor, PML. *Cell (Cambridge, Mass.)* **66**, 663–674.

Kastner, P., Perez, A., Lutz, Y., Rochette-Egly, C., Gaub, M. P., Durand, B., Lanotte, M., Berger, R., and Chambon, P. (1992). Structure, localization, and transcriptional properties of two classes of retinoic acid α fusion proteins in acute promyelocytic leukemia (APL): Structural similarities with a new family of oncoproteins. *EMBO J.* **11**, 629–642.

Kerppola, T. K., Luk, D., and Curran, T. (1993). Fos is a preferential target of glucocorticoid receptor inhibition of AP-1 activity *in vitro. Mol. Cell. Biol.* **13**, 3782–3791.

Kim, B., and Little, J. W. (1992). Dimerization of a specific DNA-binding protein on the DNA. *Science* **255**, 203–206.

Klevit, R. E., Herriott, J. R., and Horvath, S. J. (1990). Solution structure of a zinc finger domain of yeast ADRI. *Proteins* **7**, 215–226.

Klock, G., Strahle, U., and Schutz, G. (1987). Oestrogen and glucocorticoid responsive elements are closely related but distinct. *Nature (London)* **329**, 734–736.

Klocker, H., Kaspar, F., Eberle, J., Uberreiter, S., Radmayr, C., and Bartsch, G. (1992). Point mutations in the DNA binding domain of the androgen receptor in two families with Reifenstein syndrome. *Am. J. Hum. Genet.* **50**, 1318–1327.

Knegtel, R. M. A., Katahira, M., Schilthuis, J. G., Bonvin, A. M. J. J., Boelens, R., Eib, D., van der Saag, P. T., and Kaptein, R. (1993). The solution structure of the human retinoic acid receptor-β DNA-binding domain. *J. Biol. NMR Res.* **3**, 1–17.

Koken, M. H. M., Puvion-Dutilleul, F., Guillemin, M. C., Viron, A., Linares-Cruz, G., Sturrman, N., de Jong, L., Szostecki, C., Calvo, F., Chomienne, C., Degos, L., Puvion, E., and de Thé, H. (1994). The t(15;17) translocation alters a nuclear body in a retinoic acid-reversible fashion. *EMBO J.* **13**, 1073–1083.

Kraulis, P. J. (1991). MOLSCRIPT: A program to produce both detailed and schematic plots of protein structure. *J. Appl. Crystallogr.* **24**, 946–950.

Kraulis, P. J., Raine, A. R. C., Gadhavi, P. L., and Laue, E. D. (1992). Structure of the zinc-containing DNA-binding domain of GAL4. *Nature (London)* **356**, 448–450.

Kumar, V., and Chambon, P. (1988). The estrogen receptor binds tightly to its responsive element as a ligand-induced homodimer. *Cell (Cambridge, Mass.)* **55**, 145–156.

Kurokawa, R., Yu, V. C., Naar, A., Kyakumoto, S., Han, Z., Silverman, S., Rosenfeld, M. G., and Glass, C. K. (1993). Differential orientations of the DNA-binding domain and carboxy-terminal dimerization interface regulate binding site selection by nuclear receptor heterodimers. *Genes Dev.* **7**, 1423–1435.

LaBaer, J. (1989). Ph.D. Thesis, University of California, San Francisco.

Landschulz, W. H., Johnson, P. F., McKnight, S. L. (1988). The leucine zipper: A hypothetical structure common to a new class of DNA binding proteins. *Science* **240**, 1759–1764.

Laudet, V., Hanni, C., Coll, J., Catzflis, F., and Stehelin, D. (1992). Evolution of the nuclear receptor gene superfamily. *EMBO J.* **11**, 1003–1013.

Lee, M. S., Gippert, G. P., Soman, K. V., Case, D. A., and Wright, P. E. (1989). Three-dimensional solution structure of a single zinc finger DNA-binding domain. *Science* **245**, 635–637.

Lee, M. S., Kliewer, S. A., Provencal, J., Wright, P. E., and Evans, R. M. (1993). Structure of the retinoid X receptor α DNA binding domain: A helix required for homodimeric DNA binding. *Science* **260**, 1117–1121.

Lohnes, D., Dierich, A., Ghyselinck, N., Kastner, P., Lampron, C., LeMeur, M., Lufkin, T., Mendelsohn, C., Nakshatri, H., and Chambon, P. (1992a). Retinoid receptors and binding proteins. *In* "Transcriptional Regulation in Cell Differentiation and Development" (P. Rigby, R. Krumlauf, and Grosveld), pp. 69–76. Company of Biologists, Cambridge, UK.

Lohnes, D., Kastner, P., Dierich, A., Mark, M., LeMeur, M., and Chambon, P. (1992b). Function of retinoic acid receptor γ in the mouse. *Cell (Cambridge, Mass.)* **73,** 643–658.

Lufkin, T., Lohnes, D., Mark, M., Dierich, A., Gorry, P., Gaub, M.-P., LeMeur, M., and Chambon, P. (1993). High postnatal lethality and testis degeneration in retinoic acid receptor α mutant mice. *Proc. Natl. Acad. Sci. U.S.A.* **90**, 7725–7229.

Luisi, B. F., Xu, W. X., Otwinowski, Z., Freedman, L. P., Yamamoto, K. R., and Sigler, P. B. (1991). Crystallographic analysis of the interaction of the glucocorticoid receptor with DNA. *Nature (London)* **352**, 497–505.

Mader, S., Kumar, V., de Verneuil, H., and Chambon, P. (1989). Three amino acids of the oestrogen receptor are essential to its ability to distinguish an oestrogen from a glucocorticoid-responsive element. *Nature (London)* **338**, 271–274.

Marcelli, M., Zoppi, S., Grino, P. B., Griffin, J. E., Wilson, J. D., and McPhaul, M. J. (1991). A mutation in the DNA-binding domain of the androgen receptor gene causes complete testicular feminization in a patient with receptor-positive androgen resistance. *J. Clin. Invest.* **87**, 1123–1126.

Marks, M. S., Levi, B.-Z., Segars, J. H., Driggers, P. H., Hirschfeld, S., Nagata, T., Appella, E., and Ozato, K. (1992). H-2RIIBP expressed from a baculovirus vector binds to multiple hormone response elements. *Mol. Endocrinol.* **6**, 219–230.

Marmorstein, R., Carey, M., Ptashne, M., and Harrison, S. C. (1992). DNA recognition by GAL4: Structure of a protein/DNA complex. *Nature (London)* **356**, 408–414.

Meyer, M.-E., Pornon, A., Ji, J., Bocquel, M.-T., Chambon, P., and Gronemeyer, H. (1990). Agonistic and antagonistic activities of RU486 on the functions of the human progesterone receptor. *EMBO J.* **9**, 3923–3932.

Miller, J., McLachlan, A. D., and Klug, A. (1985). Repetitive zinc-binding domains in the protein transcription factor IIIA from *Xenopus* oocytes. *EMBO J.* **4**, 1609–1614.

Miner, J. N., and Yamamoto, K. R. (1992). Regulatory cross-talk at composite response elements. *Trends Biochem. Sci.* **16**, 423–426.

Morellet, N., Jullian, N., De Rocquigny, H., Maigret, B., Darlix, J.-L., and Roques, B. P. (1992). Determination of the structure of the nucleocapsid protein NCp7 from the human immunodeficiency virus type 1 by ^{1}H NMR. *EMBO J.* **11**, 3059–3065.

Mowszowicz, I., Lee, H.-J., Chen, H. T., Mestayer, C., Portois, M.-C., Cabrol, S., Mauvais-Jarvis, P., and Chang, C. (1993). A point mutation in the second zinc finger of the DNA binding domain of the androgen receptor gene causes complete androgen insensitivity in two siblings with receptor-positive androgen resistance. *Mol. Endocrinol.* **7**, 861–869.

Naar, A. M., Boutin, J.-M., Lipkin, S. M., Yu, V. C., Holloway, J. M., Glass, C. K., and Rosenfeld, M. G. (1991). The orientation and spacing of core DNA-binding motifs dictate selective transcriptional responses to three nuclear receptors. *Cell (Cambridge, Mass.)* **65**, 1267–1279.

Nagpal, S., Friant, S., Nakshatri, H., and Chambon, P. (1993). RARs and RXRs: Evidence for two autonomous transactivation functions (AF-1 and AF-2) and heterodimerization *in vivo*. *EMBO J.* **12**, 2349–2360.

Neuhaus, D., Nakaseko, Y., Schwabe, J. W. R., and Klug, A. (1992). Solution structures of two zinc fingers domains from SWI5, obtained using two-dimensional ^{1}H NMR spectroscopy. A zinc finger structure with a third strand of β-sheet. *J. Mol. Biol.* **228**, 637–651.

Nordeen, S. K., Suh, B. J., Kuhnel, B., and Hutchison, C. A. (1992). Structural determinants of a glucocorticoid receptor recognition element. *Mol. Endrocrinol.* **4**, 1866–1873.

Omichinski, J. G., Clore, G. M., Schaad, O., Felsenfeld, G., Trainor, C., Appella, E., Stahl, S. J., and Gronenborn, A. M. (1993). NMR structure of a specific complex of Zn-containing DNA binding domain of GATA-1. *Science* **261**, 438–446.

Pakdel, F., and Katzenellenbogen, B. S. (1992). Human estrogen receptor mutants with altered estrogen and antiestrogen ligand discrimination. *J. Biol. Chem.* **267**, 3429–3437.

Pavletich, N. P., and Pabo, C. O. (1991). Zinc finger–DNA recognition: Crystal structure of a Zif268–DNA complex at 2.1 Å. *Science* **252**, 809–817.

Payvar, F., DeFranco, D., Firestone, G. L., Edgar, B., Wrange, O., Okret, S., Gustafsson, J. A., and Yamamoto, K. R. (1983). Sequence-specific binding of the glucocorticoid receptor to MTV-DNA at sites within and upstream of the transcribed region. *Cell (Cambridge, Mass.)* **35**, 381–392.

Perlmann, T., Rangarajan, P. N., Umesono, K., and Evans, R. M. (1993). Determinants for selective RAR and TR recognition of direct repeat HREs. *Genes Dev.* **7**, 1411–1422.

Perutz, M. F. (1990). Frequency of abnormal human haemoglobin caused by C→T transition in CpG dinucleotides. *J. Mol. Biol.* **213**, 203–206.

Picard, D., Salser, S. J., and Yamamoto, K. R. (1988). A movable and regulable inactivation function within the steroid binding domain of the glucocorticoid receptor. *Cell (Cambridge, Mass.)* **54**, 1073–1080.

Picard, D., Khursheed, B., Garabedian, M. J., Fortin, M. G., Lindquist, S., and Yamamoto, K. R. (1990). Reduced levels of hsp90 compromise steroid receptor action *in vivo. Nature (London)* **348**, 166–168.

Pinsky, L., Trifiro, M., Kaufman, M., Beitel, L. K., Mahtre, A., Kazemi-Esfarjani, P., Sabbaghian, N., Lumbroso, R., Alvarado, C., Vasiliou, A., and Gottlieb, B. (1992). Androgen resistance due to mutation of the androgen receptor. *Clin. Invest. Med.* **15**, 456–472.

Pratt, W. B., Jolly, D. J., Pratt, D. V., Hollenberg, S. M., Giguère, V., Cadepond, F. M., Schweizer-Groyer, G., Catelli, M.-G., Evans, R. M., and Baulieu, E.-E. (1988). A region of the steroid binding domain determines formation of the non-DNA-binding 9S glucocorticoid receptor complex. *J. Biol. Chem.* **263**, 267–273.

Qian, X., Jeon, C., Yoon, H., Agarwal, K., and Weiss, M. A. (1993). Structure of a new nucleic-acid-binding motif in eukaryotic transcriptional elongation factor TFIIS. *Nature (London)* **365**, 277–279.

Quigley, C. A., Evans, B. A. J., Simental, J. A,. Marschke, K. B., Sar, M., Lubahn, D. B., Davies, P., Hughes, I. A., Wilson, E. M., and French, F. S. (1992). Complete androgen insensitivity due to deletion of exon C of the androgen receptor highlights the functional importance of the second zinc finger of the androgen receptor *in vivo. Mol. Endocrinol.* **6**, 1103–1112.

Ratajcak, T., Wilkinson, S. P., Brockway, M. J., Hahnel, R., Moritz, R. L., Begg, G. S., and Simpson, R. J. (1989). The interaction site for tamoxifen aziridine with the bovine estrogen receptor. *J. Biol. Chem.* **264**, 13453–13459.

Reichel, H., Koeffler, M. D., and Norman, A. W. (1992). The role of the vitamin D endocrine system in health and disease. *N. Engl. J. Med.* **320**, 980–991.

Rhodes, D., and Klug, A. (1993). Zinc fingers. *Sci. Am.* **268**, 56–68.

Rut, A. R., Hewison, M., Kristjansson, K., Luisi, B., Hughes, M. R., and O'Riordan, J. L. H. (1994). A stereochemical model for vitamin D resistant rickets. Submitted for publication.

Saijo, T. M., Ito, M., Takeda, E., Huq, A. H. M. M., Naito, E., Yokota, I., Sone, T., Pike, J. W., and Kuroda, Y. (1991). A unique mutation in the vitamin D receptor gene in three Japanese patients with vitamin D-dependent rickets. Type II: Utility of single stranded conformation polymorphism analysis for heterozygous carrier detection. *Am. J. Hum. Genet.* **49**, 668–673.

Saunders, P. T. K., Padayachi, T., Tincello, D. G., Shalet, S. M., and Wu, F. C. W. (1992). Point mutatutions detected in the androgen receptor of three men with partial androgen insensitivity syndome. *Clin. Endocrinol.* **37**, 214–220.

Schena, M., and Yamamoto, K. R. (1988). Mammalian glucocorticoid receptor derivatives enhance transcription in yeast. *Science* **241**, 965–967.

Schena, M., Freedman, L. P., and Yamamoto, K. R. (1989). Mutations in the glucocorticoid receptor zinc finger region that distinguish interdigitated DNA binding and transcriptional enhancement activities. *Genes Dev.* **3**, 1590–1601.

Schena, M., Lloyd, A. M., and Davis, R. W. (1991). A steroid-inducible gene expression system for plant cells. *Proc. Natl. Acad. Sci. U.S.A.* **88,** 10421–10425.

Schmitt, J., and Stunnenberg, H. G. (1993). The glucocorticoid receptor hormone binding domain mediates transcriptional activation *in vitro* in the absence of ligand. *Nucleic Acids Res.* **21,** 2673–2681.

Schule, R., and Evans, R. M. (1992). Cross-coupling of signal transduction pathways: Zinc finger meets leucine zipper. *Trends Genet. Sci.* **7,** 377–380.

Schwabe, J. W. R., Neuhaus, D., and Rhodes, D. (1990). Solution structure of the DNA-binding domain of the oestrogen receptor. *Nature (London)* **348,** 458–461.

Schwabe, J. W. R., Chapman, L., Finch, J. T., and Rhodes, D. (1993a). The crystal structure of the oestrogen receptor DNA-binding domain bound to DNA: How receptors discriminate between their response elements. *Cell (Cambridge, Mass.)* **75,** 1–12.

Schwabe, J. W. R., Chapman, L., Finch, J. T., Rhodes, D., and Neuhaus, D. (1993b). DNA recognition by the oestrogen receptor: From solution to the crystal. *Structure* **1,** 187–204.

Seagraves, W. A. (1991). Something old, some things new: The steroid receptor superfamily in Drosophila. *Cell* **67,** 225–228.

Segars, J. H., Marks, M. S., Hirschfield, S., Driggers, P. H., Martinez, E., Grippo, J. F., Wahli, W., and Ozzto, K. (1993). Inhibition of estrogen-responsive gene activation by the retinoid X receptor β: Evidence for multiple inhibitory pathways. *Mol. Cell. Biol.* **13,** 2258–2268.

Sone, T., Kerner, S. A., Saijo, T., Takeda, E., and Pike, J. W. (1991). Mutations in the DNA binding domain of the vitamin D receptor associated with hereditary resistance of 1, 25-dihydroxyvitamin D3. *In* "Norman A. W., Bouillon, R., Thomasset, M. (eds.) "Vitamin D: Gene Regulation, Structure-Function Analysis, and Clinical Applications," pp. 84–85. de Gruyter, New York.

Sone, T., Marx, S. J., Marx, J., Liberman, U. A., and Pike, J. W. (1990). A unique point mutation in the human vitamin D receptor chromosomal gene confers hereditary resistance to 1,25-dihydroxyvitamin D3. *Mol. Endocrinol.* **4,** 623–631.

Strahle, U., Boshart, M., Klock, G., Stewart, F., and Schutz, G. (1989). Glucocorticoid and progesterone-specific effects are determined by differential expression of the respective hormone receptors. *Nature (London)* **339,** 629–632.

Summers, M. F., South, T. L., Kim, B., and Hare, D. S. (1991). High resolution structure of an HIV zinc fingerlike domain via a new NMR-based distance geometry approach. *Biochemistry* **29,** 329–340.

Towers, T. L., Luisi, B. F., Asianov, A., and Freedman, L. P. (1993). DNA target selectivity by the vitamin D_3 receptor: Mechanism of dimer binding to an asymmetric repeat element. *Proc. Natl. Acad. Sci. U.S.A.* **90,** 6310–6314.

Truss, M., Chalepakis, G., and Beato, M. (1990). Contacts between steroid hormone receptors and thymines in DNA: An interference method. *Proc. Natl. Acad. Sci. U.S.A.* **87,** 7180–7184.

Tsukiyama, T., Ueda, H., Hirose, S., and Niwa, O. (1992). Embryonal long terminal repeat-binding protein is a murine homolog of FTZ-F1, a member of the steroid receptor superfamily. *Mol. Cell. Biol.* **12,** 1286–1291.

Umesono, K., and Evans, R. M. (1989). Determinants of target gene specificity for steroid/thyroid hormone receptors. *Cell (Cambridge, Mass.)* **57,** 1139–1146.

Umesono, K., Giguère, V., Glass, C. K., Rosenfeld, M. G., and Evans, R. M. (1988). Retinoic acid and thyroid hormone induce gene expression through a common responsive element. *Nature (London)* **336,** 262–265.

Umesono, K., Murakami, K. K., Thompson, C. C., and Evans, R. M. (1991). Direct repeats as selective response elements for the thyroid hormone, retinoic acid and vitamin D_3 receptors. *Cell (Cambridge, Mass.)* **65,** 1255–1266.

Weis, K., Rambaud, S., Lavau, C., Jansen, J., Carvalho, T., Carmo-Fonseca, M., Lamond, A., and Dejean, A. (1994). Retinoic acid regulates aberrant nuclear localization of PML–RARα in acute promyelocytic leukemia cells. *Cell* **76,** 345–356.

Wilson, T. E., Paulsen, R. E., Padgett, K. A., and Milbrandt, J. (1992). Participation of non-zinc finger residues in DNA binding by two orphan receptors. *Science* **256,** 107–110.

Wooster, R., Mangion, J., Eeles, R., Smith, S., Dowsett, M., Averill, D., Barrett-Lee, P., Easton, D. F., Ponder, B. A., and Stratton, M. R. (1992). A germline mutation in the androgen receptor gene in two brothers with breast cancer and Reifenstein syndrome. *Nat. Genet.* **2,** 132–134.

Yagi, H., Ozono, K., Miyake, H., Nagashima, K., Kuroume, T., and Pike, J. W. (1993). A new point mutation in the deoxyribonucleic acid-binding domain of the vitamin D receptor in a kindred with hereditary 1,25-dihydroxyvitamin D-resistant rickets. *J. Clin. Endrocrinol. Metab.* **76,** 509–515.

Yoshinaga, S. K., and Yamamoto, K. R. (1992). Signalling and regulation by a mammalian glucocorticoid receptor in *Drosophila* cells. *Mol. Endocrinol.* **5,** 844–853.

Yoshinaga, S. K., Peterson, C. L., Herskowitz, I., and Yamamoto, K. R. (1992). Roles of SWI1, SWI2, and SWI3 proteins for transcriptional enhancement by steroid receptors. *Science* **258,** 1598–1604.

Zechel, C., Shen, X-Q, Chambon, P., and Gronemeyer, H. (1994a). Dimerization interfaces formed between the DNA binding domains determine the cooperative binding of RXR/RAR and RXR/TR heterodimers to DR5 and DR4 elements. *EMBO J.* **13,** 1414–1424.

Zechel, C., Shen, X-Q., Chen, J.-Y., Chen, Z.-P., Chambon, P., and Gronemeyer, H. (1994b). The dimerization interfaces formed between the DNA binding domains of RXR, RAR and TR determine the binding specificity and polarity of the full-length receptors to direct repeats. *EMBO J.* **13,** 1425–1433.

Zhang, X.-K., Hoffmann, B., Tran, B.-V., Graupner, G., and Pfahl, M. (1992). Retinoid X receptor is an auxiliary protein for thyroid hormone and retinoic acid receptors. *Nature (London)* **355,** 441–446.

Zhang, X.-K., Lehmann, J., Hoffmann, B., Dawson, M. I., Cameron, J., Graupner, G., Hermann, T., Tran, P., and Pfahl, M. (1992). Homodimer formation of retinoid X receptor induced by 9-*cis* retinoic acid. *Nature (London)* **358,** 587–591.

Zoppi, S., Marcelli, M., Deslypere, J.-P., Griffin, J. E., Wilson, J. D., and McPaul, M. J. (1992). Amino acid substitutions in the DNA-binding domain of the human androgen receptor are a frequent cause of receptor-binding positive androgen resistance. *Mol. Endocrinol.* **6,** 409–415.

Function/Activity of Specific Amino Acids in Glucocorticoid Receptors

S. STONEY SIMONS, JR.[1]

Steroid Hormones Section/LMCB, NIDDK
National Institutes of Health
Bethesda, Maryland 20892

I. Introduction

The functions initially ascribed to the glucocorticoid receptor were an ability to bind cognate ligands, to bind DNA, and to regulate the transcription of responsive genes. Furthermore, early results indicated that these activities were conveniently segregated to distinct regions of the receptor protein. This domain structure of the receptor was

---

[1] Address correspondence to Building 8, Room B2A-07, NIDDK/LMCB, NIH, Bethesda, MD 20892 (Phone: 301-496-6796; FAX: 301-402-3572).

confirmed once the receptor had been cloned (Evans, 1988; Beato, 1989). However, as subsequent studies have uncovered additional transcriptionally active regions and other activities, such as stabilization by sodium molybdate, association with heat shock proteins, and receptor dimerization, the compartmentalization of activities began to break down with many activities overlapping (Chakraborti *et al.*, 1992) (Fig. 1). Thus, the prospects of understanding glucocorticoid receptor activity at the level of each amino acid seemed to diminish. Fortunately, the availability of receptors altered by molecular biological manipulations, coupled with NMR and X-ray analysis of purified fragments, has now made an analysis of the role of specific amino acids in various glucocorticoid receptor structures and functions appear to be an achievable objective.

This new technology does, however, have inherent limitations that restrict the interpretations of the data. For example, NMR and X-ray analyses give static pictures of pure proteins in what may be viewed as an artificial environment. One of the most interesting interactions of glucocorticoid receptors to be examined by these techniques is the binding of receptor dimers to glucocorticoid response elements (GREs), which are specific DNA sequences composed of the two palindromic half-sites that are required for steroid-induced gene transcription (Beato, 1989). The binding of truncated receptors was relatively independent of the number of nucleotides between half-sites (with $n = 3$ or 4) in an X-ray study (Luisi *et al.*, 1991) but displayed an almost absolute requirement for a spacing of $n = 3$ nucleotides in gel shift assays (Chalepakis *et al.*, 1990; Dahlman-Wright *et al.*, 1990). Furthermore, though a spacing of $n = 4$ gave the best resolution in the X-ray study, the same spacing between GREs has <7% of the ability of the same construct with a spacing of $n = 3$ to support glucocorticoid-induced transcription (Chalepakis *et al.*, 1990; Dahlman-Wright *et al.*, 1991). Thus, when the half-site spacing is four nucleotides, the otherwise low-affinity complex may have been uniquely observable by X-ray crystallography owing to stabilization by singular parameters, such as crystal packing forces.

Point mutations and deletions theoretically offer the best method of examining the role of individual amino acids in the expression of biological activities but such substitutions can have unexpected repercussions. For example, protein folding is an extremely complex and poorly understood phenomenon in which nonadjacent amino acids can influence the final tertiary structure. Even partially folded protein structures do not have to involve contiguous domains (Hughson *et al.*, 1990). It has been suggested that protein stability is influenced by the equi-

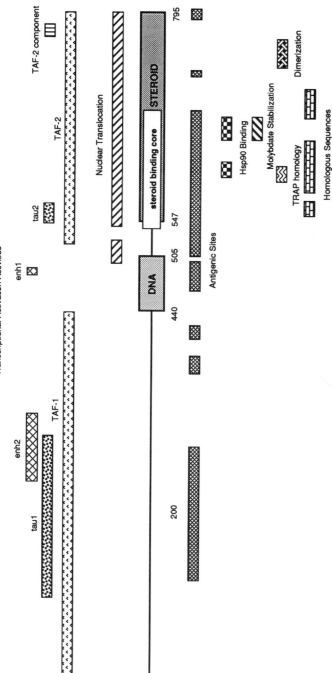

FIG. 1. Activities present in the glucocorticoid receptor. The entire rat glucocorticoid receptor is represented by the solid line and overlapping boxes, which correspond to the DNA and steroid binding domains. Selected amino acid positions, including those for the boundaries of the DNA and steroid binding domains, are indicated by the numbers below the linear diagram. Some of the major properties and activities, for which defined sequences are available, are indicated. See text for further description.

librium distribution of partially denatured states, which may reflect the conformations of newly synthesized protein emerging from the ribosome (Flanagan *et al.*, 1992) and can be radically altered by single amino acid substitutions (Flanagan *et al.*, 1993). Protein folding may also be altered by single amino acid deletions (Thomas *et al.*, 1992). Conversely, replacement by dissimilar amino acids can cause no noticeable changes as determined by conformationally sensitive monoclonal antibodies (Cunningham and Wells, 1989). The ideal amino acid to use in point mutagenesis studies would be completely "neutral." By the criteria of size, polarity, and hydrogen bonding, serine was formerly thought to be the most neutral substitution for examining the effects of the thiol group of cysteine. However, other data suggest that alanine may be a less disruptive replacement for cysteine (Weissman and Kim, 1991). In fact, based on calculations of amino acid size, protein chain conformation, and hydrophobicity, alanine is currently thought to be the best "neutral" amino acid (Cunningham and Wells, 1989). In summary, it seems impossible to determine with certainty the function of specific amino acids by the foregoing techniques without much additional physicochemical data concerning the integrity of each protein.

Despite the limitations of these techniques, they have been used to accumulate a vast amount of useful data since glucocorticoid receptors were first sequenced about 8 years ago (Hollenberg *et al.*, 1985; Miesfeld *et al.*, 1986; Danielsen *et al.*, 1986). The purpose of this article is to begin to compile what is known, at a molecular level, about the role of individual amino acids in the structure and function of the glucocorticoid receptor. The discussion of the data has been divided into two general sections. The first pertains to data where large but defined sequences of the receptor have been examined. The second section will cover the consequences of individual point mutations. In all sections, the numbering of the rat glucocorticoid receptor is used in an effort to limit confusion about the exact amino acid(s) involved. The species from which the data were originally generated is indicated when it appears to be relevant. Many of the data presented are limited to those for intact receptors since the interactions of separate segments may be required for the wild-type response (see Section IV).

II. Structure/Function Relationships in the Glucocorticoid Receptor Domains

It is very useful to view the glucocorticoid (and all other steroid receptors) by functional domains (Evans, 1988; Beato, 1989) (Fig. 1). However, there is considerable ambiguity over the precise amino acids defining the ends of each domain. Thus, the DNA binding domain has been variously described as spanning amino acids 390–521 (Baniahmad et al., 1991), 439–509 (Schüle et al., 1990), 440–505 (Evans, 1988), and 440–525 (Godowski et al., 1988). Similarly, the steroid binding domain has been cited as encompassing 574–795 (Ohara-Nemoto et al., 1990), 568–795 (Schüle et al., 1990), 547–795 (Rusconi and Yamamoto, 1987), and 546–795 (Evans, 1988). For the purposes of this review, we consider the DNA binding domain to cover amino acids 440 to 505 and the steroid binding domain to extend from 547 to 795.

A. Hypervariable, Immunogenic, or Transactivation Domain (Amino Acids 1–439)

The amino-terminal half of the receptor, up to the start of the DNA binding domain, is the largest single domain of the glucocorticoid receptor. Its name is derived from the fact that this region displayed the least homology between different receptors (Evans, 1988). Furthermore, almost all the initial antireceptor antibodies were found to have epitopes in this portion of the receptor (Fig. 1). Soon thereafter, it was determined from transient transfections of truncated receptors that this entire domain harbored transcriptional activity, which was called TAF-1 (Bocquel et al., 1989; Tasset et al., 1990), and later AF-1 (Nagpal et al., 1992).

1. Antigenic Sites

Numerous antibodies have been raised against sites in the immunogenic domain. Starting at the amino-terminal end of the domain, the monoclonal antibody 250 against the rat receptor (Okret et al., 1984) recognizes the sequence 119–273 (Rusconi and Yamamoto, 1987), while the epitope for the monoclonal antirat receptor antibody mab 49 (Westphal et al., 1982) lies between amino acids 120 and 160 (Hoeck and Groner, 1990). The rat sequence 127–235 has been used to produce the monoclonal antibody MGR3H6 (Demonacos et al., 1993; C. E. Sekeris, personal communication). Three other polyclonal antibodies have been developed against overlapping peptides of the human receptor. The sequence closest to the amino terminus, that is, 144–172 (=165–193 of rat), afforded the antibody GR135 (Hollenberg et al., 1987), whereas the peptide corresponding to 150–175 (=171–205 of rat) yielded the second antibody (Srivastava and Thompson, 1990). The third antipeptide antibody was against 171–184 (=192–205 of rat) and cross-reacts with human, rat, and mouse receptors (Antakly et al., 1990). Two polyclonal antipeptide antibodies to 245–259, and to 346–367, of the human receptor (=265–279 and 366–387, respectively, of rat) will recognize human, rat, and mouse receptors (Cidlowski *et al.*, 1990). Seven monoclonal antibodies, including mab 49, were raised against partially purified rat receptors. These antibodies appear to recognize the amino-terminal end of the receptor because they immunoprecipitated 98-kDa receptors but not the degraded 40-kDa species that were obtained from frozen samples (Westphal *et al.*, 1982, 1984) and presumably represent proteolyzed fragments lacking the amino-terminal half of the receptor (Carlstedt-Duke *et al.*, 1987). An additional seven polyclonal antibodies were obtained after immunizing rabbits with purified rat receptors. Each antibody recognized epitopes amino terminal to the chymotrypsin cleavage site at 410. Several of these antibodies cross-reacted with mouse, rabbit, and chick receptors but only two recognized human receptors (Okret, 1983).

The most widely used antibody has been BUGR-2 (Gametchu and Harrison, 1984), which recognizes rat and mouse but not human (Harmon *et al.*, 1984) or avian (Hendry *et al.*, 1993) receptors. Another monoclonal antibody (FIGR) has been independently prepared and appears to recognize the same epitope as BUGR-2 (Bodwell *et al.*, 1991). The site of BUGR-2 recognition was originally mapped to 407–423 (Rusconi and Yamamoto, 1987) and has been restricted to 407–416 (Ip *et al.*, 1991). It should be noted, though, that amino acids 407–409 are not required for antibody binding because the 42-kDa chymotryptic receptor fragment, which starts at 410 (Carlstedt-Duke *et al.*, 1987), is strongly recognized by BUGR-2 both on solid supports by Western blotting (Dahlman *et al.*, 1989; Chakraborti *et al.*, 1990; Opoku and Simons, 1993) and in solution (Chakraborti *et al.*, 1992). It is not known whether the secondary or tertiary structure of this sequence is important for recognition. However, this antigenic sequence can be fused to other regions of the receptor without a loss of reactivity with BUGR-2 (Rusconi and Yamamoto, 1987). Ip *et al.* (1991) have found that the adjacent sequence of 417–423 is recognized by a polyclonal antipeptide antibody that was produced from the mouse receptor sequence of 395–411 (=407–423 of rat) (Shea *et al.*, 1991). This polyclonal antibody only recognizes those complexes of mouse, rat, and rabbit, but not human, receptors that have been activated to the DNA binding form.

AC88 was developed as an antibody against heat shock protein 90 (hsp90) but cross-reacts with sequences in the amino-terminal half of the rat glucocorticoid receptor (Howard *et al.*, 1990). Thus, studies on hsp90 association with rat glucocorticoid receptors that employ the AC88 antibody need to be interpreted with care.

It should be realized that most antibodies will detect proteins other than that from which they were produced. Thus, the BuGR2 monoclonal antibody will immunoadsorb non-receptor proteins. Interestingly, these same non-receptor proteins were not detected by BuGR2 on Western blots (Hendry et al., 1993). Even antipeptide antibodies do not always recognize the corresponding primary sequence in proteins. Nonlinear epitopes, constituted by appropriately folded proteins, can also be recognized. Thus, an antibody that was raised against the peptide DDDED also recognizes the heat shock cognate protein hsc70, which does not contain the DDDED sequence (Imamoto et al., 1992).

2. Transcriptional Activity

This entire domain has been identified as possessing transcriptional activity and is referred to as TAF-1 (Bocquel et al., 1989; Tasset et al., 1990) or AF-1 (Nagpal et al., 1992). This activity has been limited to amino acids 98–282 (Hollenberg and Evans, 1988) or -292 (Schüle et al., 1990) of the rat receptor, as defined by deletion experiments with the human receptor in CV-1 cells, and has been called tau1 (τ1). This activity may be further restricted to 208–305, based on studies with the mouse receptor in COS-7 cells (Danielsen et al., 1987). This 208–305 sequence is also thought to reduce the nonspecific DNA binding of receptors (Danielsen et al., 1987). The same apparent transcriptional activity was localized to 237–318 of the rat receptor in CV-1 cells and called enh2, although considerable additional activity ($\sim$45%) resided between 106 and 236 (Godowski et al., 1988). A 58-bp region of human receptors (=208–264 in rat) retained 60–70% of the activity of the intact tau1 when assayed in yeast (Dahlman-Wright et al., 1994).

When a BamH1 linker was inserted at various locations of the tau1 region, the transcriptional activity in a transient transfection assay was almost always decreased or eliminated (Giguère et al., 1986). In addition to its transcriptional activity, the tau1 region is also capable of squelching transcription from the CYC1 promoter in yeast (Wright et al., 1991; McEwan et al., 1993). Thus, cotransfection of the tau1 sequence reduced the basal activity of the reporter gene, presumably by competing for the binding of limiting transcription factors. Using other tau1 constructs (=98–282 of rat) containing either an insertion (at position 141) or the DNA binding domain (434–519) in a yeast cell-free transcription system, it was found that the ability of E. coli overexpressed receptor fragments to squelch was proportional to their transactivation activity (McEwan et al., 1993). However, further deletions revealed a 41-bp sequence (=187–227 in human, =208–247 in rat) that, in yeast cells, retained full squelching activity but lost >80% of the transcriptional activity of tau1 (Dahlman-Wright et al., 1994). In a cell-free transcription assay with HeLa cell nuclear extracts, a recombinant truncated receptor containing both the tau1 sequence and the DNA binding domain (2-523) was found to be transcriptionally active (Schmitt and Stunnenberg, 1993) (see Section II,B,7).

3. *Activities at Other Sites*

An early noted feature of the rat receptor was a string of 21 glutamines at 75–96 interrupted only by an arginine at 77 (Miesfeld *et al.*, 1986). Homology comparisons with the mouse (Danielsen *et al.*, 1986) and human (Hollenberg *et al.*, 1985) receptors, which contain only 9 and 2 glutamines respectively, suggested that this sequence was unimportant. However, other studies indicate to the contrary (Lanz *et al.*, 1993). The replacement of the polyglutamine tract by polyserine was tolerated but polyalanine eliminated the transcriptional activity of the mutant receptors in a transient transfection assay. The replacement by a stretch of random amino acids, this time in a truncated receptor (1–556) lacking the steroid binding domain, was also detrimental due to an increased instability of the protein. Sequences of polyalanine, and even polyserine, longer than 23 amino acids dramatically reduced the transcriptional activity of receptors, whereas up to 80 glutamines could be inserted with little effect. Short stretches of polyalanine or polyserine were also without effect but there was an abrupt loss of activity in going from 23 to 25 alanines. This was postulated as being due to some sort of transient membrane attachment by the polyalanine that prevents a crucial posttranslational modification of the receptors, although the stability of steroid-bound receptors was not explicitly addressed (Lanz *et al.*, 1993). Each of these mutants does contain extra amino acids at one or both ends of the repetitive stretch. However, it is unlikely that these extra amino acids are significant because none of the mutants with repeats of less than 20 amino acids displayed any decreased transcriptional activity. The full-length receptor with a polyalanine ($n = 23$) substitution was also found to have dominant negative activity in that it repressed the activity of cotransfected wild-type truncated (1–556) receptors in transient transfection assays (Lanz *et al.*, 1993). Thus, glutamine appears to be uniquely neutral at this position of the receptor.

The observed molecular weight of glucocorticoid receptors on SDS gels is about 10,000 larger than that calculated from the deduced amino acid sequence (Reichman *et al.*, 1984; Miesfeld *et al.*, 1986; Urda *et al.*, 1989). Such a large difference in M_r could, theoretically, be due to glycosylation. However, studies with glycosidases, specific lectins, and [^{3}H]D-glucosamine all argue against there being glycosylation of the rat receptor in tissue culture cells (HTC cells) (Yen and Simons, 1991). A cyanogen bromide cleavage product of the mouse receptor (=120–336 of rat) has recently been found to migrate on SDS–polyacrylamide gels with an apparent molecular weight that is ~12,000 larger than the calculated 23,400. It was therefore proposed that some property of this sequence is responsible for the aberrant migration of the intact receptors (Hutchison et al., 1993). It will be interesting to see if the abnormal gel behavior of this region is due to posttranslational modifications or is intrinsic to the amino acid sequence.

Mutant receptors with the phenotype of increased nuclear binding (nt^i mutants) were found to be missing the amino-terminal 404 amino acids, as a result of aberrant splicing between exons 1 and 3 of the mRNA. These receptors may also contain an extra 41 nonreceptor amino acids at the N terminus owing to the initiation of translation from an exon 1 methionine (Dieken et al., 1990).

B. DNA BINDING DOMAIN (AMINO ACIDS 440–505)

The DNA-binding domain is required for the binding of receptors to both nonspecific and specific, or biologically active, DNA. This binding of activated receptor–steroid complexes to specific DNA sites is, in turn, required for the production of a biological response (see Sections II,B,7 and III,C) (Beato, 1989; Chalepakis et al., 1990; Dahlman-Wright et al., 1990, 1991). These biologically active DNA sites are called glucocorticoid response elements. Despite the pivotal role of GREs, there is a poor correlation between the affinity of DNA binding to GREs in cell-free assays and the magnitude of the whole-cell response (Adler et al., 1993). This argues for an involvement of chromosomal proteins in modifying the binding to DNA in nuclei, as has been postulated for many years (Simons et al., 1976; Simons, 1979; Kumar and Dickerman, 1983; Ueda et al., 1988; Okamoto et al., 1993).

The DNA binding domain, and discussions of it, are dominated by two "zinc fingers," each of which contains one zinc atom coordinated by four cysteines (Luisi et al., 1991). The base of each finger (Fig. 2) is sometimes called the "knuckle." This domain has also been the most extensively studied of all the regions of the glucocorticoid receptor. Two comprehensive reviews of this domain have appeared (Freedman, 1992; Luisi et al., this volume). Nevertheless, a precise demarcation of the boundaries has been difficult. For example, mutation of the basic amino acids at 507–514, carboxyl terminal to the second zinc finger, eliminates most of the DNA binding (Godowski et al., 1989; K. R. Yamamoto, personal communication). However, this region is not abso-

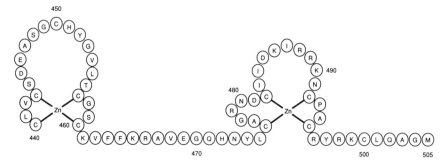

Fig. 2. Amino acids of the DNA binding domain. This sequence shows the complexed zinc ions (Zn^{2+}) and the zinc finger structures with the "knuckles" being composed of the two or five amino acids between the cysteines complexed with zinc.

lutely required for DNA binding, which can even be seen with each individual zinc finger peptide (Archer *et al.*, 1990).

1. *Antigenic Sites*

Several antibodies have been identified that recognize antigenic sites in this domain. Two monoclonal antipeptide antibodies to the DNA binding domain sequences of 465–471 and 472–495 of the rat receptor have been reported (Flach *et al.*, 1992). Three polyclonal antibodies raised against progesterone receptor sequences that are highly homologous to the sequence of 469–500 in the second zinc finger of the rat glucocorticoid receptor recognize glucocorticoid receptors (Wilson *et al.*, 1988). Another monoclonal antibody has been prepared that recognizes the D box, or proximal knuckle of the second zinc finger (476–482; Fig. 2) (Dahlman-Wright *et al.*, 1993). A peptide corresponding to positions 593–612 of the human androgen receptor (483–502 of rat receptor) afforded a monoclonal antibody (F.52.24.4) that was more effective in precipitating human glucocorticoid receptors than androgen receptors even though the glucocorticoid receptor shared only 14 out of 20 of the immunizing amino acids (Veldscholte *et al.*, 1992). The sequence of 440–795 was used to prepare a polyclonal antibody (aP1) (Hoeck *et al.*, 1989), which probably recognizes some epitope(s) in this domain.

2. *Phosphorylation*

It was originally reported that some of the amino acids in the 15-kDa DNA binding domain were phosphorylated (Dalman *et al.*, 1988; Hoeck and Groner, 1990). Subsequent studies (Benjamin *et al.*, 1990) argued that this phosphorylation was due to a contaminating phosphopeptide(s), which is consistent with ~80% of the total receptor phosphorylation being in the immunogenic domain (Bodwell et al., 1991) (see Section III,B). However, work in WCL2 cells, in which the mouse glucocorticoid receptor is overexpressed and has an atypical nuclear localization in the absence of steroid, revealed that ~5% of the total receptor phosphorylation is contained in the 15-kDa tryptic fragment (=387–517 of rat) that encompasses all of the DNA binding domain (Hutchison et al., 1993).

3. DNA Binding and Specific DNA Binding

Three amino acids, GSV, of the P box (458–462) overlapping the distal knuckle of the first zinc finger (Fig 2.) were found by mutagenesis studies to be the determinants for the specific DNA binding and biological activity of glucocorticoid receptors (Mader et al., 1989; Danielsen et al., 1989; Umesono and Evans, 1989). In fact, all the steroid receptors can be grouped into glucocorticoid or estrogen/thyroid receptor subfamilies based on the sequence of this P box (Danielsen et al., 1989; Umesono and Evans, 1989). The interaction of this knuckle with distinct hormone response elements is responsible, in part, for the unique biological activities of each subfamily of receptors.

There is some disagreement as to whether the GSV versus EGA sequence of the distal knuckle is sufficient for complete specificity of receptor binding to GRE versus estrogen response elements (ERE), respectively. Alroy and Freedman (1992) found that GSV was adequate for specificity in studies involving just the DNA binding domain fragment of 440–525 and noted that, while changing V to A caused an approximately three-fold decrease in receptor affinity for DNA, the valine was not required to maintain the specific binding of receptor to GRE versus ERE (Alroy and Freedman, 1992). In contrast, other studies found that this knuckle was not sufficient for complete specificity and that amino acids in the interfinger region (467–469) (Zilliacus et al., 1992b), and the second zinc finger (Danielsen et al., 1989), were necessary. There is general agreement, however, that sequences outside of both this knuckle and the DNA binding domain are required for the high affinity of intact, wild-type receptors for DNA (Dahlman-Wright et al., 1992; Freedman, 1992).

The determinants for receptor binding to glucocorticoid versus thyroid response elements (TREs) appear to be more complex. The five amino acids in the proximal knuckle of the second zinc finger, or D box (477–481; Fig. 2), are additionally involved in specific binding to a

TRE so that changes in both knuckles are required to shift recognition from GRE to TRE (Umesono and Evans, 1989).

4. *Cooperative DNA Binding*

In addition to simply mediating the binding to DNA, the DNA binding domain of the glucocorticoid receptor also permits the cooperative binding of receptors to DNA (Dahlman-Wright *et al.*, 1990; Baniahmad *et al.*, 1991). It does not, however, seem that all DNA binding is cooperative (Perlmann *et al.*, 1990; Cairns *et al.*, 1991).

Receptor dimerization involves the short D box sequence of 477–481. This region has also been shown to be involved in the specificity and recognition of spacing of various HREs (Umesono and Evans, 1989; Danielsen *et al.*, 1989; Dahlman-Wright *et al.*, 1992). This knuckle further participates in the cooperative DNA binding of glucocorticoid receptors because its replacement with the corresponding fragment of a different receptor afforded a protein that bound to the GRE but without cooperativity (Hard *et al.*, 1990; Freedman and Towers, 1991). Thus, converting the D box sequences of the glucorticoid receptor to those of the human thyroid β receptor in the truncated receptor (413–519) reduced the cooperativity of DNA binding to a GRE by about 100-fold (Dahlman-Wright *et al.*, 1991). Conversely, this cooperativity of DNA binding was found to be transferable when the five amino acid of the glucocorticoid receptor (477–481) were substituted for the D box of the thyroid β1 receptor (Dahlman-Wright *et al.*, 1993). Definitive evidence for the involvement of these sequences came from the observation of protein–protein contacts in the X-ray structure of the DNA binding domain complexed with DNA (Luisi *et al.*, 1991). The energetics of this dimerization was strong enough to pull a receptor off of one GRE half-site so that the normal 3-bp spacing between the two truncated receptor molecules could be maintained. However, the fact that no dimerization of the DNA binding domain was observed under the high protein conditions of NMR analysis (Hard *et al.*, 1990; Baumann *et al.*, 1993) suggests that dimerization is induced by DNA binding (see also Section III,C). It should be noted that the peptide used for this X-ray analysis, and for many other studies, contains six and two non-receptor amino acids at the amino and carboxyl termini, respectively, of the molecule (Luisi *et al.*, 1991).

5. *Interactions with Factors That Increase DNA Binding*

It is generally accepted that the DNA binding of receptors for the classical steroids (androgens, estrogens, glucocorticoids, mineralocorticoids, and progestins) does not require auxiliary factors similar to those seen for other members of the superfamily of steroid receptors, such as thyroid (Rosen et al., 1993; Sugawara et al., 1993) and ecdysone (Thomas et al., 1993) receptors. However, quantitative studies with DNA–cellulose columns uncovered the requirement of a low-molecular-weight factor (700–3000) for the DNA binding of a sub-population of activated glucocorticoid complexes. The binding of this subpopulation was inhibited by sodium arsenite, whereas methyl methanethiolsulfonate selectively prevented the DNA binding of factor-independent complexes (Cavanaugh and Simons, 1990a,b). The low-molecular-weight factor was ubiquitous, interacted with pro-gesterone and estrogen receptors (Cavanaugh and Simons, 1993), did not bind to DNA, and interacted with the receptor only after activation (Cavanaugh and Simons, 1990b). This interaction occurred at a site(s) between 410 and 795 (Cavanaugh and Simons, 1993). A much larger factor (130,000 M_r) has also been found to increase the amount of DNA binding by six-fold in gel shift assays (Kupfer et al., 1993). Because of the qualitative nature of gel shift assays, it is not yet clear whether this increased binding results from an increased affinity of all recep-tors or a recruitment of receptors that do not bind to DNA in the absence of factor. As with the preceding smaller factor, this 130,000 M_r factor had negligible affinity for DNA. In contrast to the lower M_r factor, the activity of the 130,000 M_r factor did not appear to be gener-al in that it increased the DNA binding of truncated androgen recep-tors that had been over-expressed in yeast but had no effect on the binding of either full-length androgen receptors or similarly truncated progesterone receptors. Experiments with truncated glucocorticoid re-ceptors revealed that the 130,000 M_r factor interacts with amino acids 366–406 and/or 557–630 (Kupfer et al., 1993).

6. α-Helical Structures

A solution NMR study of the DNA binding domain fragment 440–525 (with a few extra nonreceptor amino acids) revealed two regions of α-helical structure: 459–469 and 493–504 (Hard et al., 1990). These conclusions were refined to the sequences 457–469 and 486–503 by the subsequent X-ray study (Luisi et al., 1991), and a more precise NMR study (Baumann et al., 1993), in which it was found that the amino-terminal α-helix (457–469) lies in the major groove of DNA and provides most of the specifying contacts (Luisi et al., 1991). Two α-helixes were now seen in the region of the second zinc finger at 486–491 and 493–503. An α-helix that had been predicted to extend as far as Lys-514 (Godowski et al., 1989) was not observed. However, it is possible that the disorder in the X-ray structure of the carboxyl-terminal 16 amino acids (starting at 512) may have disrupted some of the normal tertiary structure.

7. *Transcriptional Activity*

Most studies of transcriptional activity have involved genes that are induced. The minimal sequence reported to retain transcriptional activity (~10% of wild type) is 407–556, although data from deletion experiments suggested that 440–525 may be sufficient (Miesfeld *et al.*, 1987). Glucocorticoids can also repress gene expression, although the number of examples is much smaller (Beato *et al.*, 1989), and the same 407–556 segment is sufficient to inhibit expression from constructs containing the bovine prolactin promoter (Miesfeld *et al.*, 1988). A slightly smaller fragment (406–523) was reported to be inactive in a cell-free transcriptional assay for induction (Cao *et al.*, 1993). This inactivity could reflect differences between the whole-cell and cell-free transcription assays, because cell-free assays have yet to be shown to be steroid dependent (Tsai *et al.*, 1990; Schmitt and Stunnenberg, 1993), or result from the missing carboxyl-terminal sequence (see below).

BamH1 linker insertion mutants in the DNA binding domain almost always decreased or eliminated the transcriptional activity of receptors in a transient transfection assay (Giguère *et al.*, 1986; Hollenberg *et al.*, 1987), presumably due to disruption of DNA binding. In contrast, the insertion of two amino acids at each end of the DNA binding domain (after positions 439 and 505) in a hybrid receptor of the estrogen receptor containing the DNA binding domain of the glucocorticoid receptor did not dramatically alter receptor activity (Green *et al.*, 1988).

A transcriptionally active region, called enh1, in the distal half of second zinc finger (484–492) has been defined genetically for truncated receptors in yeast and CV-1 cells (Schena *et al.*, 1989). Mutations within this region, at 488–491, affected transactivation but not DNA binding (Schena *et al.*, 1989). These results are consistent with the transactivation activity of the 407–566 fragment (Miesfeld *et al.*, 1987). However, it should be noted that this fragment also contains half of another transactivation domain called tau2 (Schüle *et al.*, 1990) (see Section II,D,8). Multiple changes in the enh1 region, and sequences just amino terminal to it, were found to reduce the specificity of the receptor for GRE versus ERE sites (Danielsen *et al.*, 1989).

Other examples of mutant receptors with no transcriptional activity in the presence of DNA binding are uncommon, although the results with antiglucocorticoids, which bind to GREs but are transcriptionally inactive (Miller *et al.*, 1984; Richard-Foy *et al.*, 1987; Webster *et al.*,

1988; Bocquel *et al.*, 1993), suggests that they should exist (see also Section III,C). I488* and I491* (=I490* in Giguère *et al.*, 1986) are carboxyl-terminal truncations at BamH1 linkers inserted at the designated positions of human receptors [=507 and 510 (or 509) of rat receptors]. These mutants had wild-type binding to DNA–cellulose but <1% of wild-type transcriptional activity in a transient transfection assay (Hollenberg *et al.*, 1987). Whether this is simply due to the lack of nuclear translocation sequences (see Section II,C,4), thereby rendering these receptors incapable of nuclear binding in intact cells, remains to be determined.

A carboxyl-terminal truncated construct of the human receptor containing an internal deletion, corresponding to 1–9/405–550 of the rat, is fully active in causing the constitutive lysis of dexamethasone-resistant human leukemic lymphoblasts (ICR 27 cells) (Nazareth *et al.*, 1991). Further studies revealed that an even smaller fragment (=418–484 of rat receptors) missing most of the second zinc finger was sufficient for constitutive cell lysis (E. B. Thompson, personal communication). The significance of this observation, considering the usual requirement of DNA binding for transcriptional activity, is not yet clear because it has not been established how cell killing is evoked. Nevertheless, new protein and RNA synthesis was required (E. B. Thompson, personal communication). In this respect, it should be remembered that a single zinc finger appears able to bind to a GRE (Archer *et al.*, 1990, vs. Freedman *et al.*, 1988) and thus could be sufficient for transactivation.

C. Hinge Region/Nuclear Translocation Domain (Amino Acids 506–546)

The sequences between the DNA and steroid binding domains are often called the hinge region. The boundaries of this domain depend on those of the adjacent domains as opposed to any primary biological response. In fact, most of the activities that have been described in this domain would be in the DNA binding domain if the carboxyl-terminal end of the latter domain were extended to 525 (Godowski *et al.*, 1988).

1. Antigenic Sites

The only antibody that has been reported to uniquely recognize this region of the receptor is the polyclonal antibody AP64, which was raised to the peptide 500–517 and recognizes both rat and human receptors (Urda *et al.*, 1989). The antigenic site of AP64 has subsequently been restricted to 506–514 (Miyashita *et al.*, 1993). This antibody recognizes only activated complexes, owing to an uncovering of the antigenic site during activation, and is capable of blocking the DNA binding of activated complexes (Urda *et al.*, 1989).

A very high titer polyclonal antibody (aP1) has been raised against the carboxyl-terminal half of the rat receptor (440–795) and recognizes rat, mouse, and human receptors (Hoeck *et al.*, 1989). At least one epitope has been localized to the region of 510–560 (Hoeck and Groner, 1990). Another high-titer polyclonal antibody (EP) has been elicited from the human sequence 499–597 (=518–615 of rat) (Demonacos *et al.*, 1993; C. E. Sekeris, personal communication).

2. *Steroid Binding*

It has been reported that deletion of 489–531 of human receptor (=508–549 in rat) gives a species that does not bind steroid (Hollenberg *et al.*, 1987). Whether this lack of binding is due instead to an instability of the deletion mutant receptor is not known.

3. *DNA Binding*

No contact of any of these sequences with DNA has yet been identified. Nevertheless, replacement of the basic amino acids of 507 through 514 with the shorter sequence of Pro-Asp-Leu, to give the mutant LS10 (Godowski *et al.*, 1989), virtually abolished the DNA binding of full-length receptors (K. R. Yamamoto, personal communication).

4. *Nuclear Translocation*

Basic amino acid sequences required for nuclear translocation were initially discovered in studies with nucleoplasmin (Dingwall *et al.*, 1982) and the SV40 T antigen (Lanford and Butel, 1984). This nuclear translocation sequence has been further described as a bipartite sequence of basic amino acids separated by 10 spacer amino acids (Robbins *et al.*, 1991). Whole-cell immunofluorescence studies were used to identify a nuclear localization sequence (NL1) in the rat glucocorticoid receptor, which was between 497 and 524 (Picard and Yamamoto, 1987) and contained the complete bipartite sequence (Robbins *et al.*, 1991). However, deletion of the steroid binding domain (524–795) of a receptor–LexA fusion protein afforded only an ~50:50 distribution of cytoplasmic to nuclear receptors, whereas the intact protein gave much more nuclear binding (Godowski *et al.*, 1988). This suggested that more than 497–524 is required for full nuclear translocation and retention (see Section II,D,5).

Fusion of the SV40 T antigen nuclear localization sequence to the amino terminus of the rat glucocorticoid receptor gave constitutive nuclear localization, but addition of an agonist steroid was still required for transcriptional activation (Picard et al., 1988). Thus, nuclear localization is not sufficient for transcriptional activation (Picard et al., 1988). Deletion of part of the nuclear translocation sequence of human receptors (510–533 in rat) did not alter the ability of the synthetic glucocorticoid dexamethasone (Dex) to induce transcription from a mouse mammary tumor virus (MMTV)/CAT reporter gene in transient transfections (Hollenberg et al., 1987; Oro et al., 1988), suggesting that the KTKKKIK portion of the nuclear translocation sequence is not essential. This is consistent with the report that both halves of the bipartite sequence are not required for some nuclear translocation (Robbins et al., 1991).

Nuclear binding is a two-step mechanism that involves an association with nuclear pores followed by an energy-dependent nuclear translocation (Richardson et al., 1988; Garcia-Bustos et al., 1991). Very few systems have been established that allow the nuclear translocation of macromolecules in broken-cell systems. It is notable, however, that the existing protocols do not support the cell-free nuclear translocation of glucocorticoid receptors (Miyashita et al., 1993). Thus, components in addition to those already identified are likely to be required for the nuclear binding of steroid receptors.

5. Protein–Protein Interactions

There is some speculation that the sequences just downstream of the second zinc finger may form an amphipathic helix that could participate in protein–protein interactions (Miner et al., 1991). Glucocorticoid receptors are thought to block the induction by AP-1 (jun–fos heterodimer) of responsive genes by interacting directly with jun or fos (see Section II,F,3). However, this inhibition is seen both with the truncated receptor fragment of 440–509, which lacks this potential helical sequence, and with the apo- (or non-zinc-containing) form of the 440–533 fragment, which would lack considerable tertiary structure. Thus, it seems more likely that some primary sequence, or loose secondary structure, is involved at least in the interactions with AP-1 (Kerppola et al., 1993).

6. Transcriptional Activity

No transcriptionally active elements have been identified in the hinge region, as defined in this review. For example, deletion of 491–515 of the human receptor (=510–534 of rat) has little or no effect on its transactivation capacity (Hollenberg et al., 1987; P. Herrlich, personal communication). However, the ability of this same deletion mutant to repress induction from a collagenase AP-1 site was reduced by about 70% (P. Herrlich, personal communication). Similarly, mutations in the sequence of 507–514 gave a receptor (LS10) that was capable of effecting induction from the mouse mammary tumor virus GRE but was unable to suppress transcription from a prolactin nGRE containing construct (Godowski et al., 1989). These effects could be due to some yet unidentified component, to disruption of the above postulated amphipathic helix, or to more pervasive changes in tertiary structure.

Another puzzling feature of the LS10 mutation is that the induction activity was maintained in the absence of detectable DNA binding activity in gel shift assays (K. R. Yamamoto, personal communication). Whether this apparent paradox reflects an underestimation by the gel shift assay of the DNA binding affinity, the differences between receptor binding to a GRE without and with chromosomal proteins that may contribute to the biologically active site in intact cells (Simons et al., 1976; Simons, 1979; Kumar and Dickerman, 1983; Ueda et al., 1988; Okamoto et al., 1993), or a consequence of very high levels of receptors in the transiently transfected cells remains to be investigated (see Section III,C).

D. Steroid Binding Domain (Amino Acids 547–795)

1. Antigenic Sites

Two monoclonal antipeptide antibodies have been described. They both recognize regions in the carboxyl-terminal end of the steroid binding domain of the rat receptor (721–727 and 788–795) (Flach et al., 1992). The polyclonal antibody aP1 (Hoeck et al., 1989) appears to contain at least one immunoglobulin that reacts with this region because aP1 will readily immunoprecipitate, and detect by Western blotting, the fragment of 537–673 (Chakraborti et al., 1990) (see Section II,C,1). The sequence 499–597 from the human receptor (=518–615 of rat) was used to produce a polyclonal antibody that potentially will recognize sequences in the steroid binding domain (Demonacos et al., 1993; C. E. Sekeris, personal communication).

2. Mero-receptors

For many years, the smallest segment of the glucocorticoid receptor, and of all the steroid receptors, that retained steroid binding activity was an approximately 27-kDa tryptic digestion fragment called the mero-receptor (Sherman et al., 1978). On the basis of amino acid sequencing data, which identified Lys-517 as the site of trypsin cleavage (Carlstedt-Duke *et al.,* 1987), it was proposed that the mero-receptor corresponds to 518–795 (Denis *et al.,* 1988). Thus, mero-receptor would include all of the steroid binding domain and most of the hinge region. Further analysis, however, indicated that there are two closely spaced components of the mero-receptor at 30 and 27 kDa (Reichman *et al.,* 1984), which are thought to have 781 and 761, respectively, as the carboxyl-terminal amino acids (Simons *et al.,* 1989).

3. *Steroid Binding*

The initial activity determined for the steroid binding domain was, obviously, the ability to bind glucocorticoid steroids. However, the demarcation of the boundaries of this domain has proved elusive (see beginning of Section II). A wide variety of deletions have been prepared in the steroid binding domain and most of them do eliminate steroid binding activity (e.g., Rusconi and Yamamoto, 1987; Hollenberg *et al.,* 1987, 1989; Danielsen *et al.,* 1987). Nevertheless, expression in a cell-free system of that fragment (547–795) that most closely encompasses the steroid binding domain affords a protein with an affinity for Dex that is 300-fold lower than that of the intact receptor under the same conditions (Rusconi and Yamamoto, 1987). The smallest receptor fragment that displays nearly wild-type steroid binding affinity upon expression in whole-cell (Giguère *et al.,* 1986; Hollenberg *et al.,* 1987) or cell-free systems appears to be 497–795, which had an affinity that was reduced by only 6-fold (Rusconi and Yamamoto, 1987). In contrast, trypsin digestion of native 98-kDa receptors afforded a 16-kDa fragment that is much smaller than the entire steroid binding domain but has an affinity for Dex that is reduced only by 23-fold (Simons *et al.,* 1989). This 16-kDa fragment has been assigned the sequence of 537–673, based on an analysis of the protease digestion patterns (Simons *et al.,* 1989), and appears to represent a steroid binding core sequence because it retained considerable affinity in addition to all the specificity of binding seen with the intact receptor (Chakraborti *et al.,* 1992). These apparent discrepancies in the minimum region required for steroid binding may result from large inhibitory effects by small stretches of amino acids at either end of the steroid binding domain or from differences in the preparation of the receptor fragments (see below).

Two sequences with homology to other steroid receptors (Simpson *et al.,* 1987; Wang *et al.,* 1989) and to steroid binding proteins and enzymes (Picado-Leonard and Miller, 1988) have been noted (Fig. 1) and have been suggested to be involved in steroid binding (Simpson *et al.,*

1987; Picado-Leonard and Miller, 1988). The first sequence (556–573) is at the amino terminus of the steroid binding domain (Simpson et al., 1987). The second sequence, at 683–706, was proposed to be required for steroid binding (Picado-Leonard and Miller, 1988) but may only contribute to binding as good steroid binding is seen with the 16-kDa steroid binding core fragment of 537–673, which lies outside of this region (Simons et al., 1989; Chakraborti et al., 1992).

The carboxyl terminus of the receptor has attracted considerable attention. The initial cloning of the human receptor yielded the cDNA for a second, or β, form of the receptor that lacked the carboxyl-terminal 35 amino acids and did not bind steroid (Hollenberg et al., 1985). Deletion of 769–795 of rat receptor also eliminated steroid binding (Godowski et al., 1987; Garabedian and Yamamoto, 1992). With progesterone receptors, removal of the carboxyl-terminal 42 amino acids not only eliminated the binding of agonists but also yielded receptors that displayed full agonist activity after binding antagonists (Vegeto et al., 1992). These results suggested an intriguing model in which the nature of the conformational change in the carboxyl terminus of receptors resulting from steroid binding determines the agonist versus antagonist activity of the final complex (Allan et al., 1992) (see Section II,D,8). By analogy, it was suggested that deletion of the carboxyl-terminal ~40 amino acids of the glucocorticoid receptor would similarly eliminate the binding of agonist, but not antagonist, steroids. Interestingly, this prediction has not yet been verified for glucocorticoid (Lanz et al., 1993) or estrogen (Beekman et al., 1993) receptors, which may reflect mechanistic differences between the various classes of receptors. These effects of truncations are reminiscent of those both for the ~220-amino-acid protein chloramphenicol acetyltransferase, where removal of the last nine amino acids inactivates the enzyme (Robben et al., 1993), and for the 149-amino-acid protein staphylococcal nuclease, where deletion of the carboxyl-terminal 13 amino acids gave a partially denatured protein. However, the truncated nuclease could still bind substrate with nearly wild-type affinity (Flanagan et al., 1992). Nevertheless, this sensitivity of protein structure, in conjunction with recent studies on protein folding (see Section I), raises the possibility that the preceding results with carboxyl-deleted receptors could reflect the steroid binding properties of differently folded, newly synthesized, truncated proteins. In support of this hypothesis, removal of the sequence for the carboxyl-terminal 29 amino acids from the cDNA encoding the fragment 407–795 reduced the affinity of Dex for the resulting cell-free translated receptors (407–766) by over 1000-fold (Rusconi and Yamamoto, 1987), whereas trypsin digestion of intact, prefolded receptors affords an even smaller, 16-kDa fragment (537–673) that bound steroid with the same specificity of wild-type receptors (Chakraborti *et al.*, 1992) and only a 23-fold reduction in affinity (Simons *et al.*, 1989).

The mammalian glucocorticoid receptor is transcriptionally active in yeast (Schena and Yamamoto, 1988). This important observation allows investigators to take advantage of the much simpler genetics in yeast to investigate the mechanism of steroid hormone action. However, the affinity of Dex for receptors expressed in yeast is much lower than that in mammalian cells (Garabedian and Yamamoto, 1992; Wright *et al.*, 1990). Furthermore, the specificity of binding for full-length (Schena and Yamamoto, 1988) and amino-terminal truncated (417–777 of hGR) receptors (Wright *et al.*, 1990) has been altered. Some of the diminished activity of various steroids may be due to a reduced cellular permeability, but this does not explain why RU-486, which usually is a pure antiglucocorticoid in mammalian cells, had 25% agonist activity in yeast, or why Dex-Mes, which typically has appreciable amounts of agonist activity (Simons and Thompson, 1981; Oshima and Simons, 1992), was a pure antagonist in yeast (Garabedian and Yamamoto, 1992). The tau1 transcriptional activation domain was proposed to contribute toward aberrant specificities in steroid binding even though deletion of the tau1 domain further reduced the apparent affinities of expressed truncated receptors in yeast (Wright and Gustafsson, 1992). The specificity of steroid binding and biological activity could be improved by lowering the level of expressed receptor but this did not affect the affinity of steroid binding, which was still at least 100-fold lower than that seen for receptors in a mammalian cell environment (Wright and Gustafsson, 1992). It has been suggested that yeast cells may be unable to properly process overexpressed receptors (Obourn *et al.*, 1993). Unfortunately, this reduced steroid binding affinity is not unique for receptors translated in yeast. The affinity of receptors expressed in *E. coli* was also decreased by ≥100-fold (Wright *et al.*, 1990). Thus, cellular factors and/or processes not present in yeast cells or bacteria may be involved in the binding of steroid to receptors. Alternatively, these differences could be promoter dependent and provide insight into the interactions between receptors and promoter associated transcription factors.

4. *Heat Shock Protein 90 (hsp90) Binding*

Heat shock protein 90 (hsp90) was the first nonreceptor molecule identified to be associated with all the classical steroid receptors (Joab *et al.*, 1984; Catelli *et al.*, 1985). Extensive subsequent work has established that hsp90 association is required for steroid binding to the glucocorticoid receptor, whereas disassociation of hsp90 precedes the DNA binding of receptor–steroid complexes (Meshinchi et al., 1990; Hutchison et al., 1992b). The site of hsp90 association was first proposed to be 583–633 of the mouse receptor (=595–645 of rat) owing to the similarity of this region in glucocorticoid and estrogen receptors (Danielsen et al., 1986, 1987). Other investigators extended the sequence to 583–645 (Fig. 1) based on homologies with other receptors (Pratt et al., 1988; Wang et al., 1989; O'Donnell et al., 1991; Lee et al., 1992; see Pratt, 1993, for a more recent review).

Initial deletion experiments implicated sequences in the region of 568–616 as interacting with hsp90 (Pratt et al., 1988; Howard et al., 1990). It is interesting that the sequence identified by both studies (593–616) is also homologous to the region of thyroid receptors that is thought to be required both for thyroid receptor homodimerization (Lee et al., 1992) and for association with TRAP (O'Donnell et al., 1991), which is a thyroid response auxiliary protein(s) that increases thyroid receptor binding to DNA. Thyroid receptors do not bind hsp90 (Dalman et al., 1990) and glucocorticoid receptors have not been found to heterodimerize. Thus, either the tertiary structures of this region in the two receptors are very different or the nonconserved amino acids of 593–616 may be important for specifying hsp90 binding versus thyroid receptor homodimerization and heterodimerization.

Others, also working with glucocorticoid receptor deletion mutants, have found that no particular sequence of the steroid binding domain was required for the formation of 8S complexes (Cadepond et al., 1991), which had been shown previously to contain hsp90. These results are not easily reconciled with the foregoing data, especially because the deletion mutant 532–697 (550–715 of rat) used in this study was later found to be missing most of the steroid binding domain, due to a truncation at 550 (568 of rat) (Housley et al., 1990), but still gave an 8S complex, presumably due to hsp90 binding.

All of these studies utilized deletions and identified the hsp90 binding site by loss of activity. More direct approaches have yielded similar conclusions. Thus, the addition of specific peptides to block hsp90 association with mouse receptors was used to identify two sites (=599–618 and 644–671 of rat), of which 644–671 was the most important (Dalman et al., 1991a). However, the ability of a deletion mutant $X\Delta616$–694, which is missing the sequences between 616 and 694, to be coimmunoprecipitated by anti-hsp90 antibodies indicated that the 644–671 sequence is not absolutely required for hsp90 association (Schlatter et al., 1992). The contribution of the other site could not be directly determined because deletion of 586–644, which includes this site, gave a mutant receptor with increased sensitivity to proteolysis. This instability may result from the reduced association of hsp90 (Housley *et al.*, 1990), presumably as the newly synthesized protein is being folded. When correctly folded receptors are dissociated from hsp90, the receptors are comparatively resistant to proteolysis, although some receptors have lost the ability to bind DNA as a result of a proposed "misfolding" of the steroid binding domain (Hutchison *et al.*, 1992a).

The most compelling identification of hsp90 binding sites involves hsp90 association with truncated receptors. The sequence of 540–795 was sufficient to direct the association of hsp90 with newly synthesized hybrid receptor proteins (Scherrer *et al.*, 1993). The 27-kDa meroreceptor, encompassing most or all of the steroid binding domain (see Section II,D,2), contained prebound hsp90 (Denis *et al.*, 1988). Finally, the smallest segment of the receptor shown to retain hsp90 was 537–673 (Chakraborti and Simons, 1991), which has been identified as the steroid binding core element (Simons *et al.*, 1989). Therefore, there may be multiple sites of hsp90 interaction with receptor, but those within the region of 537–673 are sufficient for a stable complex.

5. *Nuclear Translocation*

Even though the intact receptor contains the nuclear translocation sequence in the hinge region (see Section II,C,4), steroid-free glucocorticoid receptors are not nuclear. The unoccupied steroid binding domain somehow prevents the nuclear binding of glucocorticoid, but not estrogen or progestin, receptors (Ylikomi *et al.*, 1992). Using *in vivo* immunofluorescence to follow various hybrid proteins, the additional sequences required for nuclear binding were localized to 540–795 and called NL2 (Picard and Yamamoto, 1987). The activity of the NL2 sequence was dependent on steroid binding and it has not yet been possible to demonstrate any intrinsic nuclear localization activity of NL2. A similar steroid-induced, nuclear localization activity has been identified in progesterone receptors (Guiochon-Mantel *et al.*, 1989). Nonetheless, it is conceivable that other molecules, such as heat shock proteins, associate with steroid binding domain sequences either to occlude the NL1 sequence in the hinge region (Urda *et al.*, 1989) or to cause the "docking" of receptors in a cytoplasmic compartment (Dalman *et al.*, 1991b; Scherrer *et al.*, 1993).

6. *Molybdate Stabilization*

It is well known that sodium molybdate forms stable complexes with thiols (Kay and Mitchell, 1968; Kaul *et al.*, 1987). Ever since Pratt described the ability of molybdate to stabilize the steroid binding activity of receptors, and to block activation (or transformation) (Leach *et al.*, 1979), it has been suspected that molybdate exerted its effects by interacting with cysteines of the receptor. A series of indirect experiments led to the postulate that the sequence of 644–671, and especially cysteines 656 and 661, were required for molybdate stabilization (Dalman *et al.*, 1991a). Experiments with receptor fragments of wild-type and mutant receptors have supported the involvement of this region. However, they have also ruled out the involvement of Cys-656 and 661 in any of molybdate's effects (Modarress *et al.*, 1994) (see Section III,E,4).

7. *Receptor Dimerization*

The importance of sequences in the D box of the DNA binding domain for receptor dimerization has already been discussed (see Section II,B,4). A second dimerization domain has been proposed at 730–764 on the basis of homology with a sequence implicated in dimerization of estrogen receptors (Fawell *et al.*, 1990). However, the role of these sequences in the glucocorticoid receptor is not clear because mutants lacking this region are able to cause levels of transactivation similar to that seen for the wild-type receptor (Danielsen *et al.*, 1987; Hollenberg and Evans, 1988). This region may have some, albeit adverse, function though because removal of the steroid binding domain from full-length receptors that are overexpressed by a recombinant baculovirus in Sf9 cells prevents formation of insoluble aggregates, presumably due to homo- and hetero-oligomers (Alnemri and Litwack, 1993).

8. *Transactivation Activity*

In addition to the transactivation activities of sequences in the immunogenic and DNA binding domains, additional activity has been localized to the steroid binding domain. The tau2 (τ2) transactivation domain was first mapped by Evans *et al.* to 526–550 of the human receptor (Giguère *et al.*, 1986; Hollenberg *et al.*, 1987; Schüle *et al.*, 1990) (=544–568 of rat receptor). The tau2 domain is often depicted as residing at the carboxyl terminus of the hinge region, but the sequence places it mostly, if not entirely, in the steroid binding domain as defined in this review. The tau2 site is thus part of the larger TAF-2 (Bocquel *et al.*, 1989; Tasset *et al.*, 1990), or AF-2 (Nagpal *et al.*, 1992), transcriptional activation domain. The transcriptional activity of the TAF-2 sequence appearance to have been seen under cell-free conditions with the truncated receptor of 406–795 (Schmitt and Stunnenberg, 1993), although contributions due to enh1 in the DNA binding domain (Fig. 1) could not explicitly be excluded. As was first found for the intact glucocorticoid receptor (Tsai *et al.*, 1990), the induction by this truncated receptor was constitutive and unresponsive to the presence of steroid (Schmitt and Stunnenberg, 1993).

Most BamH1 linker insertion mutants in tau2 decreased or eliminated the steroid-inducible transcriptional activity in transiently transfected CV-1 cells (Giguère *et al.*, 1986), presumably because steroid binding was also disrupted. Likewise, almost all deletions in the steroid binding domain eliminate transcriptional activation, apparently due to the loss of steroid binding activity. However, once the entire steroid binding domain is deleted, the receptor becomes a constitutive transcriptional activator (Danielsen *et al.*, 1987; Hollenberg *et al.*, 1989). Although the effects of these deletions could be due to global effects on receptor structure, a less complicated explanation is that some inhibitory activity or component, such as hsp90, exists in the steroid binding domain, since it is known to be sufficient for hsp90 association (Chakraborti and Simons, 1991; Scherrer *et al.*, 1993).

A putative helix-turn-zipper motif at 590–630 of rat receptors is conserved among a large number of the members of the steroid receptor superfamily and is proposed to be involved in transcriptional activation, possibly by interacting with other transcription factors (Maksymowych *et al.*, 1992). It should be noted that Leu-600 and -602 have since been revised to Pro-600 and Phe-602 (M. J. Garabedian and K.R. Yamamoto, personal communication), which are the same residues as seen for many of the other receptors containing this motif (Maksymowych *et al.*, 1992).

A potential modulator of transcriptional activity has been identified in the carboxyl-terminal ~40 amino acids of the receptor. Deletion of these amino acids from the progesterone receptor both eliminated the binding of progestins, but not antiprogestins (see Section II,D,3), and caused the bound antiprogestins to express agonist activity (Vegeto *et al.*, 1992; Allan *et al.*, 1992). Preliminary reports suggest that the same phenomenon occurs with glucocorticoid receptors (Allan *et al.*, 1992), although others have not yet been able to confirm this (Lanz *et al.*, 1993). Nevertheless, this region may still be involved in transcriptional activation. The combination of a double point mutation (M770A,L771A) and a deletion (of 780 and 781) was inactive with the agonist Dex but afforded strong agonist activity with the antiglucocorticoid RU-486 (Lanz *et al.*, 1993).

The carboxyl-terminal 40 residues also encompass the 16 amino acids of the mouse receptor (=766–781 of the rat receptor) that have been identified as being an essential component of the TAF-2 transcriptional activity (Danielian *et al.,* 1992). The homologous sequence of the human thyroid receptor α has been found not to have any intrinsic transcriptional activity, suggesting that additional sequences are required (Saatcioglu *et al.,* 1993) and that the TAF-2 region is much larger (see also Section III,E,3).

E. PLASTICITY OF RECEPTORS

For many years, it was believed that the activity of a given protein was a property of the entire protein and deletions, rearrangements, or substitutions would destroy that activity, just as was encountered upon modifying the steroid binding domain of the glucocorticoid receptor (see Section II,D,3). However, subsequent studies have shown that substitutions and rearrangements of entire domains of the receptor are remarkably easy. In fact, the domain structure of receptors may convey considerable plasticity, a concept that may have been first articulated by Godowski *et al.* (1988). Thus, receptors retained steroid binding activity and immunoprecipitability when the antigenic site for BUGR-2 (407–423) was placed at the carboxyl terminus of truncated receptors (Rusconi and Yamamoto, 1987). The DNA binding domain did not have to be adjacent to the steroid binding domain in order for steroid-induced DNA binding to occur (Godowski *et al.,* 1988). Fully functional receptors were also obtained when the steroid binding domain was moved to the amino terminus of the receptor (Picard *et al.,* 1988; Hollenberg and Evans, 1988). When most of the amino-terminal end of human receptors (10–384 = 11–404 of rat receptor) was repeated as an insert just after the hinge region (534 of rat), the hybrid receptor was four times as active as the wild-type receptor (Hollenberg and Evans, 1988).

Domains from other steroid receptors can substitute for the corresponding segment of the glucocorticoid receptor (Green and Chambon, 1987; Danielsen *et al.,* 1989; Umesono and Evans, 1989; Schüle *et al.,* 1990). The steroid binding domain of the glucocorticoid receptor has frequently been replaced by the comparable domain of a newly isolated member of the steroid receptor superfamily to give a hybrid receptor that will activate a glucocorticoid-responsive gene only when stimulated by the appropriate ligand. Such "domain swaps" have proven very useful in identifying the ligands for new receptors (Giguère *et al.,* 1987; Issemann and Green, 1990; Christopherson *et al.,* 1992; Schmidt *et al.,* 1992).

The functional domains of the glucocorticoid receptor can even be interchanged with similar domains from nonreceptor proteins without a loss of function. Thus, portions of lexA (Godowski *et al.*, 1988), E1A (Picard *et al.*, 1988), GAL4 (Hollenberg and Evans, 1988; Schüle *et al.*, 1990), β-galactosidase (Picard *et al.*, 1988; Scherrer *et al.*, 1993), Rev (=human immunodeficiency virus type 1 regulatory protein) (Hope *et al.*, 1990), c-fos (Superti-Furga *et al.*, 1991), and dihydrofolate reductase (Israel and Kaufman, 1993) have all been fused to segments of the glucocorticoid receptor containing the steroid binding domain without destruction of the steroid binding activity. In most cases, the activity of the fusion protein was now controlled by steroid binding, just as for the intact glucocorticoid receptor. The two exceptions are of interest. The enzyme activity of the β-galactosidase fusion protein was unaffected by steroid binding (Picard *et al.*, 1988), and added steroid represses the whole-cell activity of the dihydrofolate reductase/receptor protein, apparently because of the resulting nuclear translocation of the hybrid molecule (Israel and Kaufman, 1993). The antiglucocorticoid RU-486 also caused nuclear translocation of the dihydrofolate reductase/receptor fusion protein but, surprisingly, did not inhibit the whole-cell activity. This apparent inconsistency could result from the fact that Dex, but not RU-486, additionally caused a 10-fold decrease in the absolute amount of fusion protein as determined by the whole-cell fluorescence of fluoresceinated methotrexate binding to the fusion protein. In this case, the levels of cytoplasmic hybrid enzyme remaining in the cytoplasm after the addition of RU-486 may have been sufficient for normal cellular activity.

Perhaps the ultimate example of receptor plasticity and functional independence of receptor domains is described in a report by Spanjaard and Chin (1993). No appreciable steroid-inducible activity from a reporter gene was seen when the human glucocorticoid receptor was expressed as two halves [1–506, or 1–525 of rat (= immunogenic/DNA binding domains), and methionine in front of 465–777, or 484–795 of rat, which contains all of the hinge region and steroid binding domains] in CV-1 cells. However, when the leucine zipper regions of c-jun, or c-fos, were fused to the carboxyl-terminal end of the immunogenic/DNA binding domain fragment, or to the amino-terminal end of the steroid binding domain peptide, respectively (or vice versa), strong hormonally inducible gene expression was observed. These data were most readily explained by jun and fos forming heterodimeric leucine zipper complexes to bring the two halves of the glucocorticoid receptor together in a noncovalent, but functionally active, complex that binds to the GRE of the reporter gene (Spanjaard and Chin, 1993). These data also suggest that the increased transcriptional activation attendant upon steroid binding does not occur by changes in receptor structure that are transmitted through the primary structure of the receptor. Another interpretation is that the induction results from the steroid-induced activation of TAF-2 (or tau2) sequences that are brought into proximity of the transcriptional machinery by a jun–fos-mediated association with the GRE-bound DNA binding domain. A further experiment with a hybrid protein containing just the DNA binding domain linked to jun or fos should be able to distinguish between a direct effect of steroid binding on TAF-2 activity and a through-space interaction with TAF-1 (or tau1).

The plasticity of receptor molecules is not, however, unlimited. Disulfides can be formed between Cys-656, -661, and/or -640 in the steroid binding domain by the addition of low concentrations (0.1 mM) of methyl methanethiolsulfonate. These disulfides do not undergo thiol–disulfide rearrangements (Opoku and Simons, 1993). This result was unexpected because studies with α-lactalbumin (Ewbank and Creighton, 1991) and bovine trypsin inhibitor (Kosen et al., 1992) suggest that all possible disulfides can be formed in a given molecule regardless of which disulfide is first formed. Thus, the tertiary structure of the glucocorticoid receptor may be sufficiently rigid to prevent thiol–disulfide rearrangements both within the steroid binding domain and within the receptor as a whole (Opoku and Simons, 1993).

F. Properties/Activities That Have Not Been Restricted to a Defined Domain

Several properties/activities of receptors have been restricted to portions of the receptor that span more than one of the previously discussed domains. These properties are covered in the following sections.

1. Post-translational Modifications

Phosphorylation has long been a candidate for modifying receptor function, especially after the appearance of reports that receptors are phosphorylated (Hoeck et al., 1989; Hoeck and Groner, 1990; Bodwell et al., 1991; Orti et al., 1992). Protein kinase A (PKA) has been found to increase receptor-mediated transactivation and deletion experiments suggested that the effect was mediated through the DNA binding domain, although no direct interaction of PKA and the receptor was established (Rangarajan et al., 1992). About 5% of the total phosphorylation of receptors has been localized to the DNA binding domain (Hutchison et al., 1993). However, all receptor phosphorylation was reported to occur independently of PKA (Orti et al., 1993). Furthermore, mutation of all the known phosphorylated amino acids in the receptor did not eliminate receptor activity in whole cells (Mason and Housley, 1993) (see Section III,B). This last result argues against phosphorylation being required for biological activity but does not address the possible role of low amounts of phosphorylation in the DNA binding domain or elsewhere.

An attractive candidate for posttranslational modifications of the receptor has been the formation of one or more disulfide bonds (Tienrungroj et al., 1987; Miller and Simons, 1988; Silva and Cidlowski, 1989; Chakraborti et al., 1990, 1992). Evidence for intramolecular disulfides among the transcription factors NF-κB (Matthews et al., 1992; Hayashi et al., 1993) and OxyR (Storz et al., 1990) further suggested that it would not be unusual to find disulfides among the 20 cysteines of the native rat glucocorticoid receptor. Nevertheless, an investigation of various forms of the native receptor (steroid-free and both unactivated and activated receptors bound by either an agonist or an antagonist) revealed no intramolecular disulfides (Opoku and Simons, 1993). Thus, it was concluded that the structural or functional changes of native glucocorticoid receptors that are associated with steroid binding, activation, and dissociation of hsp90 neither involved nor required the formation or reduction of stable intramolecular disulfides. Disulfide bonds were also not detected in the bacterially expressed human receptor fragment of 477–777 (=496–795 of rat), which has an affinity for Dex that is ~100 times lower than normal (Ohara-Nemoto et al., 1990).

The approximately 100-fold lower affinity of Dex and triamcinolone acetonide for glucocorticoid receptors expressed in yeast still lacks a convincing explanation (Schena and Yamamoto, 1988; Wright et al., 1990; Garabedian and Yamamoto, 1992; Wright and Gustafsson, 1992). It is theoretically possible that these differences with receptors expressed in mammalian cells could be due to yet undetected posttranslational modifications.

2. Synergism

Synergism among the steroid receptors has been defined as a response to saturating concentrations of steroids from two or more elements that is greater than the arithmetic sum of the responses from the individual elements. The synergism of steroid receptors with themselves or with other trans-acting factors at saturating concentrations of steroid has proved to be an important component of steroid-regulated gene transcription (Schüle et al., 1988a,b; Strahle et al., 1988). Linkage of the amino and the carboxyl halves of the receptor (except for the DNA binding domain) to the GAL4 DNA binding domain afforded synergism without altering the DNA binding affinity (Baniahmad *et al.,* 1991; Muller *et al.,* 1991).

The ability of truncated and/or deleted receptors to cause synergistic gene activation from reporters containing one versus two GREs in transient transfections is currently being debated. In one case, removal of the amino-terminal sequences corresponding to 1–435, including tau1, increased the magnitude of synergism (as defined here) of human receptors with luciferase receptor constructs from 1.9 to 3.5 in SK-N-MC cells, although the absolute level of induction decreased (Rupprecht *et al.,* 1993). Thus, the tau1 region appeared inhibitory for synergism. In contrast, other studies in yeast with the human receptor observed a 7.9-fold synergistic response and increased absolute levels of expression when the tau1 region was combined with the DNA binding domain but no synergism when the steroid binding domain, including tau2, was used (Wright and Gustafsson, 1991). The same study found that the tau1 region also permits cooperative DNA binding. A report by Lanz *et al.* (1993) confirmed and extended the latter conclusions for tau1 in HeLa or CV-1 cells. Experiments with both full-length receptors and constitutively active truncated receptors (1–556) lacking the steroid binding domain revealed that synergism with two GREs was stronger in the absence of the steroid binding domain, whereas synergism with transcription factors other than the glucocorticoid receptor (i.e., SP1) was much stronger with receptors containing the steroid binding domain. Thus, the steroid binding domain (and perhaps tau2) may be required for "hetero-synergism" with other transcription factors (Lanz *et al.,* 1993).

3. *Interaction with AP-1*

The ability, and often requirement, of thyroid, retinoic acid, and vitamin D receptors to heterodimerize with RXR receptors and COUP-TF has been recently established. However, no similar interactions have yet been documented for the classical steroid receptors. Thus, one of the surprises in this field has been the demonstration that a classical steroid receptor does interact with members of the jun/fos superfamily, albeit not by heterodimerization.

Two genes, collagenase and proliferin, that are induced by the jun–fos heterodimer (AP-1) have been the center of much attention. In both cases, glucocorticoids cause repression of the AP-1 induced activity via what was initially thought to be a negative glucocorticoid response element, or nGRE. It is now believed that the mechanism of glucocorticoid receptor interaction with AP-1 in the enhancer region of these two genes is not the same. With the collagenase gene (Jonat *et al.,*

1990; Yang-Yen *et al.*, 1990; Schüle *et al.*, 1990; König *et al.*, 1992; Kerppola *et al.*, 1993), and some other genes (Lucibello *et al.*, 1990; Miner *et al.*, 1991), glucocorticoid inhibition of induction by AP-1 required primarily the DNA binding domain of the receptor, although the inhibition involved protein–protein interactions as opposed to receptor binding to DNA. With the proliferin gene, AP-1 induction was repressed by glucocorticoid receptors whereas the induction by jun–jun homodimers (Diamond *et al.*, 1990) and jun–fra heterodimers (Miner and Yamamoto, 1992) was increased by receptors. In these cases, protein–protein interactions were important but now the DNA binding of receptors to the proliferin gene was also necessary (Diamond *et al.*, 1990).

Glucocorticoid, but not the closely related mineralocorticoid, receptors have been reported to repress AP-1 stimulation of genes under the control of the proliferin promoter. From a combination of deletion experiments, it was concluded that repression resulted from a specific interaction between the glucocorticoid receptor sequence 105–440 and AP-1 (Pearce and Yamamoto, 1993). It was further pointed out that such protein–protein interactions may be of wider significance in explaining different responses for receptors that recognize seemingly identical DNA elements. Similar experiments on repression of induction by glucocorticoid/mineralocorticoid hybrid receptors from a reporter construct containing five repeats of a collagenase AP-1 site indicated that both the amino terminal and DNA binding domains of glucocorticoid receptors are required (P. Herrlich, personal communication) (see Section III,C).

Yet another model for interactions with AP-1 has emerged from studies in which the receptor fragment 440–533 inhibited the binding of AP-1 to both a consensus AP-1 site and the proliferin GRE in gel shift assays (Kerppola *et al.*, 1993). This inhibition by receptor was 10 times more effective for the binding of jun–fos heterodimers than for jun–jun homodimers. Jun–jun or jun–fos binding to a consensus AP-1 site was also prevented by the smaller fragment of 440–509 and even by the apo-form of 440–553, which lacks zinc and does not bind to DNA. These results led to the hypothesis that the inhibition of AP-1 action is due to receptors preferentially binding the jun–fos heterodimers, thereby causing a shift in the species bound at equilibrium to the AP-1 site from jun–fos heterodimers to jun–jun homodimers, which are less potent activators (Kerppola *et al.*, 1993). However, it is not clear that this equilibrium model is sufficient to explain the increased induction seen with glucocorticoid receptors and jun–jun homodimers from the proliferin gene because jun–jun homodimers by themselves were only weakly active (Diamond *et al.*, 1990). Some of these apparent discrepancies may be due to differences between receptor interactions with AP-1 in *in vitro* DNA binding assays and *in vivo* bioactivity assays. In particular, whereas 440-553 blocked the DNA binding of AP-1 in a gel shift assay (Kerppola *et al.*, 1993), a slightly larger receptor fragment (405–534) was unable to repress AP-1 induction from a reporter containing the collagenase promoter (Shüle *et al.*, 1990). Also, the constitutive induction by a hybrid receptor containing a tandem repeat of the DNA/hinge region (407–556) could be repressed by AP-1 (Touray *et al.*, 1991).

With some other glucocorticoid-responsive genes, such as the transiently transfected MMTV/CAT, the effect of AP-1 was found to be cell specific with inhibition in NIH3T3 cells and synergistic activation in T cells (Maroder *et al.*, 1993). Synergism with AP-1 was found to be fully retained in a truncated human receptor corresponding to 389–795 of the rat receptor, which is missing tau1 and most of the amino-terminal domain, but abolished upon removal of 551–795.

Direct receptor–AP-1 interactions (usually only with jun but occasionally fos) (Touray *et al.*, 1991; Kerppola *et al.*, 1993) could be observed by some (Diamond *et al.*, 1990; Jonat *et al.*, 1990; Yang-Yen *et al.*, 1990), even in the absence of cross-linking agents (Touray *et al.*, 1991), but not by all (Lucibello *et al.*, 1990; Schüle *et al.*, 1990) investigators. This suggests that the interactions between the glucocorticoid receptor and AP-1 are very sensitive to changes in the environment.

4. *Interaction with Other Proteins*

Studies of another gene that is repressed by glucocorticoids have provided evidence that NF-κB can interact with the receptor (Ray and Prefontaine, 1994). Glucocorticoid repression of interleukin-6 expression is thought to be part of the anti-inflammatory response. Conversely, NF-κB has been implicated in interleukin-6 induction. In an exciting development that has implications beyond the realm of inflammation, it was found that glucocorticoid receptors and the p65 subunit of the NF-κB heterodimer physically interacted with, and blocked the action of, each other (Ray and Prefontaine, 1993). It was proposed that the DNA binding domain is necessary for this interaction, just as for AP-1 (Ray and Prefontaine, 1994).

A related study has reported that NF-IL6, a transcription factor that activates interleukin-6 gene expression, both synergizes with the glucocorticoid receptor in the induction of the rat α1-acid glycoprotein gene and physically interacts with the receptor (Nishio *et al.*, 1993). This binding to receptors was demonstrated by the ability of immobilized NF-IL6 with an intact leucine zipper to specifically adsorb ^{35}S-labeled receptors from solution. Receptors lacking the tau1 sequences (98–282) of the amino-terminal domain could still synergize with NF-IL6, consistent with this region of the receptor not being required for protein–protein interactions with NF-IL6 (Nishio et al., 1993).

Calreticulin is a poorly understood but ubiquitous calcium binding protein. Calreticulin also binds with high affinity and specificity to a sequence of intergins (KXGFFKR) (Rojiani et al., 1991) that is highly homologous to amino acids 461–466 of the rat receptor, which is a major component of the α-helix contact with the major groove of GRE DNA (Section II,B,6). It has recently been reported that increased levels of calreticulin can inhibit glucocorticoid receptor-mediated gene transcription in intact cells. This repression appears due to the binding of calreticulin to the KVFFKR sequence of receptors (Fig. 2), as determined both by the binding of a glutathione S-transferase (GST) fusion protein containing the DNA binding domain of human receptors (420–506; = 439–525 of rat) to a calreticulin affinity column and by the peptide-specific inhibition of fusion protein binding to a synthetic GRE in gel shift assays (Burns et al., 1994). This binding by calreticulin, and modification of transcriptional activity, has also been reported for androgen and retinoid receptors (Dedhar et al., 1994) and thus may be common to all members of the steroid receptor super family (Burns et al., 1994; Dedhar et al., 1994).

Transcriptional stimulation by Oct-2A, a transcription factor seen predominantly in lymphoid cells, was found to be inhibitable by glucocorticoid receptors only in nonlymphoid cells (Wieland et al., 1991). The DNA binding domain fragment (407–556) of the receptor retained inhibitory activity. However, the inability of Oct-2A to inhibit glucocorticoid receptor-mediated transcription, in addition to other data, suggested that the receptor interacts with some as yet unidentified co-activator as opposed to Oct-2A (Wieland et al., 1991).

The ubiquitous transcription factor OTF-1, or Oct-1, and Oct-2A derive from separate genes but exhibit the same DNA binding specificity (Wieland et al., 1991). Full-length glucocorticoid receptors inhibit the transcriptional activity of OTF-1 in intact cells. Unlike Oct-2A, this activity of receptors appears to involve direct interactions between receptor and OTF-1 in that receptors may be able to be cross-linked to OTF-1 in solution and could inhibit the DNA binding of OTF-1 in gel shift assays (Kutoh et al., 1992). These interactions appear to involve amino- and/or carboxyl-terminal receptor sequences because no inhibition of OTF-1 binding in the gel shift assays could be demonstrated with the overexpressed human receptor DNA binding domain (=413–519 of rat) (Kutoh et al., 1992).

Another factor that seems to interact with regions other than the DNA binding domain of the glucocorticoid receptor is GATA-1. Erythroid differentiation of mouse erythroleukemia cells by added chemicals involves the induction of several erythroid-specific genes, each of which contains GATA response elements that are activated by the transcription factor GATA-1. The inhibition of erythrodifferentiation by glucocorticoids may involve protein–protein interactions of GATA-1 and glucocorticoid receptors, as evidenced by gel shift assays using extracts that were immunodepleted of receptors, by the lack of a high-affinity GRE in the responsive genes, and by the full activity of a non-DNA binding receptor (Chang et al., 1993). The inability of various truncated receptors to repress GATA-1-induced gene expression suggested that the amino-terminal 106 amino acids of the mouse receptor (=1–118 of rat) interact with GATA-1 (Chang et al., 1993).

For a long time, it was believed that no factor was required for the DNA binding of activated glucocorticoid receptors. Other reports, however, indicate that two different factors are involved in the DNA binding of glucocorticoid receptors. The first is a low-molecular-weight factor (700–3000) that was required for binding of a subpopulation of receptors to columns of calf thymus DNA–cellulose (Cavanaugh and Simons, 1990a,b, 1993). The second factor is much larger (130,000) and is involved in the binding to GRE sequences in gel shift assays (Kupfer et al., 1993) (see also Section II,B,5).

Hsp90 is not the only heat shock protein that has been found to be associated with glucocorticoid receptors (see Section II,D,4). A 56-kDa protein that was complexed with each of several steroid receptors (Tai et al., 1986) was subsequently found to be both a heat shock protein (hsp56) (Sanchez, 1990) and an immunophilin that binds the immunosuppressive agent FK506 (Tai et al., 1992). These observations raised fascinating mechanistic possibilities concerning the antiinflammatory actions of glucocorticoids that, unfortunately, have not yet received much experimental support (Ning and Sanchez, 1993; Pratt et al., 1993). There is some debate as to how hsp56 interacts with the receptor (Pratt et al., 1993; Pratt, 1993), with cross-linking studies supporting the binding of hsp56 directly to the receptor (Rexin et al., 1991). Additional studies with the nt[i] receptor, which is missing the amino-terminal 404 amino acids (Dieken et al., 1990) but is still cross-linked to hsp56 (Rexin et al., 1988), indicate that the attachment site for hsp56 is in the carboxyl-terminal half of the receptor and probably in the steroid binding domain (Pratt et al., 1993).

III. Structure/Function Relationships for Specific Point Mutations in the Glucocorticoid Receptor Domains

All the point mutations of the glucocorticoid receptors from various species that have been found in the literature are given in the Appendix, which is meant to be used as a quick reference. More detailed discussions of the consequences of the mutations on receptor structure and function are given in the following sections. At this point, it should also be noted that the originally published sequence of the rat glucocorticoid receptor (Miesfeld *et al.*, 1986) has been revised over the years at several amino acids (i.e., positions 98, 226, 260, 276, 600, and 602; see Appendix).

A. General Considerations

A very successful and time-honored approach for identifying the function of specific amino acids has been to compare the properties of two proteins that differ only by the identity of the amino acid at the position in question. Conversely, amino acid substitutions between proteins are considered noncritical if both proteins have equivalent properties. A mutant mouse glucocorticoid receptor from an androgen-dependent tumor retained wild-type properties after having the valine corresponding to position 449 of rat changed to glycine, which is the normal amino acid in human and rat receptors (Kasai, 1990). Thus, the amino acid at position 449 can be classified as nonessential. Similarly, the amino acids of human (Hollenberg *et al.*, 1985), mouse (Danielsen *et al.*, 1986), rat (Miesfeld *et al.*, 1986a), and sheep (Yang *et al.*, 1990, 1992) glucocorticoid receptors that have changed with evolution (see Appendix) are currently viewed as relatively unimportant and nonfunctional. In contrast, the major decreases in steroid affinity for the monkey (alternately called cotton-top tamarin or marmoset) (Brandon *et al.*, 1991) and guinea pig (Keightly and Fuller, 1993) receptors preclude any conclusions as to which of the altered amino acids are unimportant, or required, for function.

As a note of caution, it should be realized that most of the differences between human, mouse, and rat receptors reside in the immunogenic, or transactivation, domain (Fig. 1). Future studies may reveal these changes to be crucial in the control of tissue-specific and/or gene-specific expression. Furthermore, an axiom of this approach is that conservative mutations to similar amino acids will have no effect when noncrucial amino acids are involved. Unfortunately, exceptions make blanket generalizations difficult and indicate that other parameters, such as protein folding and relative stability of conformers (Flanagan *et al.*, 1992, 1993), need to be considered. For example, the Cys500Gln mutation retains 50% of the original transcriptional activity in transient transfection assays, whereas the closely related mutations of Cys500Asn or Ile afford <1% activity. Likewise, Met505Cys or Leu retain >53% activity whereas Met505Thr has only 1% activity (S. Rusconi, personal communication).

B. HYPERVARIABLE, IMMUNOGENIC, OR TRANSACTIVATION DOMAIN (AMINO ACIDS 1–439)

The first, literally, modified amino acid of the glucocorticoid receptor is the amino-terminal amino acid, which has a blocked -NH$_2$ group (Eisen *et al.*, 1985; Carlstedt-Duke *et al.*, 1987). Thus, it has not yet been established directly that methionine is, in fact, the first amino acid of the receptor (see Appendix).

As with the other steroid receptors (Orti *et al.*, 1992), the steroid-free glucorticoid receptor is a phosphoprotein and steroid binding causes a two- to fourfold increase in phosphorylation (Hoeck and Groner, 1990; Bodwell *et al.*, 1991). Unlike the progesterone receptor, which is hyperphosphorylated in response to the binding of both agonists and antagonists (Sheridan *et al.*, 1988), glucocorticoid receptor phosphorylation is not increased by antisteroid binding (Hoeck *et al.*, 1989; Orti *et al.*,. 1989). The activated state of the receptor–glucocorticoid complex appears to be the substrate in a reaction that does not involve c-AMP-dependent protein kinase (Ori *et al.*, 1993).

A total of seven sites in steroid-treated mouse receptors have been identified and account for ~80% of the total receptor phosphorylation (Bodwell *et al.*, 1991). The equivalent residues in rat that are modified are Ser-134, -162, -224, -232, -246, -327, and Thr-171 (see Appendix). With the possible exception of Ser-246, none of these becomes fully phosphorylated (Bodwell *et al.*, 1991; Orti *et al.*, 1993). It is not yet known which amino acids are basally phosphorylated, and the biological importance of phosphorylation has resisted elucidation. Phosphorylation of Thr-171 and Ser-327 is unlikely to be significant because human and sheep receptors contain Ala and Pro, respectively, at these positions. Each of the seven sites has been mutated, either individually or all seven at once, to Ala (or to Asp) without any significant effect on transcriptional activation of the MMTV GRE and promoter in transiently transfected cells (Mason and Housley, 1993). Thus, phosphorylation of these sites is not required in receptor-mediated transcriptional activation. It remains to be seen if other marginally phosphorylated sites (Hutchison *et al.*, 1993) are relevant (see Section II,B,2) or if hitherto unmodified residues can be recruited for phosphorylation (Hilliard *et al.*, 1994). Other properties of receptors, such as the concentration of steroid required for half-maximal induction or the ability to cause repression versus induction from different promoters in different cells, may also be affected by phosphorylation. Finally, associated nonreceptor proteins, such as hsp90 and hsp56, may be the crucial target for phosphorylation.

Several specific sites for chemical (Met-28) and proteolytic (Phe-409 and Tyr-413) cleavage occur in this region (see Appendix). Only three mutations of the wild-type receptors have appeared for this domain (N383S, P435R, and L439P). Of these, the only one that caused any change in receptor activity was P435R, which facilitated the activation of receptors bound by the agonist triamcinolone acetonide but not the antagonist RU-486 (or RU-38,486) (see Appendix).

Almost all the differences in the number of amino acids in glucocorticoid receptors from different species arise from variations in the length of the polyglutamine stretch starting after Val-74. There are some single amino acid deletions, relative to the rat receptor, at Pro-9 through Arg-12 in various receptors and three single deletions in the guinea pig at Thr-167, Lys-176, and Leu-204. Some of the receptors have more amino acids at a particular position but, except for the 4-amino-acid insertion in sheep after the equivalent position of Leu-214 in rat, these insertions are seen in more than one species and are limited to a single amino acids (i.e., after the rat residues of Asn-243, Ser-430, and Pro-534). Not unexpectedly, many of these modifications are within the hypervariable immunogenic domain and fall outside of domains with known functions (Fig. 1).

C. DNA BINDING DOMAIN (AMINO ACIDS 440–505)

No inter- or intramolecular disulfides appear to exist in the native receptor (see Section III,E,1). However, intramolecular disulfides are readily formed during gel analysis of denatured receptors. These disulfides have been predicted to form between those cysteines that normally complex zinc. Thus, the protein-bound zinc ions may not only maintain the receptor in a conformation beneficial for DNA binding but may also prevent the facile oxidation of thiols that could inactivate the receptor (Opoku and Simons, 1993).

Most of the interest, and mutational analysis, in the DNA binding domain has centered on efforts to understand the origin of specific DNA binding. Many interesting data have been accumulated with truncated receptors or with receptors expressed in yeast (e.g., Schena *et al.*, 1989; Zilliacus *et al.*, 1992a). However, in keeping with the fore-

going activity comparisons for full-length receptors, and because of
the currently unexplained differences in steroid binding to receptors
translated in yeast (see Section II,D,3), most of these data have not
been covered here. For a more comprehensive discussion of these results, especially in comparison to other receptors, the reader is referred to two reviews (Freedman, 1992; Luisi *et al.*, this volume).

Some of the results in this domain epitomize the limitations of point
mutant studies. It was initially thought that Cys-500 was coordinated
with zinc, which was consistent with the inactivity of the C500R mutant (Schena *et al.*, 1989). However, the full activity of two other mutants (C500S and C500A) argued that Cys-500 did not bind zinc (Severne *et al.*, 1988), as was eventually confirmed by X-ray analysis (Luisi
et al., 1991). Saturation mutagenesis of Cys-500 afforded 19 mutants,
12 of which (including C500V) displayed ≤1% transcriptional activity
(S. Rusconi, personal communication). Similarly, of the five mutations
at Met-505, three (including M505T) gave ≤1% transcriptional activity (S. Rusconi, personal communication). Thus, it is difficult to predict
the effect of any given mutation.

Four studies involving the biological activity of a particularly large
number of mutants for full-length receptors in mammalian (CV-1)
cells (Hollenberg and Evans, 1988) and yeast cells (Schena *et al.*, 1989)
and for truncated receptors in CV-1 cells (Zandi *et al.*, 1993; Lanz *et al.*,
1993) have been reported. About half of the mutations caused a loss of
transactivation activity. One single-point mutant (I484L) (Lanz *et al.*,
1993) and one double mutant (H451N/S459G) (Zandi *et al.*, 1993) displayed greater than wild-type activity in the context of a truncated
receptor (407–556). Results for the same mutations in the full-length
receptor were not given but may be the same (see the following).

The P box, or distal knuckle of the first zinc finger, has been inten-
sively studied ever since it was shown to provide the determinants of
specific DNA binding and biological activity (Mader *et al.*, 1989; Danielsen *et al.*, 1989; Umesono and Evans, 1989). Several studies, including those with the truncated rat receptor sequence of 440–525, have
established that the GSV triad in 458–462 (Fig. 2) is specific for GRE
binding. The Val is pivotal, but not required for specific binding to
GREs versus EREs. Mutation of Val to Ala, which is the corresponding
amino acid of the estrogen receptor sequence (i.e., EGA), caused about
a threefold decrease in affinity of DNA binding (Alroy and Freedman,
1992). Other studies with a slightly different fragment (414–519) of
the human receptor found, *inter alia,* that changing GSV to ESA afforded low DNA binding affinity and much reduced biological activity
with a GRE, whereas swapping the GSV of the glucocorticoid receptor
for the EGA of the estrogen receptor gave a hybrid glucocorticoid re-

ceptor with intermediate DNA binding affinity and much reduced biological activity for a GRE (Zilliacus *et al.*, 1991). Somewhat surprisingly, a number of mutations of the GSV sequence were permissive for biological activity in the truncated receptors 407–556 and some 3–556 constructs (Zandi *et al.*, 1993); the effect with full-length receptors was not reported. The double mutant H451N/S459G, expressed in the 407–556 fragment, may prove to be very interesting because it displayed increased DNA binding activity and transcriptional activation (Zandi *et al.*, 1993).

A region in the second zinc finger just amino terminal to the enh1 domain has been the focus of some recent attention. Saturation mutagenesis was used to examine the role of Ile-484 in truncated rat receptors (407–556) that contain just the DNA binding domain, the enh1 sequence in the second zinc finger, and part of the tau2 activation sequence (544–574) (Lanz *et al.*, 1993). Most of the 17 mutations displayed reduced or negligible transcriptional activity in transiently transfected CV-1 cells. All receptors were stable, as seen by *in situ* immunofluorescence, and there was an excellent correlation between *in vivo* activity and *in vitro* DNA binding capacity for all but the I484C mutation, which retained wild-type DNA binding capacity but exhibited ≤5% transcriptional activity (Lanz *et al.*, 1993). It will be interesting to see whether the DNA binding of the I484C mutant is stabilized by the formation of a disulfide bond between the contacting faces of the receptor dimer. Analysis of the same mutation in the context of the 1–523 receptor, which contains the tau1 and enh1 regions but not the tau2 sequence, revealed no effect of the I484C mutation on biological activity. The activity of the I484 mutation in the context of the full-length receptor was similar to that of the 1–523 fragment, consistent with an inactivation of the tau2 region activity in the intact receptor. After comparing the NMR (Hard *et al.*, 1990) and X-ray (Luisi *et al.*, 1991) structures of the glucocorticoid receptor, it was proposed that I484 is involved in initial contacts of receptor with DNA but, upon binding, is part of a loop that "flips" 120° and correctly orients residues 486–491 for DNA–protein contacts and for protein–protein contacts of receptor dimerization (Lanz *et al.*, 1993). A recently refined NMR structure of a slightly smaller segment of the DNA binding domain (439–520) revealed that the magnitude of the "flip" is slightly smaller (i.e., about 90°) upon binding DNA (Baumann *et al.*, 1993). The fact that this flip to permit optimal protein–protein contacts is not seen in the presence of the high concentrations of protein required for NMR analysis also suggests that receptor dimerization may occur only as a consequence of DNA binding.

The flip of the zinc finger loop was also proposed to bring about the optimal positioning of tau2 (at 523–556) (Lanz *et al.*, 1993) and thus may account for the transcriptional activity of the 486–491 region, or enh1. However, the still appreciable activity of 1–523 receptors with or without the I484C mutation argues that the participation of this loop in establishing protein–protein contacts in receptor dimers is not required for biological activity.

In view of the foregoing, well-documented modular construction of the receptor, a continuing question has been whether DNA binding and transactivation activity can be separated. In addition to the I484C mutation, human receptors containing the point mutation corresponding to K461G in the rat bound DNA but were unable (<1% activity) to transactivate a MMTV/GRE-controlled reporter gene in CV-1 cells (Hollenberg and Evans, 1988; Yang-Yen *et al.*, 1990). Interestingly, the same mutant receptor displayed good transactivation activity with the same MMTV reporter construct in another cell line (HeLa) and even evoked more activity than the wild-type receptor with several promoters (interleukin-6 and thymidine kinase) lacking a GRE (Ray *et al.*, 1991). The K461G mutant also increased the activity of AP-1-induced transcription (Yang-Yen *et al.*, 1990). Thus, this mutation is detrimental for transactivation only in selected circumstances.

Recent studies have also inquired whether receptor activation and repression can be separated. The double mutant P493R/A494S in rat receptors, called LS7, was unable to activate transcription from MMTV LTR containing constructs even though its DNA binding activity was relatively unaffected (Godowski *et al.*, 1989). However, in contrast to the results with the K461G mutant, the LS7 receptors repressed transcription from both the prolactin (Godowski *et al.*, 1989) and collagenase (Yang-Yen *et al.*, 1990) promoters. Several point mutants of the human receptor (at rat positions S444G, L455V, and the double mutant Y497L/R498G) were as good, or better, transactivators as the wild type receptor but were unable to repress induction from a collagenase AP-1 promoter (P. Herrlich, personal communication). Furthermore, mutations in the region of the D box dimerization sequence (A477T and N473D/A477T) that inhibit DNA binding and activation did not affect repression (P. Herrlich, personal communication). These two mutants should be very useful in future studies of the interactions of receptors with AP-1 (see Section II,F,3). Thus, it appears clear that repression and induction involve separable portions of the receptor and that repression may not require receptor dimerization.

Three other mutations in the rat receptor (R488Q, R489K, and N491S), which are close to the positions affected in the preceding double mutant LS7, provide an interesting set of results (Schena *et al.*,

1989). R488Q was almost inactive in yeast (3% activity) but retained 28% activity in CV-1 cells. R489K displayed 10-fold less activity in yeast at 19°C than at 30°C (3 vs. 34%). N491S had high levels of activity in yeast (58%) but was virtually inactive in CV-1 cells (<0.5%). Whether or not there were also differences in the EC_{50} was not stated. The reasons for these temperature- and cell-specific variations have not been determined. However, it should be noted that all three mutations fall within the region called enh1, which has been implicated as having transcriptional activity (see Section II,B,7).

The current model of steroid hormone action requires the DNA binding of receptors before any *induced* biological activity can be evoked (see Section II,F,3 for repression without DNA binding). Thus, it should not be possible to obtain mutants that retain transactivation but little or no DNA binding activity. Nevertheless, a few such mutants have been reported in the human receptor [rat numbering = C450G, K465G, Y474G, K486G, and M505G (Hollenberg and Evans, 1988) and A477T (P. Herrlich, personal communication)]. These unexpected properties were postulated to be due to *in vitro* instability (Hollenberg and Evans, 1988) but could also be due to differences in the affinity of receptors for DNA under *in vitro* versus *in vivo* conditions, to the high concentration of transfected receptors, or to chromosomal components that could increase the affinity of receptor binding to DNA (Simons *et al.*, 1976; Simons, 1979; Kumar and Dickerman, 1983; Ueda *et al.*, 1988; Okamoto *et al.*, 1993) (see Section II,C,6).

An X-ray study of the DNA binding domain (440–525) complexed with a GRE finally provided some explanations for the preceding results at an atomic level (Luisi *et al.*, 1991). The intermolecular contacts for receptor dimerization involve the alternate amino acids Leu-475, Ala-477, Arg-479, Asp-481, and Ile-483, plus Ile-487 and Asn-491. This conclusion has been supported recently by the finding that mutation of the amino acid corresponding to Ala-477 dramatically reduced the DNA binding capacity of glucocorticoid (P. Herrlich, personal communication) and androgen (Kaspar *et al.*, 1993) receptors. Phosphate contacts at specific DNA binding sites were seen with His-451, Tyr-452, Try-474, Arg-489, Lys-490, and Arg-496. Only some of these residues, plus others, were involved in phosphate contacts at nonspecific DNA binding sites (i.e., Cys-450, Tyr-452, Arg-466, Arg-489, and Arg-496). Direct contacts between the receptor and nucleotide bases of specific DNA binding sites were seen for Lys-461, Val-462, and Arg-466. It is interesting that only Val-462 has been identified by point mutants as being important specific DNA binding, though even this amino acid was not required for specific binding (Alroy and Freedman, 1992). The structural reasoning for this is not yet clear. It has been proposed that the function of the two other amino acids of the GSV triad are to stabilize the overall tertiary structure of the receptor (Freedman, 1992).

D. Hinge Region/Nuclear Translocation Domain (Amino Acids 506–546)

Trypsin cleavage in this region is sensitive to steroid binding and appears to reflect a steroid-induced conformational change in the receptor. Thus, digestion of steroid-free receptors occurred after Lys-536 (Simons *et al.*, 1989), whereas steroid binding inhibited cleavage at this site and caused predominant cutting near Lys-517 (Carlstedt-Duke *et al.*, 1987; Simons *et al.*, 1989).

It has been suggested that a Gly-to-Ala mutation at the rat position of Pro-534 might be responsible for the 10-fold lower affinity of steroid binding to tamarin versus human receptors (Brandon *et al.*, 1991), even though this position is outside of what is often considered to be the steroid binding domain. The fact that the mouse receptor contains an Ala at almost the exact same position (Pro-534 contains an inserted A in tamarin vs. a substituted A in mouse, Appendix) may argue against this hypothesis.

E. Steroid Binding Domain (Amino Acids 547–795)

1. *Covalent Modifications*

Initial studies suggested that steroid-induced hyperphosphorylation occurred in this domain (Smith *et al.*, 1989). However, other work from the same lab argues that there is no phosphorylation of the steroid binding domain and that the earlier results were probably influenced by a contaminating phosphopeptide (Bodwell *et al.*, 1991).

Several sites of proteolysis or cyanogen bromide-induced cleavage have been identified. Cyanogen bromide causes fragmentation after Met-583, -622, -664, -684, and -770 (Carlstedt-Duke *et al.*, 1987). Trypsin digestion of receptors is, as mentioned in Section III,D, modified by steroid binding. Steroid-free receptors yield a 16-kDa fragment, which is thought to arise from scission after Arg-673 (and Lys-536), whereas steroid-bound complexes are preferentially cleaved after Lys-781 and -761 (and near Lys-517) to afford the 30- and 28-kDa mero-receptors respectively (Simons *et al.*, 1989).

The 20 cysteines in glucocorticoid receptors (see Appendix) would seem to offer ample opportunities to form inter- and/or intramolecular disulfides. However, other experiments demonstrated that there are no disulfides of either type in steroid-free receptors or in receptors bound by agonists or antagonists in the unactivated or activated state (Chakroborti *et al.*, 1990; Opoku and Simons, 1993). Similarly, even an *S. aureus* protein A fusion protein with the human equivalent of rat receptor sequences from 496–795, which was overexpressed in *E. coli* and appeared to have a relaxed tertiary structure (see Section, III,E,2), did not contain any disulfides (Ohara-Nemoto *et al.*, 1990). However, disulfides were readily formed in the denatured state during gel analysis (see Section III,C).

2. Steroid Binding and Affinity Labeling

The affinity labeling (Simons, 1987), antiglucocorticoid (Simons and Thompson, 1981) Dex-Mes covalently modifies just one cysteine (Cys-656) in the native receptor of rat (Simons *et al.*, 1987; Carlstedt-Duke *et al.*, 1988), mouse (Smith *et al.*, 1988), and human (Stromstedt *et al.*, 1990) receptors. However, when a protein A/human receptor fusion protein containing the amino acids equivalent to 496–795 of rat receptor was over-expressed in *E. coli* and covalently labeled with Dex-Mes, all five cysteines of the receptor fragment were labeled. This result, in conjunction with the about 40-fold lower affinity of this fusion protein, suggested that the tertiary structure of this sequence is less rigid and much more open than when in the context of the full-length receptor (Ohara-Nemoto *et al.*, 1990). Photoaffinity labeling by triamcinolone acetonide is seen at Met-622 and Cys-754 (Carlstedt-Duke *et al.*, 1988; Stromstedt *et al.*, 1990).

The effect of all point mutations on steroid binding can be grouped into one of four categories: elimination of binding, reduction of binding affinity, no effect, and increased binding affinity. The number of mutations that have been documented to eliminate steroid binding is actually small. E558G in mice is reported to destroy steroid binding (Danielsen *et a.*, 1986). Mutations at Cys-683 and -754 each caused a 10- to 20-fold reduction in biological potency in mouse receptors, presumably due to reduced affinity. These effects were multiplicative, though, in that the double mutation (C683,754S) afforded a receptor with about 400-fold lower potency (Chen and Stallcup, 1994). Two sets of adjacent point mutants at 770/771 and at 774/775 were first observed to have no effect on steroid binding in the mouse receptor while blocking transcriptional activation (Danielian *et al.*, 1992). Conflicting results have been obtained in rat. Receptors containing the M770A/L771A mutations and overexpressed in vaccinia virus-infected HeLa cells had a threefold reduced affinity for Dex, good DNA binding activity, and only a 50% reduction in a cell-free transcription assay that was not responsive to added steroid (Schmitt and Stunnenberg, 1993). Other investigators have found that the 770/771 and the 774/775 mutation each eradicate both steroid binding and transcriptional activation (Lanz et al., 1993). It may be that the activity of these mutants depends not only on the context of the receptor in which they are expressed but also on the system in which they are examined. If so, these mutations would be of great interest and worthy of much further attention.

Several other mutations have been isolated that were reported to have no transcriptional activity (see Section III,E,3), presumably due to a lack of steroid binding activity. In at least one case, the mutation (C742G) rendered the receptor protein unstable (Byravan et al., 1991), so that the effect on steroid binding probably was indirect. In other cases, there may be a dissociation of steroid binding activity and transcriptional activity, as was initially proposed for the foregoing 770/771 and 774/775 double mutants in mice (Danielian et al., 1992). In fact, three individual mutations (L584S, D659G, and E706K) were found to have a decreased binding affinity for, but no biological activity with, Dex in COS-7 cells, although defects in other steps prior to transcriptional activation were also considered as possible (Garabedian and Yamamoto, 1992). One such example is the activation-labile (act[l]) phenotype (Schmidt et al., 1980), which has been found to be the result of a L771F mutation (Ashraf and Thompson, 1993; Powers et al., 1993).

The largest group of altered receptors is where the mutation causes a decrease in the steroid binding affinity. Such mutations have invariably been accompanied by decreased transcriptional activation, although the correlation is often poor in the few instances where the two properties have been directly compared (e.g., Hurley et al., 1991; Malchoff et al., 1993). This discrepancy may reflect differences in the conditions of the two assays or may be indicative of postreceptor binding defects, as has been speculated earlier.

The early data on modifications of the steroid binding domain led to the suspicion that all mutations were detrimental to binding (Giguère et al., 1986; Danielsen et al., 1986; Hollenberg et al., 1989; Hurley et al., 1991). Although mutations that do not affect steroid binding are not common, there is evidence that they do exist. A total of 8 out of the 32 point mutations prepared by Stallcup et al. are thought to have wild-type affinity because their biological activity was unchanged. These mutations are at positions 568, 618, 622, (three different mutations at this position), 640, 656, and 793 (M. Stallcup, personal communication) (see Appendix).

Perhaps the least common are those mutations that increase the affinity of receptor for steroids. Only two examples have been reported so far in mammalian cells and both involve mutation of the Cys-656 that is affinity labeled by Dex-Mes (Simons *et al.*, 1987). Thus in rat, but not in mouse (M. Stallcup, personal communication), C656S afforded receptors with a 3- to 4-fold increased affinity and somewhat increased specificity in steroid binding (Chakraborti *et al.*, 1991). For human receptors expressedin COS-7 cells, the C656S mutation also did not increase the affinity. This discrepancy could be attributed to a lower affinity of the wild type human versus rat receptors in COS-7 cells, while the affinity of both C656S mutants was the same (J. Carlstedt-Duke, personal communication). The C656G mutation in rat receptors caused a 9-fold increase in affinity for Dex and a $\geq$10-fold increase in the specificity of binding Dex versus progesterone or aldosterone (Chakraborti *et al.*, 1991). At the same time, C656G was $\geq$6 times more potent than the wild-type receptor in eliciting transcriptional activation. For these reasons, C656G has been called a "super" receptor (Chakraborti *et al.*, 1991). This super receptor was also proportionately more active in yeast (Garabedian and Yamamoto, 1992). F620S afforded a receptor with increased affinity for Dex and triamcinolone acetonide in yeast (Garabedian and Yamamoto, 1992), where the affinity of the wild-type receptor for these steroids is low (see Section II,D,3), but not in CV-1 cells. These results have been interpreted as supporting the participation of some nonreceptor factor(s) in steroid binding (Garabedian and Yamamoto, 1992).

The tamarin receptor has 33 amino acid substitutions compared to the human receptor (see Appendix), and a 10-fold lower affinity (Brandon *et al.*, 1991). Four of these substitutions are in, or very close to, the steroid binding domain (534, 634, 636, and 788). The Ala-634 is the same residue as found in mouse, and thus is probably a benign mutation, whereas Arg-788 is a conservative replacement for the lysine seen with all other receptors (see Appendix). At position 534, the tamarin has an extra amino acid (Ala), which is the same amino acid as found in mouse at this position (see Appendix). However, the effect of this added amino acid on protein structure is almost impossible to predict as Stallcup *et al.* have found that the mutation of adjacent amino acids can cause $\geq$10-fold differences in biological activity, for example, 559 versus 560, 572 versus 573, 640 versus 641, and 792 versus 793 (M. Stallcup, personal communication). Ser-636 is another prime candidate for the decreased affinity of tamarin receptors (Brandon *et al.*, 1991). A similar analysis to understand why the guinea pig receptor has a 20-fold lower affinity than the mouse receptor (Kraft *et al.*, 1979) is much more difficult because, among other reasons, there are 30 amino acid differences just in the steroid binding domains of guinea pig (Keightley and Fuller, 1993) and mouse receptors (see Appendix). One interesting difference is the C656W mutation that was found in guinea pigs. The same mutation in the human receptor had no effect on transcriptional activity (Keightley and Fuller, 1993). Similarly, replacement of Cys-656 by tyrosine in the rat receptor did not change the cell-free affinity of receptors expressed in COS-7 cells (P. K. Chakraborti and S. S. Simons, Jr., unpublished results), although it did reduce the transcriptional activity of the receptor in a yeast system (Garabedian and Yamamoto, 1992).

The multiple mutations of Cys-656, ranging from the sterically small glycine, which created a "super" receptor, to the very bulky tyrosine and tryptophan, which did not alter steroid binding activity, raise interesting questions regarding the function of this cysteine. The fact that it is highly conserved among glucocorticoid receptors (see Appendix), is not seen in any of the other classical steroid receptors (Lopez *et al.*, 1990), and is the only cysteine in the steroid binding domain that is affinity-labeled by Dex-Mes would appear to suggest an important role for this amino acid. Nevertheless, it is clearly dispensable for all functions so far examined. Further studies are required to resolve this issue.

Investigations by two independent laboratories of receptors with the activation-labile (act[l]) phenotype, which was initially defined as a loss of steroid binding activity under the cell-free conditions used to activate receptor–steroid complexes (Schmidt *et al.*, 1980), have yielded some unexpected results. Ashraf and Thompson examined the original act[l] cells (4R4) and found that the phenotype is due to the mutation L771F. However, the same mutant receptor could also produce a receptor-minus (r⁻) phenotype when expressed in a different cellular environment, that is, in ICR27 versus 4R4 cells (Ashraf and Thompson, 1993). This is a very interesting result, in keeping with the evidence that other nonreceptor factors can influence the binding activity of receptors (see Section II,D,3). It should be kept in mind, though, that the r⁻ phenotype strictly indicates that the level of receptors is below some arbitrary, low amount. Thus, any mechanism that reduced the level of L771F receptor protein in ICR27 cells, such as gene dosage, would produce the r⁻ phenotype (J. M. Harmon, personal communication). Harmon and co-workers also found the L771F mutation in ICR27 cells, which afforded the loss of steroid binding at 37°C in transiently transfected cells, but not at 0°C in cell extracts, typical of the act[l] phenotype. Another cell line with the act[l] phenotype (3R7) contained a second receptor with a different mutation (C440Y), where C440Y bound steroid but not DNA (Powers et al., 1993). Thus, more than one genetic composition can afford similar phenotypes.

3. Transactivation Activity

As discussed earlier, most mutations that reduce steroid binding also reduce transactivation (see Section III,E,2). However, the transcriptional inactivity of receptor mutants with high affinities for steroid (i.e., L584S, D659G, and E706K) suggested that it is possible to separate the two activities (Garabedian and Yamamoto, 1992). If this explanation is correct, it implies that the sequences of the steroid binding domain that are responsible for transcriptional activity encompass a much larger area than either the tau2 region (Giguère et al., 1986; Hollenberg et al., 1987; Schüle et al., 1990) or the sequences immediately around 765–779, which are thought to be part of the TAF-2 transactivation domain (Danielian et al., 1992) (see Section II,D,8 and Fig. 1). Another potential example of dissociated steroid binding and transactivation is the D641V mutation of human receptors. This mutation (659 in rat) caused a threefold decrease in steroid binding affinity, a sevenfold decrease in Dex sensitivity in the patient, but a complete insensitivity in transiently transfected COS-7 cells (Hurley et al., 1991). Alternatively, as proposed by the authors, this discrepancy may reflect some difference in expression in the transiently transfected COS-7 cells versus the affected patient.

4. Other Activities

It has been theorized that molybdate stabilization involves Cys-656 and -661 of the rat receptor (Dalman et al., 1991a). However, other experiments with receptors containing a double-point mutation at these positions (C656,661S) (Chakraborti et al., 1992) have demonstrated that these cysteines are not required for molybdate stabilization (Modarress et al., 1994). No reports of the effect of point mutations on hsp90 binding, hsp56 binding, or receptor dimerization have been noted.

IV. INTERDOMAIN INTERACTIONS

No interdomain disulfide linkages have been observed (Chakraborti et al., 1990; Ohara-Nemoto et al., 1990; Opoku and Simons, 1993). In fact, some experiments argue that there are no stable disulfides in steroid-free, unactivated, or activated complexes of agonists or antagonists (Opoku and Simons, 1993). Despite the fact that there is no evidence for any covalent bonds between domains of the receptor, and that the activities of the various domains are quite localized, there are several reports of inter- (or trans-) domain interactions.

A. BETWEEN IMMUNOGENIC AND DNA BINDING DOMAINS

The mutation of Lys-461, at the distal base of the first zinc finger (Fig. 2), to Tyr is reported to eliminate the transactivation activity of TAF-1 in truncated receptors (3–556) without affecting the DNA binding (Zandi et al., 1993). It seems unlikely that Lys-461 could be interacting directly with the TAF-1 region to augment its activity because the X-ray data indicate that Lys-461 is contacting DNA (Luisi et al., 1991). However, this conceivably could change in the context of the intact 98-kDa protein. Similarly, mutations at 488, 489, or 491 of rat receptor do not affect DNA binding but do decrease the constitutive transcriptional activity of the carboxyl-terminus-truncated receptors 1–556 (Schena et al., 1989).

Conversely, comparisons of the DNA binding of intact receptors with those missing most of the immunogenic domain (1–409) indicated that the amino-terminal half of the receptor increases the specificity of DNA binding (Eriksson and Wrange, 1990). This activity may be encoded in the sequence of 208–305, which was previously proposed to be responsible for decreasing the nonspecific DNA binding of receptors (Danielsen et al., 1987).

B. BETWEEN STEROID BINDING AND DNA BINDING DOMAINS

Although there is agreement that regions outside of the DNA binding domain are required for wild-type binding affinity to GREs (Dahlman-Wright et al., 1992; Freedman, 1992; Hutchison et al., 1992a), a reexamination of the issue (Eriksson and Wrange, 1990) with receptors overexpressed in yeast led to the conclusion that the contribution of amino-terminal sequences was quite small and that the major effect was due to sequences in the steroid binding domain (Dahlman-Wright et al., 1992), possibly in the region from 730–764 that was first implicated in the dimerization of estrogen receptors (Fawell et al., 1990). Alternatively, given the known peculiarities of glucocorticoid receptors expressed in yeast (see Section II,D,3), it is conceivable that the differences in the relative importance of the amino-terminal and carboxyl-terminal half of the receptors reflect yet to be determined modifications of receptors expressed in the normal target cells (Eriksson and Wrange, 1990) versus yeast cells (Dahlman-Wright *et al.*, 1992). An early report (Hollenberg *et al.*, 1987) on an internal deletion of the human receptor from 533–696 (551–714 of rat GR), which was later found to be a truncation from 550–777 (568–795 of the rat) (Housley *et al.*, 1990), suggests that the steroid binding domain may be inhibitory for DNA binding, even in the activated state. Thus this truncated receptor, which is missing most of the steroid binding domain, was constitutively active and displayed a higher DNA binding affinity and transcriptional activity than the full-length receptor in CV-1 cells (Hollenberg *et al.*, 1987).

The inability of about half of the activated human L cell receptors to bind to DNA has been ascribed to a "misfolding" of the steroid binding domain that somehow inhibits DNA binding. This "misfolding" appears to be restricted to the steroid binding domain because trypsin digestion of these receptors generated 15-kDa DNA binding domain fragments that were then capable of binding to DNA (Hutchison *et al.*, 1992a).

I484 is in the second zinc finger (Fig. 2). Nonetheless, mutation of this residue to any amino acid other than leucine, valine, methionine, or tyrosine drastically reduced or eliminated the transcriptional activity of the constitutively active 407–556 fragment. Although this could easily be accounted for by a concomitant loss of DNA binding activity, one mutant (I484C) retained wild-type DNA binding activity, even in the context of the full-length receptor (Lanz *et al.*, 1993). Thus, it was postulated that residue Ile-484 somehow interacts with the tau2 region of the steroid binding domain to allow optimal activity of this transactivating domain (Lanz *et al.*, 1993) (see Section III,C).

C. OTHERS

Deletion of most of the hinge region, corresponding to 518–549 from the human receptor, was found to eliminate steroid binding (Hollenberg *et al.*, 1987). Whether this represents a true interaction between domains or just the removal of some crucial amino acids at the aminoterminal end of the steroid binding domain remains to be determined. Other deletions that affect less of the steroid binding domain, that is, removal of 1–546 (Rusconi and Yamamoto, 1987) and of 5–516 (Segard-Maurel *et al.*, 1992), also reduced steroid binding affinity by ≥300-fold. Mutations of 507–514 in the hinge region eliminated DNA binding (K. R. Yamamoto, personal communication) and transcriptional repression but not induction (Godowski *et al.*, 1989) (see Section II,C,6). Further studies of this mutation (LS10) promise to yield very interesting results.

V. CONCLUSIONS

A tremendous amount of information concerning structure/function relationships has appeared since the cloning of the first full-length glucocorticoid receptor in December of 1985 (Hollenberg *et al.*, 1985). During the elapsed time, molecular biologists have been generating modified receptors faster than many people can fully characterize them and certainly faster than most people can assimilate all the results. These trends will only accelerate as more detailed information about the action of receptors on a molecular level becomes available. It is hoped that this review has begun a compilation of some of this information in a manner that makes its retrieval easier, and thus its utility greater.

VI. APPENDIX

The single-letter code for the amino acid sequence for the rat receptor is listed in its entirety. Those amino acids with an asterisk have been reassigned to a different amino acid since the original publication of the sequence. All numbering refers to the rat receptor. The sequences of the other receptors were aligned for maximum homology to the rat receptor by means of the program "BestFit" on the University of Wisconsin GCG package. Only the differences from the rat receptor for the mouse, human, sheep, guinea pig, and tamarin or monkey are given. A dash means that this amino acid has been deleted, and a question mark indicates that the amino acid for this position is unknown for the receptor in question. The listing of more than one amino acid at a single position (e.g., the corresponding position of rat 214 in the sheep receptor) means that all of these amino acids have replaced the single amino acid of the rat receptor. When describing the mutations at a given position, the species in which the original mutation was characterized is given as a single letter followed by a colon (R: = rat, M: = mouse, H: = human, S: = sheep, G: = guinea pig, and T: = tamarin or monkey). Some mutations are referred to by giving the wild-type amino acid, then the position, followed by the new amino acid (e.g., H451N).

Position	____ Glucocorticoid receptors from ____						Description of mutations
	Rat	Mouse	Human	Sheep	Guinea-pig	Tamarin	
1	M			?			R: amino terminus is blocked (Eisen *et al.*, 1985; Carlstedt-Duke *et al.*, 1987)
2	D			?			
3	S			?	L		
4	K			?			
5	E			?			
6	S			?			
7	L			?	V		
8	A		T	?	T	T	
9	P		-	?	S	-	
10	P			?	S		
11	G			?	-		
12	R			?	-	K	
13	D		E	?	K	E	
14	E			?			
15	V		N	?		N	
16	P			?		S	
17	G	S	S	?	S	S	
18	S			?			
19	L		V	?	V	V	
20	L			?			
21	G		A	?		T	
22	Q	R		?	S		
23	G		E	?	E	E	
24	R			?			
25	G			?	R		
26	S		D	?	N	N	
27	V			?			
28	M			?	I		R: cyanogen bromide cleaves at this methionine (Carlstedt-Duke *et al.*, 1987)
29	D			?			
30	F	L		?			
31	Y			?		C	
32	K			?			
33	S	T	T	?	T	I	
34	L			?	V		
35	R			?			
36	G			?			
37	G			?			
38	A			?			
39	T			?			
40	V			?		L	
41	K			?			
42	V			?			
43	S			?			
44	A			?		V	
45	S			?			
46	S			?			
47	P			?		T	
48	S			?			
49	V		L	?	L	L	
50	A			?			
51	A		V	?			
52	A			?			
53	S			?	A		
54	Q			?			
55	A		S	?	S	S	
56	D			?			
57	S			?			
58	K			?			
59	Q			?			
60	Q		R	?	R		*(continued)*

Position	Rat	Mouse	Human	Sheep	Guinea-pig	Tamarin	Description of mutations
61	R			?			
62	I		L	?	L	L	
63	L			?			
64	L		V	?	V	V	
65	D			?			
66	F			?			
67	S		P	?	P	P	
68	K			?			
69	G			?			
70	S			?			
71	T	A	V	?	G	V	
72	S			?			
73	N			?			
74	V	-	-	?	-		
75	Q	-	-	?	-	-	
76	Q	-	-	?	-	-	
77	R	-	-	?	-	-	
78	Q	-	-	?	-	-	
79	Q	-	-	?	-	-	
80	Q	-	-	?	-	-	
81	Q	-	-	?	-	-	
82	Q	-	-	?	-	-	
83	Q	-	-	?	-	-	
84	Q	-	-	?	-	-	
85	Q	-	-	?	-	-	
86	Q	A	-	?	-	-	
87	Q		-	?	-	-	
88	Q		-	?	-	-	
89	Q		-	?	-	-	
90	Q		-	?	-	-	
91	Q		-	?	-	-	
92	Q		-	?	-	-	
93	Q		-	?	-	-	
94	Q		A	?	A	-	
95	Q	P		?			
96	Q			?			
97	P			?			
98	G*	D	D	?	D	D	R: = D not G (S. Rusconi, personal communication)
99	L			?			
100	S			?			
101	K			?			
102	A			?			
103	V			?			
104	S			?			
105	L			?			
106	S			?			
107	M			?			
108	G			?			
109	L			?			
110	Y			?			
111	M			?			
112	G			?			
113	E			?			
114	T			?			
115	E			?			
116	T			?			
117	K			?			
118	V			?			
119	M			?			
120	G			?			
121	N			?			
122	D			?			

Position	Glucocorticoid receptors from						Description of mutations
	Rat	Mouse	Hu-man	Sheep	Guinea-pig	Tam-arin	
123	L			?			
124	G			?			
125	Y		F	?	F	F	
126	P			?			
127	Q			?			
128	Q			?			
129	G			?			
130	Q			?			
131	L		I	?	I	I	
132	G		S	?	S	S	
133	L			?			
134	S			?	P		M: is phosphorylated in steroid treated cells (Bodwell et al., 1991) but mutation to Ala has little to no effect on biological activity (Mason & Housley, 1993)
135	S			?			
136	G			?			
137	E			?			
138	T			?			
139	D			?			
140	F		L	?		L	
141	R		K	?		Q	
142	L			?			
143	L			?			
144	E			?			
145	E			?			
146	S			?			
147	I			?			
148	A			?			
149	N			?			
150	L			?			
151	N			?	S		
152	R			?			
153	S			?			
154	T			?			
155	S			?			
156	V	R		?			
157	P			?			
158	E			?			
159	N			?			
160	P			?			
161	K			?			
162	S			?	N		M: is phosphorylated in steroid treated cells (Bodwell et al., 1991) but mutation to Ala has little to no effect on biological activity (Mason & Housley, 1993)
163	S			?			
164	T		A	A	A	A	
165	S	P					
166	A		T			S	
167	T	A	A	A	-	S	
168	G		V	V	V	V	
169	C		S	S	S	S	
170	A				G		
171	T			A	A	A	M: is phosphorylated in steroid treated cells (Bodwell et al., 1991) but mutation to Ala has little to no effect on biological activity (Mason & Housley, 1993)
172	P						
173	T					K	
174	E						
175	K						
176	E						
177	F						
178	P						

(continued)

Position		Glucocorticoid receptors from					Description of mutations
	Rat	Mouse	Human	Sheep	Guinea-pig	Tamarin	
179	K	Q					
180	T						
181	H			Q			
182	S						
183	D						
184	A	P	V	V	L	V	
185	S						
186	S						
187	E						
188	Q						
189	Q			E			
190	N		H				
191	R		L	L	L	L	
192	K						
193	S		G	G		G	
194	Q						
195	T	P		K	A		
196	G						
197	T			S			
198	N						
199	G						
200	G						
201	S		N		N	N	
202	V			M		A	
203	K						
204	L				-		
205	Y		H		F	C	
206	P	T	T	T		T	
207	T				P	A	
208	D						
209	Q						
210	S						
211	T						
212	F						
213	D						
214	L	I	I	IWRKK	I	I	
215	L						
216	K	Q	Q	Q		Q	
217	D						
218	L						
219	E						
220	F						
221	S						
222	A		S	S	S	S	
223	G			E			
224	S						M: is phosphorylated in steroid treated cells (Bodwell et al., 1991) but mutation to Ala has little to no effect on biological activity (Mason & Housley, 1993)
225	P						
226	S*	G	G		G	G	R: = G not S (M.J. Garabedian and K.R. Yamamoto, personal communication)
227	K						
228	D	E	E	E	E	E	
229	T				R		
230	N			S	S		
231	E			D		Q	
232	S						M: is phosphorylated in steroid treated cells (Bodwell et al., 1991) but mutation to Ala has little to no effect on biological activity (Mason & Housley, 1993)
233	P						
234	W						
235	R						

Position	Rat	Mouse	Human	Sheep	Guinea-pig	Tamarin	Description of mutations
236	S				P		
237	D						
238	L			I			
239	L						
240	I				M		
241	D						
242	E						
243	N		NC	NC	SC	NC	
244	L						
245	L						
246	S						M: is phosphorylated in steroid treated cells (Bodwell *et al.*, 1991) but mutation to Ala has little to no effect on biological activity (Mason & Housley, 1993)
247	P						
248	L						
249	A						
250	G						
251	E						
252	D					E	
253	D						
254	P		S	S		S	
255	F						
256	L						
257	L						
258	E						
259	G						
260	N*	D		S			R: = D not N (M.J. Garabedian and K.R. Yamamoto, personal communication)
261	T	V	S	S	S	S	
262	N						
263	E						
264	D						
265	C						
266	K						
267	P						
268	L						
269	I			L			
270	L						
271	P						
272	D						
273	T			A			
274	K						
275	P						
276	K*						R: = R not K (M.J. Garabedian and K.R. Yamamoto, personal communication)
277	I						
278	K	Q					
279	D						
280	T		N	N	N	N	
281	G						
282	D						
283	T		L	L	G	L	
284	I		V			V	
285	L						
286	S			P			
287	S						
288	P				S	S	
289	S			N	N		
290	S		N			N	
291	V						
292	A		T	P	P	T	
293	L				Q		

(continued)

Position	Rat	Mouse	Human	Sheep	Guinea-pig	Tamarin	Description of mutations
294	P						
295	Q						
296	V						
297	K						
298	T				I		
299	E				G		
300	K						
301	D		E	E	E	E	
302	D						
303	F						
304	I						
305	E						
306	L						
307	C						
308	T						
309	P						
310	G						
311	V						
312	I						
313	K						
314	Q						
315	E						
316	K						
317	L						
318	G					S	
319	P		T			T	
320	V						
321	Y						
322	C						
323	Q						
324	A						
325	S						
326	F						
327	S		P	P		P	M: is phosphorylated in steroid treated cells (Bodwell et al., 1991) but mutation to Ala has little to no effect on biological activity (Mason & Housley, 1993)
328	G						
329	T		A	A	A	A	
330	N						
331	I						
332	I						
333	G						
334	N						
335	K						
336	M						
337	S						
338	A						
339	I						
340	S						
341	V					I	
342	H						
343	G						
344	V						
345	S						
346	T						
347	S						
348	G						
349	G						
350	Q						
351	M						
352	Y						
353	H						

Position	Rat	Mouse	Human	Sheep	Guinea-pig	Tamarin	Description of mutations
354	Y						
355	D						
356	M						
357	N						
358	T						
359	A						
360	S						
361	L						
362	S						
363	Q						
364	Q						
365	Q						
366	D						
367	Q						
368	K						
369	P						
370	V		I	I	I	I	
371	F						
372	N		K				
373	V						
374	I						
375	P						
376	P						
377	I						
378	P						
379	V						
380	G						
381	S						
382	E						
383	N						H: N to S has no effect on dose-response curve or ability to transactivate (Karl *et al.*, 1993)
384	W						
385	N						
386	R						
387	C						
388	Q						
389	G						
390	S						
391	G						
392	E		D	D		D	
393	D						
394	S	N	N		N	N	
395	L						
396	T						
397	S						
398	L						
399	G						
400	A		T	T	T	T	
401	L	M			V		
402	N						
403	F						
404	P	A		S			
405	G						
406	R						
407	S		T			T	
408	V						
409	F						R: Major chymotrypsin cleavage after this amino acid in activated complexes (Carlstedt-Duke *et al.*, 1987) but no cleavage in fusion protein of protein A with the DNA-binding domain of the human receptor (=390-519 of rat) (Dahlman *et al.*, 1989)
410	S						
411	N						

(continued)

Position	Glucocorticoid receptors from						Description of mutations
	Rat	Mouse	Human	Sheep	Guinea-pig	Tamarin	
412	G						
413	Y						R: Major chymotrypsin cleavage after this amino acid in activated complexes (Carlstedt-Duke *et al.*, 1987); only cleavage site in fusion protein of protein A with the DNA-binding domain of the human receptor (=390-519 of rat) (Dahlman *et al.*, 1989)
414	S						
415	S						
416	P						
417	G		S			S	
418	M			L			
419	R						
420	P						
421	D						
422	V						
423	S						
424	S						
425	P						
426	P						
427	S						
428	S						
429	S						
430	S		ST			ST	
431	A	T			T		
432	A		T		T	T	
433	T						
434	G						
435	P						H: P to R facilitated activation of agonist (triamcinolone acetonide), but not antagonist (RU 486), bound receptors (Segard-Maurel *et al.*, 1992)
436	P						
437	P						
438	K						
439	L						R: L to P has no effect on activity in vitro or in vivo (Luisi *et al.*, 1991)
440	C						R: co-ordinated with zinc (Kellenbach *et al.*, 1991; Luisi *et al.*,1991); C to A does not eliminate zinc binding but does dramatically reduce (<10%) DNA binding (Archer *et al.*, 1990). H: C to G may change effect of receptor on IL-1α induction from repression to induction (Ray *et al.*, 1991); H: C to Y eliminates DNA binding and transactivation without affecting steroid binding activity and, in combination with L771F, is responsible for an activation labile phenotype (Powers *et al.*,1993)
441	L						
442	V						
443	C						R: co-ordinated with zinc (Kellenbach *et al.*, 1991; Luisi *et al.*, 1991)
444	S						
445	D						
446	E						
447	A				L		
448	S						
449	G	V					R: some point mutants between 449 and 455 had little effect on biological activity in truncated receptor of 407-556 (Zandi *et al.*, 1993). M: G to V has wild type properties (Kasai, 1990)
450	C						H: C to G gave receptor with ≥40% transactivation but ≤10% DNA binding activity of wild type receptor (Hollenberg and Evans, 1988)
451	H						R: some point mutants between 451 and 462 eliminate or reduce biological activity in truncated receptor of 407-556 (Zandi *et al.*, 1993); double mutation H451N/S459G in truncated receptor of 407-556 has increased affinity for DNA and transactivation activity (Zandi *et al.*, 1993)
452	Y						R: part of interchain hydrophobic interactions holding 3° structure of DNA binding domain together (Luisi *et al.*, 1991)

Position	Glucocorticoid receptors from						Description of mutations
	Rat	Mouse	Human	Sheep	Guinea-pig	Tam-arin	
453	G						
454	V						
455	L						H: L to V causes 25-100% increase in transactivation while eliminating repression (P. Herrlich, personal communication)
456	T						
457	C						R: co-ordinated with zinc (Kellenbach et al., 1991; Luisi et al., 1991)
458	G						H: involved in specificity of DNA binding, along with other amino acids (Zilliacus et al., 1992b); G to E had no effect on biological activity with MMTV promoter (Umesono and Evans, 1989).
459	S						H: involved in specificity of DNA binding, along with other amino acids (Zilliacus et al., 1992b). R: double mutation H451N/S459G in truncated receptor of 407-556 has increased affinity for DNA and transactivation activity (Zandi et al., 1993)
460	C						R: co-ordinated with zinc (Kellenbach et al., 1991; Luisi et al., 1991)
461	K						R: makes base contacts in binding to specific DNA sites (Luisi et al., 1991). H: K to G causes receptor to induce, as opposed to repress, AP-1 regulated gene expression (Lucibello et al., 1990; Yang-Yen et al., 1990) and DNA binding is retained but transactivation is lost (Hollenberg and Evans, 1988)
462	V						H and R: involved in specificity of DNA binding, along with other amino acids (Zilliacus et al., 1992b; Luisi et al., 1991)
463	F						R: part of interchain hydrophobic interactions holding 3° structure of DNA binding domain together (Luisi et al., 1991)
464	F						R: part of interchain hydrophobic interactions holding 3° structure of DNA binding domain together (Luisi et al., 1991)
465	K						H: K to G gave receptor with ≥40% transactivation but ≤10% DNA binding activity of wild type receptor (Hollenberg and Evans, 1988)
466	R						R: makes base contacts in binding to specific DNA sites. Mutation of R to K or G is inactive in vivo (Luisi et al., 1991)
467	A						
468	V						R: this region (468-472) can tolerate insertion of tandem repeats of 3-9bp without loss of biological activity and of 23bp with partial loss (Zandi et al., 1993)
469	E						
470	G					GR	M: insertion of R causes a 2 fold decrease in transcriptional activation (Kasai, 1990)
471	Q						
472	H						
473	N			?			
474	Y			?			R: part of interchain hydrophobic interactions holding 3° structure of DNA binding domain together (Luisi et al., 1991). H: Y to G gave receptor with ≥40% transactivation but ≤10% DNA binding activity of wild type receptor (Hollenberg and Evans, 1988)
475	L			?			
476	C			?			R: co-ordinated with zinc (Kellenbach et al., 1991; Luisi et al., 1991)
477	A			?			H: A to T prevents DNA binding and transactivation from a gene regulated by a single, but not a multiple, copy of GRE but does not affect repression (P. Herrlich, personal communication)
478	G			?			
479	R			?			H: thought to form inter-receptor salt bridge with D481 and thus play a major role in receptor dimerization (Dahlman-Wright et al., 1993), although mutation to G has no major effect (Hollenberg and Evans, 1988)
480	N			?			
481	D			?			H: thought to form inter-receptor salt bridge with R479 and thus play a major role in receptor dimerization (Dahlman-Wright et al., 1993), although mutation to G has no major effect (Hollenberg and Evans, Cell, 1988)
482	C			?			R: co-ordinated with zinc (Kellenbach et al., 1991; Luisi et al., 1991)
483	I			?			
484	I			?			R: most non-conservative point mutants eliminate both DNA binding and transactivation (Lanz et al., 1993)
485	D			?			
486	K			?			H: K to G gave receptor with ≥40% transactivation but ≤10% DNA binding activity of wild type receptor (Hollenberg and Evans, 1988)

(continued)

Position	Glucocorticoid receptors from						Description of mutations
	Rat	Mouse	Human	Sheep	Guinea-pig	Tamarin	
487	I			?			
488	R			?			R: R to Q caused normal DNA binding and fair activity in CV-1 cells but <3% transcriptional activation in yeast (Schena et al., 1989)
489	R			?			R: R to K caused normal DNA binding and good activity in CV-1 cells but <10% transcriptional activation, and temperature sensitive, in yeast (Schena et al., 1989)
490	K			?	E		
491	N			?			R: N to S caused normal DNA binding and good activity in yeast but <1% transcriptional activation in CV-1 cells (Schena et al., 1989)
492	C			?			H: C to S may cause incorrect folding of DNA binding domain (Zilliacus et al., 1992a); R: co-ordinated with zinc (Kellenbach et al., 1991; Luisi et al., 1991)
493	P			?			R: P493R/A494S double mutant caused normal DNA binding but poor transcriptional activation (Godowski et al., 1989)
494	A			?			See comments for position 493
495	C			?			H: C to S may cause incorrect folding of DNA binding domain (Zilliacus et al., 1992a); C495W/R498Q double mutation eliminates trans-activation and repression (P. Herrlich, personal communication). R: co-ordinated with zinc (Kellenbach et al., 1991; Luisi et al., 1991)
496	R			?			M: R to H is biologically inactive and has greatly reduced nuclear translocation (like nt⁻) (Danielsen et al., 1986 and 1987). H: R to S has no effect on transactivation or repression (P. Herrlich, personal communication)
497	Y			?			R: part of interchain hydrophobic interactions holding 3° structure of DNA binding domain together (Luisi et al., 1991). H: double mutation of Y497L and R498G causes 25-100% increase in transactivation while eliminating repression (P. Herrlich, personal communication)
498	R			?			H: C495W/R498Q double mutation eliminates transactivation and repression; Y497L/R498G double mutation causes 25-100% increase in transactivation while eliminating repression (P. Herrlich, personal communication)
499	K			?			
500	C			?			R: C to S, A, or M has little or no effect (Severne et al., 1988; S. Rusconi, personal communication) but many others give inactive receptors; C to R is biologically inactive (Schena et al., 1989). H: only C in 481-777 fragment overexpressed in E.coli that is not labeled by Dex-Mes (Ohara-Nemoto et al., 1990)
501	L			?			
502	Q			?			
503	A			?			
504	G			?			
505	M			?			R: M to L or C has little effect but others eliminate activity (S. Rusconi, personal communication). H: M to G gave receptor with 10% transactivation but 1% DNA binding activity of wild type receptor (Hollenberg and Evans, 1988)
506	N			?			
507	L			?			
508	E			?	Q		
509	A			?			
510	R			?			
511	K			?			
512	T			?			
513	K			?			
514	K			?			
515	K			?			
516	I			?			
517	K			?			R: trypsin cuts activated complexes after this amino acid (Carlstedt-Duke et al., 1987)
518	G			?			
519	I			?			
520	Q			?			
521	Q			?			
522	A			?			
523	T			?			
524	A		T	?	T	T	
525	G			?			

Position	Glucocorticoid receptors from						Description of mutations
	Rat	Mouse	Human	Sheep	Guinea-pig	Tamarin	
526	V			?			
527	S			?			
528	Q			?			
529	D		E	?	N	E	
530	T			?			
531	S			?			
532	E			?			
533	N			?			
534	P	A	PG	?		PA	T: added A is one of 4 mutations in steroid binding domain proposed to be responsible for 10 fold lower affinity in monkey (Brandon *et al.*, 1991)
535	N			?			
536	K			?			R: trypsin is thought to cleave after this site in steroid-free receptors to form 16 kDa core binding fragment (Simons *et al.*, 1989)
537	T			?			
538	I			?			
539	V			?			
540	P			?			
541	A			?			
542	A		T	?	T	T	
543	L			?			
544	P			?			
545	Q			?			
546	L			?			
547	T			?			
548	P			?			
549	T			?			
550	L			?			
551	V			?			
552	S			?			
553	L			?			M: L553G/L554G double mutant eliminates transcriptional activity (M.R. Stallcup, personal communication)
554	L			?			
555	E			?			M: E to A requires 2-3 fold higher steroid concentrations for biological activity (M.R. Stallcup, personal communication)
556	V			?			M: V to G requires 4-100 fold higher steroid concentrations for biological activity (M.R. Stallcup, personal communication)
557	I			?			M: I to G requires 4-100 fold higher steroid concentrations for biological activity (M.R. Stallcup, personal communication)
558	E			?			M: E to G eliminates steroid binding activity (Danielsen *et al.*, 1986)
559	P			?			M: P to A requires ≥100 fold higher steroid concentrations for biological activity, presumable due to decreased steroid binding affinity (Byravan *et al.*, 1991)
560	E			?			M: E to A requires 2-3 fold higher steroid concentrations for biological activity (M.R. Stallcup, personal communication)
561	V			?			M: V to G requires 4-100 fold higher steroid concentrations for biological activity (M.R. Stallcup, personal communication)
562	L			?	I		M: L to A requires 4-100 fold higher steroid concentrations for biological activity (M.R. Stallcup, personal communication)
563	Y			?	H		
564	A			?	S		
565	G			?			
566	Y			?			
567	D			?			M: D to A requires 4-100 fold higher steroid concentrations for biological activity (M.R. Stallcup, personal communication)
568	S			?			M: S to A has no effect on biological activity (M.R. Stallcup, personal communication)
569	S			?	T		
570	V			?	S		
571	P			?			
572	D			?			M: D to G eliminates transcriptional activity (M.R. Stallcup, personal communication)

(continued)

Position	Glucocorticoid receptors from						Description of mutations
	Rat	Mouse	Human	Sheep	Guinea-pig	Tamarin	
573	S			?			M: S to A requires 4-100 fold higher steroid concentrations for biological activity (M.R. Stallcup, personal communication)
574	A		T	?	T	T	
575	W			?			
576	R			?			
577	I			?			
578	M			?			
579	T			?			
580	T			?			
581	L			?			R: L to P causes no activity in yeast or COS-7 cells (Garabedian and Yamamoto, 1992) M: L to F requires 4-100 fold higher steroid concentrations for biological activity (M.R. Stallcup, personal communication)
582	N			?			
583	M			?			R: cyanogen bromide cleaves at this methionine (Carlstedt-Duke et al., 1987)
584	L			?			R: L to S causes no activity in yeast or COS-7 cells and decreased affinity for Dex in COS-7 cells (Garabedian and Yamamoto, 1992)
585	G			?			
586	G			?			
587	R			?			
588	Q			?			
589	V			?			
590	I			?			
591	A			?			
592	A			?			
593	V			?			
594	K			?			
595	W			?			
596	A			?			
597	K			?			
598	A			?			
599	I			?			
600	L*	P	P	?	P	P	R: = P not L (M.J. Garabedian and K.R. Yamamoto, personal communication)
601	G			?			
602	L*	F	F	?	F	F	R: = F not L (M.J. Garabedian and K.R. Yamamoto, personal communication)
603	R			?	K		
604	N			?			
605	L			?			
606	H			?			
607	L			?			
608	D			?			
609	D			?			
610	Q			?			
611	M			?			
612	T			?			
613	L			?			
614	L			?			
615	Q			?			
616	Y			?			
617	S			?			
618	W			?			M: W to A has no effect on biological activity (M.R. Stallcup, personal communication)
619	M			?			H: M to L causes 2-3 fold reduced affinity for Dex (J. Carlstedt-Duke, personal communication)
620	F			?			R: F to S causes increased affinity for Dex and triamcinolone acetonide in yeast, but not in CV-1 cells (Garabedian and Yamamoto, 1992)
621	L			?			
622	M			?			R: M to P causes reduced transcriptional activity in yeast (Garabedian and Yamamoto, 1992); photolabeled by triamcinolone acetonide (rat and human) (Carlstedt-Duke et al., 1988; Stromstedt et al., 1990) and by R5020 (rat) (Stromstedt et al., 1990); cyanogen bromide

Position	Rat	Mouse	Human	Sheep	Guinea-pig	Tamarin	Description of mutations
							cleaves at this methionine (Carlstedt-Duke et al., 1987). M: M to L, C, or S has no effect on biological activity (Chen and Stallcup, 1994). H: M to L has no significant effect on affinity (J. Carlstedt-Duke, personal communication)
623	A			?			
624	F			?			
625	A			?			
626	L			?			
627	G			?			
628	W			?			M: W to A requires ≥100 fold higher steroid concentrations for biological activity (M.R. Stallcup, personal communication)
629	R			?			M: R to A eliminates transcriptional activity (M.R. Stallcup, personal communication)
630	S			?			
631	Y			?			
632	R			?	K		
633	Q			?			
634	S	A		?		A	T: S to A is one of 4 mutations in steroid binding domain proposed to be responsible for 10 fold lower affinity in monkey (Brandon et al., 1991)
635	S			?	N		
636	G		A	?		S	T: G to S is one of 4 mutations in steroid binding domain proposed to be responsible for 10 fold lower affinity in monkey (Brandon et al., 1991)
637	N			?	S		
638	L			?			
639	L			?			
640	C			?			R: forms an intramolecular disulfide with C656 or C661 (Chakraborti et al., 1992); C to S causes 3 fold loss in binding affinity (Chakraborti et al., 1991). H: labeled by Dex-Mes in 481-777 fragment overexpressed in E.coli (Ohara-Nemoto et al., 1990). M: C to A has no effect on biological activity (Chen and Stallcup, 1994).
641	F			?			M: F to A requires 4-100 fold higher steroid concentrations for biological activity (M.R. Stallcup, personal communication)
642	A			?			
643	P			?			M: P to A requires ≥100 fold higher steroid concentrations for biological activity (M.R. Stallcup, personal communication)
644	D			?			
645	L			?			
646	I			?			
647	I			?			
648	N			?			
649	E			?			
650	Q			?			
651	R			?			
652	M			?			
653	S	T	T	?		T	
654	L			?			
655	P			?			
656	C			?	W		Covalently labeled by Dex-Mes (R: Simons et al., 1987; Carlstedt-Duke et al., 1988. M: Smith et al., 1988. H: Stromstedt et al., 1990). R: forms an intramolecular disulfide with C661 or C640 (Chakraborti et al., 1992); forms specific complex with arsenite (Lopez et al., 1990; Chakraborti et al., 1992); C to Y has no effect on affinity of receptors from COS-7 cells (Chakraborti and Simons, unpublished results) but reduces transcriptional activity in yeast (Garabedian and Yamamoto, 1992); C to G or S yields "super" receptor (Chakraborti et al., 1991). H: C to W has no effect on transcriptional activity (Keightly and Fuller, 1993); C to S has no effect on affinity (J. Carlstedt-Duke, personal communication). M: C to S has no effect on biological activity (M.R. Stallcup, personal communication)
657	M			?			
658	Y			?			

(continued)

Position	Glucocorticoid receptors from						Description of mutations
	Rat	Mouse	Human	Sheep	Guinea-pig	Tam-arin	
659	D			?			R: D to G causes no activity in yeast or COS-7 cells and decreased affinity for Dex in COS-7 cells (Garabedian and Yamamoto, 1992). H: D to V causes 3 fold decreased steroid affinity and 7 fold reduced glucocorticoid sensitivity (Hurley et al., 1991)
660	Q			?			
661	C			?			R: forms an intramolecular disulfide with C656 or C640 (Chakraborti et al., 1992); forms specific complex with arsenite (Lopez et al., 1990; Chakraborti et al., 1992); C to R is transcriptionally active in yeast only with deacylcortivazol (Garabedian and Yamamoto, 1992); C to S causes 4 fold loss in binding affinity (Chakraborti et al., 1991). H: labeled by Dex-Mes in 481-777 fragment overexpressed in E.coli (Ohara-Nemoto et al., 1990). M: C to G requires 2-3 fold higher steroid concentrations for biological activity (Chen and Stallcup, 1994).
662	K			?	R		
663	H			?	Y		
664	M			?			R: cyanogen bromide cleaves at this methionine (Carlstedt-Duke et al., 1987)
665	L			?			
666	F		Y	?	Y	Y	
667	V	I		?			
668	S			?			
669	S	T		?			
670	E			?			
671	L			?			R: L to S causes reduced transcriptional activity in yeast or COS-7 cells and decreased affinity for Dex in COS-7 cells (Garabedian and Yamamoto,1992)
672	Q		H	?	K	H	
673	R			?			Trypsin is thought to cleave after this site in steroid-free receptors to form 16 kDa core binding fragment (Simons et al., 1989)
674	L			?			
675	Q			?			
676	V			?			
677	S			?			
678	Y			?			
679	E			?			
680	E			?			
681	Y			?			
682	L			?			M: L to F requires 4-100 fold higher steroid concentrations for biological activity (M.R. Stallcup, personal communication)
683	C			?			H: labeled by Dex-Mes in 481-777 fragment overexpressed in E.coli (Ohara-Nemoto et al., 1990); C to S has no effect on affinity (J. Carlstedt-Duke, personal communication). M: C to S or A requires 4-100 fold higher steroid concentrations for biological activity (Chen and Stallcup, 1994).
684	M			?			R: cyanogen bromide cleaves at this methionine (Carlstedt-Duke et al., 1987). M: M to I requires 4-100 fold higher steroid concentrations for biological activity (M.R. Stallcup, personal communication)
685	K			?			
686	T			?			
687	L			?			
688	L			?			
689	L			?			
690	L			?			
691	S			?			
692	S			?			
693	V			?			
694	P			?			
695	K			?			
696	E		D	?		D	
697	G			?			
698	L			?			

Position	Glucocorticoid receptors from						Description of mutations
	Rat	Mouse	Human	Sheep	Guinea-pig	Tam-arin	
699	K			?			
700	S			?			
701	Q			?			
702	E			?			
703	L			?			
704	F			?			
705	D			?			
706	E			?			R: E to K causes no activity in yeast or COS-7 cells and decreased affinity for Dex in COS-7 cells (Garabedian and Yamamoto, 1992)
707	I			?			
708	R			?			
709	M			?			
710	T			?			
711	Y			?			
712	I			?			
713	K			?			
714	E			?			
715	L			?			R: L to V has no effect in yeast (Garabedian and Yamamoto, 1992)
716	G			?			
717	K			?			
718	A			?			
719	I			?			
720	V			?			
721	K			?			
722	R			?			
723	E			?			
724	G			?			
725	N			?			
726	S			?			
727	S			?			
728	Q			?			
729	N			?			
730	W			?			
731	Q			?			
732	R			?			
733	F			?			
734	Y			?			
735	Q			?			
736	L			?			
737	T			?			
738	K			?			
739	L			?			
740	L			?			
741	D			?			
742	S			?			
743	M			?	L		
744	H			?			
745	E	D		?			
746	V			?	I		
747	V			?			H: V to I in patient with 1° cortisol resistance causes 2-fold decreased affinity and 4-fold decreased EC_{50} (Malchoff et al., 1993)
748	E			?	G		
749	N			?			
750	L			?			
751	L			?			
752	T	S	N	?	N	N	
753	Y			?	I		
754	C			?			Photolabeled by triamcinolone acetonide (rat and human (Carlstedt-Duke et al., 1988) and by R5020 (rat) (Stromstedt et al., 1990). M: C to G requires >100 fold higher steroid concentrations for biological activity, apparently due to decreased protein stability and degradation to a 68 kDa species (no 98 kDa seen) (Byravan et al.,

(*continued*)

Position	Rat	Mouse	Human	Sheep	Guinea-pig	Tamarin	Description of mutations
							1991); C to S requires 4-100 fold more steroid for biological activity (Chen and Stallcup, 1994). H: labeled by Dex-Mes in 481-777 fragment overexpressed in E. coli (Ohara-Nemoto et al., 1990); C to S causes 2-3 fold lower affinity (J. Carlstedt-Duke, personal communication)
755	F			?			
756	Q			?	K		
757	T			?			
758	F			?			
759	L			?			
760	D			?			
761	K			?			
762	T	S		?			R: T to I causes reduced transcriptional activity in yeast (Garabedian and Yamamoto, 1992)
763	M			?			
764	S			?	N		
765	I			?			
766	E			?			
767	F			?			
768	P			?			
769	E			?			
770	M			?			R: cyanogen bromide cleaves at this methionine (Carlstedt-Duke et al., 1987). M: double mutation M770A/L771A has no major affect on steroid binding but eliminates transcriptional activation (Danielian et al., 1992). However, in rat receptors, this mutation has either a three fold lower affinity for steroid (Schmitt and Stunnenberg, 1993) or no affinity at all (Lanz et al., 1993)
771	L			?			M: double mutation M770A/L771A has no major affect on steroid binding but eliminates transcriptional activation (Danielian et al., 1992). R: double mutation M770A/L771A has either a three fold lower affinity for steroid (Schmitt and Stunnenberg, 1993) or no affinity at all (Lanz et al., 1993) H: L to F gives rise to an activation labile or r⁻ phenotype, depending on the cellular environment (Ashraf and Thompson, 1993); this same mutation gives receptors with unstable binding at 37° but not 0°C and, in combination with C440Y, will also give an activation labile phenotype (Powers et al., 1993).
772	A			?			
773	E			?			M: E to A does not affect steroid binding but decreases transcriptional activation by ≈40% (Danielian et al., 1992)
774	I			?			M: I774A/I775A double mutant had no major affect on steroid binding but eliminated transcriptional activation (Danielian et al., 1992). However, Lanz et al. (1993) found that these mutations in rat receptors did not bind steroid.
775	I			?			See comments for position 774
776	T			?			
777	N			?			
778	Q			?			
779	I			?	L		
780	P			?			
781	K			?			
782	Y			?			M: Y to N needs 3 to 4 fold more Dex for equal biological activity, presumably due to 3-4 fold lower affinity (Danielsen et al., 1986)
783	S			?			
784	N			?			
785	G			?			
786	N			?	D		
787	I			?			
788	K			?		R	T: K to R is one of 4 mutations in steroid binding domain proposed to be responsible for 10 fold lower affinity in monkey (Brandon et al., 1991)
789	K			?			
790	L			?			

sition	Glucocorticoid receptors from						Description of mutations
	Rat	Mouse	Human	Sheep	Guinea-pig	Tamarin	
791	L		?				
792	F		?				M: F to A requires 4-100 fold higher steroid concentrations for biological activity and has altered steroid binding specificity (M.R. Stallcup, personal communication)
793	H		?				M: H to L has no effect on biological activity (M.R. Stallcup, personal communication)
794	Q		?				
795	K		?				

ACKNOWLEDGMENTS

I thank Drs. David Brandon, Jan Carlstedt-Duke, Peter Fuller, Michael Garabedian, Jeff Harmon, Peter Herrlich, Paul Housley, Bill Pratt, Anaradha Ray, Sandro Rusconi, Constantin Sekeris, Mike Stallcup, Brad Thompson, and Keith Yamamoto for providing unpublished data. I am also indebted to Drs. Clayton Collier, David Jackson, Jeff Harmon, Bill Pratt, Anaradha Ray, and Mike Stallcup for their critical comments concerning this review.

REFERENCES

Adler, A. J., Scheller, A., and Robins, D. M. (1993). The stringency and magnitude of androgen-specific gene activation are combinatorial functions of receptor and non-receptor binding site sequences. *Mol. Cell. Biol.* **13,** 6326–6335.

Allan, G. F., Leng, X., Tsai, S. Y., Weigel, N. L., Edwards, D. P., Tsai, M.-J., and O'Malley, B. W. (1992). Hormone and antihormone induce distinct conformational changes which are central to steroid receptor activation. *J. Biol. Chem.* **267,** 19513–19520.

Alnemri, E. S., and Litwack, G. (1993). The steroid binding domain influences intracellular solubility of the baculovirus overexpressed glucocorticoid and mineralocorticoid receptors. *Biochemistry* **32,** 5387–5393.

Alroy, I., and Freedman, L. P. (1992). DNA binding analysis of glucocorticoid receptor specificity mutants. *Nucleic Acids Res.* **20,** 1045–1052.

Antakly, T., Raquidan, D., O'Donnell, D., and Katnick, L. (1990). Regulation of glucocorticoid receptor expression: I. Use of a specific radioimmunoassay and antiserum to a synthetic peptide of the N-terminal domain. *Endocrinology (Baltimore)* **126,** 1821–1828.

Archer, T. K., Hager, G. L., and Omichinski, J. G. (1990). Sequence-specific DNA binding by glucocorticoid receptor "zinc finger peptides." *Proc. Natl. Acad. Sci. U.S.A.* **87,** 7560–7564.

Ashraf, J., and Thompson, E. B. (1993). Identification of the activation-labile gene: A single point mutation in the human glucocorticoid receptor presents as two distinct receptor phenotypes. *Mol. Endocrinol.* **7,** 631–642.

Baniahmad, C., Muller, M., Altschmied, J., and Renkawitz, R. (1991). Co-operative binding of the glucocorticoid receptor DNA binding domain is one of at least two mechanisms for synergism. *J. Mol. Biol.* **222,** 155–165.

Baumann, H., Paulsen, K., Kovacs, H., Berglund, H., Wright, A. P., Gustafsson, J.-A., Hard, T. (1993). Refined solution structure of the glucocorticoid receptor DNA-binding domain. *Biochemistry* **32,** 13463–13471.

Beato, M. (1989). Gene regulation by steroid hormones. *Cell (Cambridge, Mass.)* **56**, 335–344.

Beato, M., Chalepakis, G., Schauer, M., and Slater, E. P. (1989). DNA regulatory elements for steroid hormones. *J. Steroid Biochem.* **32**, 737–748.

Beekman, J. M., Allan, G. F., Tsai, S. Y., Tsai, M.-T., and O'Malley, B. W. (1993). Transcriptional activation by the estrogen receptor requires a conformational change in the ligand binding domain. *Mol. Endocrinology* **7**, 1266–1274.

Benjamin, W. S. van der W., Hendry, W. J., III, and Harrison, R. W., III (1990). The mouse glucocorticoid receptor DNA-binding domain is not phosphorylated *in vivo*. *Biochem. Biophys. Res. Commun.* **166**, 931–936.

Bocquel, M. T., Kumar, V., Stricker, C., Chambon, P., and Gronemeyer, H. (1989). The contribution of the N- and C-terminal regions of steroid receptors to activation of transcription is both receptor and cell-specific. *Nucleic Acids Res.* **17**, 2581–2595.

Bocquel, M. T., Ji, J., Ylikomi, T., Benhamou, B., Vergezac, A., Chambon, P., and Gronemeyer, H. (1993). Type II antagonists impair the DNA binding of steroid hormone receptors without affecting dimerization. *J. Steroid Biochem. Mol. Biol.* **45**, 205–215.

Bodwell, J. E., Orti, E., Coull, J. M., Pappin, D. J. C., Smith, L. I., and Swift, F. (1991). Identification of phosphorylated sites in the mouse glucocorticoid receptor. *J. Biol. Chem.* **266**, 7549–7555.

Brandon, D. D., Markwick, A. J., Flores, M., Dixon, K., Albertson, B. D. and Loriaux, D. L. (1991). Genetic variation of the glucocorticoid receptor from a steroid-resistant primate. *J. Mol. Endocrinol.* **7**, 89–96.

Burns, K., Duggan, B., Atkinson, E. A., Famulski, K. S., Nemer, M., Bleackley, R. C., and Michalak, M. (1994). Modulation of gene expression by calreticuline binding to the glucocorticoid receptor. *Nature* **367**, 476–480.

Byravan, S., Milhon, J., Rabindran, S. K., Olinger, B., Garabedian, M. J., Danielson, M., and Stallcup, M. R. (1991). Two point mutations in the hormone binding domain of the receptor that dramatically reduce its function. *Mol. Endocrinol.* **5**, 752–758.

Cadepond, F., Schweizer-Groyer, G., Segard-Maurel, I., Jibard, N., Hollenberg, S. M., Giguère, V., Evans, R. M., and Baulieu, E.-E. (1991). Heat shock protein 90 as a critical factor in maintaining glucocorticosteroid receptor in a nonfunctional state. *J. Biol. Chem.* **266**, 5834–5841.

Cairns, W., Cairns, C., Pongratz, I., Poellinger, L., and Okret, S. (1991). Assembly of a glucocorticoid receptor complex prior to DNA binding enhances its specific interaction with a glucocorticoid response element. *J. Biol. Chem.* **266**, 11221–11226.

Cao, X., Preiss, T., Slater, E. P., Westphal, H. M., and Beato, M. (1993). Expression and functional analysis of steroid receptor fragments secreted from *Staphylococcus aureus*. *J. Steroid Biochem. Mol. Biol.*, **44**, 1–11.

Carlstedt-Duke, J., Stromstedt, P.-E., Wrange, O., Bergman, T., Gustafsson, J.-A., and Jornvall, H. (1987). Domain structure of the glucocorticoid receptor protein. *Proc. Natl. Acad. Sci. U.S.A.* **84**, 4437–4440.

Carlstedt-Duke, J., Stromstedt, P.-E., Persson, B., Cederlund, E., Gustafsson, J.-A., and Jornvall, H. (1988). Identification of hormone-interacting amino acid residues within the steroid-binding domain of the glucocorticoid receptor in relation to other steroid hormone receptors. *J. Biol. Chem.* **263**, 6842–6846.

Catelli, M. G., Binart, N., Jung-Testas, I., Renoir, J. M., Baulieu, E. E., Feramisco, J. R., and Welch, W. J. (1985). The common 90-kd protein component of non-transformed "8S" steroid receptors is a heat-shock protein. *EMBO J.* **4**, 3131–3135.

Cavanaugh, A. H., and Simons, S. S., Jr. (1990a). Glucocorticoid receptor binding to calf thymus DNA. 1. Identification and characterization of a macromolecular factor involved in receptor–steroid complex binding to DNA. *Biochemistry* **29**, 989–996.

Cavanaugh, A. H., and Simons, S. S., Jr. (1990b). Glucocorticoid receptor binding to calf thymus DNA. 2. Role of a DNA-binding activity factor in receptor heterogeneity and a multistep mechanism of receptor activation. *Biochemistry* **29**, 996–1002.

Cavanaugh, A. H., and Simons, S. S., Jr. (1994). Factor-assisted DNA binding as a possible general mechanism for steroid receptors. Functional heterogeneity among activated receptor–steroid complexes. *J. Steriod Biochem. Mol. Biol.* **48**, 433–446.

Chakraborti, P. K., and Simons, S. S., Jr. (1991). Association of heat shock protein 90 with the 16 kDa steroid binding core fragment of rat glucocorticoid receptors. *Biochem. Biophys. Res. Commun.* **176**, 1338–1344.

Chakraborti, P. K., Hoeck, W., Groner, B., and Simons, S. S., Jr. (1990). Localization of the vicinal dithiols involved in steroid binding to the rat glucocorticoid receptor. *Endocrinology (Baltimore)* **127**, 2530–2539.

Chakraborti, P. K., Garabedian, M. J., Yamamoto, K. R., and Simons, S. S., Jr. (1991). Creation of "super" glucocorticoid receptors by point mutations in the steroid binding domain. *J. Biol. Chem.* **266**, 22075–22078.

Chakraborti, P. K., Garabedian, M. J., Yamamoto, K. R., and Simons, S. S., Jr. (1992). Role of cysteines 640, 656, and 661 in steroid binding to rat glucocorticoid receptors. *J. Biol. Chem.* **267**, 11366–11373.

Chalepakis, G., Schauer, M., Cao, X., and Beato, M. (1990). Efficient binding of glucocorticoid receptor to its responsive element requires a dimer and a DNA flanking sequence. *DNA Cell Biol.* **9**, 355–368.

Chang, T.-J., Scher, B. M., Waxman, S., and Scher, W. (1993). Inhibition of mouse GATA-1 function by the glucocorticoid receptor: Possible mechanism of steroid inhibition of erythroleukemia cell differentiation. *Mol. Endocrinol.* **7**, 528–542.

Chen, D., and Stallcup, M. R. (1994). The hormone binding role of two cysteines near the C-terminus of the mouse glucocorticoid receptor. *J. Bio. Chem.* **269**, 7914–7918.

Christopherson, K. S., Mark, M. R., Bajaj, V., and Godowski, P. J. (1992). Ecdysteroid-dependent regulation of genes in mammalian cells by a *Drosophila* ecdysone receptor and chimeric transactivators. *Proc. Natl. Acad. Sci. U.S.A.* **89**, 6314–6318.

Cidlowski, J. A., Bellingham, D. L., Powell-Oliver, F. E., Lubahn, D. B., and Sar, M. (1990). Novel antipeptide antibodies to the human glucocorticoid receptor: Recognition of multiple receptor forms *in vitro* and distinct localization of cytoplasmic and nuclear receptors. *Mol. Endocrinol.* **4**, 1427–1437.

Cunningham, B. C., and Wells, J. A. (1989). High-resolution epitope mapping of hGH–receptor interactions by alanine-scanning mutagenesis. *Science* **244**, 1081–1085.

Dahlman, K., Stromstedt, P.-E., Rae, C., Jornvall, H., Flock, J.-I., Carlstedt-Duke, J., and Gustafsson, J.-A. (1989). High level expression in *Escherichia coli* of the DNA-binding domain of the glucocorticoid receptor in a functional form utilizing domain-specific cleavage of a fusion protein. *J. Biol. Chem.* **264**, 804–809.

Dahlman-Wright, K., Siltala-Roos, H., Carlstedt-Duke, J., and Gustafsson, J.-A. (1990). Protein–protein interactions facilitate DNA binding by the glucocorticoid receptor DNA-binding domain. *J. Biol. Chem.* **265**, 14030–14035.

Dahlman-Wright, K., Wright, A., Gustafsson, J.-A., and Carlstedt-Duke, J. (1991). Interaction of the glucocorticoid receptor DNA-binding domain with DNA as a dimer is mediated by a short segment of five amino acids. *J. Biol. Chem.* **266**, 3107–3112.

Dahlman-Wright, K., Wright, A. P., and Gustafsson, J. A. (1992). Determinants of high-affinity DNA binding by the glucocorticoid receptor: Evaluation of receptor domains outside the DNA-binding domain. *Biochemistry* **31**, 9040–9044.

Dahlman-Wright, K., Grandien, K., Nilsson, S., Gustafsson, J. A., and Carlstedt-Duke, J. (1993). Protein–protein interactions between the DNA-binding domains of nuclear receptors: Influence on DNA-binding. *J. Steroid Biochem. Mol. Biol.* **45**, 239–250.

Dahlman-Wright, K., Almlof, T., McEwan, I. J., Gustafsson, J.-A., and Wright, A. P. H. (1994). Delineation of a small region within the major transactivation domain of the human glucocorticoid receptor that mediates transactivation of gene expression. *Proc. Natl. Acad. Sci. U.S.A.* **91**, 1619–1623.

Dalman, F. C., Sanchez, E. R., Lin, A. L.-Y., Perini, F., and Pratt, W. B. (1988). Localization of phosphorylation sites with respect to the functional domains of the mouse L cell glucocorticoid receptor. *J. Biol. Chem.* **263**, 12259–12267.

Dalman, F. C., Koenig, R. J., Perdew, G. H., Massa, E., and Pratt, W. B. (1990). In contrast to the glucocorticoid receptor, the thyroid hormone receptor is translated in the DNA binding state and is not associated with hsp90. *J. Biol. Chem.* **265**, 3615–3618.

Dalman, F. C., Scherrer, L. C., Taylor, L. P., Akil, H., and Pratt, W. B. (1991a). Localization of the 90-kDa heat shock protein-binding site within the hormone-binding domain of the glucocorticoid receptor by peptide competition. *J. Biol. Chem.* **266**, 3482–3490.

Dalman, F. C., Sturzenbecker, L. J., Levin, A. A., Lucas, D. A., Perdew, G. H. Petkovitch, M., Chambon, P., Grippo, J. F., and Pratt, W. B. (1991b). Retinoic acid receptor belongs to a subclass of nuclear receptors that do not form "docking" complexes with hsp90. *Biochemistry* **30**, 5605–5608.

Danielian, P. S., White, R., Lees, J. A., and Parker, M. G. (1992). Identification of a conserved region required for hormone dependent transcriptional activation by steroid hormone receptors. *EMBO J.* **11**, 1025–1033.

Danielsen, M., Northrop, J. P., and Ringold, G. M. (1986). The mouse glucocorticoid receptor: Mapping of functional domains by cloning, sequencing and expression of wild-type and mutant receptor proteins. *EMBO J.* **5**, 2513–2522.

Danielsen, M., Northrop, J. P., Jonklaas, J., and Ringold, G. M. (1987). Domains of the glucocorticoid receptor involved in specific and nonspecific deoxyribonucleic acid binding, hormone activation, and transcriptional enhancement. *Mol. Endocrinol.* **1**, 816–822.

Danielsen, M., Hinck, L., and Ringold, G. M. (1989). Two amino acids within the knuckle of the first zinc finger specify DNA response element activation by the glucocorticoid receptor. *Cell (Cambridge, Mass.)* **57**, 1131–1138.

Dedhar, S., Rennie, P. S., Shago, M., Hagesteijn, C.-Y.L., Yang, H., Filmus, J., Hawley, R. G., Bruchovsky, N., Cheng, H., Matusik, R. J., and Giguere, V. (1994). Inhibition of nuclear hormone receptor activity by calreticulin. *Nature* **367**, 480–483.

Demonacos, C., Tsawdaroglou, N. C., Djordjevic-Markovic, R., Papalopoulou, M., Galanopoulos, V., Papadogeorgaki, S., and Sekeris, C. E. (1993). Import of the glucocorticoid receptor into rat liver mitochondria *in vivo* and *in vitro*. *J. Steroid Biochem. Mol. Biol.* **46**, 401–413.

Denis, M., Gustafsson, J.-A., and Wilström, A.-C. (1988). Interaction of the $M_r = 90,000$ heat shock protein with the steroid-binding domain of the glucocorticoid receptor. *J. Biol. Chem.* **263**, 18520–18523.

Diamond, M. I., Miner, J. N., Yoshinaga, S. K., and Yamamoto, K. R. (1990). Transcription factor interactions: Selectors of positive or negative regulation from a single DNA element. *Science* **249**, 1266–1272.

Dieken, E. S., Meese, E. U., and Miesfeld, R. L. (1990). nti Glucocorticoid receptor transcripts lack sequences encoding the amino-terminal transcriptional modulatory domain. *Mol. Cell. Biol.* **10**, 4574–4581.

Dingwall, C., Sharnick, S. V., and Laskey, R. A. (1982). A polypeptide domain that species migration of nucleoplasmin into the nucleus. *Cell (Cambridge, Mass.)* **30**, 449–458.

Eisen, L. P., Reichman, M. E., Thompson, E. B., Gametchu, B., Harrison, R. W., and Eisen, H. J. (1985). Monoclonal antibody to the rat glucocorticoid receptor. Relationship between the immunoreactive and DNA-binding domain. *J. Biol. Chem.* **260**, 11805–11810.

Eriksson, P., and Wrange, O. (1990). Protein–protein contacts in the glucocorticoid receptor homodimer influence its DNA binding properties. *J. Biol. Chem.* **265**, 3535–3542.

Evans, R. M. (1988). The steroid and thyroid hormone receptor superfamily. *Science* **240**, 889–895.

Ewbank, J. J., and Creighton, T. E. (1991). The molten globule protein conformation probed by disulphide bonds. *Nature (London)* **350**, 518–520.

Fawell, S. E., Lees, J. A., White, R., and Parker, M. G. (1990). Characterization and colocalization of steroid binding and dimerization activities in the mouse estrogen receptor. *Cell (Cambridge, Mass.)* **60**, 953–962.

Flach, H., Kaiser, U., and Westphal, H. M. (1992). Monoclonal antipeptide antibodies to the glucocorticoid receptor. *J. Steroid Biochem. Mol. Biol.* **42**, 467–474.

Flanagan, J. M., Kataoka, M., Shortle, D., and Engelman, D. M. (1992). Truncated staphylococcal nuclease is compact but disordered. *Proc. Natl. Acad. Sci. U.S.A.* **89**, 748–752.

Flanagan, J. M., Kataoka, M., Fujisawa, T., and Engelman, D. M. (1993). Mutations can cause large changes in the conformation of a denatured protein. *Biochemistry* **32**, 10359–10370.

Freedman, L. P. (1992). Anatomy of the steroid receptor zinc finger region. *Endocr. Rev.* **13**, 129–145.

Freedman, L. P., and Towers, T. L. (1991). DNA binding properties of the vitamin D3 receptor zinc finger region. *Mol. Endocrinol.* **5**, 1815–1826.

Freedman, L. P., Luisi, B. F., Korszun, Z. R., Basavappa, R., Sigler, P. B., and Yamamoto, K. R. (1988). The function and structure of the metal coordination sites within the glucocorticoid receptor DNA binding domain. *Nature (London)* **334**, 543–546.

Gametchu, B., and Harrison, R. W. (1984). Characterization of a monoclonal antibody to the rat liver glucocorticoid receptor. *Endocrinology (Baltimore)* **114**, 274–279.

Garabedian, M. J., and Yamamoto, K. R. (1992). Genetic dissection of the signaling domain of a mammalian steroid receptor in yeast. *Mol. Biol. Cell* **3**, 1245–1257.

Garcia-Bustos, J., Heitman, J., and Hall, M. N. (1991). Nuclear protein localization. *Biochim. Biophys. Acta* **1071**, 83–101.

Giguère, V., Hollenberg, S. M., Rosenfeld, M. G., and Evans, R. M. (1986). Functional domains of the human glucocorticoid receptor. *Cell (Cambridge, Mass.)* **46**, 645–652.

Giguère, V., Ong., E. S., Segui, P., and Evans, R. M. (1987). Identification of a receptor for the morphogen retinoic acid. *Nature (London)* **330**, 624–629.

Godowski, P. J., Rusconi, S., Miesfeld, R., and Yamamoto, K. R. (1987). Glucocorticoid receptor mutants that are constitutive activators of transcriptional enhancement. *Nature (London)* **325**, 365–368.

Godowski, P. J., Picard, D., and Yamamoto, K. R. (1988). Signal transduction and transcriptional regulation by glucocorticoid receptor–LexA fusion proteins. *Science* **241**, 812–816.

Godowski, P. J., Sakai, D. D., and Yamamoto, K. R. (1989). Signal transduction and transcriptional regulation by the glucocorticoid receptor. *UCLA Symp. Mol. Cell. Biol.* **95**, 197–210.

Green, S., and Chambon, P. (1987). Oestradiol induction of a glucocorticoid-responsive gene by a chimaeric receptor. *Nature (London)* **325**, 75–78.

Green, S., Kumar, V., Theulaz, I., Wahli, W., and Chambon, P. (1988). The N-terminal DNA-binding "zinc finger" of the oestrogen and glucocorticoid receptors determines target gene specificity. *EMBO J.* **7**, 3037–3044.

Giochon-Mantel, A., Loosfelt, H., Lescop, P., Sar, S., Atger, M., Perrot-Applanat, M., and Milgrom, E. (1989). Mechanisms of nuclear localization of the progesterone receptor: Evidence for interaction between monomers. *Cell (Cambridge, Mass.)* **57**, 1147–1154.

Hard, T., Kellenbach, E., Boelens, R., Maler, B. A., Dahlman, K., Freedman, L. P., Carlstedt-Duke, J., Yamamoto, K. R., Gustafsson, J.-A., and Kaptein, R. (1990). Solution structure of the glucocorticoid receptor DNA-binding domain. *Science* **249**, 157–160.

Harmon, J. M., Eisen, H. J., Brower, S. T., Simons, S. S., Jr., Langley, C. L., and Thompson, E. B. (1984). Identification of human leukemic glucocorticoid receptors using affinity labeling and anti-human glucocorticoid receptor antibodies. *Cancer Res.* **44**, 4540–4547.

Hayashi, T., Ueno, Y., and Okamoto, T. (1993). Oxidoreductive regulation of nuclear factor kB. Involvement of a cellular reducing catalyst thioredoxin. *J. Biol. Chem.* **268**, 11380–11388.

Hendry, W. J., III, Hakkak, R., and Harrison, R. W. (1993). An analysis of autologous glucocorticoid receptor protein regulation in AtT-20 cells also reveals differential specificity of the BuGR2 monoclonal antibody. *Biochim. Biophys. Acta* **1178**, 176–188.

Hilliard, G. M. IV, Crook, R. G., Weigel, N. L., and Pike, J. W. (1994). 1,25-Dihydroxyvitamin D_3 modulates phosphorylation of serine 205 in the human vitamin D receptor: site-directed mutagenesis of this residue promotes alternative phosphorylation. *Biochemistry* **33**, 4300–4311.

Hoeck, W., and Groner, B. (1990). Hormone-dependent phosphorylation of the glucocorticoid receptor occurs mainly in the amino-terminal transactivation domain. *J. Biol. Chem.* **265**, 5403–5408.

Hoeck, W., Rusconi, S., and Groner, B. (1989). Down-regulation and phosphorylation of glucocorticoid receptors in cultured cells. *J. Biol. Chem.* **264**, 14396–14402.

Hollenberg, S. M., and Evans, R. M. (1988). Multiple and cooperative trans-activation domains of the human glucocorticoid receptor. *Cell (Cambridge, Mass.)* **55**, 899–906.

Hollenberg, S. M., Weinberger, C., Ong, E. S., Cerelli, G,. Oro, A., Lebo, R., Thompson, E. B., Rosenfeld, M. G., and Evans, R. M. (1985). Primary structure and expression of a functional human glucocorticoid receptor cDNA. *Nature (London)* **318**, 635–641.

Hollenberg, S. M., Giguère, V., Segui, P., and Evans, R. M. (1987). Colocalization of DNA-binding and transcriptional activation functions in the human glucocorticoid receptor. *Cell (Cambridge, Mass.)* **49**, 39–46.

Hollenberg, S. M., Giguère, V., and Evans, R. M. (1989). Identification of two regions of the human glucocorticoid receptor hormone binding domain that block activation. *Cancer Res.* **49**, 2292s–2294s.

Hope, T. J., Huang, X., McDonald, D., and Parslow, T. G. (1990). Steroid–receptor fusion of the human immunodeficiency virus type 1 Rev transactivator: Mapping cryptic functions of the arginine-rich motif. *Proc. Natl. Acad. Sci. U.S.A.* **87**, 7787–7791.

Housley, P. R., Sanchez, E. R., Danielsen, M., Ringold, G. M., and Pratt, W. B. (1990). Evidence that the conserved region in the steroid binding domain of the glucocorticoid receptor is required for both optimal binding of hsp90 and protection from proteolytic cleavage. *J. Biol. Chem.* **265,** 12778–12781.

Howard, K. J., Holley, S. J., Yamamoto, K. R., and Distelhorst, C. W. (1990). Mapping the HSP90 binding region of the glucocorticoid receptor. *J. Biol. Chem.* **265,** 11928–11935.

Hu, L.-M., Bodwell, J., Hu, J.-M., Orti, E., and Munck, A. (1994). Glucocorticoid receptors in ATP-depleted cells. Dephosphorylation, loss of hormone binding, hsp90 dissociation, and ATP-dependent cycling. *J. Biol. Chem.* **269,** 6571–6577.

Hughson, F. M., Wright, P. E., and Baldwin, R. L. (1990). Structural characterization of a party folded apomyoglobin intermediate. *Science* **249,** 1544–1548.

Hurley, D. M., Accili, D, Stratakis, C. A., Karl, M., Vamvakopoulos, N., Rorer, E., Constantine, K., Taylor, S. I., and Chrousos, G. P. (1991). Point mutation causing a single amino acid substitution in the hormone binding domain of the glucocorticoid receptor in familial glucocorticoid resistance. *J. Clin. Invest.* **87,** 680–686.

Hutchison, K. A., Czar, M. J., and Pratt, W. B. (1992a). Evidence that the hormone-binding domain of the mouse glucocorticoid receptor directly represses DNA binding activity in a major portion of receptors that are "misfolded" after removal of hsp90. *J. Biol. Chem.* **267,** 3190–3195.

Hutchison, K. A., Czar, M. J., Scherrer, L. C., and Pratt, W. B. (1992b). Monovalent cation selectivity for ATP-dependent association of the glucocorticoid receptor with hsp70 and hsp90. *J. Biol. Chem.* **267,** 14047–14053.

Hutchison, K. A., Dalman, F. C., Hoeck, W., Groner, B., and Pratt, W. B. (1993). Localization of the ~12 kDa M_r discrepancy in gel migration of the mouse glucocorticoid receptor to the major phosphorylated cyanogen bromide fragment in the transactivating domain. *J. Steroid Biochem. Mol. Biol.* **46,** 681–686.

Imamoto, N., Matsuoka, Y., Kurihara, T., Kohno, K., Miyagi, M., Sakiyama, F., Okada, Y., Tsunasawa, S., and Yoneda, Y. (1992). Antibodies against 70-kD heat shock cognate protein inhibit mediated nuclear import of karyophilic proteins. *J. Cell Biol.* **119,** 1047–1061.

Ip, M. M., Shea, W. K., Sykes, D., and Young, D. A. (1991). The truncated glucocorticoid receptor in the P1798 mouse lymphosarcoma is associated with resistance to glucocorticoid lysis but not to other glucocorticoid-induced functions. *Cancer Res.* **51,** 2786–2796.

Israel, D. I., and Kaufman, R. J. (1993). Dexamethasone negatively regulates the activity of a chimeric dihydrofolate reductase/glucocorticoid receptor protein. *Proc. Natl. Acad. Sci. U.S.A.* **90,** 4290–4294.

Issemann, I., and Green, S. (1990). Activation of a member of the steroid hormone receptor superfamily by peroxisome proliferators. *Nature (London)* **347,** 645–650.

Joab, I., Radanyi, C., Renoir, M., Buchou, T., Catelli, M.-G., Binart, N., Mester, J., and Baulieu, E.-E. (1984). Common non-hormone binding component in nontransformed chick oviduct receptors of four steroid hormones. *Nature (London)* **308,** 850–853.

Jonat, C., Rahmsdorf, H. J., Park, K.-K., Cato, A. C. B., Gebel, S., Ponta, H., and Herrlich, P. (1990). Antitumor promotion and antiinflammation: Down-modulation of Ap-1 (Fos/Jun) activity by glucocorticoid hormone. *Cell (Cambridge, Mass.)* **62,** 1189–1204.

Karl, M., Lamberts, S. W., Detera-Wadleigh, S. D., Encio, I. J., Stratakis, C. A., Hurley, D. M., Accili, D., and Chrousos, G. P. (1993). Familial glucocorticoid resistance caused by a splice site deletion in the human glucocorticoid receptor gene. *J. Clin. Endocrinol. Metabl.* **76,** 683–689.

Kasai, Y. (1990). Two naturally-occurring isoforms and their expression of a glucocorticoid receptor gene from an androgen-dependent mouse tumor. *FEBS J.* **274**, 99–102.

Kaspar, F., Klocker, H., Denninger, A., Cato, A. C. (1993). A mutant androgen receptor from patients with Reifenstein syndrome: Identification of the function of a conserved alanine residue in the D box of steroid receptors. *Mol. Cell. Biol.* **13**, 7850–7858.

Kaul, B. B., Enemark, J. H., Merbs, S. L., and Spence, J. T. (1987). Molybdenum(VI)-dioxo, molybdenum(V)-oxo, and molybdenum(IV)-oxo complexes with 2,3:8,9-dibenzo-1,4,7,10-tetrathiadecane. Models for the molybdenum binding site of the molybdenum cofactor. *J. Am. Chem. Soc.* **107**, 2885–2891.

Kay, A., and Mitchell, P. C. H. (1968). Molybdenum–cysteine complex. *Nature (London)* **219**, 267–268.

Keightley, M.-C., and Fuller, P. J. (1994). Unique sequences in the guinea-pig glucocorticoid receptor induce constitutive transactivation and decrease steroid sensitivity. *Mol. Endocrinol.* **8**, 431–439.

Kellenbach, E., Maler, B. A., Yamamoto, K. R., Boelens, R., and Kaptein, R. (1991). Identification of the metal coordinating residues in the DNA binding domain of the glucocorticoid receptor by [113]Cd-^{1}H heteronuclear NMR spectroscopy. *FEBS Lett.* **291**, 367–370.

Kerppola, T. K., Luk, D., and Curran, T. (1993). Fos is a preferential target of glucocorticoid receptor inhibition of AP-1 activity *in vitro. Mol. Cell. Biol.* **13**, 3782–3791.

König, H., Ponta, H., Rahmsdorf, H. J., and Herrlich, P. (1992). Interference between pathway-specific transcription factors: Glucocorticoids antagonize phorbol ester-induced AP-1 activity without altering AP-1 site occupation *in vivo. EMBO J.* **11**, 2241–2246.

Kosen, P. A., Marks, C. B., Falick, A. M., Anderson, S., and Kuntz, I. D. (1992). Disulfide bond-coupled folding of bovine pancreatic trypsin inhibitor derivatives missing one or two disulfide bonds. *Biochemistry* **31**, 5705–5717.

Kumar, S. A., and Dickerman, H. W. (1983). Specificity of nucleic acid structure for binding steroid receptors. In "Biochem. Actions Hormones" (G. Litwack, ed.), Vol. 10, pp. 259–301. Academic Press, New York.

Kupfer, S. R., Marschke, K. B., Wilson, E. M., and French, F. S. (1993). Receptor accessory factor enhances specific DNA binding of androgen and glucocorticoid receptors. *J. Biol. Chem.* **268**, 17519–17527.

Kutoh, E., Strömstedt, P. E., and Poellinger, L. (1992). Functional interference between the ubiquitous and constitutive octamer transcription factor 1 (OTF-1) and the glucocorticoid receptor by direct protein–protein interaction involving the homeo subdomain of OTF-1. *Mol. Cell. Biol.* **12**, 4960–4969.

Lanford, R. E., and Butel, J. S. (1984). Construction and characterization of an SV40 mutant defective in nuclear transport of T antigen. *Cell (Cambridge, Mass.)* **37**, 801–813.

Lanz, R., Hug, M., Gola, M., Tallone, T., Wieland, S., and Rusconi, S. (1994). Active, interactive and inactive steroid receptor mutants. *Steroids* **59**, 148–152.

Leach, K. L., Dahmer, M. K., Hammond, N. D., Sando, J. J., and Pratt, W. B. (1979). Molybdate inhibition of glucocorticoid receptor inactivation and transformation. *J. Biol. Chem.* **254**, 11884–11890.

Lee, J. W., Gulick, T., and Moore, D. D. (1992). Thyroid hormone receptor dimerization function maps to a conserved subregion of the ligand binding domain. *Mol. Endocrinol.* **6**, 1867–1873.

Lopez, S., Miyashita, Y., and Simons, S. R., Jr. (1990). Structurally based, selective interaction of arsenite with steroid receptors. *J. Biol. Chem.* **265,** 16039–16042.

Lucibello, F. C., Slater, E. P., Jooss, K. U., Beato, M., and Muller, R. (1990). Mutual transrepression of fos and the glucocorticoid receptor: Involvement of a functional domain in fos which is absent in fosB. *EMBO J.* **9,** 2827–2834.

Luisi, B. F., Xu, W. X., Otwinowski, Z., Freedman, L. P., Yamamoto, K. R., and Sigler, P. B. (1991). Crystallographic analysis of the interaction of the glucocorticoid receptor with DNA. *Nature (London)* **352,** 497–505.

Mader, S., Kumar, V., de Verneuil, H., and Chambon, P. (1989). Three amino acids of the oestrogen receptor are essential to its ability to distinguish an oestrogen from a glucocorticoid-responsive element. *Nature (London)* **338,** 271–274.

Maksymowych, A. B., Hsu, T.-C., and Litwack, G. (1992). A novel, highly conserved structural motif is present in all members of the steroid receptor superfamily. *Receptor* **2,** 225–2409.

Malchoff, D. M., Brufsky, A., Reardon, G., McDermott, P., Javier, E. C., Bergh, C.-H., Rowe, D., and Malchoff, C. D. (1993). A mutation of the glucocorticoid receptor in primary cortisol resistance. *J. Clin. Invest.* **91,** 1918–1925.

Maroder, M., Farina, A. R., Vacca, A., Felli, M. P., Meco, D., Screpanti, I., Frati, L., and Gulino, A. (1993). Cell-specific bifunctional role of jun oncogene family members on glucocorticoid receptor-dependent transcription. *Mol. Endocrinol.* **7,** 570–584.

Mason, S. A., and Houseley, P. R. (1993). Site-directed mutagenesis of the phosphorylation sites in the mouse glucocorticoid receptor. *J. Biol. Chem.* **268,** 21501–21504.

Matthews, J. R., Wakasugi, N., Virelizier, J.-L., Yodoi, J., and Hay, R. T. (1992). Thioredoxin regulates the DNA binding activity of NF-kappaB by reduction of a disulphide bond involving cysteine 62. *Nucleic Acids Res.* **20,** 3821–3830.

McEwan, I. J., Wright, A. P. H., Dahlman-Wright, K., Carlstedt-Duke, J., and Gustafsson, J.-A. (1993). Direct interaction of the tau1 transactivation domain of the human glucocorticoid receptor with the basal transcriptional machinery. *Mol. Cell. Biol.* **13,** 399–407.

Meshinchi, S., Sanchez, E. R., Martell, K. J., and Pratt, W. B. (1990). Elimination and reconstitution of the requirement for hormone in promoting temperature-dependent transformation of cytosolic glucocorticoid receptors to the DNA-binding state. *J. Biol. Chem.* **265,** 4863–4870.

Miesfeld, R., Rusconi, S., Godowski, P. J., Maler, B. A., Okret, S., Wikström, A.-C., Gustafsson, J.-A., and Yamamoto, K. R. (1986). Genetic complementation of a glucocorticoid receptor deficiency by expression of cloned receptor cDNA. *Cell (Cambridge, Mass.)* **46,** 389–399.

Miesfeld, R., Godowski, P. J., Maler, B. A., and Yamamoto, K. R. (1987). Glucocorticoid receptor mutants that define a small region sufficient for enhancer activation. *Science* **236,** 423–427.

Miesfeld, R., Sakai, D., Inoue, A., Schena, M., Godowski, P. J., and Yamamoto, K. R. (1988). Glucocorticoid receptor sequences that confer positive and negative transcriptional regulation. *UCLA Symp. Mol. Cell. Biol.* **75,** 193–200.

Miller, N. R., and Simons, S. S., Jr. (1988). Steroid binding to hepatoma tissue culture cell glucocorticoid receptors involves at least two sulfhydryl groups. *J. Biol. Chem.* **263,** 15217–15225.

Miller, P. A., Ostrowski, M. C., Hager, G. L., and Simons, S. S., Jr. (1984). Covalent and noncovalent receptor–glucocorticoid complexes preferentially bind to the same regions of the long terminal repeat of murine mammary tumor virus proviral DNA. *Biochemistry* **23,** 6883–6889.

Miner, J. N., and Yamamoto, K. R. (1992). The basic region of AP-1 specifies glucocorticoid receptor activity at a composite response element. *Genes Dev.* **6**, 2491–2501.

Miner, J. N., Diamond, M. C., and Yamamoto, K. R. (1991). Joints in the regulatory lattice: Composite regulation by steroid receptor–AP1 complexes. *Cell Growth, Differ.* **2**, 525–530.

Miyashita, Y., Miller, M., Yen, P. M., Harmon, J. M., Hanover, J. A., and Simons, S. S., Jr. (1993). Glucocorticoid receptor binding to rat liver nuclei occurs without nuclear transport. *J. Steroid Biochem. Mol. Biol.* **46**, 309–320.

Modarress, K. J., Cavanaugh, A. H., Chakraborti, P. K., and Simons, S. S. Jr. (1994). Metal oxyanion stabilization of the rat glucocorticoid receptor is independent of thiols. Submitted for publication.

Muller, M., Baniahmad, C., Kaltschmidt, C., and Renkawitz, R. (1991). Multiple domains of the glucocorticoid receptor involved in synergism with the CACCC box factor(s). *Mol. Endocrinol.* **5**, 1498–1503.

Nagpal, S., Saunders, M., Kastner, P., Durand, B., Nakshatri, H., and Chambon, P. (1992). Promoter context- and response element-dependent specificity of the transcriptional activation and modulating functions of retinoic acid receptors. *Cell (Cambridge, Mass.)* **70**, 1007–1019.

Nazareth, L. V., Harbour, D. V., and Thompson, E. B. (1991). Mapping the human glucocorticoid receptor for leukemic cell death. *J. Biol. Chem.* **266**, 12976–12980.

Ning, Y.-M., and Sanchez, E. R. (1993). Potentiation of glucocorticoid receptor-mediated gene expression by the immunophilin ligands FK506 and rapamycin. *J. Biol. Chem.* **268**, 6073–6076.

Nishio, Y., Isshiki, H., Kishimoto, T., and Akira, S. (1993). A nuclear factor for interleukin-6 expression (NF-IL6) and the glucocorticoid receptor synergistically activate transcription of the rat alpha 1-acid glycoprotein gene via direct protein–protein interaction. *Mol. Cell. Biol.* **13**, 1854–1862.

Obourn, J. D., Koszewski, N. J., and Notides, A. C. (1993). Hormone- and DNA-binding mechanisms of the recombinat human estrogen receptor. *Biochemistry* **32**, 6229–6236.

O'Donnell, A. L., Rosen, E. D., Darling, D. S., and Koenig, R. J. (1991). Thyroid hormone receptor mutations that interfere with transcriptional activation also interfere with receptor interaction with a nuclear protein. *Mol. Endocrinol.* **5**, 94–99.

Ohara-Nemoto, Y., Stromstedt, P.-E., Dahlman-Wright, K., Nemoto, T., Gustafsson, J.-A., and Carlstedt-Duke, J. (1990). The steroid-binding properties of recombinant glucocorticoid receptor: A putative role for heat shock protein hsp90. *J. Steroid Biochem. Mol. Biol.* **37**, 481–490.

Okamoto, K., Hirano, H., and Isohashi, F. (1993). Molecular cloning of rat liver glucocorticoid–receptor translocation promoter. *Biochem. Biophys. Res. Commun.* **193**, 848–854.

Okret, S. (1983). Comparison between different rabbit antisera against the glucocorticoid receptor. *J. Steroid Biochem.* **19**, 1241–1248.

Okret, S., Wikström, A.-C., Wrange, O., Andersson, B., and Gustafsson, J.-A. (1984). Monoclonal antibodies against the rat liver glucocorticoid receptor. *Proc. Natl. Acad. Sci. U.S.A.* **81**, 1609–1613.

Opoku, J., and Simons, S. S., Jr. (1993). Absence of intramolecular disulfides in the structure and function of native rat glucocorticoid receptors. *J. Biol. Chem.* **269**, 503–510.

Oro, A. E., Hollenberg, S. M., and Evans, R. M. (1988). Transcriptional inhibition by a glucocorticoid receptor–beta-galactosidase fusion protein. *Cell (Cambridge, Mass.)* **55**, 1109–1114.

Orti, E., Mendel, D. B., Smith, L. I., and Munck, A. (1989). Agonist-dependent phosphorylation and nuclear dephosphorylation of glucocorticoid receptors in intact cells. *J. Biol. Chem.* **264,** 9728–9731.

Orti, E., Bodwell, J. E., and Munck, A. (1992). Phosphorylation of steroid hormone receptors. *Endoc. Rev.* **13,** 105–128.

Orti, E., Hu, L.-M., and Munck, A. (1993). Kinetics of glucocorticoid receptor phosphorylation in intact cells. Evidence for hormone-induced hyperphosphorylation after activation and recycling of hyperphosphorylated receptors. *J. Biol. Chem.* **268,** 7779–7784.

Oshima, H., and Simons, S. S., Jr. (1992). Modulation of transcription factor activity by a distant steroid modulatory element. *Mol. Endocrinol.* **6,** 416–428.

Pearce, D., and Yamamoto, K. R. (1993). Mineralocorticoid and glucocorticoid receptor activities distinguished by nonreceptor factors at a composite response element. *Science* **259,** 1161–1165.

Perlmann, T., Eriksson, P., and Wrange, O. (1990). Quantitative analysis of the glucocorticoid receptor–DNA interaction at the mouse mammary tumor virus glucocorticoid response element. *J. Biol. Chem.* **265,** 17222–17229.

Picado-Leonard, J., and Miller, W. L. (1988). Homologous sequences in steroidogenic enzymes, steroid receptors, and a steroid binding protein suggest a consensus steroid-binding sequence. *Mol. Endocrinol.* **2,** 1145–1150.

Picard, D., and Yamamoto, K. R. (1987). Two signals mediate hormone-dependent nuclear localization of the glucocorticoid receptor. *EMBO J.* **6,** 3333–3340.

Picard, D., Salser, S. J., and Yamamoto, K. R. (1988). A movable and regulable inactivation function within the steroid binding domain of the glucocorticoid receptor. *Cell (Cambridge, Mass.)* **54,** 1073–1080.

Powers, J. H., Hillman, A. G., Tang, D. C., and Harmon, J. M. (1993). Cloning and expression of mutant glucocorticoid receptors from glucocorticoid-sensitive and -resistant human leukemic cells. *Cancer Res.* **53,** 4059–4063.

Pratt, W. B. (1993). Role of heat-shock proteins in steroid receptor function. *In* "Steroid Hormone Action: Frontiers in Molecular Biology" (M. G. Parker, ed.), pp. 64–93. Oxford Univ. Press, Oxford.

Pratt, W. B., Jolly, D. J., Pratt, D. V., Hollenberg, S. M., Giguère, V., Cadepond, F. M., Schweizer-Groyer, G., Catelli, M.-G., Evans, R. M., and Baulieu, E.-E. (1988). A region in the steroid binding domain determines formation of the non-DNA-binding, 9S glucocorticoid receptor complex. *J. Biol. Chem.* **263,** 267–273.

Pratt, W. B., Czar, M. J., Stancato, L. F., and Owens, J. K. (1993). The hsp56 immunophilin component of steroid receptor heterocomplexes: Could this be the elusive nuclear localization signal-binding protein? *J. Steroid Biochem. Mol. Biol.* **46,** 269–279.

Rangarajan, P. N., Umesono, K., and Evans, R. M. (1992). Modulation of glucocorticoid receptor function by protein kinase A. *Mol. Endocrinol.* **6,** 1451–1457.

Ray, A., and Prefontaine, K. E. (1993). Physical association and functional antagonism between the p65 subunit of NF-κB and the glucocorticoid receptor. *Proc. Natl. Acad. Sci. U.S.A.* **91,** 752–756.

Ray, A., LaForge, K. S., and Sehgal, P. B. (1991). Repressor to activator switch by mutations in the first Zn finger of the glucocorticoid receptor: Is direct DNA binding necessary? *Proc. Natl. Acad. Sci. U.S.A.* **88,** 7086–7090.

Reichman, M. E., Foster, C. M., Eisen, L. P., Eisen, J. H., Torain, B. F., and Simons, S. R., Jr. (1984). Limited proteolysis of covalently labeled glucocorticoid receptors as a probe of receptor structure. *Biochemistry* **23,** 5376–5384.

Rexin, M., Busch, W., Segnitz, B., and Gehring, U. (1988). Tetrameric structure of the nonactivated glucocorticoid receptor in cell extracts and intact cells. *FEBS Lett.* **241,** 234–238.

Rexin, M., Busch, W., and Gehring, U. (1991). Protein components of the nonactivated glucocorticoid receptor. *J. Biol. Chem.* **266,** 24601–24605.

Richard-Foy, H., Sistare, F. D., Riegel, A. T., Simons, S. S., Jr., and Hager, G. L. (1987). Mechanism of dexamethasone 21-mesylate antiglucocorticoid action: II. Receptor–antiglucocorticoid complexes do not interact productively with mouse mammary tumor virus long terminal repeat chromatin. *Mol. Endocrinol.* **1,** 659–665.

Richardson, W. D., Mills, A. D., Dilworth, S. M., Laskey, R. A., and Dingwall, C. (1988). Nuclear protein migration involves two steps: Rapid binding at the nuclear envelope followed by slower translocation through nuclear pores. *Cell (Cambridge, Mass.)* **52,** 655–664.

Robben, J. Van der Schueren, J., and Volckaert, G. (1993). Carboxyl terminus is essential for intracellular folding of chloramphenicol acetyltransferase. *J. Biol. Chem.* **268,** 24555–24558.

Robbins, J., Wilworth, S. M., Laskey, R. A., and Dingwall, C. (1991). Two interdependent basic domains in nucleoplasmin nuclear targeting sequence: Identification of a class of bipartite nuclear targeting sequence. *Cell (Cambridge, Mass.)* **64,** 615–623.

Rojiani, M. V., Finlay, B. B., Gray, V., Dedhar, S. (1991). *In vitro* interaction of a polypeptide homologous to human Ro/SS-A antigen (calreticulin) with a highly conserved amino acid sequence in the cytoplasmic domain of integrin alpha subunits. *Biochemistry,* **30,** 9859–9866.

Rosen, E. D., Beninghof, E. G., and Koenig, R. J. (1993). Dimerization interfaces of thyroid hormone, retinoic acid, vitamin D, and retinoid X receptors. *J. Biol. Chem.* **268,** 11534–11541.

Rupprecht, R., Arriza, J. L., Spengler, D., Reul, J. M., Evans, R. M., Holsboer, F., and Damm, K. (1993). Transactivation and synergistic properties of the mineralocorticoid receptor: Relationship to the glucocorticoid receptor. *Mol. Endocrinol.* **7,** 597–603.

Rusconi, S., and Yamamoto, K. R. (1987). Functional dissection of the hormone and DNA binding activities of the glucocorticoid receptor. *EMBO J.* **6,** 1309–1315.

Saatcioglu, F., Bartunek, P., Deng, T., Zenke, M., and Karin, M. (1993). A conserved C-terminal sequence that is deleted in v-erbA is essential for the biological activities of c-erbA (the thyroid hormone receptor). *Mol. Cell. Biol.* **13,** 3675–3685.

Sanchez, E. R. (1990). Hsp56: A novel heat shock protein associated with untransformed steroid receptor complexes. *J. Biol. Chem.* **265,** 22067–22070.

Schena, M., and Yamamoto, K. R. (1988). Mammalian glucocorticoid receptor derivatives enhance transcription in yeast. *Science* **241,** 965–967.

Schena, M., Freedman, L. P., and Yamamoto, K. R. (1989). Mutations in the glucocorticoid receptor zinc finger region that distinguish interdigitated DNA binding and transcriptional enhancement activities. *Genes Dev.* **3,** 1590–1601.

Scherrer, L. C., Picard, D., Massa, E., Harmon, J. M., Simons, S. S., Jr., Yamamoto, K. R., and Pratt, W. B. (1993). Evidence that the hormone binding domain of steroid receptors confers hormonal control on chimeric proteins by determining their hormone-regulated binding to heat-shock protein 90. *Biochemistry* **32,** 5381–5386.

Schlatter, L. K., Howard, K. J., Parker, M. G., and Distelhorst, C. W. (1992). Comparison of the 90-kilodalton heat shock protein interaction with *in vitro* translated glucocorticoid and estrogen receptors. *Mol. Endocrinol.* **6,** 132–140.

Schmidt, A., Endo, N., Rutledge, S. J., Vogel, R., Shinar, D., and Rodan, G. A. (1992).

Identification of a new member of the steroid hormone receptor superfamily that is activated by a peroxisome proliferator and fatty acids. *Mol. Endocrinol.* **6**, 1634–1641.

Schmidt, T. J., Harmon, J. M., and Thompson, E. B. (1980). "Activation-labile" glucocorticoid–receptor complexes of a steroid-resistant variant of CEM-C7 human lymphoid cells. *Nature (London)* **286**, 507–510.

Schmitt, J., and Stunnenberg, H. G. (1993). The glucocorticoid receptor hormone binding domain mediates transcriptional activation *in vitro* in the absence of ligand. *Nucleic Acids Res.* **21**, 2673–2681.

Schüle, R., Muller, M., Kaltschmidt, C., and Renkawitz, R. (1988a). Many transcription factors interact synergistically with steroid receptors. *Science* **242**, 1418–1420.

Schüle, R., Muller, M., Otsuka-Murakami, H., and Renkawitz, R. (1988b). Cooperativity of the glucocorticoid receptor and the CACCC-box binding factor. *Nature (London)* **332**, 87–90.

Schüle, R., Rangarajan, P., Kliewer, S., Ransone, L. J., Bolado, J., Yang, N., Verma, I. M., and Evans, R. M. (1990). Functional antagonism between oncoprotein c-jun and the glucocorticoid receptor. *Cell (Cambridge, Mass.)* **62**, 1217–1226.

Segard-Maurel, I., Jibard, N., Schweizer-Groyer, G., Cadepond, F., and Baulieu, E.-E. (1992). Mutations in the "zinc fingers" or in the N-terminal region of the DNA binding domain of the human glucocorticoid receptor facilitate its salt-induced transformation, but do not modify hormone binding. *J. Steroid Biochem. Mol. Biol.* **41**, 727–732.

Severne, Y., Wieland, S., Schaffner, W., and Rusconi, S. (1988). Metal binding "finger" structures in the glucocorticoid receptor defined by site-directed mutagenesis. *EMBO J.* **7**, 2503–2508.

Shea, W. K., Cowens, J. W., and Ip, M. M. (1991). New site-directed polyclonal antibody maps N-terminus of occluded region of the non-transformed glucocorticoid receptor oligomer to within BUGR epitope. *J. Steroid Biochem. Mol. Biol.* **39**, 433–447.

Sheridan, P. L., Krett, N. L., Gordon, J. A., and Horwitz, K. B. (1988). Human progesterone receptor transformation and nuclear down-regulation are independent of phosphorylation. *Mol. Endocrinol.* **2**, 1329–1342.

Sherman, M. R., Pickering, L. A., Rollwagen, F. M., and Miller, L. K. (1978). Meroreceptors: Proteolytic fragments of receptors containing the steroid-binding site. *Fed. Proc., Fed. Am. Soc. Exp. Biol.* **37**, 167–173.

Silva, C. M., and Cidlowski, J. A. (1989). Direct evidence for intra- and intermolecular disulfide bond formation in the human glucocorticoid receptor. *J. Biol. Chem.* **264**, 6638–6647.

Simons, S. S., Jr. (1979). Factors influencing association of glucocorticoid receptor–steroid complexes with nuclei, chromatin, and DNA: Interpretation of binding data. In "Glucocorticoid Hormone Action" (J. D. Baxter, and G. G. Rousseau, eds.), pp. 161–187. Springer-Verlag, Berlin.

Simons, S. S., Jr. (1987). Selective covalent labeling of cysteines in bovine serum albumin and in HTC cell glucorticoid receptors by dexamethasone 21-mesylate. *J. Biol. Chem.* **262**, 9669–9675.

Simons, S. S., Jr., and Thompson, E. B. (1981). Dexamethasone 21-mesylate: An affinity label of glucocorticoid receptors from rat hepatoma tissue culture cells. *Proc. Natl. Acad. Sci. U.S.A.* **78**, 3541–3545.

Simons, S. S., Jr., Martinez, H. M., Garcea, R. L., Baxter, J. D., and Tomkins, G. M. (1976). Interactions of glucocorticoid receptor–steroid complexes with acceptor sites. *J. Biol. Chem.* **251**, 334–343.

Simons, S. S., Jr., Pumphrey, J. G., Rudikoff, S., and Eisen, H. J. (1987). Identification of cysteine-656 as the amino acid of HTC cell glucocorticoid receptors that is covalently labeled by dexamethasone 21-mesylate. *J. Biol. Chem.* **262**, 9676–9680.

Simons, S. S., Jr., Sistare, F. D., and Chakraborti, P. K. (1989). Steroid binding activity is retained in a 16-kDa fragment of the steroid binding domain of rat glucocorticoid receptors. *J. Biol. Chem.* **264**, 14493–14497.

Simpson, R. J., Grego, B., Govindan, M. V., and Gronemeyer, H. (1987). Peptide sequencing of the chick oviduct progesterone receptor form B. *Mol. Cell. Endocrinol.* **52**, 177–184.

Smith, L. I., Bodwell, J. E., Mendel, D. B., Ciardelli, T., North, W. G., and Munck, A. (1988). Identification of cysteine-644 as the covalent site of attachment of dexamethasone 21-mesylate to murine glucocorticoid receptors in WEHI-7 cells. *Biochemistry* **27**, 3747–3753.

Smith, L. I., Mendel, D. B., Bodwell, J. E., and Munck, A. (1989). Phosphorylated sites within the functional domains of the ~100-kDa steroid-binding subunit of glucocorticoid receptors. *Biochemistry* **28**, 4490–4498.

Spanjaard, R. A., and Chin, W. W. (1993). Reconstitution of ligand-mediated glucocorticoid receptor activity by trans-acting functional domains. *Mol. Endocrinol.* **7**, 12–16.

Srivastava, D., and Thompson, E. B. (1990). Two glucocorticoid binding sites on the human glucocorticoid receptor. *Endocrinology (Baltimore)* **127**, 1770–1778.

Storz, G., Tartaglia, L. A., and Ames, B. N. (1990). Transcriptional regulator of oxidative stress-inducible genes: Direct activation of oxidation. *Science* **248**, 189–194.

Strahle, U., Schmid, W., and Schutz, G. (1988). Synergistic action of the glucocorticoid receptor with transcription factors. *EMBO J.* **7**, 3389–3395.

Stromstedt, P.-E., Berkenstam, A., Jornvall, H., Gustafsson, J.-A., and Carlstedt-Duke, J. (1990). Radiosequence analysis of the human progestin receptor charged with [^{3}H]promegestone. A comparison with the glucorticoid receptor. *J. Biol. Chem.* **265**, 12973–12977.

Sugawara, A., Yen, P. M., Darling, D. S., and Chin, W. W. (1993). Characterization and tissue expression of multiple triiodothyronine receptor-auxiliary proteins and their relationship to the retinoid X-receptors. *Endocrinology (Baltimore)* **133**, 965–971.

Superti-Furga, G., Bergers, G., Picard, D., and Busslinger, M. (1991). Hormone-dependent transcriptional regulation and cellular transformation by Fos-steriod receptor fusion proteins. *Proc. Natl. Acad. Sci. U.S.A.* **88**, 5114–5118,

Tai, P.-K.K., Maeda, Y., Nakao, K., Wakim, N. G., Duhring, J. L., and Faber, L. E. (1986). A 59-kilodalton protein associated with progestin, estrogen, androgen, and glucocorticoid receptors. *Biochemistry* **25**, 5269–5275.

Tai, P.-K.K., Albers, M. W., Chang, H., Faber, L. E., and Schreiber, S. L. (1992). Association of a 59-kilodalton immunophilin with the glucocorticoid receptor complex. *Science* **256**, 1315–1318.

Tasset, D., Tora, L., Fromental, C., Scheer, E., and Chambon, P. (1990). Distinct classes of transcriptional activating domains function by different mechanisms. *Cell (Cambridge, Mass.)* **62**, 1177–1187.

Thomas, H. E., Stunnenberg, H. G., and Stewart, A. F. (1993). Heterodimerization of the *Drosophila* ecdysone receptor with retinoid X receptor and ultraspiracle. *Nature (London)* **362**, 471–475.

Thomas, P. J., Ko, Y. H., and Pedersen, P. L. (1992). Altered protein folding may be the molecular basis of most cases of cystic fibrosis. *FEBS Lett.* **312**, 7–9.

Tienrungroj, W., Meshinchi, S., Sanchez, E. R., Pratt, S. E., Grippo, J. F., Holmgren, A., and Pratt, W. B. (1987). The role of sulfhydryl groups in permitting transformation and DNA binding of the glucocorticoid receptor. *J. Biol. Chem.* **262**, 6992–7000.

Touray, M., Ryan, F., Jaggi, R., and Martin, F. (1991). Characterization of functional inhibition of the glucocorticoid receptor by Fos/Jun. *Oncogene* **6**, 1227–1234.

Tsai, S. Y., Srinivassan, G., Allan, G. F., Thompson, E. B., O'Malley, B. W., and Tsai, M.-J. (1990). Recombinant human glucocorticoid receptor induces transcription of hormone response genes *in vitro*. *J. Biol. Chem.* **265**, 17055–17061.

Ueda, K., Isohashi, F., Okamoto, K., Kokuhu, I., Kimura, K., Yoshikawa, K., and Sakamoto, Y. (1988). Tight binding of glucocorticoid–receptor complexes to histone-agarose. *Biochem. Biophys. Res. Commun.* **151**, 763–767.

Umesono, K., and Evans, R. M. (1989). Determinants of target gene specificity for steroid/thyroid hormone receptors. *Cell (Cambridge, Mass.)* **57**, 1139–1146.

Urda, L. A., Yen, P. M., Simons, S. S., Jr., and Harmon, J. M. (1989). Region-specific antiglucocorticoid receptor antibodies selectively recognize the activated form of the ligand-occupied receptor and inhibit the binding of activated complexes to deoxyribonucleic acid. *Mol. Endocrinol.* **3**, 251–260.

Vegeto, E., Allan, G. F., Schrader, W. T., Tsai, M.-J., McDonnell, D. P., and O'Malley, B. W. (1992). The mechanism of RU486 antagonism is dependent on the conformation of the carboxy-terminal tail of the human progesterone receptor. *Cell (Cambridge, Mass.)* **69**, 703–713.

Veldscholte, J., Berrevoets, C. A., Zegers, N. D., van der Kwast, T. H., Grootegoed, J. A., and Mulder, E. (1992). Hormone-induced dissociation of the androgen receptor-heat-shock protein complex: Use of a new monoclonal antibody to distinguish transformed from nontransformed receptors. *Biochemistry* **31**, 7422–7430.

Wang, L.-H., Tsai, S. Y., Cook, R. G., Beattie, W. G., Tsai, M.-J., and O'Malley, B. W. (1989). COUP transcription factor is a member of the steroid receptor superfamily. *Nature (London)* **340**, 163–166.

Webster, N. J. G., Green, S., Jin, J. R., and Chambon, P. (1988). The hormone-binding domains of the estrogen and glucocorticoid receptors contain an inducible transcription activation function. *Cell (Cambridge, Mass.)* **54**, 199–207.

Weissman, J. S., and Kim, P. S. (1991). Reexamination of the folding of BPTI: Predominance of native intermediates. *Science* **253**, 1386–1393.

Westphal, H. M., Moldenhauer, G., and Beato, M. (1982). Monoclonal antibodies to the rat liver glucocorticoid receptor. *EMBO J.* **11**, 1467–1471.

Westphal, H. M., Mugele, K., Beato, M., and Gehring, U. (1984). Immunochemical characterization of wild-type and variant glucocorticoid receptors by monoclonal antibodies. *EMBO J.* **3**, 1493–1498.

Wieland, S., Döbbeling, U., and Rusconi, S. (1991). Interference and synergism of glucocorticoid receptor and octamer factors. *EMBO J.* **10**, 2513–2521.

Wilson, E. M., Lubahn, D. B., French, F. S., Jewell, C. M., and Cidlowski, J. A. (1988). Antibodies to steroid receptor deoxyribonucleic acid binding domains and their reactivity with the human glucocorticoid receptor. *Mol. Endocrinol.* **2**, 1018–1026.

Wright, A. P. H., and Gustafsson, J.-A. (1991). Mechanism of synergistic transcriptional transactivation by the human glucocorticoid receptor. *Proc. Natl. Acad. Sci. U.S.A.* **88**, 8283–8287.

Wright, A. P. H., and Gustafsson, J.-A. (1992). Glucocorticoid-specific gene activation by the intact human glucocorticoid receptor expressed in yeast. Glucocorticoid specificity depends on low level receptor expression. *J. Biol. Chem.* **267**, 11191–11195.

Wright, A. P. H., Carlstedt-Duke, J., and Gustafsson, J.-A. (1990). Ligand-specific trans-activation of gene expression by a derivative of the human glucocorticoid receptor expressed in yeast. *J. Biol. Chem.* **265,** 14763–14769.

Wright, A. P. H., McEwan, I. J., Dahlman-Wright, K., and Gustafsson, J.-A. (1991). High level expression of the major transactivation domain of the human glucocorticoid receptor in yeast cells inhibits endogenous gene expression and cell growth. *Mol. Endocrinol.* **5,** 1366–1372.

Yang, K., Jones, S. A., and Challis, J. R. (1990). Changes in glucocorticoid receptor number in the hypothalamus and pituitary of the sheep fetus with gestational age and after adrenocorticotropin treatment. *Endocrinology (Baltimore)* **126,** 11–17.

Yang, K., Hammond, G. L., and Challis, J. R. (1992). Characterization of an ovine glu-cocorticoid receptor cDNA and developmental changes in its mRNA levels in the fetal sheep hypothalamus, pituitary gland and adrenal. *J. Mol. Endocrinol.* **8,** 173–180.

Yang-Yen, H.-F., Chambard, J.-C., Sun, Y.-L., Smeal, T., Schmidt, T. J., Drouin, J., and Karin, M. (1990). Transcriptional interference between c-jun and the glucocorticoid receptor: Mutual inhibition of DNA binding due to direct protein–protein interac-tion. *Cell (Cambridge, Mass.)* **62,** 1205–1215.

Yen, P. M., and Simons, S. S., Jr. (1991). Evidence against posttranslational glycosyla-tion of rat glucocorticoid receptors. *Receptor* **1,** 191–205.

Ylikomi, T., Bocquel, M. T., Berry, M., Gronemeyer, H., and Chambon, P. (1992). Cooper-ation of proto-signals for nuclear accumulation of estrogen and progesterone recep-tors. *EMBO J.* **11,** 3681–3694.

Zandi, E., Galli, I., Döbbeling, U., and Rusconi, S. (1993). Zinc finger mutations that alter domain interactions in the glucocorticoid receptor. *J. Mol. Biol.* **230,** 124–136.

Zilliacus, J., Dahlman-Wright, K., Wright, A., Gustafsson, J.-A., and Carlstedt-Duke, J. (1991). DNA binding specificity of mutant glucocorticoid receptor DNA-binding domains. *J. Biol. Chem.* **266,** 3101–3106.

Zilliacus, J., Dahlman-Wright, K., Carlstedt-Duke, J., and Gustafsson J.-A. (1992a). Zinc coordination scheme for the C-terminal zinc binding site of nuclear hormone recep-tors. *J. Steroid Biochem. Mol. Biol.* **42,** 131–139.

Zilliacus, J., Wright, A. P. H., Norinder, U. L. F., Gustafsson, J.-A., and Carlstedt-Duke, J. (1992b). Determinants for DNA-binding site recognition by the glucocorticoid receptor. *J. Biol. Chem.* **267,** 24941–24947.

Genetic Diseases of Steroid Metabolism

PERRIN C. WHITE[1]

Division of Pediatric Endocrinology
Cornell University Medical College
New York, New York 10021

I. INTRODUCTION

Steroid hormones are essential for normal sexual development, ability to withstand stress, and regulation of fluid and electrolyte balance. The hormones mediating these functions are grouped into sex steroids, glucocorticoids and mineralocorticoids. Synthesis of these different classes of steroids, and the appropriate regulation thereof, requires the precisely controlled expression of several different biosynthetic enzymes in various tissues. Deficiency of any of these enzymes results in disordered synthesis of one or more classes of hormones. This may adversely affect health.

[1]Present address: Department of Pediatrics, The University of Texas Southwestern Medical Center at Dallas, Dallas, Texas 75235.

131

With few exceptions, the enzymes required for biosynthesis and metabolism of steroid hormones have been identified. Over the past decade, the complementary DNA (cDNA) and genomic gene encoding each enzyme has been cloned. For many of the corresponding deficiency syndromes, causative mutations have been documented. This article reviews these studies. Because of space limitations, biochemical, physiological, and clinical studies of many of these enzymes are reviewed relatively briefly so that genetic studies may be considered in somewhat more detail. It is also not possible to consider enzymes primarily involved in degradation and excretion of steroid hormones.

Most of the proteins mediating steroid biosynthesis fall into two major structural classes. Whereas individual enzymes are considered later in the article, it is appropriate to begin with a review of the distinguishing features of each type of enzyme.

II. Structures and Functions of Steroid-Metabolizing Enzymes

A. Cytochromes P450

1. Basic Characteristics

Cytochromes P450 are enzymes with molecular weights of approximately 50,000 containing heme (iron chelated to protoporphyrin IX). Whereas some bacterial P450s are soluble, all eukaryotic P450s characterized thus far are membrane bound, being located either in the endoplasmic reticulum ("microsomal" enzymes) or in the inner membranes of mitochondria. Of the enzymes reviewed in this article, steroid 17α-hydroxylase, 21-hydroxylase, and aromatase are microsomal P450s, whereas cholesterol desmolase, 11β-hydroxylase, and aldosterone synthase are mitochondrial enzymes. An extensive review of cytochrome P450 enzymes was published by Schenkman and Greim (1993).

P450s are "mixed-function oxidases." They utilize molecular oxygen and reducing equivalents (i.e., electrons) provided by NADPH to catalyze oxidative conversions of an extremely wide variety of mostly lipophilic substrates.

The reducing equivalents are not accepted directly from NADPH but from accessory electron transport proteins instead. Microsomal P450s utilize a single accessory protein, NADPH-dependent cytochrome P450 reductase. This is a flavoprotein containing one molecule each of FAD and FMN (Porter and Kasper, 1985, 1986). Mitochondrial P450s require two proteins; NADPH-dependent adrenodoxin reductase donates electrons to adrenodoxin, which in turn transfers them to the P450. Adrenodoxin (or ferredoxin) reductase is also a flavoprotein but contains only a molecule of FAD (Hanukoglu *et al.*, 1987; Nonaka *et al.*, 1987; Sagara *et al.*, 1987; Hanukoglu and Gutfinger, 1989). Adrenodoxin contains nonheme iron complexed with sulfur (Okamura *et al.*, 1987; Marg *et al.*, 1992). Only one isoform of each accessory protein has been documented in mammals (Morel *et al.*, 1988; Picado-Leonard *et al.*, 1988; Solish *et al.*, 1988).

2. Evolutionary Relationships

At this time, over 200 different P450 enzymes have been analyzed using molecular genetic methods (Nelson *et al.*, 1993). This "super-family" of enzymes may be divided into a number of gene families. Members of each family are >35% identical in amino acid sequence, whereas members of different families of P450s are as little as 15–20% identical. This relatively low degree of sequence identity between different P450 families suggests that the corresponding genes have evolved separately for longer than one billion years in some cases.

When their coding sequences are aligned, the genes in a given family have identical numbers and positions of introns. Remarkably, most families of P450 genes have little or no similarity in intron–exon organization, and there is no apparent correspondence between functional domains of these enzymes and the positions of introns. This is in contrast to the situation in other superfamilies of genes (e.g., immunoglobulins) in which each functional domain of a protein is encoded by a separate exon. It is not known what mechanisms account for the different organizations of various P450 gene families. It is unlikely that this phenomenon represents loss of different introns from a common ancestor gene, because in that case any two families would be expected to have at least a few introns in common (White *et al.*, 1986). It is possible that the positions of introns in each gene were modified by activation of cryptic splice donor or acceptor sites, thus allowing intronic sequences to become coding sequences during the course of evolution. A contemporary example of this phenomenon, discussed in more detail in Section IV,D, is a 21-hydroxylase deficiency mutation that activates a cryptic splice site and incorporates 19 additional nucleotides into mRNA (Higashi *et al.*, 1988b). Alternatively, introns may have been inserted into primordial P450 genes by viruses, retrotransposons, etc.

A consistent nomenclature has been proposed for P450 enzymes (Nelson *et al.*, 1993). Each family is designated by "CYP" followed by a unique number that sometimes (but not always) has mnemonic value. A family may be divided into subfamilies designated by capital letters, the members of which are >55% identical in sequence. Each gene is then designated by a number; pseudogenes have a "P" appended. For example, steroid 11β-hydroxylase is "CYP11B1."

3. Identification of Important Structural Features

Alignment of the amino acid sequences of many P450s (which, in many cases, are predicted from the nucleotide sequences of the corresponding cDNA clones) have identified a small number of strongly conserved residues that are presumed to be important for catalytic function (Fig. 1) (Nelson and Strobel, 1988, 1989; Black, 1992). The basic three-dimensional structure of P450 enzymes has been deduced from X-ray crystallographic studies of two bacterial P450s. The first of these, P450cam (CYP101, camphor 5-exo-hydroxylase from Pseudomonas putida), is a soluble molecule that bears little similarity in primary structure (roughly 15%) to eukaryotic P450s (Poulos et al., 1987; Poulos, 1991; Poulos and Raag, 1992). It most closely resembles mitochondrial P450s in its requirement for a two-protein electron transport chain consisting of an iron–sulfur protein resembling adrenodoxin and an NADPH-dependent reductase of this protein. The other, P450BM-3 (CYP102 from Bacillus megaterium), is a complex protein consisting of a P450-like N-terminal domain (MW 55,000) and a C-terminal domain (MW 66,000) that is 35% identical to eukaryotic cytochrome P450 reductase. Thus, this P450 requires no accessory protein and is able to directly utilize NADPH to oxidize fatty acids. The P450 domain is about 25–30% identical to the eukaryotic CYP4 and CYP52 families (Ruettinger et al., 1989). Because mammalian P450s are functional in a similar manner when expressed as fusion proteins with cytochrome P450 reductase in yeast (Murakami et al., 1987), CYP102 seems to be a good model for such P450s.

The P450 domain of CYP102 was released by tryptic digestion and subjected to crystallographic analysis (Ravichandran et al., 1993). Based on these analyses and earlier functional studies, several conclusions may be drawn.

a. Heme Binding. The heme iron is critical for catalytic function of P450s. Of its six coordination positions, four interact with the protoporphyrin ring. Whereas other "b"-type cytochromes usually have the fifth position (i.e., one of the two axial positions) coordinated to a basic amino acid side chain such as histidine, P450s utilize a cysteine sulfhydryl as the ligand at the fifth position. This cysteine is located in a relatively highly conserved "heme binding peptide" near the C terminus of P450s, the consensus sequence of which is FXXGX (R/H) XCXG, where X is not an absolutely conserved residue.

```
                                                 A
CYP102                    TIKEMPQPKTFGELKNLPLLNTDKPVQALMKIADELG   37
CYP21    MLLLGLLLLPLLAGARLLWNWW--KLRSLHLPPLAPGFLHLLQPDLPIYLLGLT----QKFG   56
CYP11B1    (25) GTRAARVPRTVLPFEAMPRRPGNRWLRLLQIWREQGYEDLHLEVHQTFQELG   76
                                 *     S                         *


    1-1       1-2     B      1-5        B'           C
EIFKFEAPG-RVTRYLSSQRLIKEACD--ESRFDKNLSQA-LKFVRDFAGDGLFTSWTHEKNWKKAHNILL   104
PIYRLHL-GLQDVVVLNSKRTIEEAMVKKWADFAGRPEPLTYKLVSKNYPDLSLGDY--SLLWKAHKKLTR   124
PIFR-YDLGGAGMVCVM-LPEDVEKLQQVDSLHPHRMSLEPWVAYRQHRGHKCVGVFLLNGPEWRFNRLRLN   145
 *  S    *    S        *       S          S                *


          D            3-1          E
PSFSQQAMKGYHAMMVDIAVQLVQKWERL----NADEHIEVPE--DMTRLTLDTIGLCGFN--YRFNSFYR   167
SALLL-GIRDSMEPVVEQLTQEFCERMRA---QPGTPVAIEEE---FSLLTCSIICYLTFGDKIK-DDNLM   187
PEVLSPKAVQRFLPMVDAVARDFSQALKKKVLQNARGSLTLDVQPSIFHYTIEASNLALFGERLGLVGHSP   216
                                                *


          F                      G
DQPHPFITSMVRALDEAMNKLQRANP----DDPAYDENKRQFQEDIKVMNDLVDK-IIADRKASGEQSDDL   233
PAYYKCIQEVLKTWSHWSIQIVDVIPFLRFFPNPGLRRLKQAIEKRDHIVEMQLRQ-HKESLVAG-QWRDM   256
SSASLNFLHALEVMFKSTVQLMFMPRSLSRWTSPKVWKEHFEAWDCIFQYGDNCIQKIYQEL-AFSRPQQY   286
    S


H                                I                     J
LTHMLNGK-----DPETGEPLDDENIRYQIITFLIAGHETTSGLLSFALYFLVKNPHVLQKAAEEAARVLV   299
MDYMLQQGVAQPSMEEGSGQLLEGHVHMAAVDLLIGGTETTANTLSWAVVFLLHHPEIQQRLQEELDHELG   326
TSIV--------AELLLNAELSPDAIKANSMELTAGSVDTTVFPLLMTLFELARNPNVQQALRQESLAAAA   349
                         *         **   *     *     *     *
                                  proton transfer


    J'         K              1-4   2-1    2-2    1-3    K'
DPVPSYKQVK-QLKYVGMVLN---EARLRLWPTAPAFSLYAKEDTVLGGEYPLEKGDELMVLIPQLHRDKT   365
PGASSSRVPY-KDRARLPLLNATIREVLRLRPVVPLALPHRTTRPSSISGYDIPEGTVIIPNLQGAHLDET   396
----SISEHPQKATTELPLLRAALKETLRLYPV-GLFLERVASSDLVLQNYHIPAGTLVRVFLYSLGRNPA   415
    *            *     ***  *      S            *        *    S
         accessory protein binding


                                    L        3-3
IWGDDVEEFRPERFENPS----AIPQHAF--KPFGNGQRACIGQQFALHEATLVLGMMLKHFDFEDHTNYE   430
VWERP-HEFWPDRFLEPG-----KNSRA---LAFGCGARVCLGEPLARLELFVVLTRLLQAFTLLPSGDAL   458
L------FPRPERYNPQRWLDIKGSGRNFYHVPFGFGMRQCLGRRLAEVEMLLLLHHVLKHLQVETLTQED   480
    * *                * * * *   *     *     *     *     *
                            heme binding


    4-1   4-2     3-2
LDIKETLTLKPEGFVVKAKSKKIPLGGIPS          471
PSLQPLPHCSVILKMQPFQVRLQPRGMGAHSPGQNQ   494
IKMVYSFILRPSMCPLLTFRAIN                504
    S     S
```

FIG. 1. Alignments of amino acid sequences of human steroid-metabolizing cytochromes P450 (CYP21, 21-hydroxylase; CYP11B1, 11β-hydroxylase) (Mornet et al., 1989; White et al., 1986) with the sequence of CYP102 from B. megaterium. The entire sequences are displayed except for CYP11B1, where the mitochondrial signal peptide (25) is not shown. Features of the secondary structure of CYP102 determined by X-ray crystallography are marked in bold type above the aligned sequences. α helices are denoted by letters, and β strands by numbers. Labeling of these elements is as described in Ravichandran et al., (1993). Residues conserved between at least two of the three enzymes are printed in bold letters, and residues conserved in all three are marked with an asterisk. Regions of presumed functional importance are noted; possible substrate binding regions are denoted by an underline and the letter S.

b. Oxygen and Water Binding. The ligand at the other axial position of heme is either a water or an oxygen molecule. When an oxygen molecule is bound, it is parallel to the axis of coordination with the iron atom. Whereas the crystal structure of CYP101 (P450cam) suggested that the relatively highly conserved "I" helix in the middle of the peptide was required for O_2 binding (Poulos *et al.*, 1987), the crystal structure of CYP102 is not consistent with this idea. However, an H_2O molecule is consistently present in a groove in the I helix adjacent to a strongly conserved acidic (aspartate or glutamate) residue and a completely conserved threonine (E267 and T268, respectively, in CYP102). Mutation of this threonine in several P450s destroys or drastically decreases enzymatic activity (Raag *et al.*, 1991; Curnow *et al.*, 1993). During catalysis, it is proposed that two protons are donated in succession to the water molecule by the carboxyl group of the acidic residue, transferred to the hydroxyl of the conserved threonine and finally donated to O_2 bound to reduced (Fe^{2+}) heme. This allows the distal oxygen atom of the bound O_2 molecule to be released as water while the proximal atom hydroxylates the substrate.

c. Substrate Binding. Like most P450 substrates, steroids are relatively hydrophobic molecules. Thus, it is likely that the substrate binding site(s) will consist primarily of hydrophobic amino acid residues.

A priori, it was possible that the substrate binding sites of steroid-metabolizing P450s would more closely resemble each other in sequence than the substrate binding sites of other P450s such as xenobiotic-metabolizing enzymes. Comparisons of the sequences of 21- and 17-hydroxylase and cholesterol desmolase (CYP21, CYP17, and CYP11A) identified two highly conserved areas, one near the N terminus (Q53-R60 in CYP21) and the other toward the C terminus (L342-V358 in CYP21) (White, 1987; Picado-Leonard and Miller, 1988).

The crystal structure of CYP102 confirms that part of the first of these indeed corresponds to a portion of a deep pocket constituting the substrate binding site (β sheet 1-1, residues E38–A44 in CYP102). However, the second conserved area corresponds to helix K (L311–W325 of CYP102), which does not interact with substrate. Instead, this region forms part of the docking site for the accessory electron transport protein. The remainder of segments that form the substrate binding pocket are widely distributed in the peptide (remainder of β sheet 1, B' and F helices) and the sequence conservation among steroid-metabolizing P450s is not particularly strong in these regions.

d. Binding to Accessory Proteins. Microsomal and mitochondrial P450s accept electrons from cytochrome P450 reductase or adrenodoxin, respectively. In either case, chemical modification studies suggest that basic amino acids (usually lysine) on the P450 interact with acidic residues on the accessory protein. The crystallographic studies of CYP102 suggest a docking site for reductase formed in part by helices B, C, D, J', and K. Helix K, as mentioned, was previously thought to interact with substrate. Support for the idea that it is instead required for redox interactions (with cytochrome P450 reductase or adrenodoxin, depending on the type of P450) comes from mutagenesis studies of CYP11A, wherein modification of either of two lysine residues in helix K destroys enzymatic activity without affecting substrate binding (Wada and Waterman, 1992).

Only two basic residues (R323 and R398 in CYP102, one in helix K and the other in the heme binding peptide) are completely conserved in all eukaryotic P450s, suggesting that other positively charged residues that are not completely conserved may be necessary for binding to the accessory protein (Shimizu et al., 1991).

e. Membrane Interactions. Microsomal P450s are anchored to the membrane of the endoplasmic reticulum by an N-terminal segment of approximately 20 amino acid residues. This segment also functions as a stop transfer signal during protein synthesis, preventing the nascent protein from being released from the endoplasmic reticulum (Kuroiwa et al., 1990, 1991; Sato et al., 1990; Sakaguchi et al., 1992). Although most of the P450 is outside the membrane, mutagenesis studies suggest that regions of some P450s corresponding to helix E on CYP102 may also interact with the membrane (Monier et al., 1988; Tusie-Luna et al., 1990).

Mitochondrial P450s are synthesized with a 20- to 40-residue N-terminal peptide that directs the newly synthesized protein to mitochondria. This peptide, which carries a few positively charged residues (Furuya et al., 1987; Kumamoto et al., 1987) but otherwise has no consensus sequence, is cleaved by a protease in the mitochondrial matrix (Omura and Ito, 1991). The P450 is bound to the mitochondrial inner membrane on the matrix side, but the nature and extent of the interactions with the membrane have not yet been well defined.

B. Short-Chain Dehydrogenases

1. Basic Characteristics

Most short-chain dehydrogenases catalyze reversible reactions. In the dehydrogenase direction, a hydride (i.e., a proton plus two electrons) is removed from the substrate and transferred to an electron acceptor, which depending on the enzyme is NAD^+ or $NADP^+$. If, for example, the reaction involves a hydroxylated substrate, the reaction converts the hydroxyl to a keto group; NADH or NADPH and a proton are also produced. Oxoreductase reactions reverse this process. Unlike oxidations mediated by cytochromes P450, no accessory protein is required for these reactions.

Most short-chain dehydrogenases contain 250–300 amino acid residues. Some enzymes are found in cytosol whereas others are located mainly in the endoplasmic reticulum. These enzymes differ markedly from other dehydrogenases such as alcohol dehydrogenase in vertebrates; the latter type of enzyme is larger (>350 residues) and has zinc at the active site, usually liganded to cysteine and/or histidine residues (Jornvall *et al.*, 1990).

More than 20 enzymes of the short-chain dehydrogenase type have been identified thus far. In addition to enzymes involved in steroid and prostaglandin metabolism in eukaryotes and bacteria, this class includes alcohol dehydrogenase in *Drosophila,* a number of bacterial enzymes involved in sugar metabolism (e.g., ribitol dehydrogenase in *Enterobacter*), and enzymes involved in nitrogen fixation (e.g., nodG protein in *Rhizobium*). A 3β-hydroxysteroid dehydrogenase is encoded by vaccinia virus and increases virulence *in vivo* (Moore and Smith, 1992).

Three types of short-chain dehydrogenases will be considered in detail in this review: 3β-, 11β-, and 17β-hydroxysteroid dehydrogenase.

2. *Identification of Important Structural Features*

The amino acid sequences of various short-chain dehydrogenases are typically 20–30% identical to each other. Several alignments of multiple sequences have been made (Fig. 2) (Persson *et al.*, 1991; Tannin *et al.*, 1991; Krozowski, 1992). Although there are minor differences between these alignments, there are at least five residues that are absolutely conserved in all of these enzymes. These include two glycines and an aspartate near the N terminus and a tyrosine and lysine toward the C terminus.

Two short-chain dehydrogenases, *Drosophila* alcohol dehydrogenase (Gordon *et al.*, 1992) and steroid 3α, 20β-dehydrogenase from *Streptomyces versicolor* (Ghosh *et al.*, 1991), have been subjected to X-ray crystallography. These data, combined with alignments of multiple sequences and with mutagenesis studies (Ensor and Tai, 1991; Obeid and White, 1992; Chen *et al.*, 1993), permit tentative conclusions to be drawn regarding the functional significance of various structural features.

```
                              βA          αA          βB
3α,20β HSD                    MNDLSGKTVIITGGARGLGAEAARQAVAAGARVVLA    36
11β-HSD      MAFMKYLLPILGLFMAYYYYSANEEFRPEMLQGKKVIVTGASKGIGREMAYHLAKMGAHVVVT    64
3β-HSD                        M-TGWSCLVTGAGGFLGQRIIRLLVKEKELKEI-    32
                              *    **    *
                                cofactor binding
```

```
          αC         βC            αD              βD
DVLDEE--GAATARE-LG-DAARYQHLDVTIEEDW-------QRVVAYAREEFGSVDGLVNNAGISTGMFLETES    100
ARSKETLQKVVSHCQVLG---AASAHYIAGTMEDMTFAE---QFVVAQGKLMGG-LDMLILNHITNTSLNLFHDD    132
-RVLDKAFGPELREEFSKLQNKTKLTVLEGDILDEPFLKRACQDVSVIIHTACI-IDVFGVTHRE-SIMNVNVKG    104
                    *      * *         *        *      S
```

```
          αE              βE               αF
VERFRKVVDINLTGVFIGMKTVIPANEDAGGGSIVNISSAAGLMGLALTSSYGASKWGVRGLSKLAAVELGTDRI    175
IHHVRKSMEVNFLSYVVLTVAALPMLKQS-NGSIVVVSSLAGKVAYPMVAAYSASKFALDGFFSSIRKEYSVSRV    206
TQLLLEACVQASVPVFIYTSSIEVAGPNS-YKEIIQNGHEEEPLENTWPAPYPHSKKL--AEKAVLAANGWNLKN    176
                *          S                    *  **
                                catalysis
```

```
βF            αG                        βG
RVNSVHPGMTYTPMTAETGIRQGEGNYPNTPMGRVGNEPGEIAGAVVKLLSDTSSYVTGAELAVDGGWTTGPTVK    249
NVSITLCVLGLIDTETAMKAVSGIVHMQAAPKEECALEIIKGGALRQEEVYYDSSLWTTLLIRNPCRKILEFLYS    281
GGTLYTCALRPMYIYGEGSRFLSASINEALNNNGILSSVGKFSTVNPVYVGNVAWAHILALRALQDPKKAPSIRG    251
      S                                           S                   S
```

```
YVMGQ                255
TSYNMDRFINK          292
QFYYISDDTPHQSY....373
```

FIG. 2. Alignments of amino acid sequences of hydroxysteroid dehydrogenases (HSD). Human 11β- and 3β-hydroxysteroid dehydrogenases (Tannin *et al.*, 1991; Lachance *et al.*, 1990) are aligned with the sequence of 3α, 20β-hydroxysteroid dehydrogenase from *S. hydrogenans*. Features are labeled as in Fig. 1, but in accordance with the nomenclature of Ghosh *et al.*, (1991), both α helices and β strands are denoted by letters.

a. Overall Structure. The structure of 3α, 20β-dehydrogenase (Ghosh *et al.*, 1991) will be discussed as it is more similar in sequence than alcohol dehydrogenase is to 11β- and 17β-hydroxysteroid dehydrogenase (Tannin *et al.*, 1991). The protein consists of seven β strands (designated A–G in the order in which they occur in the polypeptide) alternating with six α helices (B–G). These are organized into a single domain consisting of a single β sheet with the β strands in the order CBADEFG. Helices C, B, and G are on one side of the β sheet and helices D, E, and F on the other, with all helices parallel to the β sheet. In crystals, the protein is organized into tetramers. Other short-chain dehydrogenases in solution also form dimers or tetramers and an oligomeric structure may be required for activity (Ghosh *et al.*, 1991).

b. Cofactor Binding. The NADP$^+$/NADPH binding domain, like most domains that bind nucleotide cofactors, has a βαβαβ secondary structure. In contrast to the remainder of the enzyme, this region has some sequence homology to the cofactor binding site in zinc-containing dehydrogenases. It is located in short-chain dehydrogenases at the N terminus and includes the consensus sequence (I/L/V) T<u>G</u>XXXG (I/L/V) <u>G</u>; "X" signifies residues that are not strongly conserved and the underlined G's are completely conserved glycines. This sequence is located in the first βα segment and apparently binds the nicotinamide ring of the cofactor. It is uncertain whether acidic residues invariably bind the cofactor as in zinc-containing dehydrogenases; D37 in 3α, 20β-dehydrogenase does form a hydrogen bond with the cofactor, but this residue is not completely conserved in short-chain dehydrogenases. In contrast, D60 is completely conserved and D82 very strongly conserved, but according to the crystal structure neither residue is located near either the cofactor or the steroid binding site. Mutagenesis of D110 in 11β-hydroxysteroid dehydrogenase, which corresponds to D82, has only a minor effect on enzymatic activity (Obeid and White, 1992).

c. Steroid Binding. In 3α, 20β-dehydrogenase, a deep cleft presumed to be the steroid binding site is formed by parts of β strands D, E, and F, a loop between βE and αF, and two C-terminal segments, one from the same monomer and one from an adjacent monomer. With the exception of the βE–αF loop, these regions are not highly conserved among short-chain dehydrogenases. This is especially true for the C-terminal regions (after βF) of these enzymes. Thus, the substrate binding sites are probably highly variable in structure and may involve interactions between multiple monomeric subunits of each enzyme.

d. Catalytic Mechanism. Residues corresponding to Y152 and K156 of 3α, 20β-dehydrogenase are completely conserved in all short-chain dehydrogenases. Mutagenesis of the corresponding residues in 11β-hydroxysteroid dehydrogenase (Obeid and White, 1992), *Drosophila* alcohol dehydrogenase (Chen *et al.*, 1993), or 15-hydroxyprostaglandin dehydrogenase (Ensor and Tai, 1991) destroys or drastically decreases enzymatic activity. Both residues are located near the pyridine ring of the cofactor in the cleft presumed to be the steroid binding site. Y152 points into this cleft and the phenolic hydroxyl comes to within a few angstroms of the steroid. There is marked bridging of electron density between this hydroxyl and the ε-amino group of K156, which is located on the opposite side of the phenolic side chain of Y152 from the substrate.

It has been proposed that K156 lowers the apparent pK_a of the phenolic hydroxyl of Y152 from 10.0, its value in solution, into the physiological range (Obeid and White, 1992; Chen *et al.*, 1993). Thus, the ε-amino group of K156 and the hydroxyl of Y152 are both charged in the absence of substrate and cofactor. The deprotonated hydroxyl is then able to remove a proton from the hydroxyl group of the substrate, facilitating transfer of a hydride from the substrate to the oxidized cofactor.

III. Pathways of Steroid Biosynthesis

A. Adrenal

The two most physiologically important steroids secreted by the adrenal cortex are cortisol, the principal glucocorticoid in humans, and aldosterone, the most potent mineralocorticoid (Fig. 3).

Cortisol is synthesized from cholesterol in the zona fasciculata of the adrenal cortex (New *et al.*, 1989). Synthesis of cortisol requires five enzymatic steps: cleavage of the cholesterol side chain by cholesterol desmolase (CYP11A) to yield pregnenolone, 3β-dehydrogenation to progesterone, and successive hydroxylations at the 17α, 21, and 11β positions, which are mediated by three distinct cytochrome P450 enzymes (CYP17, CYP21, and CYP11B1). A "17-deoxy" pathway is also active in the zona fasciculata, in which 17α-hydroxylation does not occur and the final product is normally corticosterone.

The same 17-deoxy pathway is active in the adrenal zona glomerulosa, which contains no 17α-hydroxylase activity. However, corticosterone is not the final product in the zona glomerulosa; instead, corticosterone is successively hydroxylated and oxidized at the 18 position to yield aldosterone. In humans and rodents, the last three steps of aldosterone synthesis (deoxycorticosterone to aldosterone) are mediated by a distinct 11β-hydroxylase isozyme, aldosterone synthase (CYP11B2).

The zonae fasciculata and glomerulosa are regulated differently. Cortisol synthesis is regulated mainly by corticotropin (ACTH), which occupies a cell-surface receptor, stimulating the synthesis of cAMP. This causes short-term stimulation of cholesterol desmolase activity, probably by increasing transport of cholesterol into mitochondria, and long-term increases in transcription of genes encoding steroidogenic enzymes. The latter effects require cAMP-dependent protein kinase (Wong *et al.*, 1989; Black *et al.*, 1993; Parissenti *et al.*, 1993). In contrast, the main trophic stimuli for aldosterone biosynthesis are potassium and angiotensin II. The latter hormone acts through a cell-surface receptor to mobilize intracellular Ca^{2+}, activating protein kinase C. This also results in increases in levels of steroidogenic enzymes (Naseeruddin and Hornsby, 1990).

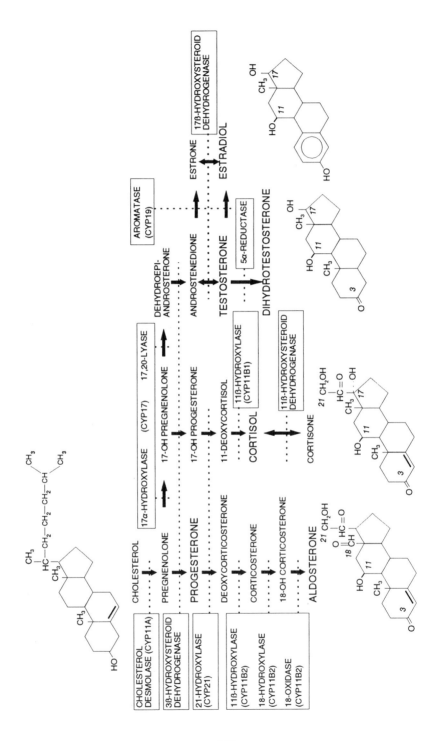

The inherited inability to synthesize cortisol is called congenital adrenal hyperplasia (White *et al.*, 1987a,b). Because cortisol is not synthesized efficiently, there is poor feedback inhibition of the hypothalamus and anterior pituitary leading to increased secretion of corticotropin. Stimulation of the adrenal cortex by corticotropin leads to overproduction and accumulation of cortisol precursors proximal to the blocked metabolic step. Depending on the defect, these precursors may produce signs and symptoms of androgen or mineralocorticoid excess. Conversely, glucocorticoid, mineralocorticoid, and sex steroid synthesis may be defective in various types of congenital adrenal hyperplasia leading to signs and symptoms of deficiencies of these types of steroids. These will be discussed along with the individual syndromes.

B. Testis

Testosterone, the predominant androgen secreted by the testis, is synthesized in Leydig cells (Griffin and Wilson, 1992). The main trophic stimulus for this process is luteinizing hormone (LH), which acts via cyclic AMP as a second messenger. Cholesterol is converted to pregnenolone by cholesterol desmolase (CYP11A). Pregnenolone is successively metabolized to 17α-hydroxypregnenolone and dehydroepiandrosterone by 17α-hydroxylase/17, 20-lyase (CYP17), and dehydroepiandrosterone is converted to androstenedione by 3β-hydroxysteroid dehydrogenase. All three of the enzymes involved in these steps are identical to those active in the adrenal cortex and are encoded by the same genes. The 17, 20-lyase activity of CYP17 is stronger in the testis than in the adrenal cortex, possibly due to higher levels of cytochrome b5 in testis (Ishii-Ohba *et al.*, 1984; Kominami *et al.*, 1992; Lin *et al.*, 1993).

Conversion of androstenedione to testosterone requires 17-

Fig. 3. Pathways of steroid biosynthesis. The pathways for synthesis of progesterone and mineralocorticoids (aldosterone), glucocorticoids (cortisol), androgens (testosterone and dihydrotestosterone), and estrogens (estradiol) are arranged from left to right. Only a subset of these reactions is active in any single steroid-secreting tissue. The enzymatic activities catalyzing each bioconversion are written in boxes. For those activities mediated by specific cytochromes P450, the systematic name of the enzyme ("CYP" followed by a number) is listed in parentheses. CYP11B2 and CYP17 have multiple activities. Other bioconversions are mediated by different enzymes in various tissues. Reversible steps are denoted by double-headed arrows. The planar structures of cholesterol, aldosterone, cortisol, dihydrotestosterone, and estradiol are placed near the corresponding labels.

ketosteroid reductase (17β-hydroxysteroid dehydrogenase), an activity present in gonads but not in the adrenal gland (Wilson *et al.*, 1988).

C. Ovary

The ovaries of women of reproductive age are unique among steroid-synthesizing tissues in their cyclic variations in the quantity and relative proportions of steroids synthesized (Carr, 1992). The primary follicle contains granulosa cells that synthesize estrogens with estradiol secretion exceeding that of estrone. Estradiol and estrone are synthesized, respectively, from testosterone and androstenedione by aromatase (CYP19). The androgens that are utilized for this reaction are synthesized by the same enzymes as are active in the testes, but the main stimulus for estrogen secretion is follicle-stimulating hormone (FSH). FSH has an especially pronounced effect on aromatase levels, which increase greatly in the mature midcycle follicle. Androgens are synthesized in granulosa cells and also in thecal cells, which are recruited into the follicle from the surrounding stroma as the follicle matures. The preovulatory surge of LH initiates the conversion of the follicle into the corpus luteum, which secretes large amounts of progesterone and estrogen.

D. Placenta

The human placenta synthesizes large amounts of progesterone from cholesterol (Casey *et al.*, 1992). This requires the same cholesterol desmolase enzyme as in other steroidogenic tissues but a unique isozyme of 3β-hydroxysteroid dehydrogenase (Lachance *et al.*, 1990; Lorence *et al.*, 1990). Estrogens are also synthesized in large amounts. Most of the necessary androgen precursors are synthesized by the fetal adrenal gland as dehydroepiandrosterone sulfate, which is desulfated in the placenta by steroid sulfatase and converted to androstenedione by 3β-hydroxysteroid dehydrogenase. Aromatase converts this to estrone, which is converted to estradiol by the 17-ketosteroid reductase activity of 17β-hydroxysteroid dehydrogenase. Estriol, a 16α-hydroxylated metabolite of estradiol, is produced by the same pathway from 16α-hydroxylated androgen metabolites originating from the fetal liver. This metabolite is of significance as a way of monitoring fetal well-being; falling levels may signify fetal distress.

E. Metabolism in Target Tissues

In addition to steroid biosynthesis in classic endocrine tissues, several biosynthetic enzymes are expressed elsewhere in the body and synthesize biologically active hormones. In many cases, this sort of metabolism allows for local modulation of levels of active steroids synthesized by classic endocrine tissues. For example, aromatase is present in breast adipose tissue and presumably increases the local concentration of estrogens available to this highly estrogen-responsive tissue (Mahendroo et al., 1991; Miller, 1991; Thijssen et al., 1991). Depending on whether the hormone is synthesized in the same cells or merely the same tissue as the target, this type of steroid action is termed "autocrine" or "paracrine." Three examples that will be discussed in this review are 17β-hydroxysteroid dehydrogenase, 11β-hydroxysteroid dehydrogenase, and steroid 5α-reductase.

IV. Molecular Biology of Steroidogenic Enzymes

A. Cholesterol Desmolase

1. Biochemistry

Cholesterol desmolase (side-chain cleavage enzyme, P450scc, CYP11A) converts cholesterol to pregnenolone. This is accomplished through three oxidations: 22R- and 20R-hydroxylations and oxidative cleavage of the 20,22 carbon–carbon bond. These reactions require a total of three O_2 molecules and six electrons. Like other mitochondrial cytochrome P450 enzymes, CYP11A receives these electrons from NADPH through the accessory proteins, adrenodoxin reductase and adrenodoxin. The three oxidations are normally performed in succession without release of hydroxylated intermediates from the enzyme.

CYP11A is synthesized as a precursor protein with a MW of 60,000. As is the case with other mitochondrial proteins encoded by nuclear genes, an amino-terminal peptide is required for transport to the mitochondrial inner membrane. This peptide is removed in the mitochondria to yield the mature protein of 56,000.

2. Genetic Analysis

Comparison of cDNA and the actual amino-terminal protein sequence has confirmed that CYP11A is synthesized with a presequence of 39 amino acid residues. The mature protein has 482 amino acid residues (Chung et al., 1986).

The single human CYP11A gene is located on chromosome 15q23-24 (Sparkes et al., 1991) and consists of nine exons spread over approximately 20 kb (Morohashi et al., 1987).

3. Congenital Lipoid Adrenal Hyperplasia

A deficiency of cholesterol desmolase activity prevents synthesis of all steroids in the adrenals and gonads. This disorder is usually found in children of consanguineous marriages and is thus presumed to be inherited in an autosomal recessive manner. Unlike other forms of congenital adrenal hyperplasia, there are no elevated levels of precursor steroids circulating in blood. Instead, large numbers of cytoplasmic lipid droplets containing cholesteryl esters are observed in pathological specimens of adrenal cortex (thus the name of this disorder). Even though the gonads also normally synthesize large amounts of steroids, such droplets are usually not prominent in these glands (Degenhart, 1984).

Affected individuals are unable to synthesize cortisol or aldosterone. Aldosterone deficiency leads to an inability to conserve sodium in the renal distal tubule and collecting duct, resulting in hyponatremia, hypovolemia, and hyperkalemia. If untreated, patients typically die within the first few weeks of life from shock.

The inability to synthesize sex steroids disrupts sexual development. Males are born with female-appearing external genitalia but have inguinal testes. Although the testes are atrophic owing to the lack of androgens, they are able to secrete Muellerian inhibitory factor during embryogenesis and so an affected male does not have a cervix, uterus, or Fallopian tubes. Females appear normal at birth but remain sexually infantile (Muller et al., 1991).

Patients with this disorder should be treated with replacement of glucocorticoids and a mineralocorticoid. Sex hormone replacement should be instituted in adolescence if the development of secondary sexual characteristics is desired. In general, genetic males as well as females should be assigned a female sex role because the external genitalia are usually female in appearance. Inguinal testes in affected males should be removed because of the risk of tumor development in undescended testes.

Despite treatment, most patients reported with this disorder have died during childhood (Degenhart, 1984). However, a few patients have survived for decades (Hauffa et al., 1985).

Biochemical analysis has been carried out on affected adrenal glands from two patients. Tissue homogenates from one patient were unable to synthesize pregnenolone from cholesterol but were able to synthesize pregnenolone from 20-hydroxycholesterol, suggesting that 20-hydroxylation was defective in this patient (Degenhart *et al.*, 1972). Another patient had decreased levels of CYP11A protein (Degenhart, 1984).

Although these studies suggest that lipoid adrenal hyperplasia is caused by mutations in the *CYP11A* gene, genetic studies have failed to confirm this hypothesis in two patients in whom sequence analysis has been performed. In one case, the presence of normal CYP11A cDNA was documented in the testis (Lin *et al.*, 1991a). This suggests that defects in one or more other genes may have the same phenotype as (putative) cholesterol desmolase deficiency itself. These genes might encode factors that regulate activity of cholesterol desmolase.

Although mutations in *CYP11A* have not been found in humans, a rabbit model of congenital lipoid adrenal hyperplasia exists in which a deletion of all or most of *CYP11A* has been documented (Pang *et al.*, 1992; Yang *et al.*, 1993). Therefore, it is likely that at least some cases of this disorder in humans will be caused by *CYP11A* mutations.

B. 17α-HYDROXYLASE/17,20-LYASE

1. *Biochemistry*

Steroid 17α-hydroxylase (CYP17, P450c17) catalyzes conversion of pregnenolone to 17α-hydroxypregnenolone. In rats, this enzyme also converts progesterone to 17α-hydroxyprogesterone, but progesterone is not a good substrate for the human or bovine enzymes. CYP17 also catalyzes an oxidative cleavage of the 17,20 carbon–carbon bond, converting 17α-hydroxypregnenolone and 17α-hydroxyprogesterone to dehydroepiandrosterone and androstenedione, respectively. A pair of electrons and a molecule of O_2 are required for each hydroxylation or lyase reaction. CYP17 is a microsomal cytochrome P450 and accepts these electrons from NADPH-dependent cytochrome P450 reductase.

In the human or bovine adrenal cortex, 17α-hydroxylase activity is required for the synthesis of cortisol, but most rodents do not express CYP17 in the adrenal cortex and secrete corticosterone as their primary glucocorticoid. Both 17α-hydroxylase and 17,20-lyase activities are required for androgen and estrogen biosynthesis. As significant amounts of 17α-hydroxylated steroids are secreted by the human adrenal cortex but not by the gonads, it is apparent that the 17, 20-lyase activity of CYP17 is stronger in the gonads than in the adrenals, relative to 17α-hydroxylase activity. Tissue-specific variations in lyase activity may reflect the activity of the electron transport protein, cytochrome b5; this hemoprotein enhances lyase activity *in vitro* (Kominami *et al.*, 1992).

2. *Genetic Analysis*

Based on analysis of cDNA, CYP17 has 508 amino acid residues, predicting a protein of 57 kDa. It is most closely related (36% sequence identity) to steroid 21-hydroxylase (CYP21) (Bradshaw *et al.*, 1987; Chung *et al.*, 1987). When expressed in cultured cells, this enzyme has both 17α-hydroxylase and 17, 20-lyase activity, confirming earlier biochemical studies (Zuber *et al.*, 1986).

In humans, a single copy of the *CYP17* gene is located on chromosome 10q24.3 (Sparkes *et al.*, 1991; Fan *et al.*, 1992). It consists of eight exons spread over 6.7 kb. When the coding sequences of *CYP17* and *CYP21* are aligned based on the predicted amino acid sequence, the last seven exons of the two genes are located in corresponding positions, indicating a relatively close evolutionary relationship (Picado-Leonard and Miller, 1987).

3. *Congenital Adrenal Hyperplasia due to 17α-Hydroxylase Deficiency*

Steroid 17α-hydroxylase deficiency is a relatively rare cause of congenital adrenal hyperplasia, accounting for about 1% of cases. As with other enzymatic defects causing congenital adrenal hyperplasia, 17α-hydroxylase deficiency is inherited as an autosomal recessive disorder. It is characterized biochemically by poor synthesis of cortisol and elevated levels of 17-deoxy steroids (Yanase *et al.*, 1991; Peter *et al.*, 1993). One of these steroids is deoxycorticosterone, which, along with its 18-hydroxy and 19-nor metabolites, acts as a mineralocorticoid. Thus, patients with 17α-hydroxylase deficiency often have hypertension and other signs of mineralocorticoid excess, in particular hypokalemia, which is usually mild (Fraser *et al.*, 1987; Biglieri and Kater, 1991a,b). Aldosterone secretion is usually suppressed in untreated patients owing to the oversecretion of other mineralocorticoid agonists, but some patients with paradoxical hypersecretion of aldosterone have been reported (Yamakita *et al.*, 1989; Cottrell *et al.*, 1990; Shima *et al.*, 1991). Hypertension is also seen in patients with 11β-hydroxylase deficiency, which is a more frequent cause of congenital adrenal hyperplasia. Whereas patients with 11β-hydroxylase deficiency have signs of androgen excess, patients with severe (classic) 17α-hydroxylase deficiency are unable to synthesize sex steroids, a problem closely resembling that seen in lipoid adrenal hyperplasia. Males are born with female-appearing external genitalia indistinguishable in appearance from congenital lipoid adrenal hyperplasia or androgen insensitivity (testicular feminization), although the latter condition is readily distinguished by elevated levels of androgens. Females appear normal at birth but remain sexually infantile. The ovaries have poor follicular development and in rare cases appear as streak gonads (Araki et al., 1987; Malcolm et al., 1992).

Some patients with mild 17α-hydroxylase deficiency are able to synthesize at least some sex steroids so that males have partially virilized (ambiguous) genitalia and females develop at least some signs of puberty (de Lange and Doorenbos, 1990). Some (mostly female) patients have been reported to have isolated 17α-hydroxylase deficiency with normal 17, 20-lyase activity.

Conversely, a few patients have been thought to have isolated 17, 20-lyase deficiency (Yanase et al., 1992). In at least one case, this was associated with cytochrome b5 deficiency in a patient with methemoglobinemia (Hegesh et al., 1986).

4. Genetic Basis of 17α-Hydroxylase Deficiency

Congenital adrenal hyperplasia due to deficiency of 17α-hydroxylase and 17,20-lyase activities is caused by mutations in CYP17 (Table I) (Kagimoto et al., 1988; Yanase et al., 1988, 1989, 1990; Winter et al., 1989; Lin et al., 1991b) as reviewed by Yanase et al., (1991) and Winter (1991). Depending on whether consanguinity is present, each kindred has had one or two unique mutations. The only exception has been in several Canadian and Dutch patients who were all found to be homozygous for an insertion of four nucleotides in codon 480. Subsequent investigation revealed that the patients were members of the same small Mennonite (Anabaptist) sect, members of which originated in the Netherlands, suggesting that the presence of this mutation reflected a founder effect (M. Kagimoto et al., 1988; K. Kagimoto et al., 1989; Imai et al., 1992).

Complete deficiency is associated with frameshifts or nonsense mutations (Kagimoto et al., 1988; Yanase et al., 1988, 1990; Winter et al., 1989). In one kindred, several exons were replaced by a segment of the E. coli lac operon; the mechanism by which this rearrangement occurred is not known but it may have involved integration into the genome by a virus (Biason et al., 1991).

Partial deficiency is associated with missense mutations in CYP17 or, in one case, deletion of a single codon (ΔF53 or 54) maintaining the reading frame of translation. These mutations have been intro-

TABLE I

Name	E/I	Activity (% nl)	Effect	Clinical severity	Ethnic group (number[a])/ other mutation	References
Steroid 17 α-hydroxylase/17,20-lyase deficiency (CYP17, chromosome 10q24–25)						
Insertion/deletion	E2–E3	0	No enzyme		Italian	Biason et al. (1991)
W17X	E1	0	Nonsense	+++	Japanese	Yanase et al. (1988)
ΔF53	E1	0		++	Japanese	Yanase et al. (1989)
S106P	E2	0		+++	Japanese	Lin et al. (1991b)
H120 + 7 nt	E2	0	Frameshift	+++	Japanese	Yanase et al. (1990)
R239X	E4	0	Nonsense		American/P342T	Ahlgren et al. (1992)
P342T	E6	40		++	American/R239X	Ahlgren et al. (1992)
Q461X	E8	0	Nonsense		Swiss/R496C	Yanase et al. (1992)
I480 + 4 nt	E8	0	Frameshift	+++	Mennonite (>3)	Kagimoto et al. (1988); Imai et al. (1992)
R496C	E8	10		Lyase	Swiss/Q461X	Yanase et al. (1992)
3β-Hydroxysteroid dehydrogenase deficiency (HSD3B2, chromosome 1p13)						
E142K	E4	0		+++	American/W171X	Simard et al. (1993)
W171X	E4	0	Nonsense	+++	Swiss (2)	Rheaume et al. (1992)
					American (2)[b]	
P186+c	E4	0	Frameshift	+++	American/W171X Dutch/Y256N	Rheaume et al. (1992)
A245P	E4	10	Unstable enzyme	++	Turkish	Simard et al. (1993)
V248 ΔS5nt	E4	0	Frameshift	+++	Mexican	Morishima et al. (1987)
Y253N	E4	0		+++	Dutch/P186+c	Simard et al. (1993)

Steroid 21-hydroxylase deficiency (CYP21, chromosome 6p21.3)

					SW[c]	SV[c]	NC[c]	
Deletion		0	No enzyme	+++	20	9	11	White et al. (1984, 1988)
Conversion		0	No enzyme	+++	8	9	3	Donohoue et al. (1986); Higashi et al. (1988a)
+L10	E1	100	Normal polymorphism	N	0	5	11	Tusie-Luna et al. (1991)
P30L	E1	30–60	? Orientation in ER	+				Higashi et al. (1988b)
"i2g"	I2	0–5	Abnormal splicing	++/+++	22	25	12	
K102R	E3	100	Normal polymorphism	N				
P105L	E3	?			Swedish			Wedell et al. (1992)
Δ8 nt	E3	0	Frameshift	+++	8	4	2	Amor et al. (1988)
I172N	E4	1	? Insertion in ER	++	3	28	5	Amor et al. (1988)
I235N	E6	0	? Substrate binding	+++	5	11	9	Higashi et al. (1988b)
V236E								
M238K								
S268T	E7	100	Normal polymorphism	N				Donohoue et al. (1990); Tusie-Luna et al. (1991)
V281L	E7	20–50	? Insertion in ER / ? Heme binding	+	1	7	34	Speiser et al. (1988)
G291S	E7	?			Swedish			Wedell et al. (1992)
F306+t	E7	0	Frameshift	+++	1	2	0	Wedell and Luthman (1993a)
i7c	I7	?0	Abnormal splicing	+++	Swedish			
Q318X	E8	0	Nonsense	+++	5	8	4	Globerman et al. (1988a)
R339H	E8	30–70		+	Turkish			Helmberg et al. (1992b)
R356W	E8	0	? Electron transfer	+++	11	9	4	Chiou et al. (1990)
W406X	E9	0	Nonsense	+++	Swedish			Wedell and Luthman (1993a)

(continued)

TABLE I (Continued)

Name	E/I	Activity (% nl)	Effect	Clinical severity	Ethnic group (number[a])/ other mutation	References
P453S	E10	30–70		+		Owerbach et al. (1992b); Helmberg et al. (1992b)
R483P	E10	?			Swedish	Wedell and Luthman (1993b)
R483 Δ1 nt	E10	0	Frameshift	+++	Swedish	Wedell et al. (1992)
Steroid 11β-hydroxylase deficiency (CYP11B1, chromosome 8q22)						
32Δc	E1	0	Frameshift	+++	American[d]	Curnow et al. (1993)
K174X	E3	0	Nonsense	+++	American/R384Q	Curnow et al. (1993)
T318M	E5	0	Proton transfer	+++	Yemenite	Curnow et al. (1993)
Q338X	E6	0	Nonsense	+++	Sikh	Curnow et al. (1993)
Q356X	E6	0	Nonsense	+++	Afro-American	Curnow et al. (1993)
R374Q	E6	0		+++	Lebanese	Curnow et al. (1993)
R384Q	E7	0		+++	American/K174X	Curnow et al. (1993)
N394+2 nt	E7	0	Frameshift	+++	Turkish	Helmberg et al. (1992a)
V441G	E8	0	? 2° structure	+++	American	Curnow et al. (1993)
R448H	E8	0	Heme binding	+++	Moroccan Jews (7)	White et al. (1991)
Aldosterone synthase deficiency (CYP11B2, chromosome 8q22)						
V35 Δ5 nt	E1	0	Frameshift	+++	Amish	Mitsuuchi et al. (1993)
R181W	E3, E7	0–100[f]		0/++	Iranian Jews (7)	Pascoe et al. (1992a)
V386A[e]						

Steroid 5α-reductase deficiency (SRD5A2, chromosome 2p23)

Mutation	Exon/Intron	Activity	Effect		Population	Reference
Deletion		0	No enzyme	+++	New Guinean	Andersson et al. (1991)
G34R	E1	2	Alters K_m for steroid, pH optimum	+++	Vietnamese; Mexican/G115D; Sicilian/R171S	Thigpen et al. (1992b)
L55Q	E1	Very low[g]			Jordanian/Q56R	Thigpen et al. (1992b)
Q56R	E1	Very low[g]			Jordanian/L55Q	Thigpen et al. (1992b)
L110 Δ2 nt	E2	0	Frameshift		Maltese/R171S	Thigpen et al. (1992b)
G115D	E2	Very low[g]			Mexican/G34R	Thigpen et al. (1992b)
Q126R	E2	Very low[g]			Brazilian/D164V	Thigpen et al. (1992b)
D164V	E3	Unstable[g]			Brazilian/Q126R	Thigpen et al. (1992b)
R171S	E3				Sicilian/G34R; Maltese/Δ2 nt	Thigpen et al. (1992b)
G183S	E3	8			Afro-Brazilian	Thigpen et al. (1992b)
G196S	E4	Low[g]	Alters NADPH K_m	+	American	Thigpen et al. (1992b)
E197D	E4	Unstable[g]			American[d]	Thigpen et al. (1992b)
A207D	E4	Very low[g]			Austrian/R246Q	Thigpen et al. (1992b)
L224P	E4	0			Native American	Thigpen et al. (1992b)
R227X	E4	Very low[g]	Nonsense		Mexican Brazilian	Thigpen et al. (1992b)
H231R	E4	Very low[g]			Afro-American[d]	Thigpen et al. (1992b)
I4,g→t	I4	Very low[g]	Destroys splice donor		Pakistani	Thigpen et al. (1992b)
R246W	E5	Very low[g]	Alters NADPH K_m		Dominican Brazilian	Thigpen et al. (1992a)
R246Q	E5	Unstable[g]	Unstable enzyme		Pakistani (2); Austrian/A207D; Afro-American[d]	Thigpen et al. (1992b)

[a] Single families unless otherwise noted

[b] One compound heterozygote each with P186+c and E142K.

[c] SW, SV, and NC denote the percentage of alleles carrying each mutation seen in patients with the salt-wasting, simple virilizing, and nonclassic forms of 21-hydroxylase deficiency, respectively.

[d] Heterozygote; second mutation not known.

[e] Homozygosity for both mutations is necessary to cause the disease.

[f] Intact 11β-hydroxylase activity, no 18-oxidase activity.

[g] In affected fibroblasts; recombinant enzyme not tested.

duced into cDNA and expressed in cultured cells. Certain mutations affect hydroxylase and lyase activities differently; for example, ΔF53 affects lyase activity more than hydroxylase activity (Yanase *et al.*, 1989), whereas P342T decreases the two activities to the same extent.

These *in vitro* activities may be correlated with clinical signs of the disorder in affected individuals. A female patient homozygous for ΔF53 had irregular menses, indicating some degree of cyclic estrogen biosynthesis. A male patient with ambiguous genitalia was a compound heterozygote for P342T and a nonsense mutation. Because P342T decreases *in vitro* 17α-hydroxylase and 17, 20-lyase activities to about 40% of normal and the nonsense mutation completely prevents synthesis of the enzyme, this suggests that the patient may have had roughly 20% of normal enzymatic activity. If true, it implies that greater than 20% of normal activity is required to synthesize sufficient androgens to permit normal male sexual development (Ahlgren *et al.*, 1992). On the other hand, 50% of normal activity must be sufficient for normal development because obligate heterozygous males are asymptomatic.

This reasoning is based on several simplifications, such as assuming that synthesis of the partially active mutant enzyme is not subject to a compensatory increase in the adrenals and testes due to stimulation by corticotropin and luteinizing hormone, respectively. Conversely, it is also assumed that the enzyme is not subjected to pseudosubstrate inhibition by increased levels of precursor steroids. These simplifications notwithstanding, it is apparent that even relatively mild impairment of this enzyme is associated with inadequate male sexual development.

In addition to allelism (Yanase *et al.*, 1989; Ahlgren *et al.*, 1992), other factors must influence the relative degree of compromise of hydroxylase and lyase activities in this disorder, considering that all mutations affect both activities to some extent (Yanase *et al.*, 1992). In support of this idea, patients have been reported who changed with age from having apparently "pure" lyase deficiency to having combined deficiency (Zachmann *et al.*, 1992). Lyase activity may be influenced by activities of cytochrome b5 (Shinzawa *et al.*, 1985; Kominami *et al.*, 1992; Sakai *et al.*, 1993) and cytochrome P450 reductase (Hall, 1991; Lin *et al.*, 1993); variations in expression of these proteins might affect lyase activity *in vivo*.

C. 3β-HYDROXYSTEROID DEHYDROGENASE

1. *Biochemistry*

Conversion of   5-3β-hydroxysteroids (pregnenolone, 17-hydroxypregnenolone, dehydroepiandrosterone) to Δ4-3-ketosteroids (progesterone, 17-hydroxyprogesterone, androstenedione) is mediated by 3β-hydroxysteroid dehydrogenase. These conversions are mediated by several isozymes in the endoplasmic reticulum. These enzymes utilize NAD$^+$ as an electron acceptor. In addition to 3β-hydroxysteroid dehydrogenase activity, these enzymes mediate a Δ5- Δ4-ene isomerase reaction that transfers a double bond from the B ring to the A ring of the steroid so that it is in conjugation with the 3-keto group. These enzymes are also able to mediate the reverse, 3-ketosteroid reductase reaction using NADH as an electron donor. The only 3β-hydroxysteroid dehydrogenase enzyme to actually be purified is the isozyme present in human placenta (Thomas *et al.*, 1988; Luu-The *et al.*, 1989). The protein has a MW of 42,000. Antibodies to this enzyme were used to isolate the corresponding cDNA, following which cDNA clones encoding other isozymes were isolated and characterized.

Thus far, two isozymes have been identified in humans by cloning of cDNA. The human type I enzyme is expressed in the placenta, skin, and adipose tissue (Luu-The *et al.*, 1989; Lorence *et al.*, 1990; Labrie *et al.*, 1991), whereas the type II enzyme is expressed in the adrenal gland and the gonads (Lachance *et al.*, 1991; Rheaume *et al.*, 1991). These isozymes contain 372 and 371 amino acids, respectively, and are 94% identical in their predicted sequence. These and all other 3β-hydroxysteroid dehydrogenase isozymes contain nucleotide cofactor binding domains and conserved tyrosine and lysine residues that are characteristic of short-chain dehydrogenases. However, they are less than 15% identical in amino acid sequence to other enzymes of this type, suggesting a very distant evolutionary relationship.

The two human recombinant enzymes differ in their kinetic properties when expressed in cultured mammalian cells; the type I isozyme has K_m values for steroids of approximately 0.2 μM for both dehydrogenase (using pregnenolone or dehydroepiandrosterone as substrates) and reductase (dihydrotestosterone as substrate) reactions, whereas the type II enzyme has 10-fold higher K_m values. These isozymes have similar apparent V_{max} values so that the first-order rate constants (V_{max}/K_m) of the type I enzyme are three to six times higher than those of the type II enzyme. The higher K_m of the type II enzyme may reflect its expression in the adrenal glands and gonads, where the levels of steroid precursors are much higher than elsewhere in the body.

2. Genetic Analysis

Genomic blot analysis suggests that there are as many as six *HSDB3* genes in humans, all of which are located on chromosome 1p11-13 (Berube *et al.*, 1989). At this time, only two of these genes have been studied in detail, and thus it is not known whether any of the others are active genes. Each of the characterized genes contains four exons spanning 8 kb; the first exon is entirely untranslated. *HSD3B1* encodes the type I isozyme expressed in placenta and skin (Lachance *et al.*, 1990), whereas *HSD3B2* encodes the type II isozyme expressed in the adrenals and gonads (Lachance *et al.*, 1991).

3. 3β-Hydroxysteroid Dehydrogenase Deficiency

Deficiency of 3β-hydroxysteroid dehydrogenase interferes with synthesis of mineralocorticoids, glucocorticoids, and sex steroids. Thus, severely affected patients are liable to succumb in the neonatal period to shock from lack of glucocorticoids and from hyponatremia, hypovolemia, and hyperkalemia caused by a lack of mineralocorticoids. Because testosterone cannot be synthesized, affected males are born with ambiguous genitalia (Mendonca *et al.*, 1987; Heinrich *et al.*, 1993). However, affected individuals secrete large amounts of dehydroepiandrosterone, which is a weak androgen. Thus, affected females are also sometimes born with ambiguous genitalia. As is the case with other forms of congenital adrenal hyperplasia, the internal genital structures correspond to the chromosomal sex.

Although classic 3β-hydroxysteroid dehydrogenase deficiency is rare, comprising fewer than 1% of cases of congenital adrenal hyperplasia, a mild or nonclassic form of the disease has been postulated to be a frequent cause of hirsutism and/or menstrual irregularities (particularly polycystic ovary syndrome) in women. However, reliable diagnostic criteria for the nonclassic disorder are difficult to establish, consisting mainly of altered ratios of Δ5 to Δ4 steroids (Mathieson *et al.*, 1992; Schram *et al.*, 1992; Azziz *et al.*, 1993; Barnes *et al.*, 1993). At this time, the proportion of patients with the putative nonclassic disorder that actually have mutations in the *HSD3B2* gene is not known.

Patients with the classic form of the disease have mutations in the *HSD3B2* gene that abolish activity (Table I) (Rheaume *et al.*, 1992; Simard *et al.*, 1993). As this gene is expressed in the adrenals and gonads, these mutations account fully for the pathogenesis of the disease.

Thus far, patients in six ostensibly unrelated kindreds have been analyzed; patients had impaired aldosterone synthesis in all but one kindred. Four chromosomes in three kindreds carried the same nonsense mutation, W171X, and two chromosomes in separate kindreds carried the same frameshift mutation, P186+c. These mutations are both expected to completely prevent synthesis of an intact enzyme.

The reason for the presence of these mutations in multiple kindreds is not immediately apparent. It is unlikely that it reflects founder effects due to unsuspected common ancestry (the kindreds are widely separated geographically) or the presence of mutational hotspots (because neither of the two recurrent *HSD3B2* mutations is a CpG→TpG, which is the most characteristic type of recurrent point mutation). Humans carry several cross-hybridizing genes in addition to *HSD3B1* and *HSD3B2* that have not yet been analyzed. It is likely that at least some of these are pseudogenes. Perhaps W171X and P186+c are normally present in one of these putative pseudogenes and are transferred to *HSD3B2* in gene conversion events. Such a mechanism accounts for the vast majority of mutations causing the most common form of congenital adrenal hyperplasia, 21-hydroxylase deficiency (discussed in the next section).

One patient was homozygous for a complex frameshift mutation in which codons 248–249 were changed from GTCCGA to AACTA. Three missense mutations were detected; two, E142K and Y253N, were found in patients with the salt-wasting form of the disease and completely destroyed enzymatic activity when mutant cDNA was expressed in cultured cells (Simard *et al.*, 1993).

One male with relatively unimpaired aldosterone synthesis was homozygous for A245P. When expressed in intact cells, cDNA carrying this mutation had about 10% of normal activity, but the mutant enzyme was unstable in cell lysates unless glycerol was added (Simard *et al.*, 1993).

The mechanisms by which these missense mutations affect enzymatic function are not known. E142 and Y253 are completely conserved in 3β-hydroxysteroid dehydrogenase enzymes from various species, but the functions of these residues are not understood. A245 is not completely conserved, but A245P may change the secondary structure of the enzyme.

Because the *HSD3B1* gene is intact, steroid biosynthesis in the placenta should be normal during gestation. Moreover, 3β-hydroxysteroid dehydrogenase activity in the skin and adipose tissue should remain intact postnatally, suggesting that complete 3β-hydroxysteroid dehydrogenase deficiency should occur extremely rarely if at all.

D. 21-Hydroxylase

1. *Biochemistry*

Steroid 21-hydroxylase (P450c21, CYP21) is a microsomal cytochrome P450 enzyme that converts 17-hydroxyprogesterone to 11-

deoxycortisol and progesterone to deoxycorticosterone. The human enzyme normally contains 494 amino acid residues and has a MW of about 52,000. Human recombinant CYP21 has K_m values for 17-hydroxyprogesterone and progesterone of 1.2 and 2.8 μM, respectively, and the apparent V_{max} for 17-hydroxyprogesterone is twice that of progesterone (Tusie-Luna et al., 1990).

2. Genetic Analysis

The structural gene encoding human CYP21 (CYP21, CYP21A2, or CYP21B) and a pseudogene (CYP21P, CYP21A1P, or CYP21A) are located in the HLA major histocompatibility complex on chromosome 6p21.3 about 30 kb apart, adjacent to and alternating with the C4B and C4A genes encoding the fourth component of serum complement (Carroll et al., 1985; White et al., 1985) (Fig. 4). Both the CYP21 and C4 genes are transcribed in the same direction. CYP21 overlaps a gene on the opposite DNA strand (OSG or XB) that encodes a putative extracellular matrix protein similar to tenascin (Bristow et al., 1993). CYP21P overlaps a truncated copy of this gene (XA) that does not encode a functional protein.

CYP21 and CYP21P each contain 10 exons spaced over 3.1 kb. Their nucleotide sequences are 98% identical in exons and about 96% identical in introns (Higashi et al., 1986; White et al., 1986).

3. Congenital Adrenal Hyperplasia Due to 21-Hydroxylase Deficiency: Clinical Features

Steroid 21-hydroxylase deficiency is by far the most frequent cause of congenital adrenal hyperplasia, accounting for more than 90% of cases of the severe, "classic" form of the disease. Classic 21-hydroxylase deficiency occurs in about one of 10,000 births. A mild, "nonclassic" form occurs in about 1% of the general population and in approximately 2–3% of Jews of Eastern European origin (Speiser et al., 1989; Zerah et al., 1990).

Cortisol deficiency leads to overproduction and accumulation of cortisol precursors, particularly 17-hydroxypregnenolone and 17-hydroxyprogesterone. These are metabolized by CYP17 (17α-hydroxylase/17, 20-lyase) to dihydroepiandrosterone and androstenedione, respectively, which are then converted to testosterone. Thus, signs of androgen excess are a prominent feature of this disorder. In classic 21-hydroxylase deficiency, these signs include ambiguous genitalia in females (due to exposure to excessive levels of androgens in utero), menstrual irregularity, male body habitus, and a male pattern of facial and body hair. Androgens also cause rapid postnatal somatic growth in both sexes as

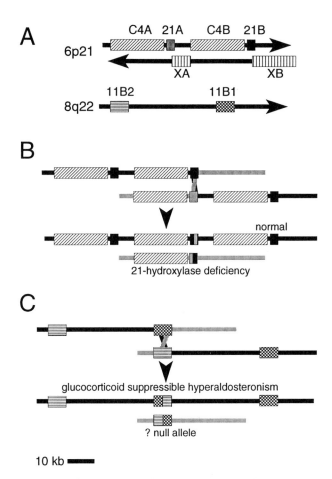

Fɪɢ. 4. Diagrams of the chromosomal regions containing the 21-hydroxylase and 11β-hydroxylase genes. (A) *CYP21P* (21A) and *CYP21* (21B) are adjacent to and alternate with the *C4A* and *C4B* genes on chromosome 6p21. The *XA* and *XB* genes are on the opposite chromosomal strand. The *CYP11B1* and *CYP11B2* are on chromosome 8q22. (B) Unequal crossing-over involving the *CYP21* genes produces two daughter chromosomes. One carries a duplication and contains a normal *CYP21* gene; it is thus a normal variant. The other carries a deletion and represents a 21-hydroxylase deficiency allele. (C) The duplication produced by unequal crossing-over involving the *CYP11B* genes generates a chimeric gene that is abnormally regulated, thus causing glucocorticoid-suppressible hyperaldosteronism. The deleted chromosome might be a null allele (i.e., causing either 11β-hydroxylase or aldosterone synthase deficiencies depending on the other allele) but no such chromosomes have as yet been detected.

well as accelerated skeletal maturation, leading to premature closure
of the epiphyses and short adult stature. Patients with nonclassic disease may be asymptomatic (nonclassic disease has been detected biochemically in screening programs and in close relatives of known patients) or they may have any of the signs of androgen excess seen in
classic patients with the exception of ambiguous genitalia in females.

About two-thirds of patients with the classic form of the disease
have a parallel defect in their ability to synthesize the mineralocorticoid hormone aldosterone. Such patients are said to have "salt-
wasting" disease; patients who are able to synthesize aldosterone normally have "simple virilizing" disease. If not promptly diagnosed and
treated, salt wasting may rapidly lead to shock and death (White *et al.*,
1987a,b; New *et al.*, 1989).

4. *Genetic Analysis of 21-Hydroxylase Deficiency*

The disorder is inherited as a monogenic autosomal recessive trait
located within the *HLA* major histocompatibility complex. There are a
number of associations between the different forms of 21-hydroxylase
deficiency and specific combinations of HLA antigens, or haplotypes.
In particular, the unusual haplotype A3;Bw47;DR7 is associated with
about 10% of classic (i.e., salt wasting or simple virilizing) 21-
hydroxylase deficiency alleles, whereas HLA-B14;DR1 is associated
with 70% of alleles for nonclassic disease (Dupont *et al.*, 1984; New *et
al.*, 1989). Based on these associations and on studies of pedigrees
containing patients with the different forms of the disease, it appears
that these forms are inherited as allelic variants.

All mutations causing 21-hydroxylase deficiency that have been described thus far are apparently the result of either of two types of
recombinations between *CYP21*, the normally active gene, and the
CYP21P pseudogene: unequal crossing-over during meiosis resulting
in a complete deletion of *C4B* and *CYP21* (White *et al.*, 1984), and gene
conversion events resulting in the transfer to *CYP21* of deleterious
mutations normally present in *CYP21P*.

These deleterious mutations include (Higashi *et al.*, 1986; White *et
al.*, 1986) an A→G substitution 13 nucleotides (nt) before the end of
intron 2 that results in aberrant splicing of pre-mRNA, an 8-nt deletion in exon 3, and a 1-nt insertion in exon 7, each of which shifts the
reading frame of translation, and a nonsense mutation in codon 318 of
exon 8. There are also eight missense mutations in *CYP21P*, seven of
which have been observed in patients with 21-hydroxylase deficiency.

Because particular mutations occur in many unrelated kindreds,
each mutation, and the degree of enzymatic compromise it causes, may

be correlated with the different clinical forms of 21-hydroxylase deficiency (i.e., salt wasting, simple virilizing, and nonclassic disease) (Table I) (Higashi *et al.*, 1991; Mornet *et al.*, 1991; Speiser *et al.*, 1992; Wedell *et al.*, 1992).

a. *Deletions and Large Gene Conversions.* Large deletions involving *C4B* and *CYP21* comprise 20–25% of alleles in patients with classic 21-hydroxylase deficiency (Mornet *et al.*, 1991; Speiser *et al.*, 1992). Many deleted alleles are associated with the HLA haplotype A3;Bw47;DR7 (White *et al.*, 1984). Deletions usually extend from somewhere between exons 3 and 8 of *CYP21P* through *C4B* to the corresponding point in *CYP21*, yielding a single remaining *CYP21* gene in which the 5' end corresponds to *CYP21P* and the 3' end to *CYP21* (White *et al.*, 1988). Deleterious mutations within the *CYP21P* portion render such a gene incapable of encoding an active enzyme. All patients who carry homozygous deletions suffer from the salt-wasting form of the disorder.

In most studies, these deletions have been detected by genomic blot hybridization as absence (or diminished intensity in heterozygotes) of gene-specific fragments produced by digestion with several restriction enzymes. Large gene conversions, in which multiple mutations are transferred from *CYP21P* to *CYP21*, are also detected by this approach. Large conversions account for 5–10% of alleles in classic 21-hydroxylase deficiency (Mornet *et al.*, 1991; Speiser *et al.*, 1992).

Deletions have also been directly documented by resolving very large restriction fragments using pulsed-field gradient electrophoresis (Collier *et al.*, 1989).

b. *Nonsense and Frameshift Mutations.* Two other mutations normally found in *CYP21P* completely prevent synthesis of an intact enzyme and cause salt-wasting 21-hydroxylase deficiency if they occur in *CYP21*: the nonsense mutation in codon 318 (Q318X) (Globerman *et al.*, 1988a) and the 8-nt deletion in exon 3 (Higashi *et al.*, 1988b). The 1-nt insertion in exon 7 of *CYP21P* has not yet been identified as an independent mutation in any patient with 21-hydroxylase deficiency.

c. *A or C→G Mutation in Intron 2.* The nucleotide 13 bp before the end of intron 2 is A or C in normal individuals. Mutation to G constitutes the single most frequent allele causing classic 21-hydroxylase deficiency (Higashi *et al.*, 1991; Mornet *et al.*, 1991; Speiser *et al.*, 1992; Owerbach *et al.*, 1992a).

This mutation causes aberrant splicing of intron 2 with retention of 19 nucleotides normally spliced out of mRNA, resulting in a shift in the translational reading frame (Higashi *et al.*, 1988a, 1991). Almost all the mRNA is aberrantly spliced, but in cultured cells a small amount of normally spliced mRNA is detected. If no other mutations were present, a small amount of normal enzyme might thus be synthesized.

Although it is not known what proportion of mRNA is normally spliced in the adrenal glands of patients with this mutation, most (but not all) patients who are homozygous or hemizygous for this mutation have the salt-wasting form of the disorder, indicating that they have insufficient enzymatic activity to permit adequate aldosterone synthesis (Higashi *et al.*, 1991; Mornet *et al.*, 1991; Owerbach *et al.*, 1992a; Speiser *et al.*, 1992).

d. *Arg-356→Trp (R356W).* This mutation abolishes enzymatic activity (Chiou *et al.*, 1990; Higashi *et al.*, 1991). It is located in a region of the gene encoding the K helix of the enzyme; this location suggests that the mutation affects interactions with the cytochrome P450 reductase, but this has not yet been demonstrated.

e. *Ile-Val-Glu-Met-235-238→Asn-Glu-Glu-Lys(I235N/V236E/M238K).* This cluster of three missense mutations also abolishes enzymatic activity (Tusie-Luna *et al.*, 1990; Higashi *et al.*, 1991). Interference with substrate binding has been suggested (based on sequence conservation with cholesterol side-chain cleavage enzyme, another cytochrome P450) (Higashi *et al.*, 1988b) but not experimentally demonstrated.

f. *Ile-172→Asn (I172N).* This mutation, the only one specifically associated with the simple virilizing form of the disease, results in an enzyme with about 1% of normal activity as measured by the first-order rate constant, V_{max}/K_m. The mutant enzyme is apparently not properly localized in microsomes (Tusie-Luna *et al.*, 1990). The mutated amino acid is located in a region of the protein that may normally interact with the membrane of the endoplasmic reticulum, and the isoleucine residue normally at this position is strongly conserved in many different P450 enzymes (Amor *et al.*, 1988). Mutation of this hydrophobic residue to a polar residue might disrupt such an interaction, weakening the association of the enzyme with the endoplasmic reticulum.

Because aldosterone is normally secreted at a rate 100–1000 times lower than that of cortisol, it is apparent that 21-hydroxylase activity would have to decrease to very low levels before it became rate-limiting for aldosterone synthesis. Apparently, as little as 1% of normal activity allows adequate aldosterone synthesis to prevent significant salt wasting in most cases, although one patient with I172N has been reported to have an elevated ratio of plasma renin to aldosterone, consistent with mild salt wasting (Amor *et al.*, 1988).

g. Val-281→Leu (V281L). V281L occurs in all or nearly all patients with nonclassic 21-hydroxylase deficiency who carry the HLA haplotype B14;DR1, an association that is presumably due to a founder effect (Speiser *et al.,* 1988). In certain populations (such as Jews of Eastern European origin) this is a very common genetic polymorphism with a gene frequency of more than 10% (Speiser *et al.,* 1985). This mutation results in an enzyme with 50% of normal activity when 17-hydroxyprogesterone is the substrate but only 20% of normal activity for progesterone (Tusie-Luna *et al.,* 1990; Wu and Chung, 1991). The mutant enzyme is not normally localized in the endoplasmic reticulum (Tusie-Luna *et al.,* 1990).

h. Pro-30→Leu (P30L). This mutation, which is found in approximately one-sixth of alleles in patients with nonclassic disease, yields an enzyme with properties very similar to that of the V281L enzyme when expressed in cultured cells (Tusie-Luna *et al.,* 1991). However, enzymatic activity is rapidly lost when the cells are lysed, suggesting that the enzyme is relatively unstable. Preliminary evidence suggests that patients carrying this mutation tend to have more severe signs of androgen excess than patients carrying V281L.

As is the case with other microsomal P450 enzymes, CYP21 is anchored to the membrane of the endoplasmic reticulum mainly by a hydrophobic "tail" at the amino terminus. Most P450 enzymes have one or more proline residues separating this tail from the remainder of the polypeptide, which are predicted to create a turn in the polypeptide chain. P30L may abolish this turn, leading to an enzyme that is not oriented properly with respect to the membrane of the endoplasmic reticulum.

i. Other Mutations. A few mutations that are apparently not gene conversions (i.e., these mutations have not been observed in the *CYP21P* genes sequenced thus far) have been reported. The most frequent of these is P453S, which occurs in a number of different populations. This suggests that *CYP21P* may carry P453S as a frequent polymorphism and that this mutation is transferred to *CYP21* in the same way as the other mutations frequently causing 21-hydroxylase deficiency (Helmberg *et al.,* 1992b; Owerbach *et al.,* 1992b; Wedell *et al.,* 1992). Other mutations include P105L, G291S, R339H, W406X, R483P, and a frameshift at R483 (Helmberg *et al.,* 1992b; Wedell *et al.,* 1992; Wedell and Luthman, 1993a,b). Each of these has been reported on only one chromosome. P105L, G291S, and R483P have not as yet been documented to affect enzymatic activity and thus have not been confirmed as actual 21-hydroxylase deficiency mutations.

5. *Prenatal Diagnosis and Neonatal Screening*

Prenatal diagnosis of 21-hydroxylase deficiency is important because an affected female fetus can be treated by administering dexamethasone to the mother, which suppresses excessive secretion of androgens by the fetal adrenal cortex. This may ameliorate masculinization of the external genitalia, often the most distressing sign of this disease (Karaviti *et al.*, 1992; Dorr and Sippell, 1993; Forest *et al.*, 1993; Speiser *et al.*, 1993). Because a limited number of mutations account for almost all 21-hydroxylase deficiency alleles, prenatal diagnosis can be accomplished by direct genotyping of chorionic villus biopsy samples.

Because a single gene conversion may transfer several mutations into a single *CYP21* gene, it is critical to genotype both parents to establish segregation of all mutations and ensure that both affected chromosomes can be identified. Moreover, several percent of 21-hydroxylase deficiency mutations in patients are *de novo* deletions or gene conversions (Hejtmancik *et al.*, 1992; Speiser *et al.*, 1992; Collier *et al.*, 1993; Tajima *et al.*, 1993).

The prevalence of classic 21-hydroxylase deficiency is similar to that of phenylketonuria, an inborn error of metabolism for which neonatal screening is mandated so that appropriate therapy can be instituted. Neonatal screening for 21-hydroxylase deficiency has been advocated to minimize the risk of infant death from undiagnosed salt-wasting crises (Pang *et al.*, 1985; Valentino *et al.*, 1990; Listernick *et al.*, 1992; Allen, 1993). It is not certain if this is a cost-effective strategy except in limited populations, such as Yupik Eskimos in which the disease has a high prevalence (Pang *et al.*, 1985). Neonatal screening is currently accomplished by measuring 17-hydroxyprogesterone levels in dried blood samples on filter paper, but molecular methods may eventually become superior in accuracy and cost when combined with screening for other inherited disorders.

E. 11β-HYDROXYLASE AND ALDOSTERONE SYNTHASE

1. *Biochemistry*

Humans have distinct 11β-hydroxylase isozymes, CYP11B1 (also called P450XIB1, P450c11, 11β-hydroxylase) and CYP11B2 (P450XIB2, P450c18, P450cmo, P450aldo, or aldosterone synthase), that are responsible for cortisol and aldosterone biosynthesis, respectively. Both CYP11B1 and CYP11B2 11β-hydroxylate 11-deoxycorticosterone to corticosterone and 11-deoxycortisol to cortisol. CYP11B2 also 18-hydroxylates and then 18-oxidizes corticosterone to aldosterone and cortisol to 18-oxocortisol. In contrast, the CYP11B1 isozyme does not synthesize detectable amounts of aldosterone from corticosterone or 18-hydroxycorticosterone (Kawamoto *et al.*, 1990a; Curnow *et al.*, 1991; Ogishima *et al.*, 1991).

These isozymes are mitochondrial cytochromes P450. The human isozymes are synthesized with 503 amino acid residues, but a signal peptide is cleaved in mitochondria to yield the mature protein of 479 residues. The sequences of the proteins are 93% identical (Mornet *et al.*, 1989). Despite their identical sizes, CYP11B1 and CYP11B2 have apparent molecular weights of 51,000 and 49,000, respectively, as determined by SDS–polyacrylamide gel electrophoresis (Curnow *et al.*, 1991; Ogishima *et al.*, 1991). A similar phenomenon is observed in the rat (Lauber and Muller, 1989); it is not known if this reflects consistent differences in posttranslational modifications or secondary structure.

2. *Genetic Analysis*

CYP11B1 and CYP11B2 are encoded by two genes (Mornet *et al.*, 1989) on chromosome 8q21-q22 (Chua *et al.*, 1987). Each consists of nine exons spread over approximately 7 kb. The location of introns in each gene is identical to that seen in the *CYP11A* gene encoding cholesterol desmolase. The predicted protein sequences of the CYP11B isozymes are each about 36% identical to that of CYP11A.

CYP11B1 is expressed at high levels in normal adrenal glands (Mornet *et al.*, 1989). *CYP11B2* is normally expressed at levels too low to be detected by hybridization of Northern blots but transcripts can be detected using a more sensitive assay wherein RNA is reverse-transcribed and then amplified with the polymerase chain reaction (RT–PCR) (Curnow *et al.*, 1991). Transcription of *CYP11B2* is dramatically increased in aldosterone-secreting tumors (Kawamoto *et al.*, 1990a; Curnow *et al.*, 1991).

CYP11B2 is located to the left of *CYP11B1* if the genes are pictured as being transcribed from left to right; the genes are located 30–40 kb apart (Lifton *et al.*, 1992b; Pascoe *et al.*, 1992b). Transcription of *CYP11B1* is appropriately regulated by cAMP (the second messenger for ACTH) (Kawamoto *et al.*, 1990b). In primary cultures of human zona glomerulosa cells, angiotensin II markedly increases levels of both *CYP11B1* and *CYP11B2* transcripts. ACTH increases *CYP11B1* mRNA levels more effectively than angiotensin II does, but it has no effect on *CYP11B2* transcription (Curnow *et al.*, 1991).

3. Congenital Adrenal Hyperplasia Due to
11β-Hydroxylase Deficiency

Steroid 11β-hydroxylase deficiency comprises 5–8% of cases of congenital adrenal hyperplasia, occurring in about one of 100,000 births in the general Caucasian population (Zachmann *et al.*, 1983).

Patients with this disorder are unable to convert 11-deoxycortisol to cortisol. As occurs in 21-hydroxylase deficiency, elevated levels of ACTH cause steroid precursors of cortisol to accumulate. These are shunted into the pathway for androgen biosynthesis, leading to signs of androgen excess.

A parallel defect usually exists in the synthesis of 17-deoxy steroids, so that deoxycorticosterone is not converted to corticosterone and instead accumulates. Because deoxycorticosterone and some of its metabolites have mineralocorticoid activity, elevated levels may cause hypertension and hypokalemia. About two-thirds of untreated patients become hypertensive, sometimes early in life (Hague and Honour, 1983; Rösler *et al.*, 1992). This clinical feature distinguishes 11β-hydroxylase deficiency from 21-hydroxylase deficiency, in which poor aldosterone synthesis causes renal salt wasting in the majority of patients.

A relatively high frequency (1/5000 to 1/7000) of 11β-hydroxylase deficiency has been reported in Israel among Jewish immigrants from Morocco, a relatively inbred population. Almost all affected alleles in this group carry the same mutation of CYP11B1, R448H (White *et al.*, 1991). This residue is located in the "heme binding peptide" (C450 is the ligand to the heme iron) and is completely conserved in all eukaryotic P450 enzymes examined thus far (see references in Nelson *et al.*, 1993).

Although R448H abolishes normal enzymatic activity (Curnow *et al.*, 1993), patients who are homozygous for this mutation differ in the severity of signs and symptoms of androgen and mineralocorticoid excess. Thus, other epigenetic or nongenetic factors probably influence the clinical phenotype of the disorder.

Other sporadic mutations of CYP11B1 causing this disease have been identified (Helmberg *et al.*, 1992a; Curnow *et al.*, 1993). Mutations that prevent synthesis of the enzyme include nonsense mutations W116X, K174X, Q338X, and Q356X and frameshift mutations in codons 32 (1-nt deletion) and 394 (2-nt insertion).

In addition to R448H, other missense mutations identified thus far are all in regions of known functional importance and abolish enzymatic activity (Curnow *et al.*, 1993). T318M modifies the absolutely conserved threonine in the I helix that is thought to be critical for proton transfer to bound molecular oxygen. R374Q also mutates a completely conserved residue located in the K helix; this mutation may affect binding of adrenodoxin. R384Q is in a region that corresponds to a β strand in CYP102 that forms part of the substrate binding pocket. Almost all P450s have a basic residue (H or R) at this or the immediately adjacent position. Finally, V441G is adjacent to the highly conserved heme binding region, and this mutation may change the secondary structure of the protein.

4. *Aldosterone Synthase (Corticosterone Methyloxidase) Deficiency*

Corticosterone methyloxidase (CMO) deficiency is an autosomal recessive inherited defect of aldosterone biosynthesis (Rösler, 1984). Patients with this disorder are subject to potentially fatal electrolyte abnormalities as neonates and a variable degree of hyponatremia and hyperkalemia combined with poor growth in childhood, but they may have no symptoms as adults. Two types of this disorder have been described; patients with CMO I deficiency have low serum levels of 18-hydroxycorticosterone whereas patients with CMO II deficiency have high levels of this aldosterone precursor and thus an elevated ratio of 18-hydroxycorticosterone to aldosterone.

Only one kindred with CMO I deficiency has been studied in detail. Patients in this kindred are homozygous for a 5-nt deletion in *CYP11B2* leading to a frameshift (Mitsuuchi *et al.*, 1993).

CMO II deficiency is apparently rare in the general population, but it has been found at an increased frequency among Jews of Iranian origin. In this population, the disease is genetically linked to a unique Msp I polymorphism in *CYP11B1* (Globerman *et al.*, 1988b), but affected individuals in fact have normal *CYP11B1* genes. Instead, they are homozygous for two missense mutations in *CYP11B2* (Pascoe *et al.*, 1992a): R181W and V386A. Individuals homozygous for either one of the mutations alone are asymptomatic.

When normal and mutant CYP11B2 are expressed in cultured cells, the R181W mutant has normal 11β-hydroxylase activity, decreased 18-hydroxylase activity, and undetectable 18-oxidase activity. The V386A mutant has slightly decreased activity as compared with the normal enzyme. No differences could be demonstrated between the enzymes carrying R181W alone and in combination with V386A, but the studies on patients suggest that the double mutant enzyme must have even more severely compromised 18-oxidase activity than the enzyme carrying R181W alone.

Taken together, these studies suggest that CMO I and CMO II deficiencies represent allelic variants. Complete loss of CYP11B2 activity leads to CMO I deficiency, a trace of activity is associated with CMO II deficiency, and a presumably greater level of activity (still undetectable *in vitro*) leads to normal aldosterone biosynthesis *in vivo*.

5. *Glucocorticoid-Suppressible Hyperaldosteronism*

Glucocorticoid-suppressible hyperaldosteronism, also called dexamethasone-suppressible hyperaldosteronism or glucocorticoid-remediable aldosteronism, is inherited as an autosomal dominant disorder (New *et al.*, 1980). It is distinguished from primary aldosteronism by suppression of (usually moderate) aldosterone hypersecretion within 48 hr of dexamethasone administration, indicating that aldosterone synthesis is inappropriately regulated by ACTH (Sutherland *et al.*, 1966; New and Peterson, 1967). This feature and the high urinary concentration of 18-oxocortisol (Gomez-Sanchez *et al.*, 1988; Rich *et al.*, 1992), a 17α-hydroxylated analog of aldosterone that cannot be produced in the zona glomerulosa because of the zone's lack of 17α-hydroxylase activity, suggest that aldosterone is being synthesized in the zona fasciculata. This implies that *CYP11B2* must be inappropriately regulated in a manner similar to the normal regulation of *CYP11B1* (Ulick *et al.*, 1990). This disorder is caused by intergenic recombinations that juxtapose the promoter of *CYP11B1* with coding sequences of *CYP11B2* (Lifton *et al.*, 1992a,b; Miyahara *et al.*, 1992; Pascoe *et al.*, 1992b). These are unequal crossovers that generate a chromosome carrying three *CYP11B* genes, the middle one of which is a chimera with 5' and 3' ends corresponding to *CYP11B1* and *CYP11B2*, respectively. This hybrid gene is expressed in the zona fasciculata and its transcription is stimulated mainly by corticotropin. A single copy of such an abnormally regulated gene should be sufficient to cause the disease, consistent with the known mode of inheritance as an autosomal dominant disorder. The encoded hybrid protein (the amino-terminal region is that of CYP11B1) retains the ability to synthesize aldosterone from deoxycorticosterone (Pascoe *et al.*, 1992b) as long as the crossover point is before exon 5, which is the case with all 17 affected kindreds studied thus far.

The other chromosome involved in the unequal crossover carries a deletion and thus has a single *CYP11B* gene. Because this gene has a *CYP11B2* promoter, it should not be expressed in the zona fasciculata, but it might also lack 18-oxidase activity depending on the breakpoint. Thus, the chromosome carrying the deletion might represent a null allele, but such chromosomes have not yet been detected among patients with either 11β-hydroxylase or corticosterone methyloxidase deficiencies. This may mean that such deletions are rare compared to those causing 21-hydroxylase deficiency, or that the single chimeric gene is expressed at sufficiently high levels that it does not cause 11β-hydroxylase deficiency and is thus not ascertained.

F. AROMATASE

1. *Biochemistry*

Aromatase (CYP19, P450arom) converts androgens such as androstenedione and testosterone to the corresponding estrogens, estrone and estradiol (Kellis and Vickery, 1987). This process requires hydroxylation of the C19 methyl group, further oxidation of this position to an aldehyde, and a third oxidation of this carbon leading to its release as formic acid. In the last oxidation, the 2β hydrogen is lost from the A ring of the steroid leading to the aromatization of the A ring. This entire process requires three molecules of oxygen and six electrons donated by three molecules of NADPH. As a microsomal cytochrome P450, aromatase receives these electrons via cytochrome P450 reductase.

2. *Genetic Analysis*

Based on analysis of cDNA, human CYP19 consists of 503 amino acids. It is only distantly related to other microsomal P450 in its sequence. The coding sequence of the single *CYP19* gene consists of nine exons spread over 32 kb (Means *et al.*, 1989; Harada *et al.*, 1990). The positions of introns do not correspond to those of any other P450 family. Human *CYP19* is unique among P450 genes in having several tissue-specific promoters, which with one exception are each 5' of a different untranslated first exon. Thus, transcription in the ovary is mainly initiated from a promoter immediately upstream of the start of the coding sequence (exon 2). In the placenta, two alternative untranslated first exons are noted in CYP19 transcripts. One is located 9 kb upstream of exon 2, whereas the other, which is included in the majority of transcripts in the placenta, is at least 40 kb farther upstream (Means *et al.*, 1991). In adipose tissue, at least one additional untranslated exon 1 is utilized (Mahendroo *et al.*, 1991). With a span of greater than 70 kb, *CYP19* is easily the largest P450 gene characterized thus far. This gene is located at chromosome 15q21.1 in humans.

3. *Defects in Placental Estrogen Biosynthesis and Fetal Survival*

Because placental estrogens are required to maintain gestation, it might be predicted that complete aromatase deficiency would act as an embryonic lethal disorder. Indeed, only one case of aromatase deficiency has been reported. A mutation in *CYP19* has been demonstrated in this case (Shozu *et al.*, 1991; Harada *et al.*, 1992a,b). This mutation affects the splice donor site at the beginning of intron 6, causing the first 87 nt of intron 6 to be included in mRNA. This results in 29 additional amino acid residues being included in the protein without changing the reading frame of translation of the mRNA. The mutant recombinant protein had a trace (0.3%) of activity, suggesting that the patient did not have complete aromatase deficiency.

It should be pointed out that a complete deficiency of cholesterol desmolase should prevent synthesis of both progestins and estrogens by the placenta, whereas 17α-hydroxylase/17,20-lyase deficiency should, like aromatase deficiency, prevent synthesis of estrogens. Thus, these deficiencies should have adverse effects on fetal survival similar to those of aromatase deficiency. Indeed, these defects are in their severe forms very rare compared to 21- and 11β-hydroxylase deficiencies, other defects causing congenital adrenal hyperplasia that do not affect estrogen biosynthesis. Although no mutations causing cholesterol desmolase deficiency have been identified, nonsense mutations and frameshifts causing 17α-hydroxylase deficiency have been documented, and these presumably lead to a profound loss of enzymatic activity. If these defects indeed adversely affect fetal survival, it is not clear why certain pregnancies are nevertheless carried to term. Apparently certain women are able to maintain their pregnancies without progestins or estrogens from the fetus.

G. 17β-Hydroxysteroid Dehydrogenase (17-Ketosteroid Reductase)

1. *Biochemistry*

The 17β-hydroxy forms of both androgens and estrogens (e.g., testosterone and estradiol) are more active than the corresponding 17-keto forms (androstenedione and estrone). The synthesis of active sex steroids in the testes and ovaries thus requires 17-ketosteroid reductase activity. Conversely, 17β-hydroxysteroid dehydrogenase activity represents a mechanism by which effects of sex steroids in other tissues may be modulated.

It is not yet certain how many different enzymes exist with 17β-hydroxysteroid dehydrogenase activity. One cytosolic isozyme in human placenta efficiently interconverts estrone and estradiol but has poor activity toward androgens (Tremblay *et al.*, 1989). Another isozyme is found in placental microsomes and acts as a 17β-hydroxy-steroid dehydrogenase on both androgens and estrogens. In addition, it has a 20α-hydroxysteroid dehydrogenase activity for progestins, which may be important in regulating levels of this class of hormones during pregnancy (Wu et al., 1993). Both of these isozymes utilize NAD+ or NADH as cofactors.

2. Genetic Analysis

Complementary DNA and genomic clones encoding the cytosolic iso-zyme of 17-hydroxysteroid dehydrogenase have been isolated. This en-zyme is a short-chain dehydrogenase with 327 amino acids and a MW of 34,000. The gene is transcribed in placenta from at least two differ-ent promoters so that one type of transcript has an unusually long (814 nucleotides) 5' untranslated region (Labrie et al., 1989). The gene is expressed at extremely high levels in placenta and also at significant levels in a wide range of tissues, including ovary, testis, prostate, breast, and adipose (Luu-The et al., 1990); both promoters are active in most tissues although the relative abundance of the corresponding transcripts varies. The gene is located along with an adjacent pseu-dogene on chromosome 17q11-12. The sequences of the gene and pseu-dogene are 89% identical; each consists of six exons over 3.2 kb (Trem-blay et al., 1989; Luu-The et al., 1990). Polymorphisms of this gene have been identified (Normand et al., 1993) but they are of uncertain functional significance.

The microsomal placental isozyme is also a short-chain dehy-drogenase, but its predicted amino acid sequence is only 23% identical to that of the cytosolic isozyme (Wu et al., 1993). When the predicted sequences of the two isozymes are aligned, the microsomal isozyme is distinguished by an amino-terminal extension of 80 amino acid resi-dues that includes a cluster of basic residues followed by a hydrophobic region. This extension presumably functions to anchor the enzyme to the membrane of the endoplasmic reticulum. Although the microso-mal isozyme catalyzes the interconversion of testosterone and an-drostenedione, expression of this isozyme in the testis has not yet been documented.

3. 17-Ketosteroid Reductase Deficiency

Deficiency of 17-ketosteroid reductase in the testes leads to an in-ability to synthesize testosterone from androstenedione. Males with this disorder are born with ambiguous genitalia (Imperato-McGinley et al., 1987; Wilson et al., 1987, 1988; Wit et al., 1988), whereas females have no apparent impairment of fertility. This is a rare disorder but a large number of cases have been documented in an inbred Arab population in Gaza (Rösler, 1992).

At puberty, affected males become virilized with growth of the testes and phallus and, in some cases, spontaneous adoption of a male gender identity despite being reared as females. Such patients usually require surgery to correct perineal hypospadias and severe chordee before they can function sexually as males. Moreover, reconstruction of a scrotum and orchidopexy or orchiectomy (depending on age at diagnosis) are required because the testes will otherwise remain undescended.

After puberty, affected patients have relatively normal levels of testosterone and dihydrotestosterone at the expense of dramatically elevated levels of androstenedione and gonadotropins. The testes also produce abnormally high levels of estrogens, particularly estrone (thus, the 17-ketosteroid deficiency affects both androgen and estrogen metabolism in these patients). High estrogen levels produce gynecomastia in about half of patients (Rösler, 1992).

The enzyme that is defective in this disorder has not yet been identified. Whereas the cytosolic placental isozyme has poor activity for androgens, the microsomal placental isozyme has an appropriate substrate specificity. The pubertal changes seen in affected males are presumably due in part to increased levels of circulating precursors (i.e., androstenedione) that allow synthesis of adequate levels of testosterone. The enzyme mediating this conversion could be a mutant enzyme with decreased but not absent activity or, alternatively, another isozyme present in other tissues. Such an isozyme might be induced at puberty by gonadotropins or other steroid hormones. Aromatase may also be induced abnormally in the testes, explaining the increased secretion of estrogens (Rösler, 1992).

H. 11β-Hydroxysteroid Dehydrogenase

1. Biochemistry

The interconversion of cortisol and cortisone (in rodents, corticosterone and 11-dehydrocorticosterone) is catalyzed by 11β-hydroxysteroid dehydrogenase. Studies of the mineralocorticoid (Type I) receptor have suggested that this activity may be of considerable physiological importance (Funder et al., 1990). In vivo, cortisol is a weak mineralocorticoid compared to aldosterone, yet in vitro the Type I receptor has about the same affinity for cortisol, corticosterone, and aldosterone (Arriza et al., 1987). However, cortisone and 11-dehydrocorticosterone are not ligands for this receptor. Thus, it has been suggested that 11β-hydroxysteroid dehydrogenase normally functions to confer specificity for aldosterone on the Type I receptor by converting cortisol and corticosterone to their inactive metabolites, cortisone and 11-dehydrocorticosterone (Stewart *et al.*, 1987; Edwards *et al.*, 1988; Funder *et al.*, 1988).

An enzyme with 11β-hydroxysteroid dehydrogenase activity has been purified from rat liver microsomes (Lakshmi and Monder, 1988) and is expressed in a wide range of rat tissues (Monder and Lakshmi, 1990). It is a glycoprotein with a molecular weight of 34,000 that requires $NADP^+$ as a cofactor. Although the purified enzyme functions only as a dehydrogenase, the recombinant enzyme expressed from cloned cDNA exhibits both 11β-dehydrogenase and the reverse oxoreductase activities (conversion of 11-dehydrocorticosterone to corticosterone) (Agarwal *et al.*, 1989, 1990). The reason for the apparent loss of oxoreductase activity during purification has not been determined.

There is accumulating evidence for the existence of one or more additional 11β-hydroxysteroid dehydrogenases, particularly in the kidney. If an 11β-hydroxysteroid dehydrogenase isozyme is to confer ligand specificity on the renal Type I receptor, it should be expressed in the distal tubule and collecting duct, the primary site of mineralocorticoid action. 11β-Hydroxysteroid dehydrogenase activity is present throughout the renal tubule, but studies using two different polyclonal antibodies to the liver enzyme found immunoreactive 11β-hydroxysteroid dehydrogenase only in the proximal tubules (Rundle *et al.*, 1989). The 11β-hydroxysteroid dehydrogenase activity in the distal tubule of the kidney appears to be NAD^+ (rather than $NADP^+$) dependent and has a much smaller apparent K_m for the steroid substrate than the 11β-hydroxysteroid dehydrogenase enzyme isolated from liver (100 nM vs. 1 μM) (Rusvai and Naray-Fejes-Toth, 1993). A similar enzyme has been detected in placenta (Walker *et al.*, 1992; Brown *et al.*, 1993).

2. *Genetic Analysis*

Cloned cDNA encoding the rat liver enzyme has been isolated (Agarwal *et al.*, 1989). Sequence analysis revealed that 11β-hydroxysteroid dehydrogenase is structurally related to other short-chain dehydrogenases. The MW of the enzyme predicted from the nucleotide sequence of the cDNA is 32,000, in contrast to the observed MW of the purified enzyme, 34,000. The difference is presumed to result from glycosylation, because when the recombinant enzyme is expressed in cultured cells, addition of tunicamycin (a glycosylation inhibitor) decreases the MW of the recombinant enzyme from 34,000 to 32,000 (Agarwal *et al.*, 1990).

The single human gene for this enzyme (*HSD11*) is located on chromosome 1 and contains six exons with a total length of over 9 kb (Tannin *et al.*, 1991).

The corresponding mRNA has been detected in a wide range of rat (Agarwal *et al.*, 1989; Krozowski *et al.*, 1990) and human tissues (Tannin *et al.*, 1991), including liver, lung, testis, colon, and kidney.

At least four different mRNA species hybridizing with cloned 11β-hydroxysteroid dehydrogenase cDNA have been found in rat kidney (Krozowski *et al.*, 1990), suggesting the presence of one or more alternative forms of 11β-hydroxysteroid dehydrogenase that are products of either alternate splicing of *HSD11* pre-mRNA or the use of a different cap site.

Alternative transcripts have been observed in the kidney that originate in the first intron (Krozowski *et al.*, 1992; Moisan *et al.*, 1992; Obeid *et al.*, 1993). The first codon of the second exon (codon 31 in human or 27 in rat) is an ATG (methionine) in a good sequence context for initiation of translation. These alternative transcripts thus encode a truncated protein with a MW of 26,000. Although functionally important domains (particularly the nucleotide cofactor binding domain near the N terminus) are preserved, this truncated protein is enzymatically inactive, and it is not apparent what function, if any, it mediates (Obeid *et al.*, 1993).

3. *Apparent Mineralocorticoid Excess*

If 11β-hydroxysteroid dehydrogenase confers steroid specificity on the Type I receptor, a defect in this activity should cause Type I receptor sites to be occupied mostly by cortisol, plasma levels of which are two to three orders of magnitude higher than those of aldosterone. Cortisol should act as a mineralocorticoid causing increased sodium retention and plasma volume. Such a phenomenon apparently occurs as an inborn error of metabolism called apparent mineralocorticoid excess (AME), a form of juvenile hypertension characterized by severe hypokalemic alkalosis in the presence of low plasma renin activity and subnormal levels of aldosterone and other known mineralocorticoids (New *et al.*, 1977). These symptoms respond to spironolactone administration or to a low-sodium diet, suggesting that they are mediated by the mineralocorticoid (Type I) receptor. Metyrapone is also effective, indicating that the steroid occupying the receptor is 11β-hydroxylated. Administration of hydrocortisone (cortisol) or cosyntropin (ACTH1-24) exacerbates the hypertension (Oberfield *et al.*, 1983).

AME patients excrete subnormal amounts of urinary cortisone metabolites compared to cortisol metabolites. These patients convert cortisol to cortisone at a lower than normal rate as measured by decreased excretion of tritiated water after administration of tritiated cortisol labeled at the 11α position (Ulick *et al.*, 1979). However, cortisone to cortisol conversion in these patients appears normal.

These findings suggest that cortisol acts as a mineralocorticoid in AME patients (Stewart *et al.*, 1988). The poor conversion of cortisol to cortisone implies that 11β-hydroxysteroid dehydrogenase, the activity that normally catalyzes this step, is defective. A similar phenomenon also occurs in licorice intoxication; glycyrrhizic acid, the hypertensive agent in licorice, is known to be a competitive inhibitor of 11β-hydroxysteroid dehydrogenase (Stewart *et al.*, 1987; Monder *et al.*, 1989).

Mutations in the *HSD11* gene (encoding the hepatic isozyme of 11β-hydroxysteroid dehydrogenase) could not be detected in patients with AME (Nikkila *et al.*, 1993). In light of the evidence for a different 11β-hydroxysteroid dehydrogenase isozyme in the renal distal tubule and cortical collecting duct, this result is not unexpected, and further genetic studies await isolation of clones encoding the additional isozyme(s).

If the enzyme encoded by *HSD11* is not responsible for 11β-hydroxysteroid dehydrogenase activity in the distal renal tubule, what then is its physiological function? This enzyme is present at very high levels in human liver, the main site of cortisone reduction, so it may function primarily as a reductase *in vivo*. Because *HSD11* is widely expressed, it has also been suggested that the encoded enzyme functions to protect the glucocorticoid (Type II) receptor in various tissues from excessive concentrations of cortisol (Whorwood *et al.*, 1992, 1993).

4. *Cortisone Oxoreductase Deficiency*

A defect in the oxoreductase reaction catalyzed by 11β-hydroxysteroid dehydrogenase (conversion of cortisone to cortisol) has been postulated in two reports of female patients presenting with signs of androgen excess; dehydrogenase activity in these patients was apparently normal (Taylor *et al.*, 1984; Phillipou and Higgins, 1985). Both reports involve families with two affected sibs, suggesting a familial disorder.

Although the liver 11β-hydroxysteroid dehydrogenase enzyme has strong reductase activity *in vivo*, mutations in the corresponding gene were not detected in one kindred with putative oxoreductase deficiency (Nikkila *et al.*, 1993), and so the genetic basis of this disorder remains unclear.

I. 5α-REDUCTASE

1. *Biochemistry*

Steroid 5α-reductase, an enzyme located in the endoplasmic reticulum, converts testosterone to dihydrotestosterone, a more potent androgen. The source of electrons for this reaction is NADPH. Because enzymes with this activity could not be purified by biochemical methods, molecular cloning techniques have been essential in studying this activity.

Humans have two 5α-reductase isozymes that are 50% identical in amino acid sequence. These isozymes contain 255 or 254 amino acids, respectively, and have MWs of 29,000.

One isozyme is expressed at low levels in peripheral tissues, particularly the skin. It has a K_m for substrate in the micromolar range and a broad pH optimum centered around pH 7 (Andersson *et al.*, 1989; Andersson and Russell, 1990). The second isozyme is expressed exclusively in male genital structures, including the skin of the genitalia and the prostate (Andersson *et al.*, 1991). It has a K_m for substrate in the nanomolar range and a narrow pH optimum of 4.5–5.5. The second isozyme is extremely sensitive (IC_{50} of 30 nM) to competitive inhibition by finasteride, a 4-azasteroid that is now used clinically to treat benign prostatic hypertrophy, whereas the first isozyme is 30-fold less sensitive to this inhibitor (Cuevas *et al.*, 1993).

Despite their similar sizes, these enzymes are not related structurally to the short-chain dehydrogenases or to any known family of oxidoreductases.

2. *Genetic Analysis*

The gene encoding the type 1 isozyme expressed in peripheral tissues (*SRD5A1*) is located on chromosome 5p15, with a pseudogene (*SRD5A1P*) on Xq24-qter (Jenkins *et al.*, 1991). The type 2 isozyme specific for male genital structures is encoded by the *SRD5A2* gene located on chromosome 2p23 (Thigpen *et al.*, 1992b). Each gene consists of five exons; the introns are located in corresponding positions in the genes.

3. *5α-Reductase Deficiency*

As mentioned in the discussions of congenital adrenal hyperplasia due to cholesterol desmolase, 17α-hydroxylase, and 3β-hydroxysteroid dehydrogenase deficiencies, testosterone is required for normal development of male genitalia. This androgen is sufficient to permit normal development of Wolffian ducts into the seminal vesicles, vas deferens, and epididymis. Development and subsequent growth of the external genitalia, urethra, and prostate require dihydrotestosterone, a stronger ligand for the androgen receptor.

Deficiency of 5α-reductase activity thus interferes with male sexual development. Affected males have a variable degree of genital ambiguity ranging from mild hypospadias to nearly female-appearing genitalia. Patients often become more normally virilized during adolescence and may assume male sex roles.

Biochemical studies of this disorder have been facilitated by the ability to grow genital skin (labia majora, foreskin, scrotum) fibroblasts from biopsies. Such cultures from normal subjects express type 2 (acid pH optimum, finasteride inhibitable) 5α-reductase activity, whereas cultures from affected individuals have absent or altered activity. Such alterations have included changes in K_m for steroid or NADPH.

The *SRD5A1* gene is normal in this disorder (Jenkins *et al.*, 1992). Mutations have been detected in the *SRD5A2* genes in most patients (Andersson *et al.*, 1991; Thigpen *et al.*, 1992a,b). About half of the patients studied are homozygous for a given mutation, and there is known consanguineous parentage in most of these cases. Over 40% of patients are compound heterozygotes, which suggests that the carrier frequency is relatively high, but this has not been directly determined.

Several patients from Papua New Guinea have a (nearly) complete deletion of *SRD5A2* (Table I) (Andersson *et al.*, 1991). One-sixth of alleles do not carry an identifiable mutation. The remainder have point mutations, most of which are missense. R246 (CGG) may represent a mutational hotspot with CpG→TpG mutations on both strands (i.e., C→T, R246W, and G→A, R246Q) recurring in patients from Latin America, Austria, and Pakistan and in one African-American. R246W alters the K_m for NADPH, whereas R246Q leads to an unstable enzyme. Another recurrent mutation, G34R, has been generated by different nucleotide changes (GGG→AGG or CGG) in Sicilian, Mexican-American, and Vietnamese patients. This mutation alters the K_m for testosterone, increases and broadens the pH optimum so that it resembles that of the type 1 enzyme, and decreases activity in transfected cells to 2% of normal.

V. Summary

All major classes of biologically active steroid hormones (progestins, mineralocorticoids, glucocorticoids, and sex steroids) are synthesized from cholesterol through 11 different bioconversions. With the exception of 5α-reductase, all the enzymes mediating these reactions fall into two classes, cytochromes P450 and short-chain dehydrogenases. Cytochromes P450 are heme-containing membrane-bound proteins with molecular weights of approximately 50,000 that utilize molecular oxygen and electrons from NADPH-dependent accessory proteins to hydroxylate substrates. Short-chain dehydrogenases have molecular weights of 30,000–40,000, have tyrosine and lysine residues at the active site, and remove a hydride from the substrate, transferring the electrons of the hydride to NAD^+ or $NADP^+$. In most cases, this reaction is reversible so that the dehydrogenase can also function as a reductase under appropriate conditions.

Inherited disorders in enzymes required for steroid biosynthesis have varying effects. Defects that prevent cortisol from being synthesized are referred to collectively as congenital adrenal hyperplasia. Because the enzymes required for cortisol biosynthesis in the adrenal cortex are in many cases required for the synthesis of mineralocorticoids and/or sex steroids, these classes of steroids may also not be synthesized normally. Thus, cholesterol desmolase and 3β-hydroxysteroid dehydrogenase deficiencies affect synthesis of all classes of steroids in both the adrenals and gonads. Steroid 21-hydroxylase deficiency, the most common cause (>90% of cases) of congenital adrenal hyperplasia, can affect both mineralocorticoid and glucocorticoid synthesis, but androgen secretion is usually abnormally high due to shunting of accumulated precursors into this pathway. Excessive secretion of androgens and mineralocorticoids occurs in 11β-hydroxylase deficiency (the second most frequent form of congenital adrenal hyperplasia). Mineralocorticoid excess is also seen in 17α-hydroxylase deficiency, but in this disorder sex steroid synthesis is defective.

All defects that affect estrogen synthesis (deficiencies of cholesterol desmolase, 3β-hydroxysteroid dehydrogenase, 17α-hydroxylase, aromatase, and 17β-hydroxysteroid dehydrogenase) are very rare, suggesting that the inability to synthesize placental estrogens may adversely affect fetal survival.

A number of enzymes are expressed at sites of steroid action and regulate the amount of active steroid available to steroid receptors. Steroid 5α-reductase converts testosterone to the more active dihydrotestosterone. Deficiency of this activity leads to incomplete development of male genitalia; 17β-hydroxysteroid dehydrogenase deficiency has similar phenotypic effects. The interconversion of cortisol and cortisone (the latter is not an active steroid) is catalyzed by 11β-hydroxysteroid dehydrogenase; this enzymatic activity prevents cortisol from occupying the renal mineralocorticoid receptor, and deficiency of this activity may be responsible for a form of hypertension called apparent mineralocorticoid excess. All the enzymes that are not cytochromes P450 exist in multiple isoforms, permitting complex regulation of steroid metabolism in different tissues.

ACKNOWLEDGMENTS

This work is supported by Grants DK37867 and DK42169 from the National Institutes of Health.

REFERENCES

Agarwal, A. K., Monder, C., Eckstein, B., and White, P. C. (1989). Cloning and expression of rat cDNA encoding corticosteroid 11-beta-dehydrogenase. *J. Biol. Chem.* **264,** 18939–18943.

Agarwal, A. K., Tusie-Luna, M. T., Monder, C., and White, P. C. (1990). Expression of 11-beta-hydroxysteroid dehydrogenase using recombinant vaccinia virus. *Mol. Endocrinol.* **4,** 1827–1832.

Ahlgren, R., Yanase, T., Simpson, E. R., Winter, J. S., and Waterman, M. R. (1992). Compound heterozygous mutations (Arg 239–stop, Pro 342–Thr) in the CYP17 (P45017 alpha) gene lead to ambiguous external genitalia in a male patient with partial combined 17-alpha-hydroxylase/17, 20-lyase deficiency. *J. Clin. Endocrinol. Metab.* **74,** 667–672.

Allen, D. B. (1993). Newborn screening for congenital adrenal hyperplasia in Wisconsin. *Wis. Med. J.* **92,** 75–78.

Amor, M., Parker, K. L., Globerman, H., New, M. I., and White, P. C. (1988). Mutation in the CYP21B gene (Ile-172–Asn) causes steroid 21-hydroxylase deficiency. *Proc. Natl. Acad. Sci. U. S. A.* **85,** 1600–1604.

Andersson, S., and Russell, D. W. (1990). Structural and biochemical properties of cloned and expressed human and rat steroid 5-alpha-reductases. *Proc. Natl. Acad. Sci. U. S. A.* **87,** 3640–3644.

Andersson, S., Bishop, R. W., and Russell, D. W. (1989). Expression cloning and regulation of steroid 5-alpha-reductase, an enzyme essential for male sexual differentiation. *J. Biol. Chem.* **264,** 16249–16255.

Andersson, S., Berman, D. M., Jenkins, E. P., and Russell, D. W. (1991). Deletion of steroid 5-alpha-reductase 2 gene in male pseudohermaphroditism. *Nature (London)* **354,** 159–161.

Araki, S., Chikazawa, K., Sekiguchi, I., Yamauchi, H., Motoyama, M., and Tamada, T. (1987). Arrest of follicular development in a patient with 17-alpha-hydroxylase deficiency: Folliculogenesis in association with a lack of estrogen synthesis in the ovaries. *Fertil. Steril.* **47,** 169–172.

Arriza, J. L., Weinberger, C., Cerelli, G., Glaser, T. M., Handelin, B. L., Housman, D. E., and Evans, R. M. (1987). Cloning of human mineralocorticoid receptor complementary DNA: Structural and functional kinship with the glucocorticoid receptor. *Science* **237,** 268–275.

Azziz, R., Bradley, E. L., Jr., Potter, H. D., and Boots, L. R. (1993). 3-Beta-hydroxysteroid dehydrogenase deficiency in hyperandrogenism. *Am. J. Obstet. Gynecol.* **168,** 889–895.

Barnes, R. B., Ehrmann, D. A., Brigell, D. F., and Rosenfield, R. L. (1993). Ovarian steroidogenic responses to gonadotropin-releasing hormone agonist testing with nafarelin in hirsute women with adrenal responses to adrenocorticotropin suggestive of 3-beta-hydroxy-delta 5-steroid dehydrogenase deficiency. *J. Clin. Endocrinol. Metab.* **76**, 450–455.

Berube, D., Luu The, V., Lachance, Y., Gagne, R., and Labrie, F. (1989). Assignment of the human 3-beta-hydroxysteroid dehydrogenase gene (HSDB3) to the p13 band of chromosome 1. *Cytogenet. Cell Genet.* **52**, 199–200.

Biason, A., Mantero, F., Scaroni, C., Simpson, E. R., and Waterman, M. R. (1991). Deletion within the CYP17 gene together with insertion of foreign DNA is the cause of combined complete 17-alpha-hydroxylase/17, 20-lyase deficiency in an Italian patient. *Mol. Endocrinol.* **5**, 2037–2045.

Biglieri, E. G., and Kater, C. E. (1991a). 17-Alpha-hydroxylation deficiency. *Endocrinol. Metab. Clin. North Am.* **20**, 257–268.

Biglieri, E. G., and Kater, C. E. (1991b). Mineralocorticoids in congenital adrenal hyperplasia. *J. Steroid Biochem. Mol. Biol.* **40**, 493–499.

Black, S. D. (1992). Membrane topology of the mammalian P450 cytochromes. *FASEB J.* **6**, 680–685.

Black, S. M., Szklarz, G. D., Harikrishna, J. A., Lin, D., Wolf, C. R., and Miller, W. L. (1993). Regulation of proteins in the cholesterol side-chain cleavage system in JEG-3 and Y-1 cells. *Endocrinology (Baltimore)* **132**, 539–545.

Bradshaw, K. D., Waterman, M. R., Couch, R. T., Simpson, E. R., and Zuber, M. X. (1987). Characterization of complementary deoxyribonucleic acid for human adrenocortical 17-alpha-hydroxylase: A probe for analysis of 17-alpha-hydroxylase deficiency. *Mol. Endocrinol.* **1**, 348–354.

Bristow, J., Tee, M. K., Gitelman, S. E., Mellon, S. H., and Miller, W. L. (1993). Tenascin-X: A novel extracellular matrix protein encoded by the human XB gene overlapping P450c21B. *J. Cell Biol.* **122**, 265–278.

Brown, R. W., Chapman, K. E., Edwards, C. R., and Seckl, J. R. (1993). Human placental 11-beta-hydroxysteroid dehydrogenase: Evidence for and partial purification of a distinct NAD-dependent isoform. *Endocrinology (Baltimore)* **132**, 2614–2621.

Carr, B. R. (1992). Disorders of the ovary and female reproductive tract. *In* "Williams Textbook of Endocrinology" (J. D. Wilson and D. W. Foster, eds.), 8th ed., pp. 733–798. Saunders, Philadelphia.

Carroll, M. C., Campbell, R. D., and Porter, R. R. (1985). Mapping of steroid 21-hydroxylase genes adjacent to complement component C4 genes in HLA, the major histocompatibility complex in man. *Proc. Natl. Acad. Sci. U. S. A.* **82**, 521–525.

Casey, M. L., MacDonald, P. C., and Simpson, E. R. (1992). Endocrinological changes of pregnancy. *In* "Williams Textbook of Endocrinology" (J. D. Wilson and D. W. Foster, eds.), 8th ed., pp. 977–992. Saunders, Philadelphia.

Chen, Z., Jiang, J. C., Lin, Z. G., Lee, W. R., Baker, M. E., and Chang, S. H. (1993). Site-specific mutagenesis of *Drosophila* alcohol dehydrogenase: Evidence for involvement of tyrosine-152 and lysine-156 in catalysis. *Biochemistry* **32**, 3342–3346.

Chiou, S. H., Hu, M. C., and Chung, B. C. (1990). A missense mutation at Ile172–Asn or Arg356–Trp causes steroid 21-hydroxylase deficiency. *J. Biol. Chem.* **265**, 3549–3552.

Chua, S. C., Szabo, P., Vitek, A., Grzeschik, K. H., John, M., and White, P. C. (1987). Cloning of cDNA encoding steroid 11-beta-hydroxylase (P450c11). *Proc. Natl. Acad. Sci. U. S. A.* **84**, 7193–7197.

Chung, B. C., Matteson, K. J., Voutilainen, R., Mohandas, T. K., and Miller, W. L. (1986). Human cholesterol side-chain cleavage enzyme, P450scc: cDNA cloning, assignment of the gene to chromosome 15, and expression in the placenta. *Proc. Natl. Acad. Sci. U. S. A.* **83**, 8962–8966.

Chung, B. C., Picado-Leonard, J., Haniu, M., Bienkowski, M., Hall, P. F., Shively, J. E., and Miller, W. L. (1987). Cytochrome P450c17 (steroid 17-alpha-hydroxylase/17, 20-lyase): Cloning of human adrenal and testis cDNAs indicates the same gene is expressed in both tissues. *Proc. Natl. Acad. Sci. U. S. A.* **84,** 407–411.

Collier, S., Sinnott, P. J., Dyer, P. A., Price, D. A., Harris, R., and Strachan, T. (1989). Pulsed field gel electrophoresis identifies a high degree of variability in the number of tandem 21-hydroxylase and complement C4 gene repeats in 21-hydroxylase deficiency haplotypes. *EMBO J.* **8,** 1393–1402.

Collier, S., Tassabehji, M., and Strachan, T. (1993). A *de novo* pathological point mutation at the 21-hydroxylase locus: Implications for gene conversion in the human genome. *Nat. Genet.* **3,** 260–265.

Cottrell, D. A., Bello, F. A., and Falko, J. M. (1990). 17-Alpha-hydroxylase deficiency masquerading as primary hyperaldosteronism. *Am. J. Med. Sci.* **300,** 380–382.

Cuevas, M. E., Collins K., and Callard, G. V. (1993). Stage-related changes in steroid-converting enzyme activities in *Squalus* testis: Synthesis of biologically active metabolites via 3-beta-hydroxysteroid dehydrogenase/isomerase and 5-alpha-reductase. *Steroids* **58,** 87–94.

Curnow, K. M., Tusie-Luna, M. T., Pascoe, L., Natarajan, R., Gu, J. L., Nadler, J. L., and White, P. C. (1991). The product of the CYP11B2 gene is required for aldosterone biosynthesis in the human adrenal cortex. *Mol. Endocrinol.* **5,** 1513–1522.

Curnow, K. M., Slutsker, L., Vitek, J., Cole, T., Speiser, P. W., New, M. I., White, P. C., and Pascoe, L. (1993). Mutations in the CYP11B1 gene causing congenital adrenal hyperplasia and hypertension cluster in exons 6, 7, and 8. *Proc. Natl. Acad. Sci. U. S. A.* **90,** 4552–4556.

Degenhart, H. J. (1984). Prader's syndrome (congenital lipoid adrenal hyperplasia). *Pediatr. Adolesc. Endocrinol.* **13,** 125–144.

Degenhart, H. J., Visser, H. K. A., Boon, H., and O'Docherty, N. J. (1972). Evidence for deficiency of 20-alpha-cholesterol hydrolase activity in adrenal tissue of a patient with lipoid adrenal hyperplasia. *Acta Endocrinol. (Copenhagen)* **71,** 512–518.

de Lange, W. E., and Doorenbos, H. (1990). Incomplete virilization and subclinical mineralocorticoid excess in a boy with partial 17,20-desmolase/17-alpha-hydroxylase deficiency. *Acta Endocrinol. (Copenhagen)* **122,** 263–266.

Donohoue, P. A., Van Dop, C., McLean, R. H., White, P. C., Jospe, N., and Migeon, C. J. (1986). Gene conversion in salt-losing congenital adrenal hyperplasia with absent complement C4B protein. *J. Clin. Endocrinol. Metab.* **62,** 995–1002.

Donohoue, P. A., Sandrini Neto, R., Collins, M. M., and Migeon, C. J. (1990). Exon 7 NcoI restriction site within CYP21B (steroid 21-hydroxylase) is a normal polymorphism. *Mol. Endocrinol.* **4,** 1354–1362.

Dorr, H. G., and Sippell, W. G. (1993). Prenatal dexamethasone treatment in pregnancies at risk for congenital adrenal hyperplasia due to 21-hydroxylase deficiency: Effect on midgestational amniotic fluid steroid levels. *J. Clin. Endocrinol. Metab.* **76,** 117–120.

Dupont, B., Virdis, R., Lerner, A. J., Nelson, C., Pollack, M. S., and New, M. I. (1984). Distinct HLA-B antigen associations for the salt-wasting and simple virilizing forms of congenital adrenal hyperplasia due to Z1-hydroxylase deficiency. *In* "Histocompatibility Testing 1984" (E. D. Albert, M. P. Baur, and W. R. Mayr, eds.), p. 660. Springer-Verlag, Berlin.

Edwards, C. R., Stewart, P. M., Burt, D., Brett, L., McIntyre, M. A., Sutanto, W. S., de Kloet, E. R., and Monder, C. (1988). Localisation of 11-beta-hydroxysteroid dehydrogenase—Tissue specific protector of the mineralocorticoid receptor. *Lancet* **2,** 986–989.

Ensor, C. M., and Tai, H. H. (1991). Site-directed mutagenesis of the conserved tyrosine-151 of human placental NAD+-dependent 15-hydroxyprostaglandin dehydrogenase yields a catalytically inactive enzyme. *Biochem. Biophys. Res. Commun.* **176,** 840–845.

Fan, Y. S., Sasi, R., Lee, C., Winter, J. S., Waterman, M. R., and Lin, C. C. (1992). Localization of the human CYP17 gene (cytochrome P450 (17 alpha)) to 10q24.3 by fluorescence *in situ* hybridization and simultaneous chromosome banding. *Genomics* **14,** 1110–1111.

Forest, M. G., David, M., and Morel, Y. (1993). Prenatal diagnosis and treatment of 21-hydroxylase deficiency. *J. Steroid Biochem. Mol. Biol.* **45,** 75–82.

Fraser, R., Brown, J. J., Mason, P. A., Morton, J. J., Lever, A. F., Robertson, J. I., Lee, H. A., and Miller H. (1987). Severe hypertension with absent secondary sex characteristics due to partial deficiency of steroid 17-alpha-hydroxylase activity. *J. Hum. Hypertens.* **1,** 53–58.

Funder, J. W., Pearce, P. T., Smith, R., and Smith, A. I. (1988). Mineralocorticoid action: Target tissue specificity is enzyme, not receptor, mediated. *Science* **242,** 583–585.

Funder, J. W., Pearce, P. T., Myles, K., and Roy, L. P. (1990). Apparent mineralocorticoid excess, pseudohypoaldosteronism, and urinary electrolyte excretion: Toward a redefinition of mineralocorticoid action. *FASEB J.* **4,** 3234–3238.

Furuya, S., Okada, M., Ito, A., Aoyagi, H., Kanmera, T., Kato, T., Sagara, Y., Horiuchi, T., and Omura, T. (1987). Synthetic partial extension peptides of P-450 (SCC) and adrenodoxin precursors: Effects on the import of mitochondrial enzyme precursors. *J. Biochem. (Tokyo)* **102,** 821–832.

Ghosh, D., Weeks, C. M., Grochulski, P., Duax, W. L., Erman, M., Rimsay, R. L., and Orr, J. C. (1991). Three-dimensional structure of holo 3-alpha, 20-beta-hydroxysteroid dehydrogenase: A member of a short-chain dehydrogenase family. *Proc. Natl. Acad. Sci. U. S. A.* **88,** 10064–10068.

Globerman, H., Amor, M., Parker, K. L., New, M. I., and White, P. C. (1988a). Nonsense mutation causing steroid 21-hydroxylase deficiency. *J. Clin. Invest.* **82,** 139–144.

Globerman, H., Rosler, A., Theodor, R., New, M. I., and White, P. C. (1988b). An inherited defect in aldosterone biosynthesis caused by a mutation in or near the gene for steroid 11-hydroxylase. *N. Engl. J. Med.* **319,** 1193–1197.

Gomez-Sanchez, C. E., Gill, J. R., Jr., Ganguly, and Gordon, R. D. (1988). Glucocorticoid-suppressible aldosteronism: A disorder of the adrenal transitional zone. *J. Clin. Endocrinol. Metab.* **67,** 444–448.

Gordon, E. J., Bury, S. M., Sawyer, L., Atrian, S., and Gonzalez-Duarte, R. (1992). Preliminary X-ray crystallographic studies on alcohol dehydrogenase from *Drosophila. J. Mol. Biol.* **227,** 356–358.

Griffin, J. E., and Wilson, J. D. (1992). Disorders of the testes and the male reproductive tract. *In* "Williams Textbook of Endocrinology" (J. D. Wilson and D. W. Foster, eds.), 8th ed., pp. 799–852. Saunders, Philadelphia.

Hague, W. M., and Honour, J. W. (1983). Malignant hypertension in congenital adrenal hyperplasia due to 11-beta-hydroxylase deficiency. *Clin. Endocrinol. (Oxford)* **18,** 505–510.

Hall, P. F. (1991). Cytochrome P-450 C21scc: One enzyme with two actions: Hydroxylase and lyase. *J. Steroid Biochem. Mol. Biol.* **40,** 527–532.

Hanukoglu, I., and Gutfinger, T. (1989). cDNA sequence of adrenodoxin reductase. Identification of NADP-binding sites in oxidoreductases. *Eur. J. Biochem.* **180,** 479–484.

Hanukoglu, I., Gutfinger, T., Haniu, M., and Shively, J. E. (1987). Isolation of a cDNA for adrenodoxin reductase (ferredoxin-NADP+ reductase). Implications for mitochondrial cytochrome P-450 systems. *Eur. J. Biochem.* **169,** 449–455.

Harada, N., Yamada, K., Saito, K., Kibe, N., Dohmae, S., and Takagi, Y. (1990). Structural characterization of the human estrogen synthetase (aromatase) gene. *Biochem. Biophys. Res. Commun.* **166,** 365–372.

Harada, N., Ogawa, H., Shozu, M., and Yamada, K. (1992a). Genetic studies to characterize the origin of the mutation in placental aromatase deficiency. *Am. J. Hum. Genet.* **51,** 666–672.

Harada, N., Ogawa, H., Shozu, M., Yamada, K., Suhara, K., Nishida, E., and Takagi, Y. (1992b). Biochemical and molecular genetic analyses on placental aromatase (P-450AROM) deficiency. *J. Biol. Chem.* **267,** 4781–4785.

Hauffa, B. P., Miller, W. L., Grumbach, M. M., Conte, F. A., and Kaplan, S. L. (1985). Congenital adrenal hyperplasia due to deficient cholesterol side-chain cleavage activity (20, 22-desmolase) in a patient treated for 18 years. *Clin. Endocrinol. (Oxford)* **23,** 481–493.

Hegesh, E., Hegesh, J., and Kaftory, A. (1986). Congenital methemoglobinemia with a deficiency of cytochrome b5. *N. Engl. J. Med.* **314,** 757–761.

Heinrich, U. E., Bettendorf, M., and Vecsei, P. (1993). Male pseudohermaphroditism caused by nonsalt-losing congenital adrenal hyperplasia due to 3-beta-hydroxysteroid dehydrogenase (3-beta-HSD) deficiency. *J. Steroid Biochem. Mol. Biol.* **45,** 83–85.

Hejtmancik, J. F., Black, S., Harris, S., Ward, P. A., Callaway, C., Ledbetter, D., Morris, J., Leech, S. H., and Pollack, M. S. (1992). Congenital 21-hydroxylase deficiency as a new deletion mutation. Detection in a proband during subsequent prenatal diagnosis by HLA typing and DNA analysis. *Hum. Immunol.* **35,** 246–252.

Helmberg, A., Ausserer, B., and Kofler, R. (1992a). Frame shift by insertion of 2 basepairs in codon 394 of CYP11B1 causes congenital adrenal hyperplasia due to steroid 11-beta-hydroxylase deficiency. *J. Clin. Endocrinol. Metab.* **75,** 1278–1281.

Helmberg, A., Tusie-Luna, M. T., Tabarelli, M., Kofler, R., and White, P. C. (1992b). R339H and P453S: CYP21 mutations associated with nonclassic steroid 21-hydroxylase deficiency that are not apparent gene conversions. *Mol. Endocrinol.* **6,** 1318–1322.

Higashi, Y., Yoshioka, H., Yamane, M., Gotoh, O., and Fujii-Kuriyama, Y. (1986). Complete nucleotide sequence of two steroid 21-hydroxylase genes tandemly arranged in human chromosome: A pseudogene and a genuine gene. *Proc. Natl. Acad. Sci. U. S. A.* **83,** 2841–2845.

Higashi, Y., Tanae, A., Inoue, H., and Fujii-Kuriyama, Y. (1988a). Evidence for frequent gene conversion in the steroid 21-hydroxylase P-450 (C21) gene: Implications for steroid 21-hydroxylase deficiency. *Am. J. Hum. Genet.* **42,** 17–25.

Higashi, Y., Tanae, A., Inoue, H., Hiromasa, T., and Fujii-Kuriyama, Y. (1988b). Aberrant splicing and missense mutations cause steroid 21-hydroxylase [P-450 (C21)] deficiency in humans: Possible gene conversion products. *Proc. Natl. Acad. Sci. U. S. A.* **85,** 7486–7490.

Higashi, Y., Hiromasa, T., Tanae, A., Miki, T., Nakura, J., Kondo, T., Ohura, T., Ogawa, E., Nakayama, K., and Fujii-Kuriyama, Y. (1991). Effects of individual mutations in the P-450 (C21) pseudogene on the P-450 (C21) activity and their distribution in the patient genomes of congenital steroid 21-hydroxylase deficiency. *J. Biochem. (Tokyo)* **109,** 638–644.

Imai, T., Yanase, T., Waterman, M. R., Simpson, E. R., and Pratt, J. J. (1992). Canadian Mennonites and individuals residing in the Friesland region of The Netherlands share the same molecular basis of 17-alpha-hydroxylase deficiency. *Hum. Genet.* **89,** 95–96.

Imperato-McGinley, J., Akgun, S., Ertel, N. H., Sayli, B., and Shackleton, C. (1987). The coexistence of male pseudohermaphrodites with 17-ketosteroid reductase deficiency and 5-alpha-reductase deficiency within a Turkish kindred. *Clin. Endocrinol. (Oxford)* **27**, 135–143.

Ishii-Ohba, H., Matsumura, R., Inano, H., and Tamaoki, B. (1984). Contribution of cytochrome b5 to androgen synthesis in rat testicular microsomes. *J. Biochem. (Tokyo)* **95**, 335–343.

Jenkins, E. P., Hsieh, C. L., Milatovich, A., Normington, K., Berman, D. M., Francke, U., and Russell, D. W. (1991). Characterization and chromosomal mapping of a human steroid 5-alpha-reductase gene and pseudogene and mapping of the mouse homologue. *Genomics* **11**, 1102–1112.

Jenkins, E. P., Andersson, S., Imperato-McGinley, J., Wilson, J. D., and Russell, D. W. (1992). Genetic and pharmacological evidence for more than one human steroid 5-alpha-reductase. *J. Clin. Invest.* **89**, 293–300.

Jornvall, H., Persson, B., Krook, M., and Kaiser, R. (1990). Alcohol dehydrogenases. *Biochem. Soc. Trans.* **18**, 169–171.

Kagimoto, K., Waterman, M. R., Kagimoto, M., Ferreira, P., Simpson, E. R., and Winter, J. S. (1989). Identification of a common molecular basis for combined 17-alpha-hydroxylase/17,20-lyase deficiency in two Mennonite families. *Hum. Genet.* **82**, 285–286.

Kagimoto, M., Winter, J. S., Kagimoto, K., Simpson, E. R., and Waterman, M. R. (1988). Structural characterization of normal and mutant human steroid 17-alpha-hydroxylase genes: Molecular basis of one example of combined 17-alpha-hydroxylase/17, 20-lyase deficiency. *Mol. Endocrinol.* **2**, 564–570.

Karaviti, L. P., Mercado, A. B., Mercado, M. B., Speiser, P. W., Buegeleisen, M., Crawford, C., Antonian, L., White, P. C., and New, M. I. (1992). Prenatal diagnosis/treatment in families at risk for infants with steroid 21-hydroxylase deficiency (congenital adrenal hyperplasia). *J. Steroid Biochem. Mol. Biol.* **41**, 445–451.

Kawamoto, T., Mitsuuchi, Y., Ohnishi, T., Ichikawa, Y., Yokoyama, Y., Sumitomo, H., Toda, K., Miyahara, K., Kuribayashi, I., Nakao, K., Hosoda, K., Yamamoto, Y., Imura, H., and Shizuta, Y. (1990a). Cloning and expression of a cDNA for human cytochrome P-450aldo as related to primary aldosteronism. *Biochem. Biophys. Res. Commun.* **173**, 309–316.

Kawamoto, T., Mitsuuchi, Y., Toda, K., Miyahara, K., Yokoyama, Y., Nakao, K., Hosoda, K., Yamamoto, Y., Imura, H., and Shizuta, Y. (1990b). Cloning of cDNA and genomic DNA for human cytochrome P-45011 beta. *FEBS Lett.* **269**, 345–349.

Kellis, J. T. J. R., and Vickery, L. E. (1987). Purification and characterization of human placental aromatase cytochrome P-450. *J. Biol. Chem.* **262**, 4413–4420.

Kominami, S., Ogawa, N., Morimune, R., De-Ying, H., and Takemori, S. (1992). The role of cytochrome b5 in adrenal microsomal steroidogenesis. *J. Steroid Biochem. Mol. Biol.* **42**, 57–64.

Krozowski, Z. (1992). 11-Beta-hydroxysteroid dehydrogenase and the short-chain alcohol dehydrogenase (SCAD) superfamily. *Mol. Cell. Endocrinol.* **84**, C25–C31.

Krozowski, Z., Stuchbery, S., White, P., Monder, C., and Funder, J. W. (1990). Characterization of 11-beta-hydroxysteroid dehydrogenase gene expression: Identification of multiple unique forms of messenger ribonucleic acid in the rat kidney. *Endocrinology (Baltimore)* **127**, 3009–3013.

Krozowski, Z., Obeyesekere, V., Smith, R., and Mercer, W. (1992). Tissue-specific expression of an 11-beta-hydroxysteroid dehydrogenase with a truncated N-terminal domain. A potential mechanism for differential intracellular localization within mineralocorticoid target cells. *J. Biol. Chem.* **267**, 2569–2574.

Kumamoto, T., Morohashi, K., Ito, A., and Omura T. (1987). Site-directed mutagenesis of basic amino acid residues in the extension peptide of P-450 (SCC) precursor: Effects on the import of the precursor into mitochondria. *J. Biochem. (Tokyo)* **102,** 833–838.

Kuroiwa, T., Sakaguchi, M., Mihara, K., and Omura, T. (1990). Structural requirements for interruption of protein translocation across rough endoplasmic reticulum membrane. *J. Biochem. (Tokyo)* **108,** 829–834.

Kuroiwa, T., Sakaguchi, M., Mihara, K., and Omura, T. (1991). Systematic analysis of stop-transfer sequence for microsomal membrane. *J. Biol. Chem.* **266,** 9251–9255.

Labrie, F., Luu-The, V., Labrie, C., Berube, D., Couet, J., Zhao, H. F., Gagne, R., and Simard, J. (1989). Characterization of two mRNA species encoding human estradiol 17-beta-dehydrogenase and assignment of the gene to chromosome 17. *J. Steroid Biochem.* **34,** 189–197.

Labrie, F., Simard, J., Luu-The, V., Trudel, C., Martel, C., Labrie, C., Zhao, H. F., Rheaume, E., Couet, J., and Breton, N. (1991). Expression of 3-beta-hydroxysteroid dehydrogenase/delta-5-delta 4-isomerase (3-beta-HSD) and 17-beta-hydroxysteroid dehydrogenase (17-beta-HSD) in adipose tissue. *Int. J. Obes.* **15,** Suppl. 2:91–Suppl. 2:99.

Lachance, Y., Luu-The, V., Labrie, C., Simard, J., Dumont, M., de Launoit, Y., Guerin, S., Leblanc, G., and Labrie, F. (1990). Characterization of human 3-beta-hydroxysteroid dehydrogenase/delta-5-delta 4-isomerase gene and its expression in mammalian cells. *J. Biol. Chem.* **265,** 20469–20475; erratum: *J. Biol. Chem.* **267,** 3551 (1992).

Lachance, Y., Luu-The, V., Verreault, H., Dumont, M., Rheaume, E., Leblanc, G., and Labrie, F. (1991). Structure of the human type II 3-beta-hydroxysteroid dehydrogenase/delta-5-delta 4-isomerase (3-beta-HSD) gene: Adrenal and gonadal specificity. *DNA Cell Biol.* **10,** 701–711.

Lakshmi, V., and Monder, C. (1988). Purification and characterization of the corticosteroid 11-beta-dehydrogenase component of the rat liver 11-beta-hydroxysteroid dehydrogenase complex. *Endocrinology (Baltimore)* **123,** 2390–2398.

Lauber, M., and Muller, J. (1989). Purification and characterization of two distinct forms of rat adrenal cytochrome P450 (11) beta: Functional and structural aspects. *Arch. Biochem. Biophys.* **274,** 109–119.

Lifton, R. P., Dluhy, R. G., Powers, M., Rich, G. M., Cook, S., Ulick, S., and Lalouel, J. M. (1992a). A chimaeric 11-beta-hydroxylase/aldosterone synthase gene causes glucocorticoid-remediable aldosteronism and human hypertension. *Nature (London)* **355,** 262–265.

Lifton, R. P., Dluhy, R. G., Powers, M., Rich, G. M., Gutkin, M., Fallo, F., Gill, J. R., Jr., Feld, L., Ganguly, A., Laidlaw, J. C., Murnaghan, D. J., Kaufman, C., Stockigt, J. R., Ulick, S., and Lalouel, J. M. (1992b). Hereditary hypertension caused by chimaeric gene duplications and ectopic expression of aldosterone synthase. *Nat. Genet.* **2,** 66–74.

Lin, D., Gitelman, S. E., Saenger, P., and Miller, W. L. (1991a). Normal genes for the cholesterol side chain cleavage enzyme, P450scc, in congenital lipoid adrenal hyperplasia. *J. Clin. Invest.* **88,** 1955–1962.

Lin, D., Harikrishna, J. A., Moore, C. C., Jones, K. L., and Miller, W. L. (1991b). Missense mutation serine 106–proline causes 17-alpha-hydroxylase deficiency. *J. Biol. Chem.* **266,** 15992–15998.

Lin, D., Black, S. M., Nagahama, Y., and Miller, W. L. (1993). Steroid 17-alpha-hydroxylase and 17, 20-lyase activities of P450c17: Contributions of serine 106 and P450 reductase. *Endocrinology (Baltimore)* **132,** 2498–2506.

Listernick, R., Frisone, L., and Silverman, B. L. (1992). Delayed diagnosis of infants with abnormal neonatal screens. *JAMA, J. Am. Med. Assoc.* **267**, 1095–1099.

Lorence, M. C., Murry, B. A., Trant, J. M., and Mason, J. I. (1990). Human 3-beta-hydroxysteroid dehydrogenase/delta-5-4-isomerase from placenta: Expression in nonsteroidogenic cells of a protein that catalyzes the dehydrogenation/isomerization of C21 and C19 steroids. *Endocrinology (Baltimore)* **126**, 2493–2498.

Luu-The, V., Lachance, Y., Labrie, C., Leblanc, G., Thomas, J. L., Strickler, R. C., and Labrie, F. (1989). Full length cDNA structure and deduced amino acid sequence of human 3-beta-hydroxy-5-ene steroid dehydrogenase. *Mol. Endocrinol.* **3**, 1310–1312.

Luu-The, V., Labrie, C., Simard, J., Lachance, Y., Zhao, H. F., Couet, J., Leblanc, G., and Labrie, F. (1990). Structure of two in tandem human 17-beta-hydroxysteroid dehydrogenase genes. *Mol. Endocrinol.* **4**, 268–275.

Mahendroo, M. S., Means, G. D., Mendelson, C. R., and Simpson, E. R. (1991). Tissue-specific expression of human P-450AROM. The promoter responsible for expression in adipose tissue is different from that utilized in placenta. *J. Biol. Chem.* **266**, 11276–11281.

Malcolm, P. N., Wright, D. J., and Edmonds, C. J. (1992). Deficiency of 17-alpha-hydroxylase associated with absent gonads. *Postgrad. Med. J.* **68**, 59–61.

Marg, A., Kuban, R. J., Behlke, J., Dettmer, R., and Ruckpaul, K. (1992). Crystallization and X-ray examination of bovine adrenodoxin. *J. Mol. Biol.* **227**, 945–947.

Mathieson, J., Couzinet, B., Wekstein-Noel, S., Nahoul, K., Turpin, G., and Schaison, G. (1992). The incidence of late-onset congenital adrenal hyperplasia due to 3-beta-hydroxysteroid dehydrogenase deficiency among hirsute women. *Clin. Endocrinol. (Oxford)* **36**, 383–388.

Means, G. D., Mahendroo, M. S., Corbin, C. J., Mathis, J. M., Powell, F. E., Mendelson, C. R., and Simpson, E. R. (1989). Structural analysis of the gene encoding human aromatase cytochrome P-450, the enzyme responsible for estrogen biosynthesis. *J. Biol. Chem.* **264**, 19385–19391.

Means, G. D., Kilgore, M. W., Mahendroo, M. S., Mendelson, C. R., and Simpson, E. R. (1991). Tissue-specific promoters regulate aromatase cytochrome P450 gene expression in human ovary and fetal tissues. *Mol. Endocrinol.* **5**, 2005–2013.

Mendonca, B. B., Bloise, W., Arnhold, I. J., Batista, M. C., Toledo, S. P., Drummond, M. C., Nicolau, W., and Mattar, E. (1987). Male pseudohermaphroditism due to nonsalt-losing 3-beta-hydroxysteroid dehydrogenase deficiency: Gender role change and absence of gynecomastia at puberty. *J. Steroid Biochem.* **28**, 669–675.

Miller, W. R. (1991). Aromatase activity in breast tissue. *J. Steroid Biochem. Mol. Biol.* **39**, 783–790.

Mitsuuchi, Y., Kawamoto, T., Miyahara, K., Ulick, S., Morton, D. H., Naiki, Y., Kuribayashi, I., Toda, K., Hara, T., Orii, T., Yasuda, K., Miura, K., Yamamoto, Y., Imura, H., and Shizuta, Y. (1993). Congenitally defective aldosterone biosynthesis in humans: Inactivation of the P450C18 gene (CYP11B2) due to nucleotide deletion in CMO I deficient patients. *Biochem. Biophys. Res. Commun.* **190**, 864–869.

Miyahara, K., Kawamoto, T., Mitsuuchi, Y., Toda, K., Imura, H., Gordon, R. D., and Shizuta, Y. (1992). The chimeric gene linked to glucocorticoid-suppressible hyperaldosteronism encodes a fused P-450 protein possessing aldosterone synthase activity. *Biochem. Biophys. Res. Commun.* **189**, 885–891.

Moisan, M. P., Edwards, C. R., and Seckl, J. R. (1992). Differential promoter usage by the rat 11-beta-hydroxysteroid dehydrogenase gene. *Mol. Endocrinol.* **6**, 1082–1087.

Monder, C., and Lakshmi, V. (1990). Corticosteroid 11-beta-dehydrogenase of rat tissues: Immunological studies. *Endocrinology (Baltimore)* **126,** 2435–2443.

Monder, C., Stewart, P. M., Lakshmi, V., Valentino, R., Burt, D., and Edwards, C. R. (1989). Licorice inhibits corticosteroid 11-beta-dehydrogenase of rat kidney and liver: *In vivo* and *in vitro* studies. *Endocrinology (Baltimore)* **125,** 1046–1053.

Monier, S., Van Luc, P., Kreibich, G., Sabatini, D. D., and Adesnik, M. (1988). Signals for the incorporation and orientation of cytochrome P450 in the endoplasmic reticulum membrane. *J. Cell Biol.* **107,** 457–470.

Moore, J. B., and Smith, G. L. (1992). Steroid hormone synthesis by a vaccinia enzyme: A new type of virus virulence factor. *EMBO J.* **11,** 1973–1980; erratum: *EMBO J.* **11**(9), 3490 (1992).

Morel, Y., Picado-Leonard, J., Wu, D. A., Chang, C. Y., Mohandas, T. K., Chung, B. C., and Miller, W. L. (1988). Assignment of the functional gene for human adrenodoxin to chromosome 11q13–qter and of adrenodoxin pseudogenes to chromosome 20cen–q13.1. *Am. J. Hum. Genet.* **43,** 52–59.

Morishima, N., Yoshioka, H., Higashi, Y., Sogawa, K., and Fujii-Kuriyama, Y. (1987). Gene structure of cytochrome P-450 (M-1) specifically expressed in male rat liver. *Biochemistry* **26,** 8279–8285.

Mornet, E., Dupont, J., Vitek, A, and White, P. C. (1989). Characterization of two genes encoding human steroid 11-beta-hydroxylase (P-450 (11) beta). *J. Biol. Chem.* **264,** 20961–20967.

Mornet, E., Crete, P., Kuttenn, F., Raux-Demay, M. C., Boue, J., White, P. C., and Boue, A. (1991). Distribution of deletions and seven point mutations on CYP21B genes in three clinical forms of steroid 21-hydroxylase deficiency. *Am. J. Hum. Genet.* **48,** 79–88.

Morohashi, K., Sogawa, K., Omura, T., and Fujii-Kuriyama, Y. (1987). Gene structure of human cytochrome P-450 (SCC), cholesterol desmolase. *J. Biochem. (Tokyo)* **101,** 879–887.

Muller, J., Torsson, A., Damkjaer Nielsen, M., Petersen, K. E., Christoffersen, J., and Skakkebaek, N. E. (1991). Gonadal development and growth in 46,XX and 46,XY individuals with P450scc deficiency (congenital lipoid adrenal hyperplasia). *Horm. Res.* **36,** 203–208.

Murakami, H., Yabusaki, Y., Sakaki, T., Shibata, M., and Ohkawa, H. (1987). A genetically engineered P450 monooxygenase: Construction of the functional fused enzyme between rat cytochrome P450c and NADPH-cytochrome P450 reductase. *DNA* **6,** 189–197.

Naseeruddin, S. A., and Hornsby, P. J. (1990). Regulation of 11-beta- and 17-alpha-hydroxylases in cultured bovine adrenocortical cells: 3′, 5′-Cyclic adenosine monophosphate, insulin-like growth factor-I, and activators of protein kinase C. *Endocrinology (Baltimore)* **127,** 1673–1681.

Nelson, D. R., and Strobel, H. W. (1988). On the membrane topology of vertebrate cytochrome P-450 proteins. *J. Biol. Chem.* **263,** 6038–6050.

Nelson, D. R., and Strobel, H. W. (1989). Secondary structure prediction of 52 membrane-bound cytochromes P450 shows a strong structural similarity to P450cam. *Biochemistry* **28,** 656–660.

Nelson, D. R., Kamataki, T., Waxman, D. J., Guengerich, F. P., Estabrook, R. W., Feyereisen, R., Gonzalez, F. J., Coon, M. J., Gunsalus, I. C., Gotoh, O., Okuda, K., and Nebert, D. W. (1993). The P450 superfamily: Update on new sequences, gene mapping, accession numbers, early trivial names of enzymes, and nomenclature. *DNA Cell Biol.* **12,** 1–51.

New, M. I., and Peterson, R. E. (1967). A new form of congenital adrenal hyperplasia. *J. Clin. Endocrinol. Metab.* **27**, 300–305.

New, M. I., Levine, L. S., Biglieri, E. G., Pareira, J., and Ulick, S. (1977). Evidence for an unidentified steroid in a child with apparent mineralocorticoid hypertension. *J. Clin. Endocrinol. Metab.* **44**, 924–933.

New, M. I., Oberfield, S. E., Levine, L. S., Dupont, B., Pollack, M. S., Gill, J. R., and Bartter, F. C. (1980). Demonstration of autosomal dominant transmission and the absence of HLA linkage in dexamethasone suppressible hyperaldosteronism. *Lancet* **1**, 550–551.

New, M. I., White, P. C., Pang, S., Dupont, B., and Speiser, P. W. (1989). *In* "The Metabolic Basis of Inherited Disease" (C. R. Scriver, A. L. Beaudet, W. S. Sly, and D. Valle, eds.), 6th ed., pp. 1881–1918. McGraw-Hill, New York.

Nikkila, H., Tannin, G. M., New, M. I., Taylor, N. F., Kalaitzoglou, G., Monder, C., and White, P. C. (1993). Defects in the HSD11 gene encoding 11β-hydroxysteroid dehydrogenase are not found in patients with apparent mineralocorticoid excess or 11-oxoreductase deficiency. *J. Clin. Endocrinol. Metab.* **77**, 687–691.

Nonaka, Y., Murakami, H., Yabusaki, Y., Kuramitsu, S., Kagamiyama, H., Yamano, T., and Okamoto, M. (1987). Molecular cloning and sequence analysis of full-length cDNA for mRNA or adrenodoxin oxidoreductase from bovine adrenal cortex. *Biochem. Biophys. Res. Commun.* **145**, 1239–1247.

Normand, T., Narod, S., Labrie, F., and Simard, J. (1993). Detection of polymorphisms in the estradiol 17-beta-hydroxysteroid dehydrogenase II gene at the EDH17B2 locus on 17qll-q21. *Hum. Mol. Genet.* **2**, 479–483.

Obeid, J., and White, P. C. (1992). Tyr-179 and Lys-183 are essential for enzymatic activity of 11-beta-hydroxysteroid dehydrogenase. *Biochem. Biophys. Res. Commun.* **188**, 222–227.

Obeid, J., Curnow, K. M., Aisenberg, J., and White, P. C. (1993). Transcripts originating in intron 1 of the HSD11 (11-beta-hydroxysteroid dehydrogenase) gene encode a truncated polypeptide that is enzymatically inactive. *Mol. Endocrinol.* **7**, 154–160.

Oberfield, S. E., Levine, L. S., Carey, R. M., Greig, F., Ulick, S., and New, M. I. (1983). Metabolic and blood pressure responses to hydrocortisone in the syndrome of apparent mineralocorticoid excess. *J. Clin. Endocrinol. Metab.* **56**, 332–339.

Ogishima, T., Shibata, H., Shimada, H., Mitani, F., Suzuki, H., Saruta, T., and Ishimura, Y. (1991). Aldosterone synthase cytochrome P-450 expressed in the adrenals of patients with primary aldosteronism. *J. Biol. Chem.* **266**, 10731–10734.

Okamura, T., Kagimoto, M., Simpson, E. R., and Waterman, M. R. (1987). Multiple species of bovine adrenodoxin mRNA. Occurrence of two different mitochondrial precursor sequences associated with the same mature sequence. *J. Biol. Chem.* **262**, 10335–10338.

Omura, T., and Ito, A. (1991). Biosynthesis and intracellular sorting of mitochondrial forms of cytochrome P450. *In* "Methods in Enzymology" (M. Waterman and E. Johnson, eds.), Vol. 206, pp. 75–81. Academic Press, San Diego.

Owerbach, D., Ballard, A. L., and Draznin, M. B. (1992a). Salt-wasting congenital adrenal hyperplasia: Detection and characterization of mutations in the steroid 21-hydroxylase gene, CYP21, using the polymerase chain reaction. *J. Clin. Endocrinol. Metab.* **74**, 553–558.

Owerbach, D., Sherman, L., Ballard, A. L., and Azziz, R. (1992b). Pro-453 to Ser mutation in CYP21 is associated with nonclassic steroid 21-hydroxylase deficiency. *Mol. Endocrinol.* **6**, 1211–1215.

Pang, S., Spence, D. A., and New, M. I. (1985). Newborn screening for congenital adrenal hyperplasia with special reference to screening in Alaska. *Ann. N. Y. Acad. Sci.* **458,** 90–102.

Pang, S., Yang, X., Wang, M., Tissot, R., Nino, M., Manaligod, J., Bullock, L. P., and Mason, J. I. (1992). Inherited congenital adrenal hyperplasia in the rabbit: Absent cholesterol side-chain cleavage cytochrome P450 gene expression. *Endocrinology (Baltimore)* **131,** 181–186.

Parissenti, A. M., Parker, K. L., and Schimmer, B. P. (1993). Identification of promoter elements in the mouse 21-hydroxylase (Cyp21) gene that require a functional cyclic adenosine 3′, 5′-monophosphate-dependent protein kinase. *Mol. Endocrinol.* **7,** 283–290.

Pascoe, L., Curnow, K. M., Slutsker, L., Rosler, A., and White, P. C. (1992a). Mutations in the human CYP11B2 (aldosterone synthase) gene causing corticosterone methyloxidase II deficiency. *Proc. Natl. Acad. Sci. U. S. A.* **89,** 4996–5000.

Pascoe, L., Curnow, K. M., Slutsker, L., Connell, J. M., Speiser, P. W., New, M. I., and White, P. C. (1992b). Glucocorticoid-suppressible hyperaldosteronism results from hybrid genes created by unequal crossovers between CYP11B1 and CYP11B2. *Proc. Natl. Acad. Sci. U. S. A.* **89,** 8327–8331.

Persson, B., Krook, M., and Jornvall, H. (1991). Characteristics of short-chain alcohol dehydrogenases and related enzymes. *Eur. J. Biochem.* **200,** 537–543.

Peter, M., Sippell, W. G., and Wernze, H. (1993). Diagnosis and treatment of 17-hydroxylase deficiency. *J. Steroid Biochem. Mol. Biol.* **45,** 107–116.

Phillipou, G., and Higgins, B. A. (1985). A new defect in the peripheral conversion of cortisone to cortisol. *J. Steroid Biochem.* **22,** 435–436.

Picado-Leonard, J., and Miller, W. L. (1987). Cloning and sequence of the human gene for P450c17 (steroid 17-alpha-hydroxylase/17, 20-lyase): Similarity with the gene for P450c21. *DNA* **6,** 439–448.

Picado-Leonard, J., and Miller, W. L. (1988). Homologous sequences in steroidogenic enzymes, steroid receptors and a steroid binding protein suggest a consensus steroid-binding sequence. *Mol. Endocrinol.* **2,** 1145–1150.

Picado-Leonard, J., Voutilainen, R., Kao, L. C., Chung, B. C., Strauss, J. F., and Miller, W. L. (1988). Human adrenodoxin: Cloning of three cDNAs and cycloheximide enhancement in JEG-3 cells, *J. Biol. Chem.* **263,** 3240–3244; erratum: *J. Biol. Chem.* **263**(22), 11016 (1988).

Porter, T. D., and Kasper, C. B. (1985). Coding nucleotide sequence of rat NADPH-cytochrome P-450 oxidoreductase cDNA and identification of flavin-binding domains. *Proc. Natl. Acad. Sci. U. S. A.* **82,** 973–977.

Porter, T. D., and Kasper, C. B. (1986). NADPH-cytochrome P-450 oxidoreductase: Flavin mononucleotide and flavin adenine dinucleotide domains evolved from different flavoproteins. *Biochemistry* **25,** 1682–1687.

Poulos, T. L. (1991). Modeling of mammalian P450s on basis of P450cam X-ray structure. *In* "Methods in Enzymology" (M. Waterman and E. Johnson, eds.), Vol. 206, pp. 11–30. Academic Press, San Diego.

Poulos, T. L., and Raag, R. (1992). Cytochrome P450cam: Crystallography, oxygen activation, and electron transfer. *FASEB J.* **6,** 674–679.

Poulos, T. L., Finzel, B. C., and Howard, A. J. (1987). High resolution crystal structure of cytochrome P450cam. *J. Mol. Biol.* **195,** 687–700.

Raag, R., Martinis, S. A., Sligar, S. G., and Poulos, T. L. (1991). Crystal structure of the cytochrome P-450CAM active site mutant Thr252Ala. *Biochemistry* **30,** 11420–11429.

Ravichandran, K. G., Boddupalli, S. S., Hasemann, C. A., Peterson, J. A., and Deisenhofer, J. (1993). Crystal structure of hemoprotein domain of P450BM-3, a prototype for microsomal P450's. *Science* **261**, 731–736.

Rheaume, E., Lachance, Y., Zhao, H. F., Breton, N., de Launoit, Y., Trudel, C., Luu-The, V., Simard, J., and Labrie, F. (1991). Structure and expression of a new cDNA encoding the almost exclusive 3-beta-hydroxysteroid dehydrogenase/delta-5-delta-4 isomerase in human adrenals and gonads. *Mol. Endocrinol.* **5**, 1147–1157.

Rheaume, E., Simard, J., Morel, Y., Mebarki, F., Zachmann, M., Forest, M. G., New, M. I., and Labrie, F. (1992). Congenital adrenal hyperplasia due to point mutations in the type II 3-beta-hydroxysteroid dehydrogenase gene. *Nat. Genet.* **1**, 239–245.

Rich, G. M., Ulick, S., Cook, S., Wang, J. Z., Lifton, R. P., and Dluhy, R. G. (1992). Glucocorticoid-remediable aldosteronism in a large kindred: Clinical spectrum and diagnosis using a characteristic biochemical phenotype. *Ann. Intern. Med.* **116**, 813–820.

Rösler, A. (1984). The natural history of salt-wasting disorders of adrenal and renal origin. *J. Clin. Endocrinol. Metab.* **59**, 689–700.

Rösler, A. (1992). Steroid 17-beta-hydroxysteroid dehydrogenase deficiency in man: An inherited form of male pseudohermaphroditism. *J. Steroid Biochem. Mol. Biol.* **43**, 989–1002.

Rösler, A., Leiberman, E., and Cohen, T. (1992). High frequency of congenital adrenal hyperplasia (classic 11-beta-hydroxylase deficiency) among Jews from Morocco. *Am. J. Med. Genet.* **42**, 827–834.

Ruettinger, R. T., Wen, L. P., and Fulco, A. J. (1989). Coding nucleotide 5′ regulatory and deduced amino acid sequences of P-450BM-3, a single peptide cytochrome P-450: NADPH-P-450 reductase from *Bacillus megaterium*. *J. Biol. Chem.* **264**, 10987–10995.

Rundle, S. E., Funder, J. W., Lakshmi, V., and Monder, C. (1989). The intrarenal localization of mineralocorticoid receptors and 11-beta-dehydrogenase: Immunocytochemical studies. *Endocrinology (Baltimore)* **125**, 1700–1704.

Rusvai, E., and Naray-Fejes-Toth, A. (1993). A new isoform of 11-beta-hydroxysteroid dehydrogenase in aldosterone target cells. *J. Biol. Chem.* **268**, 10717–10720.

Sagara, Y., Takata, Y., Miyata, T., Hara, T., and Horiuchi, T. (1987). Cloning and sequence analysis of adrenodoxin reductase cDNA from bovine adrenal cortex. *J. Biochem. (Tokyo)* **102**, 1333–1336; errata: *J. Biochem. (Tokyo)* (**3**):539 (1989); **108**(6), 1070 (1990).

Sakaguchi, M., Tomiyoshi, R., Kuroiwa, T., Mihara, K., and Omura, T. (1992). Functions of signal and signal-anchor sequences are determined by the balance between the hydrophobic segment and the N-terminal charge. *Proc. Natl. Acad. Sci. U. S. A.* **89**, 16–19.

Sakai, Y., Yanase, T., Takayanagi, R., Nakao, R., Nishi, Y., Haji, M., and Nawata, H. (1993). High expression of cytochrome b5 in adrenocortical adenomas from patients with Cushing's syndrome associated with high secretion of adrenal androgens. *J. Clin. Endocrinol. Metab.* **76**, 1286–1290.

Sato, T., Sakaguchi, M., Mihara, K., and Omura, T. (1990). The amino-terminal structures that determine topological orientation of cytochrome P-450 in microsomal membrane. *EMBO J.* **9**, 2391–2397.

Schenkman, J. B., and Greim, H., eds. (1993). "Cytochrome P450" Springer-Verlag, Berlin.

Schram, P., Zerah, M., Mani, P., Jewelewicz, R., Jaffe, S., and New, M. I. (1992). Nonclassical 3-beta-hydroxysteroid dehydrogenase deficiency: A review of our experience with 25 female patients. *Fertil. Steril.* **58**, 129–136.

Shima, H., Kawanaka, H., Yabumoto, Y., Okamoto, E., and Ikoma, F. (1991). A case of 17-alpha-hydroxylase deficiency with chromosomal karyotype 46,XY and high plasma aldosterone concentration. *Int. Urol. Nephrol.* **23**, 611–618.

Shimizu, T., Tateishi, T., Hatano, M., and Fujii-Kuriyama, Y. (1991). Probing the role of lysines and arginines in the catalytic function of cytochrome P450d by site-directed mutagenesis. Interaction with NADPH-cytochrome P450 reductase. *J. Biol. Chem.* **266**, 3372–3375.

Shinzawa, K., Kominami, S., and Takemori, S. (1985). Studies on cytochrome P-450 (P-450 17-alpha, lyase) from guinea pig adrenal microsomes. Dual function of a single enzyme and effect of cytochrome b5. *Biochim. Biophys. Acta* **833**, 151–160.

Shozu, M., Akasofu, K., Harada, T., and Kubota, Y. (1991). A new cause of female pseudohermaphroditism: Placental aromatase deficiency. *J. Clin. Endocrinol. Metab.* **72**, 560–566.

Simard, J., Rheaume, E., Sanchez, R., Laflamme, N., de Launoit, Y., Luu-The, V., van Seters, A. P., Gordon, R. D., Bettendorf, M., Heinrich, U., Moshang, T., New, M. I., and Labrie, F. (1993). Molecular basis of congenital adrenal hyperplasia due to 3-beta-hydroxysteroid dehydrogenase deficiency. *Mol. Endocrinol.* **7**, 716–728.

Solish, S. B., Picado-Leonard, J., Morel, Y., Kuhn, R. W., Mohandas, T. K., Hanukoglu, I., and Miller, W. L. (1988). Human adrenodoxin reductase: Two mRNAs encoded by a single gene on chromosome 17cen–q25 are expressed in steroidogenic tissues. *Proc. Natl. Acad. Sci. U. S. A.* **85**, 7104–7108.

Sparkes, R. S., Klisak, I., and Miller, W. L. (1991). Regional mapping of genes encoding human steroidogenic enzymes: P450scc to 15q23-q24, adrenodoxin to 11q22; adrenodoxin reductase to 17q24-q25; and P450c17 to 10q24-q25. *DNA Cell Biol.* **10**, 359–365.

Speiser, P. W., Dupont, B., Rubinstein, P., Piazza, A., Kastelan, A., and New, M. I. (1985). High frequency of nonclassical steroid 21-hydroxylase deficiency. *Am. J. Hum. Genet.* **37**, 650–667.

Speiser, P. W., New, M. I., and White, P. C. (1988). Molecular genetic analysis of nonclassic steroid 21-hydroxylase deficiency associated with HLA-B14, DR1. *N. Engl. J. Med.* **319**, 19–23.

Speiser, P. W., New, M. I., and White, P. C. (1989). Clinical and genetic characterization of nonclassic 21-hydroxylase deficiency. *Endocr. Res.* **15**, 257–276.

Speiser, P. W., Dupont, J., Zhu, D., Serrat, J., Buegeleisen, M., Tusie-Luna, M. T., Lesser, M., New, M. I., and White, P. C. (1992). Disease expression and molecular genotype in congenital adrenal hyperplasia due to 21-hydroxylase deficiency. *J. Clin. Invest.* **90**, 584–595.

Speiser, P. W., White, P. C., Dupont, J., Zhu, D., Mercado, A. B., and New, M. I. (1994). Prenatal diagnosis of congenital adrenal hyperplasia due to 21-hydroxylase deficiency by allele-specific hybridization and Southern blot. *Hum. Genet.* **93**, 424–428.

Stewart, P. M., Wallace, A. M., Valentino, R., Burt, D., Shackleton, C. H., and Edwards, C. R. (1987). Mineralocorticoid activity of liquorice: 11-Beta-hydroxysteroid dehydrogenase deficiency comes of age. *Lancet* **2**, 821–824.

Stewart, P. M., Corrie, J. E., Shackleton, C. H., and Edwards, C. R. (1988). Syndrome of apparent mineralocorticoid excess. A defect in the cortisol–cortisone shuttle. *J. Clin. Invest.* **82**, 340–349.

Sutherland, D. J., Ruse, J. L., and Laidlaw, J. C. (1966). Hypertension, increased aldosterone secretion and low plasma renin activity relieved by dexamethasone. *Can. Med. Assoc. J.* **95**, 1109–1119.

Tajima, T., Fujieda, K., and Fujii-Kuriyama, Y. (1993). *De novo* mutation causes steroid 21-hydroxylase deficiency in one family of HLA-identical affected and unaffected siblings. *J. Clin. Endocrinol. Metab.* **77**, 86–89.

Tannin, G. M., Agarwal, A. K., Monder, C., New, M. I., and White, P. C. (1991). The human gene for 11-beta-hydroxysteroid dehydrogenase. Structure, tissue distribution, and chromosomal localization. *J. Biol. Chem.* **266**, 16653–16658.

Taylor, N. F., Bartlett, W. A., Dawson, D. J., and Enoch, B. A. (1984). Cortisone reductase deficiency: Evidence for a new inborn error in metabolism of adrenal steroids. *J. Endocrinol.* **102**, Suppl., 90.

Thigpen, A. E., Davis, D. L., Gautier, T., Imperato-McGinley, J., and Russell, D. W. (1992a). Brief report: The molecular basis of steroid 5-alpha-reductase deficiency in a large Dominican kindred. *N. Engl. J. Med.* **327**, 1216–1219.

Thigpen, A. E., Davis, D. L., Milatovich, A., Mendonca, B. B., Imperato-McGinley, J., Griffin, J. E., Francke, U., Wilson, J. D., and Russell, D. W. (1992b). Molecular genetics of steroid 5-alpha-reductase 2 deficiency. *J. Clin. Invest.* **90**, 799–809.

Thijssen, J. H., Blankenstein, M. A., Donker, G. H., and Daroszewski, J. (1991). Endogenous steroid hormones and local aromatase activity in the breast. *J. Steroid Biochem. Mol. Biol.* **39**, 799–804.

Thomas, J. L., Berko, E. A., Faustino, A., Myers, R. P., and Strickler, R. C. (1988). Human placental 3-beta-hydroxy-5-ene-steroid dehydrogenase and steroid 5–4-ene-isomerase: Purification from microsomes, substrate kinetics, and inhibition by product steroids. *J. Steroid Biochem.* **31**, 785–793.

Tremblay, Y., Ringler, G. E., Morel, Y., Mohandas, T. K., Labrie, F., Strauss, J. F., and Miller, W. L. (1989). Regulation of the gene for estrogenic 17-ketosteroid reductase lying on chromosome 17cen–q25, *J. Biol. Chem.* **264**, 20458–20462.

Tusie-Luna, M. T., Traktman, P., and White, P. C. (1990). Determination of functional effects of mutations in the steroid 21-hydroxylase gene (CYP21) using recombinant vaccinia virus. *J. Biol. Chem.* **265**, 20916–20922.

Tusie-Luna, M. T., Speiser, P. W., Dumic, M., New, M. I., and White, P. C. (1991). A mutation (Pro-30 to Leu) in CYP21 represents a potential nonclassic steroid 21-hydroxylase deficiency allele. *Mol. Endocrinol.* **5**, 685–692.

Ulick, S., Levine, L. S., Gunczler, P., Zanconato, G., Ramirez, L. C., Rauh, W., Rosler, A., Bradlow, H. L., and New, M. I. (1979). A syndrome of apparent mineralocorticoid excess associated with defects in the peripheral metabolism of cortisol. *J. Clin. Endocrinol. Metab.* **49**, 757–764.

Ulick, S. Chan, C. K., Gill, J. R., Jr., Gutkin, M., Letcher, L., Mantero, F., and New, M. I. (1990). Defective fasciculata zone function as the mechanism of glucocorticoid-remediable aldosteronism. *J. Clin. Endocrinol. Metab.* **71**, 1151–1157.

Valentino, R., Tommaselli, A. P., Rossi, R., Lombardi, G., and Varrone, S. (1990). A pilot study for neonatal screening of congenital adrenal hyperplasia due to 21-hydroxylase and 11-beta-hydroxylase deficiency in Campania region. *J. Endocrinol. Invest.* **13**, 221–225.

Wada, A., and Waterman, M. R. (1992). Identification by site-directed mutagenesis of two lysine residues in cholesterol side chain cleavage cytochrome P450 that are essential for adrenodoxin binding. *J. Biol. Chem.* **267**, 22877–22882.

Walker, B. R., Campbell, J. C., Williams, B. C., and Edwards, C. R. (1992). Tissue-specific distribution of the NAD(+)-dependent isoform of 11-beta-hydroxysteroid dehydrogenase. *Endocrinology (Baltimore)* **131**, 970–972.

Wedell, A., and Luthman, H. (1993a). Steroid 21-hydroxylase deficiency: Two additional mutations in salt-wasting disease and rapid screening of disease-causing mutations. *Hum. Mol. Genet.* **2**, 499–504.

Wedell, A., and Luthman, H. (1993b). Steroid 21-hydroxylase (P450c21): A new allele and spread of mutations through the pseudogene. *Hum. Genet.* **91**, 236–240.

Wedell, A., Ritzen, E. M., Haglund-Stengler, B., and Luthman, H. (1992). Steroid 21-hydroxylase deficiency: Three additional mutated alleles and establishment of phenotype–genotype relationships of common mutations. *Proc. Natl. Acad. Sci. U. S. A.* **89**, 7232–7236.

White, P. C. (1987). Genetics of steroid 21-hydroxylase deficiency. *Recent Prog. Horm. Res.* **43**, 305–336.

White, P. C., New, M. I., and Dupont, B. (1984). HLA-linked congenital adrenal hyperplasia results from a defective gene encoding a cytochrome P-450 specific for steroid 21-hydroxylation. *Proc. Natl. Acad. Sci. U. S. A.* **81**, 7505–7509.

White, P. C., Grossberger, D., Onufer, B. J., Chaplin, D. D., New, M. I., Dupont, B., and Strominger, J. L. (1985). Two genes encoding steroid 21-hydroxylase are located near the genes encoding the fourth component of complement in man. *Proc. Natl. Acad. Sci. U. S. A.* **82**, 1089–1093.

White, P. C., New, M. I., and Dupont, B. (1986). Structure of human steroid 21-hydroxylase genes. *Proc. Natl. Acad. Sci. U. S. A.* **83**, 5111–5115.

White, P. C., New, M. I., and Dupont, B. (1987a). Congenital adrenal hyperplasia (1). *N. Engl. J. Med.* **316**, 1519–1524.

White, P. C., New, M. I., and Dupont, B. (1987b). Congenital adrenal hyperplasia (2). *N. Engl. J. Med.* **316**, 1580–1586.

White, P. C., Vitek, A., Dupont, B., and New, M. I. (1988). Characterization of frequent deletions causing steroid 21-hydroxylase deficiency. *Proc. Natl. Acad. Sci. U. S. A.* **85**, 4436–4440.

White, P. C., Dupont, J., New, M. I., Leiberman, E., Hochberg, Z., and Rosler, A. (1991). A mutation in CYP11B1 (Arg-448–His) associated with steroid 11-beta-hydroxylase deficiency in Jews of Moroccan origin. *J. Clin. Invest.* **87**, 1664–1667.

Whorwood, C. B., Franklyn, J. A., Sheppard, M. C., and Stewart, P. M. (1992). Tissue localization of 11-beta-hydroxysteroid dehydrogenase and its relationship to the glucocorticoid receptor. *J. Steroid Biochem. Mol. Biol.* **41**, 21–28.

Whorwood, C. B., Sheppard, M. C., and Steward, P. M. (1993) Licorice inhibits 11-beta-hydroxysteroid dehydrogenase messenger ribonucleic acid levels and potentiates glucocorticoid hormone action. *Endocrinology (Baltimore)* **132**, 2287–2292.

Wilson, S. C., Hodgins, M. B., and Scott, J. S. (1987). Incomplete masculinization due to a deficiency of 17-beta-hydroxysteroid dehydrogenase: Comparison of prepubertal and peripubertal siblings. *Clin. Endocrinol. (Oxford)* **26**, 459–469.

Wilson, S. C., Oakey, R. E., and Scott, J. S. (1988). Steroid metabolism in testes of patients with incomplete masculinization due to androgen insensitivity or 17-beta-hydroxysteroid dehydrogenase deficiency and normally differentiated males. *J. Steroid Biochem.* **29**, 649–655.

Winter, J. S. (1991). Clinical, biochemical and molecular aspects of 17-hydroxylase deficiency. *Endocr. Res.* **17**, 53–62.

Winter, J. S., Couch, R. M., Muller, J., Perry, Y. S., Ferreira, P., Baydala, L., and Shackleton, C. H. (1989). Combined 17-hydroxylase and 17, 20-desmolase deficiencies: Evidence for synthesis of a defective cytochrome P450c17. *J. Clin. Endocrinol. Metab.* **68**, 309–316.

Wit, J. M., van Hooff, C. O., Thijssen, J. H., and van den Brande, J. L. (1988). *In vivo* and *in vitro* studies in a 46,XY phenotypically female infant with 17-ketosteroid reductase deficiency. *Horm. Metab. Res.* **20,** 367–374.

Wong, M., Rice, D. A., Parker, K. L., and Schimmer, B. P. (1989). The roles of cAMP and cAMP-dependent protein kinase in the expression of cholesterol side chain cleavage and steroid 11-beta-hydroxylase genes in mouse adrenocortical tumor cells. *J. Biol. Chem.* **264,** 12867–12871.

Wu, D. A., and Chung, B. C. (1991). Mutations of P450c21 (steroid 21-hydroxylase) at Cys428, Val281, and Ser268 result in complete, partial, or no loss of enzymatic activity, respectively. *J. Clin. Invest.* **88,** 519–523.

Wu, L., Einstein, M., Geissler, W. M., Chan, H. C., Elliston, K. O., and Andersson, S. (1993). Expression cloning and characterization of human 17-beta-hydroxysteroid dehydrogenase type 2, a microsomal enzyme possessing 20-alpha-hydroxysteroid dehydrogenase activity. *J. Biol. Chem.* **268,** 12964–12969.

Yamakita, N., Murase, H., Yasuda, K., Noritake, N., Mercado-Asis, L. B., and Miura, K. (1989). Possible hyperaldosteronism and discrepancy in enzyme activity deficiency in adrenal and gonadal glands in Japanese patients with 17-alpha-hydroxylase deficiency. *Endocrinol. Jpn.* **36,** 515–536.

Yanase, T., Kagimoto, M., Matsui, N., Simpson, E. R., and Waterman, M. R. (1988). Combined 17-alpha-hydroxylase/17, 20-lyase deficiency due to a stop codon in the N-terminal region of 17-alpha-hydroxylase cytochrome P-450. *Mol. Cell. Endocrinol.* **59,** 249–253.

Yanase, T., Kagimoto, M., Suzuki, S., Hashiba, K., Simpson, E. R., and Waterman, M. R. (1989). Deletion of a phenylalanine in the N-terminal region of human cytochrome P-450 (17 alpha) results in partial combined 17-alpha-hydroxylase/17, 20-lyase deficiency. *J. Biol. Chem.* **264,** 18076–18082; erratum: *J. Biol. Chem.* **264**(35); 21433 (1989).

Yanase, T., Sanders, D., Shibata, A., Matsui, N., Simpson, E. R., and Waterman, M. R. (1990). Combined 17-alpha-hydroxylase/17, 20-lyase deficiency due to a 7-basepair duplication in the N-terminal region of the cytochrome P45017 alpha (CYP17) gene. *J. Clin. Endocrinol. Metab.* **70,** 1325–1329.

Yanase, T., Simpson, E. R., and Waterman, M. R. (1991). 17-Alpha-hydroxylase/17, 20-lyase deficiency: From clinical investigation to molecular definition. *Endocr. Rev.* **12,** 91–108.

Yanase, T., Waterman, M. R., Zachmann, M., Winter, J. S., Simpson, E. R., and Kagimoto, M. (1992). Molecular basis of apparent isolated 17, 20-lyase deficiency: Compound heterozygous mutations in the C-terminal region (Arg (496)–Cys, Gln (461)–Stop) actually cause combined 17-alpha-hydroxylase/17, 20-lyase deficiency. *Biochem. Biophys. Acta* **1139,** 275–279.

Yang, X., Iwamoto, K., Wang, M., Artwohl, J., Mason, J. I., and Pang, S. (1993). Inherited congenital adrenal hyperplasia in the rabbit is caused by a deletion in the gene encoding cytochrome P450 cholesterol side-chain cleavage enzyme. *Endocrinology (Baltimore)* **132,** 1977–1982.

Zachmann, M., Tassinari, D., and Prader, A. (1983). Clinical and biochemical variability of congenital adrenal hyperplasia due to 11-beta-hydroxylase deficiency. A study of 25 patients. *J. Clin. Endocrinol. Metab.* **56,** 222–229.

Zachmann, M., Kempken, B., Manella, B., and Navarro, E. (1992). Conversion from pure 17, 20–desmolase- to combined 17, 20-desmolase/17-alpha-hydroxylase deficiency with age. *Acta Endocrinol. (Copenhagen)* **127,** 97–99.

Zerah, M., Ueshiba, H., Wood, E., Speiser, P. W., Crawford, C., McDonald, T., Pareira, J., Gruen, D., and New, M. I. (1990). Prevalence of nonclassical steroid 21-hydroxylase deficiency based on a morning salivary 17-hydroxyprogesterone screening test: A small sample study. *J. Clin. Endocrinol. Metab.* **70,** 1662–1667.

Zuber, M. X., Simpson, E. R., and Waterman, M. R. (1986). Expression of bovine 17-alpha-hydroxylase cytochrome P-450 cDNA in nonsteroidogenic (COS 1) cells. *Science* **234,** 1258–1261.

Structure, Function, and Regulation of Androgen-Binding Protein/Sex Hormone-Binding Globulin

DAVID R. JOSEPH

Department of Pediatrics
The Laboratories for Reproductive Biology
The University of North Carolina
Chapel Hill, North Carolina 27599

I. INTRODUCTION

After speculation for many years about the presence of circulating sex steroid-binding proteins, specific androgen- and estrogen-binding proteins in the human plasma β-globulin fraction were first described by Mercier *et al.* (1966), Rosner *et al.* (1966), Pearlman and Crépy (1967), Rosner and Deakins (1968), Kato and Horton (1968), and Vermeulen and Verdonck (1968). Subsequent studies showed that the two activities were characteristics of the same protein, which was found to bind DHT, testosterone, and estradiol with high affinity (Murphy, 1968; Steeno *et al.*, 1968). By comparison with CBG, which had been

197

described much earlier, the binding protein was thought to regulate the cellular availability of free sex steroids. Pearlman *et al.* (1967) showed that testosterone-binding levels in the serum of women varied during the menstrual cycle and increased dramatically during pregnancy; they postulated that the increased testosterone binding acted to lower circulating androgenicity. At that time the sex steroid-binding protein was called steroid-binding β-globulin (SBβG), sex steroid-binding protein (SBP), or sex hormone-binding globulin (SHBG). Later it was shown that the protein was produced and secreted by cultured transformed liver hepatocytes (Khan *et al.*, 1981). Proof that the serum protein was made in the liver came much later with the advent of molecular biology techniques and the demonstration of liver SHBG mRNA (Gershagen *et al.*, 1987b, 1989; Que and Petra, 1987; Hammond *et al.*, 1987, 1989).

In the early 1970s, endocrinologists in several laboratories studying the androgen receptor found high levels of receptorlike activity in the rat testis and epididymis (Ritzen *et al.*, 1971, 1973; Blaquier, 1971; Hansson, 1972; Vernon *et al.*, 1972, 1974; Danzo *et al.*, 1973; Hansson *et al.*, 1973). Initially the activity was thought to be the receptor, but soon it was realized that the size and steroid-binding properties of the protein did not fit the characteristics of the receptor and it was shown that the protein was synthesized in the testis, secreted and transported to the epididymis (French and Ritzén, 1973; Danzo *et al.*, 1974, 1977). This protein was called androgen-binding protein (ABP). At that time it was proposed that ABP may serve to increase the accumulation of androgen in the seminiferous tubule and epididymis and make it available for the nuclear androgen receptor (Tindall *et al.*, 1974b; Hansson *et al.*, 1975a, b, 1976a). Later several laboratories showed that ABP was a secretory product of the Sertoli cell of the seminiferous tubule epithelium (Tindall *et al.*, 1974a; Fritz *et al.*, 1974, 1976; Steinberger *et al.*, 1975; Hagenäs *et al.*, 1975; Kierszenbaum *et al.*, 1980). Schmidt *et al.* (1981) demonstrated that Sertoli cell cultures secreted a protein with chemical characteristics essentially the same as those of epididymal ABP. Subsequently, ABP was developed for use as a biochemical marker for Sertoli cell function and formation of the blood–testis barrier (Tindall *et al.*, 1975; Hansson *et al.*, 1975a, b; Gunsalus *et al.*, 1981, 1984).

During the early 1970s, ABP and SHBG were the subject of a wide range of studies and it became obvious that the two proteins were closely related. However, characterizations of SHBG and ABP were hampered by the inability to purify the proteins to homogeneity. During the middle of the decade a major breakthrough was made with the development of steroid affinity columns (Renoir and Mercier-Bodard, 1974; Rosner and Smith, 1975; Mickelson and Petra, 1975; Musto *et al.*, 1977). Plasma SHBG was purified to homogeneity from first human and then numerous other species. Soon afterward, rat ABP was purified to homogeneity by affinity chromatography (Musto *et al.*, 1980; Feldman *et al.*, 1981). In 1980, Danzo and co-workers developed the methodology to photoaffinity-label ABP and SHBG (Taylor *et al.*, 1980; Danzo *et al.*, 1980). This methodology facilitated experiments on the subunit structure of ABP and SHBG and later identification of the steroid-binding site. Characterizations of both purified proteins further supported previous experiments that they were physicochemically closely related. These experiments included the production of polyclonal antisera against both proteins and the development of immunoassays (Renoir *et al.*, 1977; Gunsalus *et al.*, 1978; Bordin *et al.*, 1978, 1982; Khan *et al.*, 1982; Kovacs *et al.*, 1988). Immunochemical and structural studies with human and rabbit SHBG and ABP found that the two proteins were chemically nearly indistinguishable (S.-L. Cheng *et al.*, 1984; C.-Y. Cheng *et al.*, 1984, 1985; Kotite *et al.*, 1986).

The next major breakthroughs were the determination of the complete amino acid sequence of human SHBG (Walsh *et al.*, 1986) and the cloning and sequencing of rat ABP cDNA and the deduction of the rat ABP amino acid sequence (Joseph *et al.*, 1985, 1987); both studies were completed about the same time and were presented at the First International Symposium on Steroid Hormone-Binding Proteins in Lyon (Petra *et al.*, 1986a; Joseph *et al.*, 1986). Strong evidence that they were actually the same protein came from a comparison of the rat ABP and human SHBG sequences at the symposium (Petra *et al.*, 1986b). Both proteins had a similar basic primary structure and shared 68% of the 373 residues with no gaps in the alignment. Final proof of their identity came with the cloning of the rat and human genes and cloning of fetal rat liver SHBG and human ABP cDNAs (Joseph *et al.*, 1988; Hammond *et al.*, 1989; Gershagen *et al.*, 1989; Sullivan *et al.*, 1991). It is now known that ABP and SHBG are encoded by the same gene and they share the identical amino acid sequence. ABP and SHBG are the same protein produced in separate organ systems. Because both proteins have the same amino acid sequence they are presumed to have the same functional properties, but slight functional differences may exist due to oligosaccharide differences.

Historically the extracellular proteins androgen-binding protein and sex hormone-binding globulin were defined by their tissue origin and location. SHBG has also been called testosterone/estradiol-binding globulin (TeBG) or sex steroid-binding protein (SBP), which has led to some confusion in the literature. The ABP/SHBG nomenclature served to distinguish the liver and testis proteins, but was not suitable when it was discovered that the protein was made by other tissues. What do you call this protein produced by the brain or the recombinant protein produced by heterologous cell lines *in vitro?* What is the plasma protein called if it originates in the testis or ovary instead of the liver? Moreover, this nomenclature became even more antiquated when alternate nonsecreted forms of the protein were discovered.

In this review, the liver-plasma protein will be referred to as SHBG and the testis-epididymis protein will be referred to as ABP. The protein from all other sources will be called ABP/SHBG. Also, when the subject applies to both proteins, ABP/SHBG will be used (e.g., the ABP/SHBG gene). Alternate proteins encoded by the same gene resulting from alternate mRNAs or alternatively processed RNAs will be called ABP/SHBG-like proteins. Perhaps someday the ABP and SHBG groups will agree on one common term to describe this protein, as it has been done for numerous other convergent fields. However, a consensus on a term for the plasma protein—SHBG, SBP, or TeBG—remains to be reached. A vote by participants at the First International Symposium on Steroid Hormone-Binding Proteins in Lyon (1986) split equally between SHBG and SBP; it was suggested that TeBG not be used.

In this review, I will present an updated view of the structure, regulation, and function of ABP and SHBG. An emphasis will be placed on knowledge gained from molecular biology studies and findings since the last reviews by Westphal (1986), Rosner (1990), Strel'chyonok and Avvakumov (1990), Hammond (1990), and Petra (1991). Also, topics that were not covered by these reviews in detail will be addressed and several controversial subjects will be discussed. I will also share some unpublished results from this laboratory that should be of interest. The clinical aspects of SHBG will not be covered. The clinical significance was reviewed by Englebienne (1984), Moore and Bulbrook (1988), and Selby (1990) and this subject has been a major focus of the three international symposia on steroid-binding proteins. It should be pointed out that the publications of these symposia, which were held in Lyon (1986), Turin (1987), and The Hague (1990), are an excellent source of information on all aspects of the steroid-binding proteins [*Colloq. INSERM* **149,** 1–656 (1986); *Ann. N. Y. Acad. Sci.* **538,** 1–326 (1988); *J. Steroid Biochem.* **40,** 735–849 (1991)]. The next meeting is being held at Dallas in September 1994, as part of the International Congress on Hormonal Steroids.

II. Species Distribution

SHBG has been identified in the plasma of numerous species, including humans, nonhuman primates, and numerous other mammals. A comprehensive list of species (with references) that do and do not have SHBG has been reviewed by Westphal (1986) and Petra (1991). It appears to be absent in several mammalian species, including the adult rats and mice (males and females), guinea pigs and pigs (Corvol and Bardin, 1973; Wenn *et al.*, 1977; Stupnicki and Bartke, 1976; Lea and Støa, 1972). However, male and female fetal rats do have a circulating SHBG (Carreau, 1986; Gunsalus *et al.*, 1984; Becker and Iles, 1985). It should be pointed out that the apparent lack of SHBG in some species could be due to procedural problems. For example, ammonium sulfate precipitation may not detect steroid complexes that have a rapid dissociation rate or are sensitive to high salt concentrations.

Epididymal and testicular ABP have been found in most species that have been examined; rat, rabbit, and human ABP are the best characterized (Westphal, 1986). Two species, mouse and pig, have been thought not to express ABP because steroid-binding activity was not detectable in the epididymis (Mather *et al.*, 1983; Bardin *et al.*, 1988). The belief concerning mouse ABP was reversed with the demonstration of mouse ABP with immunological techniques and the cloning and sequencing of mouse ABP cDNA (Wang *et al.*, 1989). The apparent lack of mouse ABP was due to the low levels of epididymal ABP (based on RIA and DHT-binding activity) and testicular ABP mRNA, both of which are present at 2–4% of the level in the rat. The immunocytochemical location of the mouse ABP in the testis and epididymis was indistinguishable from that of the rat. The apparent lack of pig ABP may be due to the source of pig testes, which are generally obtained from immature animals (i.e., castration of farm animals). Possibly, ABP is expressed in pig testes later in development and/or at a low level as it is in the mouse.

Steroid-binding activities in plasma and testis have been described in species as distantly related as teleost fishes and sharks (Mak and Callard, 1987; Foucher and Le Gac, 1989; Foucher *et al.*, 1991). The steroid-binding properties of these proteins are very similar to those of mammalian ABP/SHBG and the physiochemical properties also appear to be related. Because amino acid sequence data have not been obtained, it is not known if fish and shark ABP/SHBG are structurally related to the mammalian proteins. SHBG has been purified from carp plasma (Chang and Lee, 1992); both estrogen- and testosterone-binding activities were associated with the same fractions after chromatography. The estimated molecular weight of carp SHBG (194,000) differs greatly from the size of mammalian SHBG. ABP in the dogfish shark is expressed in a spermatogenic stage-specific manner and it is used as a marker of Sertoli cell function (Mak and Callard, 1987). This model is particularly useful because the different spermatogenic stages in the dogfish testis are topographically separated.

III. STRUCTURE

A. QUANTITATIVE MEASUREMENT

Numerous assays have been used to quantitate the binding of labeled sex steroids (Englebienne, 1984). The various assays were discussed and evaluated during the First International Symposium on Steroid Hormone-Binding Proteins (Hiramatsu et al., 1986; Heyns, 1986; Hammond, 1986; Rosner, 1986; Petra, 1986; Degrelle, 1986; Gunsalus et al., 1986). The methods include the use of semipermeable membranes (equilibrium dialysis and ultrafiltration), electrophoresis, column chromatography, absorption to charcoal or DEAE filters, and precipitation with ammonium sulfate to determine the amount of bound steroid. Rosner (1986) has discussed the advantages and disadvantages of each method. Today, the most widely utilized method to separate bound and free steroid employs the use of dextran-coated charcoal, because of its simplicity. Nonspecific binding is determined with an excess of unlabeled steroid. Charcoal treatments must be very brief (<2 min) when assaying binding proteins that have a very short half-life of steroid dissociation, such as rat and rabbit ABP and SHBG. Equilibrium binding constants and the number of sites are usually calculated by the method of Scatchard (1949).

Immunoassay, a more sensitive procedure than steroid binding, is also a commonly utilized method to assay ABP and SHBG (Khan et al., 1982; Hammond et al., 1985; Gunsalus et al., 1986; Petra, 1986). Several human SHBG immunometric assays are commercially available (Thomas et al., 1987; Selby, 1990), but antisera against human SHBG do not recognize ABP/SHBG from heterologous species other than some primates (Renoir et al., 1980). In general, there is an excellent correlation of values obtained with steroid-binding assays and RIA (Hammond et al., 1985; Hiramatsu et al., 1986); steroid-binding assays are much less expensive. A rat ABP RIA kit was available from the National Hormone and Pituitary Program, but it is no longer available

(Gunsalus *et al.*, 1978). This lack of an ABP RIA has hampered ABP research. The development of another ABP RIA has been delayed because of the difficulty in purification of rat ABP, ABP's intrinsic instability during storage, and because of its lack of immunogenicity. To circumvent these problems, an ABP RIA using an antiserum against a rat ABP peptide is being developed (P. Petrusz, D. R. Joseph, and F. S. French, unpublished results).

B. PURIFICATION AND SUBUNIT STRUCTURE

The development of steroid affinity column chromatography led to the purification of ABP and SHBG from several species (Westphal, 1986). Updated procedures were addressed at the First International Symposium on Steroid Hormone-Binding Proteins (1986). ABP has been purified to homogeneity from human (Cheng *et al.*, 1985), rat (Musto *et al.*, 1980; Feldman *et al.*, 1981; Kovaks *et al.*, 1988) and rabbit (Cheng and Musto, 1982). SHBG has been purified from several sources, including plasma from human and rabbit (Mickelson and Petra, 1978; Mercier-Bodard *et al.*, 1979; Petra and Lewis, 1980; Fernlund and Laurell, 1981; Strel'chyonok *et al.*, 1983; S.-L. Cheng *et al.*, 1984; Suzuki and Sinohara, 1984; Khan *et al.*, 1985; Hammond *et al.*, 1986; Griffin *et al.*, 1989). Other studies have described the SHBG purification from macaque, baboon, dog, cow, and several other species (Westphal, 1986; Petra, 1991).

The purification studies and additional experiments (Larrea *et al.*, 1981a,b; Taylor *et al.*, 1980; Khan *et al.*, 1985) revealed that ABP and SHBG were heterogeneous under denaturing conditions, consisting of heavy and light components. The ratio of the heavy and light components was generally between 3:1 to 10:1. Figures 1 and 2 demonstrate the typical banding pattern of reduced rat ABP and human SHBG fractionated using SDS–PAGE. This heterogeneity was the subject of intense debate to explain the phenomenon. Larrea *et al.* (1981a,b) showed by peptide mapping of each component of ABP that they were nearly identical, suggesting there were slight differences in primary structure or differences in carbohydrate content. Similarly, Suzuki and Sinohara (1984) found that the V8 polypeptide patterns of the two SHBG components from man, cattle, dog, and rabbit were similar, but not identical, suggesting there were structural differences between the two components. Later, it was demonstrated by Danzo and co-workers that the differences in rat ABP and human SHBG components were due to differential glycosylation (Danzo and Bell, 1988; Danzo *et al.*, 1989a, b). Enzymatic or chemical deglycosylation to re-

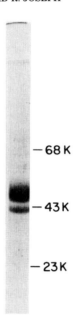

FIG. 1. Sodium dodecyl sulfate–polyacrylamide gel electrophoresis of purified ABP (20 μg). Protein was suspended in 1% sodium dodecyl sulfate, 2.5 mM EDTA, 2.75 M urea, 50 mM Tris–HCl, pH 6.7, 3% 2-mercaptoethanol, and 0.05% bromophenol blue, and heated for 10 min at 60°C. Cylindrical stacking and running gels (5 × 60 mm) contained 3.6 and 8% acrylamide, respectively. Electrophoresis was at 4°C, 1.5 mA/tube for 1 hr. Protein was stained with Coomassie blue. Molecular weight markers were bovine serum albumin (M_r = 68,000), ovalbumin (M_r = 43,000), and chymotrypsinogen-α (M_r = 23,000). From Feldman *et al.* (1981).

move N-linked and O-linked oligosaccharides yielded a single component with a molecular weight consistent with the removal of two or three oligosaccharide chains (Table I). The 48,000 and 43,000 M_r components of rat ABP were converted to a single 40,000 M_r species after deglycosylation. Human SHBG protomers (52,000 and 48,000 M_r species) were also converted to a single electrophoretic species (M_r = 42,000). Similarly, Petra *et al.* (1992) found a 39,000 M_r species after complete deglycosylation of human or rabbit SHBG. Enzymatic deglycosylation of N-linked moieties had no apparent effect on steroid binding, but removal of O-linked oligosaccharides appeared to increase slightly the affinity for DHT. Thus, denatured ABP and SHBG are heterogeneous owing to differential glycosylation. Rat ABP consists of 48,000 and 43,000 protomers, whereas the human SHBG protomers

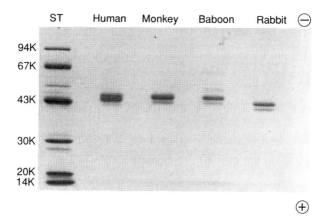

Fig. 2. Determination of heterogeneity and subunit molecular weight of SBP by sodium dodecyl sulfate gel electrophoresis in discontinuous buffers at 10°C. Stacking gel (5%) in 0.1% SDS, 0.5 M Tris–HCl, pH 6.8, and separating gel (10%) in 0.1% SDS, 1.5 M Tris–HCl, pH 8.8. Electrode buffer: 0.025 M Tris, 0.192 M glycine, 0.1% SDS, pH 8.3. Protein samples were dialyzed against buffer. Ten microliters (10 μg) SBP was added to 15 μl H_2O and 25 μl SDS buffer (0.125 M Tris–HCl, 4% SDS (w/v), 10% β-mercaptoethanol (v/v), 48% urea (w/v) and heated 5 min at 100°C. Five microliters of 0.5% bromophenol blue was added and the solution applied to wells in gels. The standard proteins (Pharmacia Fine Chemicals) were: rabbit phosphorylase b, 94K; bovine serum albumin, 67K; ovalbumin, 43K; bovine carbonic anhydrase, 30K; soybean trypsin inhibitor, 20.1K; bovine α-lactalbumin, 14.4K. From Petra et al. (1983).

are slightly larger. This difference in the size of the human and rat proteins is probably due to the presence of O-glycosylation in human SHBG at residue 7 that is not present in rat ABP. Taken together, these data support a monomer or homodimer model for native ABP and SHBG.

In vitro translation experiments also support a monomer or homodimer structure with heterogeneity due to differential glycosylation. Cell-free translation of rat testis poly(A) RNA using wheat germ extracts yielded a single 45,000 M_r protein band with SDS–PAGE after immunoprecipitation (Joseph et al., 1987). This value is very close to the predicted molecular weight of the rat ABP/SHBG precursor (unglycosylated), including the signal peptide, based on the deduced amino acid sequence from the cDNA sequence (44,539). Similarly, translation of human or monkey liver mRNA yielded a single immunoreactive 42,000 M_r species (Kottler et al., 1988, 1989).

The purification studies also estimated the molecular weights of native ABP and SHBG to be 85,000–100,000 by standard meth-

TABLE I

THE EFFECT OF DEGLYCOSYLATION ON THE APPARENT M_r
OF THE SUBUNITS OF hTeBG

Treatment	M_r, hTeBG	
	Heavy subunit	Light subunit
None	52,200 ± 1,000 (4)[a]	48,600 ± 600 (4)
Neuraminidase	50,800 ± 1,500 (3)	47,300 ± 900 (3)
N-Glycanase		44,100 ± 1,500 (3)
Neuraminidase + N-glycanase		42,600 ± 2,100 (3)
Neuraminidase + O-glycanase	49,600 ± 600 (3)	46,200 ± 1,100 (3)
		41,700 ± 1,500 (3)
3 enzymes[b]		42,900 ± 1,300 (5)
TFMSA[c]		42,000 ± 1,700 (4)

Note. Table from Danzo *et al.* (1989b).

[a] Numbers in parentheses indicate the number of replicates analyzed.

[b] N-Glycanase, neuraminidase, and O-glycanase.

[c] Trifluoromethanesulfonic acid.

odologies, including sedimentation equilibrium, native gel electrophoresis with different acrylamide concentrations (Ferguson plots), and gel filtration (Petra *et al.*, 1983, 1986c; Suzuki and Sinohara, 1984). Most data supported a homodimer structure for ABP and SHBG, however, there was some support for a monomer structure. First of all, the methods used to determine the molecular weights may lead to inaccurate estimations of size. This is especially true of sedimentation methods, where dimerization can be caused by pressure effects (Harrington, 1975). In support of a monomer model for SHBG, Englebienne *et al.* (1987) presented evidence that SHBG in undiluted serum exists as a monomer and is able to bind steroid. Furthermore, they showed that serum dilution induces the progressive dimerization of the protein, which leads to the loss of binding capacity, and the dimerization process is dependent on the steroid concentration. Following the changes in steady-state fluorescence anisotropy of labeled SHBG upon denaturation and renaturation, Casali *et al.* (1990) concluded that steroids aid in refolding and dimerization of the protein. They also concluded that monomers in the presence of guanidinium chloride have no affinity for steroids, but could not exclude the possibility that native monomers, if they exist, could bind steroid.

Although most studies by gel filtration and sedimentation equilibrium estimated the molecular weight of human SHBG to be around 100,000, Strel'chyonok *et al.* (1983) calculated a molecular weight of 51,000 by PAGE under native conditions and a molecular weight of 49,000 under denaturing conditions. In the same study, gel filtration of the native protein yielded a molecular weight of 120,000. To the contrary, Suzuki and Sinohara (1984) estimated the molecular weights of human, bovine, canine and rabbit SHBG to be 95,000–98,000 by native polyacrylamide gel electrophoresis. If purified ABP/SHBG is a dimer, the size may not reflect its native state. As a possible analogy, human corticosteroid-binding globulin exhibits monomer and dimer characteristics. In plasma, CBG appears to exist as a monomer, whereas when highly purified it can form a dimer structure (Le Gaillard *et al.*, 1975; Mickelson *et al.*, 1982). However, Strel'chyonok and Avvakumov (1990) have presented evidence for a dimeric SHBG molecule in plasma. Following the interaction of ^{125}I-labeled SHBG with immobilized SHBG, they determined that the subunits bound with a K_d of 10^{-12}– 10^{-11} M. This K_d value suggests that at physiological concentrations (approx. 10^{-7} M) SHBG exists as a dimer.

It is now widely accepted that one mole of ABP/SHBG dimer binds 1 mole of steroid. A model has been proposed in which the steroid molecule is sandwiched between two subunits (Petra *et al.*, 1986c; Petra, 1991). On the basis of fluorescent quenching and anisotropy studies, Casali *et al.* (1990) presented results to support this model; the steroid-binding interaction was linked to the protein subunit interactions. However, the assumption that one mole of steroid binds per dimer molecule is debatable. Experiments have determined the binding capacity based on the number of steroid molecules bound per molecule of protein (calculated by Scatchard analysis). Although several investigations have concluded that one molecule binds per ABP/SHBG dimer (Petra *et al.*, 1983, 1986c; Suzuki and Sinohara, 1984), Strel'chyonok *et al.* (1983) have presented evidence that one molecule of DHT binds per monomer. Furthermore, Petra *et al.* (1988) found that after storage at 4°C with steroid, two moles of steroid bound per mole of dimer. They proposed that the phenomenon supported a negative cooperative mechanism of binding.

All of these experiments suffer from the same experimental problems. First of all, the measurements of protein concentrations are difficult and may be slightly inaccurate. There could also be small errors in the determination of bound steroid by steady-state gel electrophoresis or equilibrium dialysis. Furthermore, if a fraction of the protein is denatured during purification to form a non-steroid-binding state, the methods used would underestimate the binding capacity of the native protein. Petra *et al.* (1986c) presented evidence for this phenomenon with macaque SHBG. Whereas rabbit, baboon, and human SHBG yielded approximately one binding site per dimer, macaque SHBG yielded 0.64 sites per dimer. It was suggested that this difference was due to protein denaturation. Obviously, this phenomenon could also be occurring with the other SHBGs, resulting in the underestimation of the steroid-binding capacity. Thus, although these SHBG studies utilized acceptable methodologies, because of possible errors in each measurement, the final calculations could easily be misleading. This author feels that conclusive evidence is lacking to support a one binding site per dimer model. One should keep an open mind on the subject.

C. CLONING OF THE cDNAs

Since 1985, much has been learned about the structure of ABP/SHBG from cDNA cloning studies. The cDNAs encoding human SHBG and human, rat, rabbit, and mouse testicular ABP have been isolated and sequenced (Joseph *et al.*, 1985, 1987; Gershagen *et al.*, 1987b, 1989; Hammond *et al.*, 1987, 1989; Que and Petra, 1987; Reventos *et al.*, 1988; Wang *et al.*, 1989). These studies (Joseph *et al.*, 1985, 1987; Reventos *et al.*, 1988) revealed that rat ABP is synthesized from a 1.7-kb mRNA as a 403-amino-acid residue precursor (44,539 Da), which after removal of the signal peptide forms the 373-residue subunit (41,183 Da). Reventos *et al.* (1988) identified a rat ABP cDNA that differed on one nucleotide residue from the cDNA described by Joseph *et al.* (1987), changing His-317 to an Arg residue. The analysis of partial human SHBG cDNA clones (Hammond *et al.*, 1987; Gershagen *et al.*, 1987b; Que and Petra, 1987) confirmed the amino acid sequence determined by protein analysis (Walsh *et al.*, 1986). One study (Hammond *et al.*, 1987) described a cDNA clone that encoded all of the 373-residue SHBG (40,509 Da), but lacked some of the sequence encoding the signal peptide. The full sequence of the human SHBG signal peptide was later determined from the gene sequence (Hammond *et al.*, 1989; Gershagen *et al.*, 1989). Human SHBG has a slightly smaller signal peptide (29 residues) than rat ABP. All three human SHBG cDNA cloning studies revealed a very short 3' untranslated region with an unusual poly(A) addition signal (ATTAAA) that appears to be part of the translation termination codon (TAA). It is of interest that one of the human SHBG cDNAs was isolated from a fetal liver cDNA library, suggesting that like the rat ABP, it is produced by the fetal liver (Que and Petra, 1987). Hammond *et al.* (1989) and Gershagen *et al.* (1989) characterized human ABP cDNAs and found the amino acid sequence to be identical to that of human SHBG, thus demonstrating that ABP and SHBG are products of the same gene. Similarly, Sullivan *et al.* (1991) demonstrated that fetal rat liver SHBG has an amino acid sequence identical to that of rat ABP. These last three studies also found alternate human ABP and rat SHBG mRNAs, which are discussed in the following.

Mouse ABP/SHBG also consists of 373 residues and shares 89% residue identity with rat ABP/SHBG (Wang *et al.*, 1989). As expected, rat ABP/SHBG shares much less homology with human ABP/SHBG than with mouse ABP/SHBG (68% identity). Figure 3 shows an alignment of the amino acid sequences of rat and human ABP/SHBG. Fig-

```
Rat-     LRHIDPIQSAQDSPAKYLSNGPGQEPVTVLTIDLTKISKPSSSFEFRTWD- 50
         **   * *** * **  ******* *  * * ***** * ***** ****
Human-   LRPVLPTQSAHDPPAVHLSNGPGQDPIAVMTFDLTKITKTSSSFEVRTWD- 50


Rat-     PEGVIFYGDTNTEDDWFMLGLRDGQLEIQLHNLWARLTVGFGPRLNDGRW-100
         ***********  ***********  ****** ** **** **** ****
Human-   PEGVIFYGDTNPKDDWFMLGLRDGRPEIQLHNHWAQLTVGAGPRLDDGRW-100


Rat-     HPVELKMNGDSLLLWVDGKEMLCLRQVSASLADHPQLSMRIALGGLLLPT-150
         * ** ** *** ** *** * * ***** *    ********* *
Human-   HQVEVKMEGDSVLLEVDGEEVLRLRQVSGPLTSKRHPIMRIALGGLLFPA-150


Rat-     SKLRFPLVPALDGCIRRDIWLGHQAQLSTSARTSLGNCDVDLQPGLFFPP-200
         * ** ********* *** **  ** * ** ***  ***    ** * **
Human-   SNLRLPLVPALDGCLRRDSWLDKQAEISASAPTSLRSCDVESNPGIFLPP-200


Rat-     GTHAEFSLQDIPQPHTDPWTFSLELGFKLVDGAGRLLTLGTGTNSSWLTL-250
         ** *** * ****** ** *** ** *  * * ** *** * *** *
Human-   GTQAEFNLRDIPQPHAEPWAFSLDLGLKQAAGSGHLLALGTPENPSWLSL-250


Rat-     HLQDQTVVLSSEAEPKLALPLAVGLPLQLKLDVFKVALSQGPKMEVLSTS-300
         ***** ***** * * *** ********   * **** ** *
Human-   HLQDQKVVLSSGSGPGLDLPLVLGLPLQLKLSMSRVVLSQGSKMKALALP-300


Rat-     LLRLASLWRLWSHPQGHLSLGALPGEDSSASFCLSDLWVQGQRLDIDKAL-350
         * ** * ** *** * ********** ****  ** ****** * **
Human-   PLGLAPLLNLWAKPQGRLFLGALPGEDSSTSFCLNGLWAQGQRLDVDQAL-350


Rat-     SRSQDIWTHSCPQSPSNDTHTSH-373
         **  ********** * *   **
Human-   NRSHEIWTHSCPQSPGNGTDASH-373
```

FIG. 3. Alignment of rat and human ABP/SHBG amino acid sequences. The upper and lower lines represent rat and human ABP/SHBG, respectively. Residue numbers are shown in the right margin and identical residues are indicated with an asterisk.

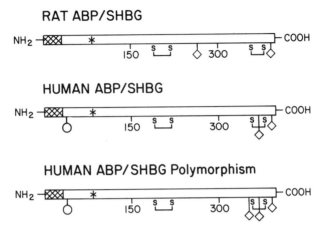

FIG. 4. Diagrams representing the precursors of rat and human ABP/SHBGs. The rectangle represents the ABP/SHBG precursor with amino acid residue numbers depicted below (403 residues). The numbering includes the signal peptide, which is indicated by crosshatching. The locations of the two disulfide bridges (characterized in hSHBG) and two sites of Asn glycosylation (Asn-X-Ser/Thr) are indicated with S–S and diamonds, respectively. Sites of O-glycosylation are depicted with a circle. The asterisk localizes the receptor-binding region.

ure 4 compares the rat and human ABP/SHBG precursor diagrammatically and depicts the locations of the disulfide linkages and potential sites of glycosylation. To clarify the numbering system, it should be pointed out that Fig. 3 represents the mature proteins and Fig. 4 depicts the precursor proteins with the signal peptides. All residue numbers that follow refer to the mature 373-residue ABP/SHBG. Several regions are highly conserved between the rat and human proteins, especially from residues 33–128. As discussed in the following, this region contains at least part of the receptor binding domain. Also four Cys residues (residues 164, 188, 333, and 361), which form two disulfide linkages (Walsh *et al.,* 1986), are conserved (Fig. 4). One potential site of N-glycosylation at residue 367 is shared by the two proteins (Fig. 4). In addition, each contains a unique site; human ABP/SHBG at residue 351 and rat ABP/SHBG at residue 244. The human ABP/SHBG sequence contains a potential O-glycosylation site at residue 7 that is absent in the rat sequence (Fig. 4). The most dissimilar region occurs at residues 132–138, which are near the putative steroid-binding site (see Section III,E). All the rat ABP/SHBG properties presented here are also conserved in mouse ABP/SHBG.

The sequence of rabbit SHBG and rabbit ABP has been presented based on amino acid sequence data (Griffin *et al.*, 1989) and partial cDNA sequence data (Wang *et al.*, 1989), however, there are numerous residue discrepancies in the two amino acid sequences. Because the amino acid sequence deduced from the nucleotide sequence should be more accurate than direct sequence data, the deduced sequence is probably more correct. This interpretation is supported by an earlier publication (Petra *et al.*, 1988) of the rabbit sequence, which had 16 differences with their latter publication (Griffin *et al.*, 1989). Griffin *et al.* noted the corrections and discussed the limitations of some of the methods used. These differences and the limitations cast doubt on the final result. Nevertheless, it is of interest that the rabbit ABP/SHBG sequence is much more closely related to human than rat or mouse (Wang *et al.*, 1989; Griffin *et al.*, 1989). Because the rabbit has circulating SHBG similar to man, this similarity may be related to its physiological function. Analysis of purified rabbit SHBG revealed that the denatured protein was smaller than primate SHBG (Fig. 2). In agreement, the amino acid sequence data revealed that rabbit SHBG lacks six residues at the N terminus that are present in rat, mouse, and human (Griffin *et al.*, 1989). Whether this represents the true N terminus of the protein awaits characterization of a full-length cDNA clone or the genomic DNA. The missing sequence could represent the product of proteolysis. Purified human SHBG was lacking the N-terminal Leu residue, which was attributed to proteolysis (Walsh *et al.*, 1986). Rabbit ABP/SHBG has potential sites of N-glycosylation identical to those in the human protein (Wang *et al.*, 1989; Griffin *et al.*, 1989), but like rat ABP/SHBG, it appears to lack the O-glycosylation site at residue 7 (Griffin *et al.*, 1989).

Characterization of the cDNAs provided conclusive evidence that the structural determinations based on biochemical studies were correct. The cloning of rat and human ABP/SHBG cDNAs and expression (transient or stable transformation) in eukaryotic cells yielded ABP/SHBG with the same chemical and steroid-binding properties as the native proteins (Sullivan *et al.*, 1991; Bocchinfuso *et al.*, 1991; Hagen *et al.*, 1992). Each cDNA yielded the same heterogeneous structure as wild type and the intact protein bound steroids with the same affinity as the wild-type protein, thus, yielding final proof that a single mRNA encodes a single ABP/SHBG peptide chain.

D. Glycosylation and Chemical Properties

Early studies on the purified proteins indicated that rat ABP and human SHBG were glycosylated (Danzo *et al.*, 1991b). It was later demonstrated that purified human SHBG contained two biantennary N-linked oligosaccharide chains and one O-linked oligosaccharide moiety (Avvakumov *et al.*, 1983; Danzo and Black, 1990a; Danzo *et al.*, 1989b). Rat and rabbit ABP also appeared to have two N-linked chains, but it was not clear if rat ABP also had an O-linked side chain (Danzo and Bell, 1988; Danzo *et al.*, 1989a). One clear difference in the two proteins is the ability to bind to concanavalin-A (Hsu and Troen, 1978; S.-L. Cheng *et al.*, 1984; C.-Y. Cheng *et al.*, 1985). Serial lectin chromatography has demonstrated that rat ABP has a greater percentage of triantennary and less biantennary glycans than human SHBG and rat ABP has a higher level of fucosylation (Danzo and Black, 1990b; Danzo *et al.*, 1991b). Interestingly, male rat plasma ABP, which is thought to originate in the testis, differs in molecular weight and carbohydrate content from epididymal ABP (Danzo *et al.*, 1987). S.-L. Cheng *et al.* (1984) demonstrated differences in concanavalin-A binding of the plasma and epididymal proteins and Danzo and Bell (1988) showed that deglycosylation of epididymal or plasma ABP resulted in the same size product. Human ABP appears to have glycosylation properties similar to those of rat ABP, but preparations contain an unknown amount of SHBG (Cheng *et al.*, 1985).

Mutagenesis experiments demonstrated that both potential N-glycosylation sites are glycosylated in rat ABP/SHBG (Joseph *et al.*, 1992). Mutant rat ABP/SHBG proteins were constructed that eliminated one or both of the two potential sites of asparagine(Asn)-linked glycosylation. Substitution of the Asn residue in either consensus sequence for Asn-linked glycosylation with an Ile residue resulted in increased mobility of the immunoreactive ABP/SHBG species. These changes were consistent with the loss of an Asn-linked oligosaccharide. Substitution of both Asn residues yielded a single immunoreactive species in the medium and cell extracts that migrated as a 39,000 M_r protein. These results confirmed the biochemical experiments indicating that the molecular weight heterogeneity of ABP/SHBG is due to differential Asn-linked glycosylation of both potential sites. The two bands observed with SDS–PAGE represent ABP/SHBG monomers containing one or two N-linked oligosaccharide moieties. Secreted wild-type ABP/SHBG and all three glycosylation mutants had essentially the same affinity for DHT ($K_d = 1–3 \times 10^{-9}$ M). Using mutagenesis (Asn to Gln conversions), Hammond and co-workers (Bocchinfuso *et al.*, 1992b) demonstrated that both potential N-glycosylation sites in human ABP/SHBG were also glycosylated. In addition, they confirmed that Thr residue 7 was a site of O-glycosylation. Western immunoblot analysis of the recombinant proteins dem-

onstrated that the subunit heterogeneity was due to differential glycosylation of the two N-linked sites and was not affected by removal of
the O-linked carbohydrate at residue 7. Like rat ABP/SHBG, mutants
lacking attached oligosaccharides were secreted and bound steroid
with apparently the same affinity as wild type. They also demonstrated by gel filtration that mutants lacking carbohydrate moieties
were all apparently homodimers, indicating that carbohydrates are
not involved in subunit association.

No strong evidence exists for posttranslational processing other
than oligosaccharide modifications. However, after chemical or enzymatic deglycosylation of ABP or SHBG, some charge heterogeneity
remained (Danzo and Bell, 1988; Danzo et al., 1989a,b). These data
suggest the presence of other modifications, such as phosphate, but the
remaining heterogeneity could be due to incomplete deglycosylation.

All ABP/SHBG sequences that have been characterized contain four
conserved Cys residues at residues 164, 188, 333, and 361. These Cys
residues in human ABP/SHBG have been shown to form disulfide
bridges in the mature protein. The locations of the two disulfide bonds
were established by peptide analysis of unreduced SHBG (Walsh et al.,
1986). Cys[164] and Cys[188] form one bridge and Cys[333] and Cys[361] form
the other bridge (Fig. 4). Mass spectrometry of rabbit SHBG confirmed
the placement of the disulfide bonds (Griffin et al., 1989). ABP/SHBGs
in other species that contain the same Cys residues are presumed to
have the same disulfide bridge structure. Mutagenesis experiments
have demonstrated the importance of these structures in function (Joseph and Lawrence, 1993). Conversion of rat ABP/SHBG Cys[188] or
Cys[333] to a Ser residue eliminated DHT-binding activity and secretion
by the COS cells (Table II). It is of interest that these Cys residues are
conserved in the ABP homologs, including laminin and protein S. In
protein S and laminin, they have been shown to exist in the same
oxidized structure as ABP/SHBG (Dählback et al., 1986, 1990) (see
Fig. 15). This conservation of Cys residues in the distantly related
homologs further supports the importance of these residues in maintaining the proper conformation for activity.

Another interesting finding from the early chemical studies of the
protein was its unusual amino acid composition. For example, rat ABP
was found to be very hydrophobic; it consisted of 20% leucine residues
(Feldman et al., 1981). The cDNA cloning experiments confirmed this
unusual composition. The hydrophobic nature of ABP/SHBG can be
better visualized in a hydrophilicity plot of the amino acid sequence.
Figure 5 shows a hydrophilicity plot comparing human and rat
ABP/SHBG. It can be easily seen that large portions of both proteins

TABLE II
ABP Mutagenesis

Mutation	Secretion[a]	DHT binding medium (%)	DHT binding *in vivo* (%)	
			3.7 nM	1.1 nM
Wild type	+	100	100	100
Asn 367 → Ile	+	94[b]	137	ND
Val 54 → Glu	−	<5	<20	<10
Met 139 → Arg	+	<5	<20	<10
Cys 188 → Ser	−	<5	<20	ND
Gly 234 → Glu	−	<5	<20	<10
Cys 333 → Ser	−	<5	<20	ND
ValIlePheTyr 57 → (Gly)$_4$	−	<5	56 ± 21[c]	20 ± 2[d]
MetArgIleAla 142 → (Gly)$_4$	−	<5	<20	<10
ArgLeuLeuThr 238 → (Gly)$_4$	−	<5	<20	<10
LeuProLeuGln 278 → (Gly)$_4$	−	<5	<20	<10
GluAspSerSer 329 → (Gly)$_4$	−	<5	<20	<10
Lys 285 → term	−	<5	<20	ND
Lys 348 → term	−	<5	<20	ND

Note. Mutant ABP cDNAs were created by site-directed mutagenesis in phage M13. ABP mutants were expressed in COS cells and analyzed for secretion and DHT-binding properties. DHT-binding activity was measured in the concentrated medium with the dextran–charcoal assay and *in vivo* directly on transformed COS cells. All data are presented as percentage of wild type. ND = not determined. Residue numbers initiate at the N terminus of the 373-residue subunit, not the precursor residues as previously used. Table from Joseph and Lawrence (1993).
[a]Immunoreactive ABP in the medium.
[b]From Joseph *et al.* (1992).
[c]Average of three experiments.
[d]Average of two experiments.

are hydrophobic, especially toward the C termini, and most of the hydrophobic regions are conserved between the two proteins. The significance of this hydrophobic characteristic is not known, but one would predict the one or more of the hydrophobic regions form the binding site for the hydrophobic ligand.

E. Functional Domains

ABP/SHBG consists of at least three functional "domains": a high-affinity steroid-binding region, the sequence that interacts with the plasma membrane receptor, and the sequence involved in subunit interaction.

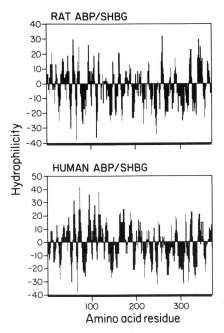

FIG. 5. Predicted hydrophilicity plot of rat and human ABP/SHBG. Hydrophilic regions are shown above the line (zero hydrophilicity) and hydrophobic regions below. The plot is according to the algorithm of Hopp and Woods (1981). The running average of the relative hydrophilicity value is given in the left margin. Amino acid residue numbers (1–373) are indicated below. The signal peptide is not included.

A common property of all studied ABP and SHBG proteins is their ability to bind sex steroids with high affinity. The steroid-binding specificities of ABP and SHBG have been reviewed in detail by Westphal (1986). There are several interesting features that are worthy of mention. Although there is considerable variation in steroid specificity between species, most bind DHT, T, and E_2 with high affinity (DHT > T > E_2). The binding of other biologically active steroids occurs at much lower affinities and probably is not biologically significant. Although the binding affinities vary between species, the K_a for DHT is generally between 1×10^9 and 1×10^{10} M. Testosterone and E_2 generally bind with slightly lower affinities, but within one order of magnitude of the DHT affinity. One exception is rabbit ABP/SHBG, which binds estrogens with an extremely low affinity (Rosner and Darmstadt, 1973; Danzo and Eller, 1975). The most significant difference between rat ABP and human SHBG is the half-life of the DHT complex, which is 5–6 min for rat ABP (Schmidt et al., 1981) and 30–

60 min for human SHBG (Hsu and Troen, 1978; Namkung *et al.*, 1989) at 0°C. The ABP/SHBG affinities for DHT and testosterone are nearly identical to or slightly lower than the affinities of the androgen receptor. However, there are several notable differences. ABP/SHBG has an extremely fast on–off rate, with a half-life of the DHT complex of 5–60 min at 0°C, depending on the species of origin. On the contrary, the androgen receptor–DHT complex is extremely stable with a half-life of greater than 24 hr. Another notable difference is the inability of ABP/SHBG to bind to several agonists and antagonists of the receptor, such as cyproterone acetate (Danzo and Eller, 1975). These differences, especially the on–off rates, undoubtedly reflect the physiological functions of SHBG and ABP.

Affinity labeling experiments to identify the steroid-binding region have led to ambiguous results, implicating various regions from residue 134 to the C terminus (Hammond *et al.*, 1987; Petra *et al.*, 1988; Grenot *et al.*, 1988, 1992; Khan and Rosner, 1990; Namkung *et al.*, 1990; Danzo *et al.*, 1991a). These experiments utilized either [3H]17β-hydroxy-4,6-androstadien-3-one (Δ-6-testosterone) or 17β-[(bromoacetyl)oxy]-5α-androstan-3-one (17β-bromoacetoxydihydrotestosterone) to affinity-label SHBG or ABP. After labeling and cleavage, the radioactive fragments were isolated and sequenced. Three laboratories reported active site labeling between residues 134 and 150. The first report was by Grenot *et al.* (1988), who identified human SHBG Met[139] as the site of photoaffinity labeling by Δ-6-testosterone. Later, in a more comprehensive study, they demonstrated that estradiol also interacts with the same site (Grenot *et al.*, 1992). Edman degradation and liquid secondary ion mass spectrometry of the labeled peptides demonstrated that Met[139] was conjugated with labeled Δ-6-testosterone or Δ-6-estradiol. Similarly, Danzo *et al.* (1991a) identified residues 141–150 of rat ABP using Δ-6-testosterone as photoaffinity ligand. Using 17β-bromoacetoxydihydrotestosterone as the alkylating agent, Namkung *et al.* (1990) identified Lys[134] as being in the steroid-binding site of human SHBG. Three other studies have identified other regions that may be involved in steroid binding. The results of Hammond *et al.* (1987) and Petra *et al.* (1988) suggested that at least a portion of the steroid-binding domain of human SHBG is located in a hydrophobic pocket near the C terminus. Khan *et al.* (1990) identified SHBG His[235] as the covalent site of attachment for 17β-bromoacetoxydihydrotestosterone. Taken together, these data indicate that several regions of the protein form the steroid-binding pocket.

Recently mutagenesis has been used to identify residues of ABP/SHBG that are required for steroid binding. In each experiment, rat or human ABP/SHBG cDNA was altered by deletion or site-directed mutagenesis and cloned into an expression vector, and the recombinant proteins were expressed in heterologous cells (COS or CHO) for analysis. Although these types of experiments can aid in the identification of essential residues for steroid binding, this method may not identify residues that are in the active site, but can also identify residues that are required for proper folding of the protein. In one study (Joseph and Lawrence, 1993), the analysis of truncated ABP/SHBG proteins revealed that removal of 26 or more residues from the C terminus eliminated secretion and DHT-binding activity. Also, alteration of amino acid residues to residues with different chemical properties (nonconservative) by site-directed mutagenesis from residue 54 to residue 333 resulted in elimination of DHT binding for 9 of 10 mutants and reduced DHT affinity for one altered protein (ABP/SHBG[Gly 54–57]) (Table II). Only 1 of the 10 mutant ABP/SHBG proteins was secreted by the COS cells. This secreted mutant ABP/SHBG[Arg139]) exhibited no detectable DHT-binding activity. Thus, the data demonstrated that nonconservative modifications of the ABP/SHBG primary sequence throughout the molecule have a detrimental effect on steroid binding and secretion. Furthermore, these data revealed that alterations of the cell membrane receptor-binding region of ABP (residues 54–57) eliminate or reduce the affinity for DHT.

Bocchinfuso *et al.* (1992a) have also used mutagenesis to aid in the identification of the steroid-binding site. In a novel experiment, they made a human–rat chimera protein that consisted of 205 human residues in the N-terminal portion and 168 residues of rat ABP/SHBG in the C-terminal portion. Interestingly, the chimera protein bound steroids with the properties of human SHBG, indicating that the binding site is present within the first 205 residues. In a similar study (Danzo and Joseph, in press), all or parts of 48 rat ABP/SHBG residues were converted to human ABP/SHBG residues flanking both sides and including the putative steroid-binding pocket (residues 139–150). Many of the resulting mutants had increased or decreased estradiol-binding affinities whereas there was little effect on DHT binding. Interestingly, the half-lives of the DHT–mutant ABP/SHBG complexes were the same as those of wild-type rat ABP/SHBG (5 min at 0°C) and not as those of the human protein (30–60 min at 0°C). The data supported a model in which distant regions come together to form the steroid-binding pocket.

Two other mutagenesis studies implicated residue 139 in the steroid-binding site. Sui *et al.* (1992) presented evidence that conversion of Met[139] to Lys or Ser decreases steroid-binding activity of human ABP/

SHBG, however, the presence of bovine SHBG in the recombinant protein preparations makes interpretation of the results ambiguous. Bocchinfuso *et al.* (1992a) demonstrated that conversion of human ABP/SHBG Met^{139} to a Trp residue reduces the steroid-binding affinity and increases the dissociation rate of DHT. Conversion of residues 133 and 136 to Asp and Gln residues (rat amino acid residues), respectively, had no effect on steroid-binding activity. The mutagenesis data, taken together with the affinity-labeling experiments and the conserved residues between rat and human ABP/SHBG, indicate that at least part of active site is located around residues 139–150, but most of the protein is required to maintain the conformation of the active site. However, the involvement of other regions in the steroid-binding site cannot be ruled out. The finding that conversion of 48 rat ABP/SHBG residues to human sequence around the putative steroid-binding pocket did not confer human DHT-binding characteristics supports the presence of another region(s) in the active site. This would explain why several regions of the protein were identified by affinity labeling.

A single study has convincingly identified at least part of the ABP/SHBG receptor-binding domain (Khan *et al.*, 1990). They digested highly purified human SHBG with trypsin and identified a 10-residue peptide that competed with labeled SHBG for the prostate receptor. This peptide corresponds to SHBG amino acid residues 48 to 57. The peptide interacted with the receptor with an affinity 10^{-5} of the affinity for SHBG ($K_i = 2 \times 10^{-9}\ M$). The nature of this peptide was confirmed by demonstrating that a synthetic peptide also competitively inhibited SHBG binding to the receptor. These data clearly demonstrated that this sequence represents at least part of the receptor-binding domain. The relatedness of this sequence with the sequence in rat ABP/SHBG and its homologs bespeaks its functional importance. Whereas rat and human ABP/SHBG share only 68% identity of amino acid residues, this decamer sequence is identical in rat and human. Furthermore, the sequences surrounding this peptide are much more highly conserved than the conservation of the overall sequence. This sequence homology is also partially conserved in the ABP/SHBG homologs (Baker *et al.*, 1987; Joseph and Baker, 1992). This region is the most highly conserved area between ABP/SHBG and vitamin K-dependent protein S, laminin, merosin, agrin, and the *Drosophila crumbs* and *slit* proteins (see Section IV,D). For example, the overall sequence homology between ABP//SHBG and protein S is 26%, but the conservation of the decamer sequence is 70% in bovine and human protein S. This conservation of sequence in the ABP/SHBG receptor-binding region suggests that the ABP/SHBG homologs may share receptors or a class of receptor.

Obviously mutagenesis and affinity-labeling experiments alone will not identify the functional domains of ABP/SHBG. The full interpretation of these results must await X-ray crystallography and other physical studies. Because glycosylated proteins are refractory to crystallization owing to heterogeneity, the finding that unglycosylated ABP/SHBG binds steroids with the same affinity as wild type should facilitate crystallization and determination of the ABP/SHBG tertiary structure. These experiments should also be able to address the hypothesis that ABP/SHBG is an allosteric protein (i.e., changes in conformation upon binding steroid modulate interaction with the receptor).

F. POLYMORPHISMS OF SHBG

Several laboratories have identified electrophoretic variants of human SHBG, which were detected by SDS–PAGE and/or isoelectric focusing (Luckock and Cavalli-Sforza, 1983; Gershagen et al., 1987a; Larrea et al., 1990; van Baelen et al., 1992); each study appears to have identified the same variant. The polymorphism was seen when analyzing serum or plasma from individuals and was sometimes not observed in pools of SHBG, which are used for most studies. Generally, the allelic frequencies of the two forms were 0.9 and 0.1, but there was considerable variation in the distribution between populations from various countries (van Baelen et al., 1992). In all families, inheritance was compatible with a simple Mendelian transmission. The variant protein (denatured and reduced) migrated as 56,000, 52,000, and 48,000 M_r components instead of the typical 52,000 and 48,000 M_r species. Figure 6 shows the banding pattern from a heterozygous individual and individuals with both homozygous alleles. The two forms were shown to have the identical N-terminal sequence and it was suggested that an additional carbohydrate chain could account for the increases mass of the variant (Gershagen et al., 1987a). Later, Power et al. (1992) demonstrated that the gene encoding the variant high-molecular-weight form contained an Asn codon in place of the typical Asp codon (SHBG amino acid residue 327). This change generated a potential site of N-glycosylation that was shown to be glycosylated in the recombinant variant SHBG (Fig. 4). Thus, the three protein components described are SHBG peptide chains with one, two, or three N-linked oligosaccharide moieties. In each study, the variant SHBG

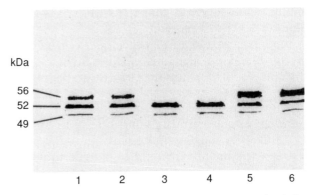

FIG. 6. SDS–PAGE of SHBG from individuals with the 1–1, the 1–2, and the 2–2 pattern. Enriched SHBG samples with known phenotype were separated on an 8–18 gradient gel (ExcelGel SDS), transferred to nitrocellulose, and immunostained. Lanes 1 and 2: samples with the 1–2 pattern (heterozygotes); lanes 3 and 4: samples with the 1–1 pattern (homozygous SHBG[1] carriers); lanes 5 and 6: samples with the 2–2 pattern (homozygous SHBG[2] carriers). From van Baelen et al. (1992).

bound steroid with the same characteristics as the typical SHBG and there appeared to be no major pathology associated with it. ABP from individuals with the variant SHBG has not been analyzed, but it would be expected to have the same properties. Thus, the only known variant of ABP/SHBG has no pathological consequences.

IV. PHYSIOLOGICAL ROLES

A. ABP: A MARKER OF SERTOLI CELL FUNCTION

Spermatogenesis is primarily regulated by the Sertoli cells of the seminiferous tubules (Bardin et al., 1988; Griswold et al., 1988; Russel and Griswold, 1993). High concentrations of androgens produced by the interstitial Leydig cells and other paracrine factors from germ cells, Leydig cells, and peritubular cells are required for Sertoli cell function and spermatogenesis (Skinner and Fritz, 1985; Mather et al., 1983). Because Sertoli cells contain nuclear androgen receptors, androgen action is at least partly direct (Sanborn et al., 1977). In addition, pituitary FSH is required for the differentiation of the Sertoli cell and mature Sertoli cell function (Fritz, 1979; Means et al., 1980; Ritzen et al., 1981). Sertoli cells maintain direct contact with the developing germ cells and secrete numerous proteins, some of which serve as regulatory factors for specific stages of spermatogenesis (Wright *et al.*, 1981; Kissinger *et al.*, 1982; DePhillip *et al.*, 1982). As a result of hormonal and paracrine influences, Sertoli cells have a cyclic function that depends on the stage of the cycle of the seminiferous epithelium (Parvinen, 1993). The concentrations of several testicular proteins have been shown to vary with the stages of the seminiferous epithelium. One of these proteins is ABP (Ritzen *et al.*, 1982).

Most ABP is secreted into the lumen of the seminiferous tubule, but a small amount appears to be secreted basally and passes to the blood (Gunsalus *et al.*, 1980). Figure 7 demonstrates the immunocytochemical location of ABP in the luminal and adluminal region of the seminiferous tubule (Fig. 7, arrows). A small amount of immunoreactive ABP is also present in the basal regions of the Sertoli cells (Fig. 7, arrowheads). Because the hormonal requirements for ABP production and secretion are similar to the hormonal requirements for spermatogenesis (Anthony *et al.*, 1984, 1984a), the protein is used widely as a marker for Sertoli cell function and development (Gunsalus *et al.*,

Fɪɢ. 7. Immunoperoxidase localization of ABP in mouse testis. Tissues were fixed and sectioned and antiserum to rat ABP (Feldman *et al.*, 1981) was used at a dilution of 1:10,000 according to the "double PAP" method. Immunoreactive ABP is present primarily in the lumen of the seminiferous tubules and is absent in interstitial tissue. It is apparent that immunoreactive ABP is also present in the apical (arrows) and basal (arrowheads) compartments of the seminiferous tubule, presumably in the cytoplasm of Sertoli cells. From Wang *et al.* (1989).

1981). It has especially been useful in monitoring hormonal effects on Sertoli cells (Ritzen *et al.*, 1981; Sanborn *et al.*, 1975; Kotite *et al.*, 1978; Bardin *et al.*, 1981) (see Section V,C).

From the testis, ABP is transported to the epididymis with the maturing sperm. ABP's function in the epididymis is unknown, but it is thought to aid in the transport of androgens to the epididymis. Because the intraluminal ABP : androgen ratio in the rat is approximately 1:1, ABP may establish and maintain the high epididymal androgen levels required for epididymal function (i.e., sperm maturation) (Turner *et al.*, 1984). Turner and Roddy (1990) and Turner and Yamamoto (1991) have shown that intraluminal ABP enhances net androgen uptake by caput epididymal tubules. Also, it has been speculated that ABP may regulate the metabolism of testosterone or DHT, but evidence is lacking. Possibly, ABP is involved in the regulation of 5α-reductase, which maintains the high level of DHT required for epididymal function.

As ABP enters the caput epididymis, it is rapidly internalized by the epithelium. Immunohistochemical experiments localized epididymal ABP in the supranuclear–Golgi region of the epididymal epithelial cells (Feldman *et al.*, 1981; Pelliniemi *et al.*, 1981). Figure 8 demonstrates immunoreactive ABP in the caput lumen and within the epithelial cells lining the lumen in the Golgi region of the cytoplasm. There is burgeoning evidence that this uptake is mediated by a receptor-mediated process. Veeramachaneni *et al.* (1990) found that ABP was apparently spared from endocytosis along with the bulk protein in the rete testis fluid entering the efferent ducts and was specifically endocytosed in the proximal portion of the caput epididymis. They later showed that the apparent sparing in the efferent ducts was actually due to the recycling of endocytosed ABP (Veeramachaneni and Amann, 1991). Gerard *et al.* (1988) have presented radioautographic evidence that ABP is specifically internalized by a receptor-mediated process in the rat epididymis principal epithelial cells. Iodinated ABP or ABP photoaffinity-labeled with ³H was internalized and was mainly present in the supranuclear region concentrated over coated structures, endosomes, multivesicular bodies, and the Golgi apparatus (Fig. 9). Interestingly, in a related study they found labeled protein in the nuclei of monkey epididymis, including the nuclear envelope (Gerard *et al.*, 1990). The mechanism of internalization seemed to involve endocytosis by the principal cells that led to labeling of the nuclear compartment. Gerard *et al.* (1992) have also demonstrated SHBG endocytosis by coated vesicles in monkey germ cells. Although no internalized SHBG was found in the nuclear area of late spermatids or sperm, spermatogonia, spermatocytes, and early

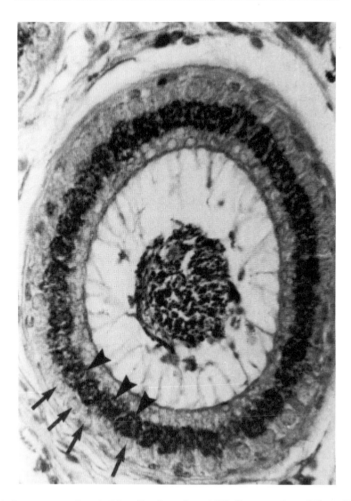

FIG. 8. Immunocytochemical localization of rat ABP. Cross section of the epididymal duct from the region of the caput, showing immunoreactive ABP (dark stain) in the lumen and all epithelial cells lining the duct. In all epithelia cells, ABP staining is confined to the supranuclear (Golgi) region (arrowheads) of the cytoplasm. Complete arrows point to nuclei in the basal region of epithelial cells. Note lack of staining in the basal region of epithelial cells and in the surrounding smooth muscle and connective tissue (× 138). From Feldman et al. (1981).

spermatids contained internalized SHBG in the nucleus. In a more recent study, Gerard et al. (1994) used transmission electron microscopy and autoradiography to demonstrate that ABP is internalized by rat germ cells. Receptor-mediated endocytosis of ABP was observed from a fluid phase or from the Sertoli cell cytoplasm. ABP was internalized by

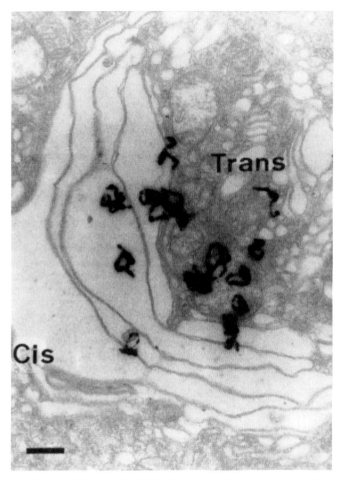

FIG. 9. Distribution of silver grains in the Golgi area 35 min after *in vivo* intraluminal injection of [125I]iodo-ABP–DHT complex (0.5 μl–20 ng ABP, 4 nCi). Clusters of silver grains are concentrated over the dilated Golgi saccules of the transface. Sections were exposed 2 months at 4°C. *Bar* = 0.5 μm. From Gerard *et al.* (1988).

spermatocytes, round spermatids, and elongated spermatids, but the intracellular cite of ABP accumulation varied throughout spermatogenesis; nuclear labeling decreased whereas endosomal labeling increased during maturation. They suggest these data indicate a role for ABP in spermatogenesis, either directly or by serving as a steroid trans-membrane carrier. Recent transgenic mouse experiments also support this hypothesis (see Section D). Obviously, the receptor(s) plays an important role in ABP's function in the testis and epididymis. Other than their binding characteristics, little is known about the characteristics of the receptors (see Section IV,C).

B. SHBG: THE MAJOR PLASMA SEX STEROID CARRIER PROTEIN

Sex steroid hormones are bound by several plasma proteins, including albumin, which binds androgens and estrogens with low affinity, and SHBG, which binds these hormones with high affinity (Siiteri *et al.*, 1982). In general, the physiological role of SHBG is poorly understood and in recent years the function has been the subject of heated debates. One long-standing theory is the "free hormone hypothesis" (Mendel, 1989, 1992). This hypothesis states that only free (unbound) steroids are available for uptake by target cells from the capillaries; only free hormones are biologically active. A corollary of this hypothesis is that SHBG acts as a reservoir or buffer of sex steroids that can be made available to cells by dissociation of bound steroid (Rosner, 1990). Ekins (1990) has reviewed in detail the measurements of free hormones and the validity of the hypothesis based on mathematical models. The rapid dissociation rate of SHBG–steroid complexes may be related to this function (Mendel, 1990). Mendel (1990, 1992) and Rosner (1990) have reviewed the evidence that supports the free hormone hypothesis. It has been shown in several studies that SHBG can inhibit the uptake and metabolism of androgen in rat prostate, prostate cultures, and prostate cell lines (Lasnitski and Franklin, 1972, 1975; Mercier-Bodard *et al.*, 1976). Damassa and Gustafson (1988) found that chronic infusions of SHBG in male rats produced marked increases in total plasma testosterone, but had little effect on the concentration of free testosterone and the maintenance of the sex accessory glands. SHBG has also been shown to reduce the metabolic clearance rate of plasma testosterone, a process that would tend to increase androgen availability (Petra *et al.*, 1985).

Several other experiments lend further support to the free hormone hypothesis. Damassa *et al.* (1991) have shown that purified SHBG inhibits androgen-induced proliferation in LNCaP prostate cells in a dose-dependent manner. The effect appeared to be due primarily to the high affinity binding of SHBG; the proliferative response induced by R1881, which binds poorly to SHBG, was not affected by added SHBG. The presence of SHBG also lowered the uptake and metabolism of DHT. In a related study, Roberts and Zirkin (1993) demonstrated that ABP inhibits androgen-dependent transcription in a mouse Sertoli cell line (MSC1). *In vivo* experiments also suggested that when ABP concentrations exceed the concentrations of testosterone in the seminiferous tubule fluid, spermatogenesis is inhibited. Both of these studies (Damassa *et al.,* 1991; Roberts and Zirkin, 1993) indicated that ABP/SHBG can inhibit androgen action, presumably by binding and sequestering androgen and lowering the level of free hormone.

It has also been proposed that SHBG acts to regulate the free androgen–estrogen balance (Anderson, 1974). This model is based on the fact that SHBG binds testosterone with a greater affinity then estradiol and estradiol binds more tightly to albumin than testosterone. In support of this model, it was found that changes in SHBG concentrations differentially affect the concentrations of testosterone and estradiol (Anderson, 1974). However, the physiological importance of the androgen–estrogen ratio is unknown. Rosner (1990) presented a critique of this model and concluded that it seems doubtful that SHBG evolved to regulate the androgen–estrogen balance.

Siiteri and Simberg (1986) have reviewed the unusual hormonal characteristics of the squirrel monkey. These monkeys have plasma levels of testosterone that are 20- to 50-fold higher than those in humans; the levels are much higher than the concentration of plasma SHBG, which has a relatively low affinity for androgens. They pointed out that the free hormone concept is untenable in this situation.

Other factors in blood and tissues may affect the steroid-binding properties of ABP and SHBG and modulate the level of free hormone. For example, Nunez and co-workers (Martin *et al.,* 1986; Felden *et al.,* 1993) have shown that fatty acids modify the steroid-binding properties of human SHBG and rat ABP. Saturated nonesterified fatty acids caused an increase in binding to human SHBG and unsaturated fatty acids inhibited the binding. Their studies indicated that fatty acids induce conformational changes in the protein that are reflected in its electrophoretic, immunological, and steroid-binding properties. They proposed that fatty acids modulate the level of free hormones; this is especially pertinent in the epididymis, which contains extremely high concentrations of fatty acids.

Other models of SHBG action propose that it selectively delivers sex steroids to specific target tissues. Pardridge (1988a, b) has proposed a model in which SHBG-bound pools of steroid are available for uptake by cells. In this model the transport of steroid hormone from plasma occurs via a mechanism of enhanced dissociation. The interactions between SHBG and the surface of the microcirculation induce conformational changes in the protein that enhance release of bound steroid. This hypothesis is based on *in vivo* studies in anesthetized rats that have shown that human SHBG selectively delivers testosterone and estradiol to specific target tissues, such as the testis and prostate. The hypothesis was the subject of intense debate at the Second International Symposium on Steroid Hormones in Turin. Pardridge (1988a) reviewed the hypothesis and Ekins and Edwards (1988) offered an opposing opinion with a critique. Ellison and Pardridge (1990) subsequently showed that SHBG reduces the testosterone availability to 5α-reductase in the rat ventral prostate. Using an aortic infusion technique, they found that the presence of SHBG causes the selective sequestration of steroid hormone within the stromal compartment, where 5α-reductase is primarily located.

Expanding evidence supports the concept that these extracellular proteins do more than simply sequester sex steroid hormones. Alternative models for the delivery of sex steroids to target tissues involve the direct interaction of SHBG with target cells. One of the first indications that steroid carrier proteins participate directly in regulating events in target tissues was an observation by Amaral et al. (1974), who demonstrated the presence of a CBG-like protein in human liver nuclei. Later, it was suggested that ABP and SHBG utilize cell-surface receptors to regulate the uptake of ligand into target tissues (Lobl, 1981). In 1984, Strel'chyonok et al. described a SHBG recognition system in human decidual endometrium plasma membranes and, in 1985, Hryb et al. proposed a similar system in human prostate with the identification of a specific plasma membrane receptor for SHBG. Hryb's model proposed that steroid delivery occurs via interaction of SHBG with the plasma membrane (Fig. 10). Only unliganded SHBG can bind to the membrane, but once bound to the receptor, SHBG can bind steroid. The model leaves the internalization of SHBG an open question. Strel'chyonok and Avvakumov (1990) suggested that the mechanism of steroid hormone delivery may not involve penetration of SHBG through the membrane and proposed a "shuttle mechanism" (Fig. 11). In this model, only the SHBG–estradiol (not SHBG–testosterone) complex can interact with the membrane.

There is, however, considerable evidence for the internalization of SHBG and ABP by target cells. Noé et al. (1992) showed that administration of human SHBG to female rats, which do not have a circulating SHBG, increased the accumulation of E_2, but not DHT, in the uterus and oviduct. The effect was tissue specific, suggesting that a receptor-mediated process may be involved. Moreover, immunohistochemistry localized human SHBG concentrated on the apical border of the endometrium and oviduct epithelial cells. They noted that the staining pattern resembled the pattern of SHBG immunoreactivity found by Bordin and Petra (1980) in the monkey prostate. Similarly, Sinnecker

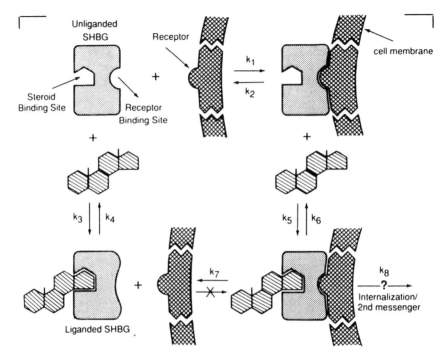

Fig. 10. Model of the interaction between steroids, SHBG, and the SHBG receptor. This model depicts our current understanding of this mystery; k_1, k_2, k_3, k_4, k_5, k_6, and k_7 are known. Although k_7 has been determined, the moment by moment concentration of the steroid–SHBG–receptor complex will be strongly influenced by its possible internalization. An understanding of the disposition of the complex awaits further experimentation. From Hryb et al. (1990).

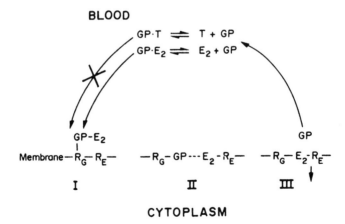

Fig. 11. A hypothetical shuttle mechanism for the SHBG-mediated estradiol (E_2) interaction with the target cell plasma membrane. GP, glycoprotein (SHBG); T, testosterone, R_G and R_E, membrane receptors for SHBG and E_2, respectively. From Strel'chyonok and Avvakumov (1990).

et al. (1988, 1990) found immunoreactive SHBG in the apical cytoplasm of the epithelium and tubule lumen of normal human endometrium and breast tissue. They concluded that either SHBG was
internalized from the plasma or it was being secreted by the epithelium. Three laboratories localized immunoreactive SHBG in the
epithelium of rat and human prostate (Bordin and Petra, 1980; Sinnecker *et al.*, 1988; Larriva-Sahd *et al.*, 1991). Porto *et al.* (1991) have
demonstrated the receptor-mediated internalization of SHBG in
MCF-7 breast cancer cells. As described earlier, expanding evidence
also supports the receptor-mediated endocytosis of ABP in germ cells
and the epididymal epithelium.

C. Membrane Receptors

Numerous studies have now shown that ABP and SHBG interact
with proteins in plasma membranes of target tissues. These studies
were last reviewed by Rosner (1990) and Strel'chyonok and Avvakumov (1990). The two laboratories have identified membrane binding proteins in the human prostate (Hryb *et al.*, 1985, 1989, 1990;
Rosner, 1990) and in human decidual endometrium (Avvakumov *et al.*,
1986; Strel'chyonok *et al.*, 1984; Strel'chyonok and Avvakumov, 1990)
and placenta membranes (Krupenko *et al.*, 1990). Rosner (1990) called
the binding protein a "receptor" because the protein has many of the
binding properties of typical membrane receptors and because SHBG
interaction with membranes appears to mediate signal transduction.
Strel'chyonok and co-workers have refrained from calling the protein a
receptor and call it a membrane recognition system for SHBG.
    Although both laboratories have described high-affinity binding in
membrane proteins from prostate and endometrium, the details of the
binding differ greatly; the largest differences were in the binding affinities, association rates, and the effects of bound steroids on SHBG
binding to membranes. In each case, two binding sites were identified
based on Scatchard analysis, but the affinities for SHBG were quite
different. Hryb *et al.* (1985) described a prostate membrane binding
site with a $K_d = 2 \times 10^{-8}$ *M*, whereas Strel'chyonok *et al.* (1984)
described a much higher affinity site ($K_d = 3 \times 10^{-12}$ *M*) in endometrium membranes. Interestingly, Hryb *et al.* (1989) found that the
solubilized prostate receptor had a higher affinity for SHBG ($K_d = 1.5
\times 10^{-9}$ *M*) than the membrane-bound receptor. The differences in
steroid effects are most interesting because the steroid-binding properties should reflect the function of the SHBG–membrane interaction.
Hryb *et al.* (1990) found that only unliganded SHBG binds to the

prostate membrane protein; once bound, SHBG is able to bind steroid. An allosteric model for SHBG binding has been proposed (Fig. 10).

The Byelorussian group found that estrogens were required for membrane binding; no binding was detected in the presence of testosterone (Fig. 11). Rosner (1990) has a different interpretation of their data, suggesting that the results are explainable if the moiety that binds to the receptor is unliganded SHBG. Therefore, the stronger the binding by steroids, the greater inhibition of binding to the receptor. The differences in their results may be technical in nature because there are major differences in their methodologies, such as binding temperatures, steroid concentrations, solubility states, and other experimental details. Comparisons of some of the data are difficult because the binding studies were done at 37 or 4°C. Strel'chyonok and Avvakumov (1990) have discussed the pitfalls of carrying out binding experiments at 37°C and suggested that it results in nonequilibrium conditions because of degradation. The differences in the apparent characteristics of the prostate and endometrium receptors may be technical, on the other hand the differences may reflect the presence of unique recognition systems in the prostate and endometrium.

Avvakumov et al. (1988) demonstrated that the carbohydrate moiety is required for membrane binding. Even though SHBG desialylation does not influence steroid binding, it does result in the complete loss of its ability to bind specifically to endometrium membranes.

During the past few years the SHBG receptor has been characterized from several other sources. An Italian group has demonstrated specific SHBG binding to several estrogen-sensitive tissues, including MCF-7 cells (breast cancer line), human premenopausal endometrium, human breast, and human liver (Fortunati et al., 1991, 1992a, b, 1993; Frairia et al., 1991, 1992). The effects of steroid hormones on binding were dependent on the steroid concentration, but in general sex steroids inhibited the binding of SHBG to the membranes. Interestingly, they found that at physiological steroid concentrations the soluble receptor cannot recognize the estradiol–SHBG complex, whereas the endometrium membrane-bound receptor can interact with both estrogen–SHBG and unliganded SHBG, but not androgen–SHBG complexes. Porto et al. (1992b) have also characterized the receptor from MCF-7 cells. They found a single class of high-affinity binding site that decreased in affinity when solubilized. Porto et al. (1992a) have also described a testicular receptor with similar characteristics and an estimated molecular weight of 174,000. This M_r value is close to the size estimated by Hryb et al. (1989) for the prostate receptor.

Specific binding sites for ABP have been described in epididymal membranes. Felden *et al.* (1992a,b) identified two types of binding sites in immature rats, which were calcium and pH dependent. Because photoaffinity-labeled ABP was used as ligand, the effect of steroid on the interaction could not be investigated. Interestingly, only one site was present in the adult rat. Similarly, another group (Krupenko *et al.*, in press) found two classes of binding using ^{125}I-labeled human SHBG as ligand, but calcium did not influence the binding. The high-affinity and low-affinity sites had K_d values of approximately 0.1 and 1 nM, respectively. The presence of physiological concentrations of DHT, testosterone, or estradiol increased the affinity of SHBG for the high-affinity receptor. Only estradiol markedly increased the affinity of SHBG for the lower-affinity site. Analysis of various regions of the epididymis revealed that the high- and low-affinity sites are differentially expressed; the high-affinity sites were found in the distal cauda and the proximal caput, where ABP is internalized (Fig. 12).

Two studies have addressed the tissue location of the receptors (Frairia *et al.*, 1992; Krupenko *et al.*, in press). Frairia *et al.* (1992)

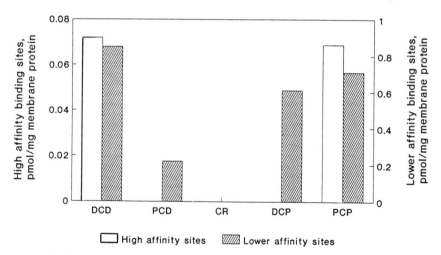

Fig. 12. Distribution of hSHBG binding sites in the epididymis. The epididymides of 10 adult (300–350 g) rats were divided into proximal caput (PCP), distal caput (DCP), corpus (CR), proximal cauda (PCD), and distal cauda (DCD) and plasma membranes were isolated from the pooled segments. The membranes were incubated with 1×10^{-11}–5×10^{-9} M [125]hSHBG that had been equilibrated with 1×10^{-7} M 5α-dihydrotestosterone in the presence or absence of 5×10^{-7} M unlabeled hSHBG. Dissociation constants and binding site concentrations were determined from Scatchard plots. From S. A. Krupenko, N. I. Krupenko, and B. J. Danzo, unpublished results.

found specific SHBG binding in membranes from endometrium, liver, and prostate, but not muscle or colon. The data indicated that the receptors are located only in sex steroid-dependent tissues, which contain estrogen receptors. Krupenko *et al.* (in press) have found that the high- and low-affinity receptor sites are differentially expressed. The epididymis, testis, and prostate contained both sites, whereas liver membranes had only the low-affinity activity. Spleen and brain were negative for both sites.

Taking together the various receptor studies, the only common property shared by the various receptors is that SHBG binding is specific. Steroids do appear to influence binding, but the effect is variable. It would appear that at least some of the differences are due to procedural variabilities between the various laboratories, but specific recognition systems may exist, especially for a tissue such as the epididymis, which rapidly internalizes large quantities of ABP.

Virtually nothing is known about the function of these membrane binding proteins in physiological processes. Understanding these mechanisms must await purification and characterization of the receptors and cloning of their cDNAs. The identified class of protein will give some insights into the type of receptor system by comparison with other receptors and membrane binding proteins. For example, if a G protein-linked receptor is identified, one would expect it is linked to a signal transduction process. In fact, there is some evidence that ABP and SHBG have hormonal activity. Rosner and co-workers have demonstrated a SHBG–DHT-dependent increase of cAMP in prostate cancer cells (Rosner *et al.*, 1988; Nakhla *et al.*, 1990). Induction of cAMP only occurred if SHBG was added to the cells before DHT or estradiol; administration of the SHBG–steroid complex had no detectable effect (Fig. 13). In another study, Nakhla *et al.* (in press) showed that estradiol, but not DHT, acts in concert with SHBG to produce a sevenfold increase in intracellular cAMP in cultured human prostate tissue. DHT, which blocks the binding of estradiol to SHBG, completely negated the effect of estradiol. They also demonstrated that this second messenger system is primarily located in the prostatic stroma and not the epithelium. An interesting phenomenon has been observed in the estradiol-induced proliferation of MCF-7 cells. Fortunati *et al.* (1993) found that even though SHBG did not alter the uptake of estradiol in the cells, SHBG did inhibit the estrogen-dependent increase in proliferation. This effect was only observed if SHBG was added to the cells before the steroid. Similarly, SHBG and estradiol increased the level of intracellular cAMP threefold when SHBG was added first (N. Fortunati, personal communication). These data appear to support the

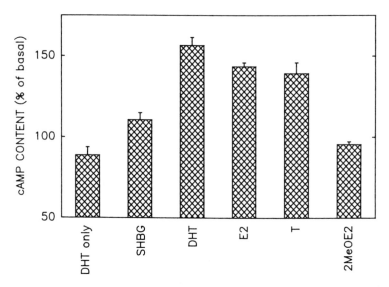

Fig. 13. Effect of steroid addition to LNCaP cells that have been preincubated with SHBG. The human prostatic carcinoma cell line, LNCaP, was incubated with highly purified human SHBG for 1 hr at 37°C, after which time the cellular accumulation of cAMP was determined. DHT, DHT added to cells *not* preincubated with SHBG; SHBG, cells preincubated with SHBG but to which steroid was not subsequently added; E2, estradiol; T, testosterone; 2MeOE2, 2-methoxyestradiol. From Nakhla *et al.* (1990).

model proposed by Rosner that is described earlier, in which only unliganded SHBG binds to the receptor.

D. Novel Functions

Before the ABP/SHBG amino acid sequence was deciphered, it was expected that the steroid-binding domain would share sequence homology with other DHT-, testosterone-, and estrogen-binding proteins, such as the nuclear receptors and the steroidogenic enzymes. To the contrary, comparisons of the amino acid sequences of these proteins with ABP/SHBG revealed no apparent similarities. However, Picado-Leonard and Miller (1988) reported that there were homologous sequences in the steroid receptors, steroidogenic enzymes, and ABP/SHBG that suggested a consensus steroid-binding sequence. They identified a 17-residue consensus sequence that they proposed was involved in steroid binding. However, computer analysis of these sequences with the ALIGN program (Dayhoff *et al.*, 1983) indicated that the apparent similarities scored no better than random chance

```
RAT ABP        6   P I Q S A Q D S P A K Y L S N G P              G Q E P V T V L T I D L T K
HUM SHBG       6   P T Q S A H D P P A V H L S N G P              G Q E P I A V M T F D L T K
HUM PROT S   249   P L N L D T K Y E L L Y L A E   Q          F A G V V L Y L K F R L P E
HUM LAMININ  303   P E Q C V V D A A L E Y V P G A H Q F G L T Q N S H F I L P T N Q S A
HUM MEROSIN  770   P C A A E S E P A L L I G S K Q F G L S R N S H I A I A F D D T K
FLY CRUMBS  1030                                           A T S L I S V T T E R
                                                      _____A_____

RAT ABP       37   I   S K P S S S F F F R T W D P E G V I F Y G D T N T E D D   W F M L G
HUM SHBG      37   I   T K T G S S F R V R T W D P E G V I T Y G D T N P F K D D   W F M L G
HUM PROT S   279   I   S R F S A E F D F R T Y D S E G V I Y G D T N P F K D D   W L L I A
HUM LAMININ  338   V R K K L S V R L S I R T L A S S G L I Y M A H Q N Q A D   Y A V L Q
HUM MEROSIN  803   V K N R L T I K L E V R T E A E S G L L F Y M A R I N H A D   F A T V Q
FLY CRUMBS  1041   E   E G Y D I N L Q F R T T L P N G V L A F G T T G E K N E P V S Y I L E

RAT ABP       71   L R D G Q L E I Q L H N L W A R L T V G F G P R L N D G R W H P V E L K M
HUM SHBG      71   L R D G R P E I Q L H N H W A Q L T V G A G P R L D D G R W H Q V E V K M
HUM PROT S   313   L R G G K I E V Q L K N E H T S K I T T G G D R W H N G L W N M V S V E E
HUM LAMININ  373   L H G G A L H F M F D L G K G R T K V S H P A L L S D G K W H T V K T D Y
HUM MEROSIN  838   L R N G L P Y F S Y D L G S G D T H T M I P T K I N D G Q W H K I K I M R
FLY CRUMBS  1077   L I N G R L N L H S S L L N K W E G V F I G S K L N D S N W H K V F V A I
                                                                              ____B

RAT ABP      108   N G D S L L L W V D G K K M L C L R Q V S A S L   A D H P Q L S M R I A L
HUM SHBG     108   E G D S V L L E V D G E E V L R L R Q V S G P L   T S K R H P I M R I A L
HUM PROT S   350   L E H S I S I K I A K E A V M D I N K P G P L F K P E N G L L E T K V Y F
HUM LAMININ  410   V K R K G F I T V D G R E S P M V T V V G D     G T M L D V E G L F Y L
HUM MEROSIN  875   S K Q E G I L V D G A S N R T I S P K K         A D I L D V V G M L Y V
FLY CRUMBS  1114   N T S H L V L S A N D E Q A I F P V G S Y E T A N N S Q P S F P   R T Y L
                   ____B____                                              ____B*____

RAT ABP      144   G G L   L L P T S K L R         F F L V P A L D G C I R D I W   L G H Q A
HUM SHBG     144   G G L   L F P A S N L R         L T L V P A L D G C L R R D S W   L D K Q A
HUM PROT S   387   A G F F R K V E S E L I         K T I N P R L D G C I R S W N L M K Q G
HUM LAMININ  444   G G L P S Q Y Q A R K I         G N I T H S I P A C I G D V T   V N S K
HUM MEROSIN  908   G G L P I N Y T T R R I         G P V T Y S I D G C V R N L H   M A E A
FLY CRUMBS  1150   G G T   I P N L K S Y L R H L T H Q P S A F V G C 1173

                                                          *
RAT ABP      176   Q L S T S     A R T S L G N   C D V D L Q P G L F F P P G T H A E F S L
HUM SHBG     176   E I S A S     A P T S L R S   C D V E S N P G I F L P P G T Q A E F N L
HUM PROT S   419   A S G I K E I I Q E K Q N K H C L V T V E K G S Y Y F D G S G Y A Q F H I
HUM LAMININ  475   Q L D K D S P V S A F T V N R C Y A V A Q E G T Y F D G S G Y A A L
HUM MEROSIN  939   P A D L E Q P T S S F H V G T C F A N A Q R G T Y F D G T G F A K

RAT ABP      209   Q D I P Q P H T D P W T F S L E L G F K L V D G A G R L L T L G T G T N S S
HUM SHBG     209   R D I P Q P H A E P W A F S L D L G L K Q A A G S G H L L A L G T P E K P S
HUM PROT S   455   D Y N V S S A E G W H V N V T L N I R P S T S T G V M L A L V S G N N T V
HUM LAMININ  509   V K E G Y K V Q S D V N I T L E F R T S S Q N G V L L G I S T A K V D A
HUM MEROSIN  972   A V G G F K V G L D L L V E F R T T T T G V L L G I S S Q K M D G

RAT ABP      247   W L T L H L Q D         Q T V V L S S E A E     P K L A L P L A V G
HUM SHBG     247   W L S L H L Q D         Q T V V L S S G S G     P G L D L P L V L G
HUM PROT S   493   P F A V S E L V D S T S E K S Q D I L L S V E N T V I Y R I Q A L S L C S D
HUM LAMININ  545   I G I E L V         G K V L M F H V N N G A G R I T P A Y E P K T A T
HUM MEROSIN 1008   M G I E M I D         E K L M F H V D N G A G R F T A V Y D A G V P G

RAT ABP      275   L P L Q L K L D V F K V A L S Q G P K M E V L S T S L L R L A S L W R L W
HUM SHBG     275   L P L Q L K L S M S R V V L S Q G S K M K A L A L P P L G L A P L L N L W
HUM PROT S   530   Q Q S H L E F R V N R N N N L E L S T P L K T E T I S H E D L Q R Q L A V L
HUM LAMININ  576   V L C D G K W H T L Q A N K S K H R I T L I V D G N A V G A E S P H T Q S
HUM MEROSIN 1039   H L C D G Q W H K V T A N K I K H R I E L T V D G N Q V E A Q S P N P A S

RAT ABP      312   S H P   Q G H     L S L G A L P       G E     D S S A S F   C L S D L
HUM SHBG     312   A K P   Q G R     L F L G A L P       G E     D S S T S F   C L N G L
HUM PROT S   567   D K A M A K A V A T V L G G L P D V P F S A   T P V N A F Y N G C M E V
HUM LAMININ  613   T S V D T N N P I Y V G G Y P A G V K Q K C L R S Q T S F   R G C L R K L
HUM MEROSIN 1076   T S A   D T N D P V F V G G F P D L K Q F G L T S I P F   R G C I
                   ____C1____                      ____C2____                   ____B*____

RAT ABP      338   W V Q G G Q R L D I D K A L S R S Q D I W T H S C P  362  ---C-TERMINUS  373
HUM SHBG     338   W A Q G G Q R L D V D Q A L N R S H E I W T H S C P  362  ---C-TERMINUS  373
HUM PROT S   602   N I N G V Q L D L D E A I S K H N D I R A H S C P   626  ---C-TERMINUS  635
HUM LAMININ  649   A L I K S P Q V Q S L D F S R A F E     L H G V F L H S C P  676  ---C-TERMINUS  680
HUM MEROSIN 1109                 R S L K L T K G     T A S H W R L I L P 1126  ---C-TERMINUS 1130
                                                              *
```

and were not significant (M. Baker, personal communication). Moreover, Fawell *et al.* (1989) demonstrated by mutagenesis that the sequence was not necessary for high-affinity steroid binding by the estrogen receptor.

What was unexpected was the similarity of ABP/SHBG and several plasma, extracellular matrix, and developmental proteins. Clues for potential functions of ABP/SHBG were presented in the description of homologies between ABP/SHBG and domains of the blood-clotting factor protein S, the extracellular matrix protein laminin, and the developmental protein *Drosophila crumbs* (Baker *et al.*, 1987; Joseph and Baker, 1992). Figure 14 shows a multiple alignment of rat and human ABP/SHBG with the homologous domains of protein S, laminin, merosin, and *crumbs* protein. The ALIGN comparison scores of laminin A chain with rat and human ABP/SHBG were 11.25 and 12.2 standard deviations, respectively, whereas the scores with *Crumbs* were 10 and 9.9 standard deviations. The probability of getting a score higher than 10 standard deviations by chance is less than 10^{-23}. The most conserved sequence is in the region that has been identified as the receptor-binding site of SHBG (Fig. 14, overlined A). Remarkably, the four ABP/SHBG Cys residues are very highly conserved in all the proteins (Fig. 14, asterisks).

Protein S is a negative coregulator in the blood coagulation pathway. Specifically, it is a cofactor of activated protein C that leads to proteolytic inactivation of factors Va and VIIIa (Dählback *et al.*, 1986, 1990; Hoskins *et al.*, 1987). It also modulates the complement system by interacting with the C4b-binding protein (Hessing, 1991). The region of protein S that binds to C4b-binding protein is also related to laminin and ABP/SHBG (Fig. 14, overlined C). It is of interest that other vitamin K-dependent factors in the blood-clotting system have a protease domain in place of the ABP/SHBG-like region.

The *Drosophila crumbs* locus encodes an EGF-like protein that is

---

FIG. 14. Multiple alignment of rat androgen-binding protein, human vitamin K-dependent protein S, human sex hormone-binding globulin, human laminin A, human merosin, and *Drosophila crumbs* protein. The progressive alignment program of Feng and Doolittle (1987) and the ALIGN program were used to construct the multiple alignment. Identities between the ABP/SHBG/protein S group and the laminin A/merosin/*crumbs* protein group are denoted with dark shaded boxes. Light-shaded boxes denote residues conserved between the ABP/SHBG/protein S group and the laminin A/merosin/*crumbs* protein group. Underlined segment A is the SHBG receptor-binding sequence; segment B is the conserved repeats in merosin; and segment C is the two segments of protein S that bind C4b-binding protein. Conserved cysteines are highlighted with an asterisk. From Joseph and Baker (1992).

important for the organization of the epithelia (Tepass *et al.*, 1990). ABP/SHBG is also related to the carboxy-terminal globular (G) domain of the A chain of laminin and merosin, a form of laminin. These two multisubunit multidomain proteins are part of basement membranes and have important roles in cell proliferation, migration, and differentiation (Beck *et al.*, 1990; Engel, 1989, 1992; Kleinman *et al.*, 1993). *In vitro*, laminin promotes the formation of cordlike structures by Sertoli cells (Hadley *et al.*, 1990). The G domain of laminin is divided into five homologous repeats of 180–190 residues (G1–G5, G5 is truncated) (Sasaki *et al.*, 1988; Ehrig *et al.*, 1990). ABP/SHBG is most homologous to the G4–G5 subdomains at the C terminus of the laminin A chain (Fig. 15; this figure is presented to demonstrate the complexity of laminin and the location of the ABP/SHBG-related domain). These two regions of ABP/SHBG correspond to the amino acid sequences encoded by exons 1–5 and 6–8, respectively. Because the G4 and G5 domains are the products of duplicated genes, it appears that the 373-residue ABP/SHBG and its homologous domain in protein S were formed by gene duplication. The G domain of laminin A is known to be multifunctional and binding sites for heparin, type IV collagen, and a plasma membrane laminin-binding protein have been identified. Synthetic peptides from the globular domain of the A chain promote cell adhesion and neurite outgrowth (Skubitz *et al.*, 1991).

   Since the original description of these homologies several other ABP homologs have been identified. An ABP/SHBG-related domain is present in the *Drosophila slit* protein, a protein that functions in glia development (Rothberg *et al.*, 1990). Agrin, an integral membrane protein that is involved in aggregation of acetylcholine receptors at neuromuscular junctions (Rupp *et al.*, 1991), has ABP/SHBG-related domains. Also, the multidomain proteins *Drosophila* segment polarity gene product *patched* (Nakano *et al.*, 1989), the cadherin-related fat tumor suppressor gene (Mahoney *et al.*, 1991), and the human basement membrane heparin sulfate proteoglycan core protein (Kallunki *et al.*, 1992) have ABP/SHBG-like domains. During preparation of this review another homolog was identified by Manfioletti *et al.* (1993). The protein encoded by growth arrest-specific gene (gas6) was shown to be closely related to vitamin K-dependent protein S. This is one of several growth arrest-specific genes whose expression is negatively regulated after serum induction. The finding defined a new member of the vitamin K-dependent protein family that is expressed in many human and mouse tissues and may be involved in the regulation of a protease cascade relevant in growth regulation.

   It is of interest that most of these ABP/SHBG domains in the homo-

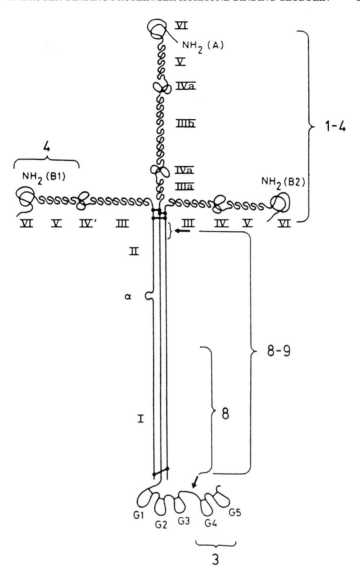

FIG. 15. Structural model of laminin. Designations of domains by roman numerals is according to Beck *et al.* (1990). Cys-rich rod domains in the short arms are designated by symbols S and the triple coiled-coil region (domain I–II) of the long arm by parallel straight lines. In the B1 chain, the α-helical coiled-coil domains are interrupted by a small Cys-rich domain α. Interchain disulfide bridges are indicated by thick bars. The primary cleavage sites of cathepsin G are marked by arrows. Regions of the molecule corresponding to fragments 1–4, 4, 8–9, 8, and 3 are indicated. From Beck *et al.* (1990).

logs are adjacent to EGF domains and/or protease–antiprotease domains, which are associated with regulatory proteins. Even though ABP's homologs have apparently unrelated functions, they all contain ABP/SHBG-like domains. ABP/SHBG and the related domains of these other proteins have evolved from the same ancestral protein and they would be expected to share functional properties. For example, the blood-clotting factor protein S and the extracellular matrix protein laminin were thought to have totally unrelated functions, but protein S has been shown to be a bone extracellular matrix protein (Maillard *et al.*, 1992) and a muscle growth factor (Gasic *et al.*, 1992). Thus, ABP/SHBG may have functions related to these other homologous proteins. Preliminary experiments (M. E. Baker, H. K. Kleinman, and D. R. Joseph, unpublished results) have demonstrated that human SHBG or protein S mimics the activity of laminin in the B16 melanoma (neural crest derived) cell attachment assay. Like laminin, SHBG increases the rate of cell attachment and spreading. This activity is thought to be mediated by the interaction of SHBG with plasma membrane receptors on the B16 melanoma cells. Because the region of ABP/SHBG that binds the plasma membrane receptor contains the most conserved residues between ABP/SHBG and its homologs, the major functional property shared by these various proteins appears to be related to receptor–ligand interaction. They may share a receptor or class of receptor.

Studies with transgenic mice also offer some clues to the function of ABP/SHBG and other ABP/SHBG gene products. Transgenic mice have been developed that carry the rat ABP/SHBG gene (Reventos *et al.*, 1993). As discussed in the following, expression of the rat ABP/SHBG gene is specific for the testis and ovary. Further analysis of these mice (hemizygous for the transgene) demonstrated that rat ABP is located in the mouse testis and epididymis (D. Joseph and P. Sullivan, unpublished results). DHT-binding assays revealed that there was a 25- to 50-fold increase in ABP activity in the testis and epididymis as compared to the normal littermates. Surprisingly, DHT-binding activity was elevated at least 25-fold in the male transgene plasma, a level comparable to that in male human plasma. ABP/SHBG activity was also elevated in female transgene plasma, at a slightly lower level. Serum testosterone analysis of the transgenic male mice revealed that the serum levels of testosterone were unchanged. As with their normal littermates, the transgenic mice exhibited a 50-fold variation in testosterone levels (0.3–15 ng/ml). Thus, it is clear that the wide fluctuations in testosterone levels in mice (Bartke *et al.*, 1973) are not due to the absence of a circulating SHBG.

Breeding of the transgenic mice to establish a homozygous transgenic line led to decreased fertility in males and females; infertility increased dramatically with age. Immunohistochemical analysis of the homozygous transgenic mice revealed large amounts of immunoreactive ABP in the testis and caput epididymis (D. R. Joseph, P. Petrusz, and D. O'Brien, unpublished results). In the epididymis, ABP was internalized by the caput epithelium in numerous vesicles. In the testis of some transgenics, the major site of immunoreactivity was the spermatocyte, specifically pachytene and spermatocytes undergoing cell division; transgenic immunoreactivity was clearly stage specific. Moreover, the immunoreactive spermatocytes were arranged in abnormal clusters and not evenly distributed around the tubule. In the testis of other transgenic mice, the distribution appeared normal. Since germ cells have been shown to internalize ABP (Gerard et al., 1994), the source of the germ cell ABP is probably the Sertoli cell, but considering the large amounts of spermatocyte ABP, a germ cell origin cannot be ruled out. The female reproductive system also contained large amounts of immunoreactive ABP in the transgenics. The major location of immunoreactive ABP was the epithelium lining the lumens of the uterus and oviduct. The source of the ABP in the oviduct and uterus is unknown, but it appears to originate from the ovary via the blood. Thus, the overexpression of ABP in the male and female reproductive system appears to cause the fertility problems. The homozygous female transgenic mice also exhibited a gait disorder in their hindlimbs, with high stepping characteristics. Interestingly, the phenotypic expression was much less pronounced in males, who exhibited abnormal walking behavior with shuffling and shaking. The onset of the disorders in males and females was gradual, beginning around puberty (35 days of age) and reaching full expression around 4–5 months. These mice appeared to suffer from a pathological problem in the central nervous system. This pathological condition may be related to the human sex-linked trait spinal and bulbar muscular atrophy, which is caused by a defect in the androgen receptor (La Spada et al., 1991). Numerous motor neurons in the brain stem and spinal cord contain androgen receptors. We (D. R. Joseph and P. Petrusz, unpublished results) have recently found increased immunoreactive ABP in motor neurons of the male and female transgenic lumbar spinal cord with three different ABP antisera. Moreover, female and male transgenics contained high levels of immunoreactive ABP in the brain cortical pyramidal cells, which send axons to the spinal cord and control motor function. These data suggest that the neurological disorder may be due to overexpression of the ABP/SHBG gene in motor neurons. Although it appears that this disorder is due to the overexpression of the rat ABP gene, the possibility exists that it is caused by the knockout of another gene during integration of the rat gene in the mouse genome.

The paramount problem contributing to our lack of knowledge of ABP and SHBG function is that there are no known natural mutants in humans or animals that affect function. This suggests that the protein is extremely important for mammalian development (i.e., mutants are lethal) or it has little or no importance. The latter would appear unlikely considering the conservation of sequence and activity during evolution. With the advent of "knockout" technology by homologous recombination it is now possible to develop mice that do not have a functional ABP/SHBG gene (Snouwaert *et al.*, 1992). Thus, the development of mice lacking the ABP/SHBG gene will give us the tool for functional studies that nature apparently would not allow.

V. STRUCTURE AND REGULATION OF THE GENE

A. STRUCTURE AND TRANSCRIPTION

To date, the rat, mouse, and human ABP/SHBG genes have been mapped; each species clearly contains a single gene. The mouse and rat ABP/SHBG genes (*Shbg* locus) were localized to chromosomes 11 and 10, respectively (Joseph *et al.*, 1991; Sullivan *et al.*, 1991). The chromosomal location of each gene was determined by analysis of restriction endonuclease polymorphisms of DNA from interspecies cell hybrids. Mouse progeny from an intersubspecies backcross were analyzed to position *Shbg* in the middle of chromosome 11 at 39 cM from the centromere (Joseph *et al.*, 1991). *Shbg* is thus closely linked to several neurological mutants, one of which, *Tr,* is associated with male sterility. The finding that the ABP/SHBG gene is also expressed in the rat brain raises the possibility that one of these mutations may be due to a defect in *Shbg*. This possibility is supported by a finding revealing that homozygous ABP/SHBG transgenic mice have a neurological defect (see IV,D). The human ABP/SHBG gene has been localized on chromosome 17 (P12–P13 region) in a region that shares considerable homology with mouse chromosome 11 (Bérubé *et al.*, 1990). *In situ* hybridization of chromosomes was used to locate the region. It was pointed out by Bérubé *et al.* that this region of chromosome 17 is a hotspot for genetic recombination, gene amplification, and integration of foreign genomes.

The rat and human ABP/SHBG genes have been cloned and sequenced (Joseph *et al.*, 1988; Hammond *et al.*, 1989; Gershagen *et al.*, 1989). The coding regions of the human and rat genes consist of 8 exons separated by unusually small introns. Each splice site flanking exons 1–8 conforms to the consensus donor and acceptor splice sequences and the sites are perfectly conserved in both species (Fig. 16, P_1 region, and Fig. 17). Interestingly, these splice sites are also conserved in the gene of vitamin K-dependent protein S, a homolog of ABP/SHBG, which contains an ABP/SHBG like domain (Gershagen *et al.*, 1991). This conservation offers further evidence that the ABP/SHBG gene and the protein S gene evolved from a common ancestor by exon shuffling.

To characterize the regulatory properties of the gene, it is imperative that the transcription start site is known. Studies that attempted to identify the start site in the rat and human ABP/SHBG gene have led to ambiguous results. Sequences upstream of the translation-initiating Met codon offered no clues to the identity of the site; there was no TATA box element in either gene. The apparent major transcription start site of the rat gene promoter was located by primer extension with two unique oligonucleotides based on sequences in exon 1, which contains the initiating Met codon (Fig. 16, promoter P_1, and Fig. 18). Both oligonucleotides identified the major site 36 bp upstream of the initiating Met codon (Joseph *et al.*, 1988). In addition, both oligonucleotides yielded a minor product 80 residues longer. These data suggested the presence of another start site or that the major product represented premature termination of reverse transcription. Because an oligonucleotide based on a gene sequence upstream of the identified major site did not hybridize with the mRNA, the longer product apparently represented extension of an mRNA containing another upstream exon.

In another study, RNase protection was used to identify the rat gene transcription start site (Sullivan *et al.*, 1993). These experiments also suggested that the major site was approximately 36 bp upstream of the Met codon. However, the RNase protection technique cannot distinguish between an RNA terminus and an RNA splice site. The sequence of the putative site, CAG, conforms to the sequence of most transcription start sites, however, in the ABP/SHBG gene this sequence also is included in a consensus acceptor splice site (TTCTCCTCAG). Taken together, the primer extension and RNase protection data suggest that the major start site is located 36 bp upstream of the initiating Met codon, however, it cannot be ruled out that another noncoding exon exists upstream. The only definitive known fact about the promoter

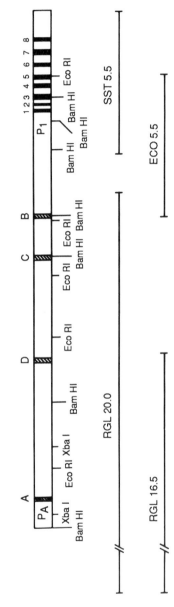

Fig. 16. Structural map of the rat ABP/SHBG gene. Exons encoding secreted testicular ABP are indicated by the black boxes numbered 1–8. Alternate exons A, B, C, and D (hatched boxes) encode unique 5′ ends of alternate ABP mRNAs. Exons B, C, and D are found only in brain ABP RNAs. The locations of exons A, B, C, and D were identified by Southern blot analysis of two genomic clones, RGL 20.0 and RGL 16.5, with oligonucleotides specific for each exon. Exons A and B were completely sequenced, and exons C and D were located only by Southern blot analysis with oligonucleotide probes. P₁ represents the promoter region that regulates rat ABP mRNA containing exons 1–8. This mRNA encodes secreted testicular ABP and fetal liver SHBG. Pₐ represents the promoter region that regulates the alternate ABP mRNA, which encodes the nonsecreted ABP/SHBG-like protein. Restriction endonuclease cleavage sites determined by sequence and Southern blot analysis are shown below the gene diagram. The solid lines below the gene diagram represent the four rat genomic clones used to define the ABP gene map. Rat genomic clones RGL 20.0 and RGL 16.5 define the 5′ end of the gene, and clones ECO 5.5 and SST 5.5 define the 3′ end of the gene. SST 5.5 is a previously described 5.5-kb *SstI* genomic fragment and ECO 5.5 is a 5.5-kb *Eco*RI fragment. The SST 5.5 clone contains all the information necessary for the Sertoli cell-specific expression of secreted ABP in transgenic mice. From Sullivan *et al.* (1993).

242

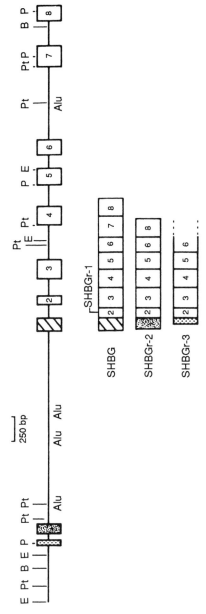

Fig. 17. Partial restriction map and organization of the human SHBG gene, and exon composition of SHBG and its related (SHBG) testicular transcripts. Exons are represented by boxes. The SHBG-r1 cDNA is identical to the SHBG cDNA, and its 5′ end is located 69 bp from the 5′ splice site of exon 2. Those containing the unique 5′ regions of the SHBG (hatched), SHBGr-2 (shaded), and SHBGr-3 (stippled) cDNAs are indicated with reference to their corresponding transcripts. The positions of *Alu* sequences are shown below the gene map. The sequence inversion in SHBGr-3 is indicated by dashed lines. Restriction endonuclease sites are shown for *Eco*RI (E), *Bam*HI (B), *Pvu*II (P), and *Pst*I (Pt). From Hammond *et al.* (1989).

and transcription start site is that it is within 1.5 kb of the initiating Met codon. This information was obtained with transgenic mice carrying the rat ABP/SHBG gene. These experiments have demonstrated that the rat gene promoter and Sertoli cell enhancer sequences are located in a 5.5-kb SstI DNA fragment (Fig. 16), within 1.5 kb upstream of the Met codon (Reventos et al., 1993). Northern hybridization blots, RNA PCR, and sequencing of the PCR products from the adult transgenic mice revealed extremely high levels of the rat ABP mRNA in the testis, but no detectable mRNA in the liver, kidney, or brain. These data demonstrated that the 5.5-kb genomic DNA fragment contains the transcription start site and an element capable of directing ABP/SHBG gene expression in the testis. Recently, ABP mRNA analyses and immunohistochemistry have demonstrated that the transgenic ovary also expresses the rat ABP/SHBG gene (see Section D).

An attempt to identify the human gene transcription start site has also led to ambiguous results (Gershagen et al., 1989). Primer extension analysis of testis and liver RNA with oligonucleotides specific for exon 1 yielded multiple bands, without a clear-cut answer to the identity of the start site. One major product mapped the site to near the starting Met residue and a longer product mapped the site 110 bp upstream. On the basis of their data, Gershagen et al. placed the apparent start sites 80 bp upstream of these sites; their rationale is not clear. The problems were compounded by the identification of an alternate exon 1 sequence, which led to the hypothesis that the promoter may be farther upstream. It would appear likely that because the N-terminal sequence of ABP/SHBG is encoded by exon 1, which is not present in the alternate transcripts, the alternate sequence is not related to the encoding of SHBG or ABP, but may encode an alternate protein (i.e., like the rat gene). This interpretation does not rule out the possibility of another noncoding 5' exon that controls the synthesis of ABP/SHBG mRNA.

Comparison of the upstream region of the rat (Joseph et al., 1988) and human (Hammond et al., 1989) ABP/SHBG genes has yielded information on important functional elements. Figure 18 aligns the region of homology between the rat (P_1 region) and human gene. It can be easily seen that approximately 600 bp upstream of the initiating Met codon are highly conserved (rat residues 897–1504 and human residues 2097–2642). The largest difference between the two sequences is the presence of a 110-bp stretch in the rat gene (residues 1124–1233) that is missing in the human gene. Regions of homology

```
R860-  TCTCAACCTTGGGTGCTTG                GGTTATAGGCAGATTTCT -896
       *  *  *  *** *  *                  **    **** *  **
H2047- TGTAATAAATGGATAAGGGAAGATAACTGAGAGGCGGGGGGCAGGTCCCT -2096

R897-  TCATAGTATCCATTAA    ACACAGAAAGACAATACTTTCTTGTGAGAGC -943
       ** ** *** ** *  *    * *** * ********* ****** ****
H2097- TCTTAATATTCACTGAATCATACACACAGACAATACCTTCTTGGGAGACA -2146

R944-  AACCACATAGGTCTGGGAAA TCTAAGGGAGGCATTCATGTCGGATCCCG -992
       ** ** ***  ** *** **  *** **  ****** ****  * *** **
H2147- GGCCTCAGAGGCTGGGAAAAGACTGGGGGAGGAGTTCAGACCAGATGCCA -2196

R993-  GGCATTCTGCCTGCATTGTCTCGATCCTCATTCCCAGCAGCCCCATAGAA -1042
       **** * ********** **** **  *  **            *
H2197- GGCACTGTGCCTGCATTTTCTCAATGAACCCTCTTTCACAGTCACCCCGT -2246

R1043- GAAGGATTATGTTCCTCATTTAACAGACAAGGTCACTGAAGCACAAAAGT -1092
       *** ***** ********** **** ***** ************** **
H2247- AAAGTATTAT TTCCTCATTTTACAG CAAGGACACTGAAGCACAAAGGT -2294

R1093- GAAGTGTG TCTCCCTAGGTCAGTCAGCAAGATTTCTGGATTCCACACCT -1141
       ******  * *** ****** ****  **
H2295- GAAGTGACTTGGCCCAAGGTCACTCAGG  GAC                  -2325

R1142- CAACATTAGCCTCGACAGAGGATGCTCGTTGCAGACGCAGGGAGAATTGT -1191

R1192- TTCCAGATCTCAAGTGTAACCTAGATAGCCTCCCCTTGTGGGAGAAGGCT -1241
                                                **** **
H2326-                                          AGAAATCT -2333

R1242- TGGAGGAGCTACACTAGATAACAGTGGAGGAGAAG    AGATGT -1282
       ******* *** *  **     ** ******** *    * **
H2334- TGGAGGACCTAGATCAGGCCCTAGAGGAGGAGAGGGGAGATGGAATATCC -2483

R1283- TCTCCAT      AGATGTCCTGGCTGCAGGGGATAGTAGTGGAAGGAC -1325
       *****    *  * **   *** *   * **** ******* ****
H2484- TCTCCCAGTTCAGAAACTTTCTCGGCAGTGGAGGATGATAGTGGAGGGAC -2533

R1326- TCTGTCCCTCATCTCATCTGCCTTCAGAGG                    -1355
       ******* *** * ***    *   ******
H2534- TCTGTCCTTCACCCCATTGATCCCCAGAGGGGTGATAGCTGAGTCTTGTG -2583

R1356-          GGCCGCATGGTCAGGGTCAGTGTCCCTATCTCTTGCCCC -1394
                ** *   *   ********** **** * ** *    **
H2584- ACTGGGCCCCTGGGCAGGGGTCAAGGGTCAGTGCCCCTGTTTCCTTTACC -2633

R1395- CCTTCTTCCCCCGGAGCAACCTTTAACCCTCCACCACCCATGTGAGAGGC̄ -1444
       ** ** ***** * ****************** ****  * ****
H2634- CCCTCCTCCCC  GGGCAACCTTTAACCCTCCACCGCCCACACGCAAGGC -2681

                                  ▼
R1445- TACCTACCCCCACTGCTTCTCCTCAGA TATTCTGAGCCACTGGGTGGAC -1493
       * *** ** * **   ******  *** *  ******** * * ******
H2682- TGCCTGCCTCTACACATTCTCCCAAGAGTTGTCTGAGCCGCCGAGTGGAC -2631

R1494- AGCTGCTAACTATG -1507
       **  *** * ****
H2632- AGTGGCTGATTATG -2645
```

Fig. 18. Comparison of the human and rat ABP/SHBG gene. An alignment of the rat and human sequences is shown beginning 600 bp upstream of the rat P$_1$ transcription start site (arrowhead). An asterisk denotes those residues that are identical between rat and human. The sequence corresponding to the SV40 TATA-like promoter element in the rat gene (TACCTA) is overlined. The translation start site (ATG) is underlined.

exist throughout the sequence and are highly conserved near the putative rat gene start site and upstream at rat residues 1043–1098. Upstream of the homologous region in both genes are species-specific repetitive elements, which share no homology.

To date no regulatory elements have been defined in either the rat or human gene. Because only 600 bp upstream of the transcription start site are homologous in the rat and human genes, it is presumed that this 600-bp region contains the most important regulatory elements. This region contains few sequences that are related to known regulatory elements. Neither gene contains a TATA box or CAAT box, but the rat gene does contain a modified TATA box (TACCTA) that has been shown to function in initiating transcription in the SV40 major late promoter (Brady *et al.*, 1982). This sequence has also been shown to bind to TFIID, the TATA box binding protein (Wiley *et al.*, 1992). This sequence is not conserved in the human ABP/SHBG gene (Hammond *et al.*, 1989). The rat gene upstream region does contain three sequences related to SP1 binding sites, but if they are SP1 sites they probably are weakly active (Joseph *et al.*, 1988). These sites are fairly well conserved in the human gene. The rat ABP/SHBG gene promoter region contains two adjacent androgen–glucocorticoid regulatory element half-sites, but these sequences are not conserved in the human gene.

Probably the most interesting regulatory sequences in the ABP/SHBG gene are the sequences that dictate tissue-specific expression. The human gene contains a sequence related to a liver-specific enhancer in the promoter region, but its activity has not been tested (Hammond *et al.*, 1989). Because this sequence is not present in the rat gene, it was suggested that the lack of this element accounts for the lack of liver expression in the adult rat. Studies of another Sertoli cell-expressed gene offer a clue to the identity of the Sertoli cell enhancer. The rat MIS gene contains a sequence (M2) that is conserved in the P-450 aromatase gene and the human and rat ABP/SHBG gene (C. Haqq and P. Donahoe, personal communication). The sequence (TGTC-CCAAGGTCA) has been shown to bind Sertoli cell nuclear proteins in band shift assays. In the rat ABP/SHBG gene, the related sequence is located 370 bp upstream of the start site. This sequence may be the Sertoli cell enhancer.

There are several other interesting features in the ABP/SHBG genes. First, the gene coding region is extremely short in the human and rat gene. All of the coding region, including seven introns, occupy only about 3 kb of DNA; most of the introns are only 200–400 bp. This unusual structure suggests that the small introns are important for function. Another interesting feature of the rat and human genes is the presence of repetitive elements. Both the human (Hammond *et al.,* 1989) and rat (Joseph *et al.,* 1988) genes contain repetitive elements in introns and upstream of the transcription start site. The human ABP/SHBG gene contains an Alu repetitive element in intron 6, whereas the rat gene contains two rodent repetitive elements in intron 5 and one inverted element downstream of the poly(A) addition signal. Although the rat and human genes are homologous to 600 bp upstream of the transcription start site, both genes contain several species-specific repetitive elements immediately 5′ to the homologous region. This similarity in structure suggests that these elements may have important functional properties. It was outlined how these elements may act to regulate RNA splicing and facilitate formation of the trans-spliced mRNA described in the following (Sullivan *et al.,* 1991). Hammond *et al.* (1989) pointed out that the exons involved in alternate splicing in the rat and human genes were each adjacent to an intron containing a repetitive element.

It is now known that the rat ABP/SHBG gene is much more complex than originally described. Studies of ABP/SHBG gene expression in the fetal rat liver and brain revealed an alternate exon 1 sequence (Sullivan *et al.,* 1991, 1993; Wang *et al.,* 1990). The localization of this exon in rat DNA revealed an alternate promoter (P_A) 15 kb upstream of the P_1 testicular start site (Fig. 16). Promoters P_A and P_1 are active in testis, fetal liver, and brain; P_1 transcripts are predominant in the testis and P_A transcripts are predominant in brain (Sullivan *et al.,* 1993).The P_A region of DNA has the characteristics of a "GC rich" promoter. The approximate transcription start site was identified by RNase protection and primer walking with RNA PCR. This promoter regulates an mRNA that encodes an ABP/SHBG-like protein, with an altered N-terminal sequence, that is not secreted by COS cells (Fig. 19, clone 1–6). The N terminus of the alternate ABP/SHBG does not appear to have a signal peptide sequence. Figure 20 demonstrates the lack of hydrophobicity of this sequence with a comparison of the rat ABP/SHBG signal peptide. It is unclear if this alternate protein binds steroid. Expression of the recombinant ABP/SHBG-like protein in COS cells yielded mostly insoluble protein, which could not be assayed. Assays for DHT-binding activity *in vivo* yielded no detectable activity, whereas the assay could detect recombinant ABP/SHBG activity in transformed COS cells (Sullivan *et al.,* 1993). Based on these data, the recombinant ABP-like protein appears not to bind DHT in COS cells.

ABP cDNA

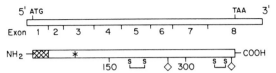

Alt-ABP cDNA (1-6)

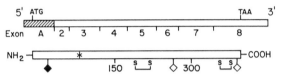

Alt-ABP cDNA (1-7)

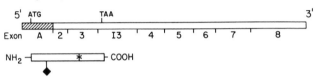

Alt-ABP cDNA (5-5)

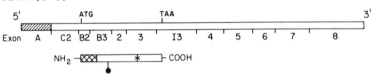

Alt-ABP (Δ6) cDNA

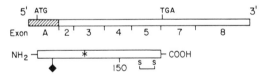

FIG. 19. Diagrams of rat ABP/SHBG and alternate cDNAs and their encoded proteins. The cDNAs are represented by open bars, with the exon sequences depicted below. The positions of the initiation Met codon (ATG) and the stop codon (TAA or TGA) are indicated for each cDNA. The hatched area in Alt-ABP cDNA represents the alternate exon 1 sequence (exon A). The rectangles below each cDNA diagram represent the encoded proteins of each cDNA with amino acid residue numbers indicated. The locations of two disulfide bridges (S–S) and sites of N-glycosylation (open diamonds) are also indicated. It is not known if these sites are glycosylated or if the disulfide bridges exist in the alternate protein (ABP cDNA 1–6). Solid diamonds and circles represent potential sites of N-linked and O-linked glycosylation, respectively. The approximate location of the plasma membrane-binding domain is indicated by an asterisk. The crosshatched area represents the signal peptide sequence in secreted ABP and the signal peptide-like sequence in the protein encoded by Alt-ABP cDNA 5.5. cDNAs with exons B, C, and D

The data do not rule out the possibility that it may exhibit binding activity when produced in its native environment.

B. ALTERNATIVE PROCESSING OF RNA TRANSCRIPTS

Alternatively processed ABP/SHBG RNA transcripts have been described in human and rat. Several alternatively processed rat P_A transcripts have been characterized from fetal liver and brain RNA (Sullivan et al., 1991, 1993; Wang et al., 1990). Some of these alternate RNAs encode truncated forms of the ABP-like protein (Fig. 19). Analysis of cDNA clones derived from fetal rat liver cDNA libraries identified two cDNAs encoded by the ABP/SHBG gene that represented alternatively spliced RNAs (Sullivan et al., 1991). One cDNA had an alternate exon 1 sequence (exon A) and also lacked testicular exon 6 DNA. This cDNA encoded a truncated form of the ABP/SHBG-like protein ($M_r = 28,000$) [Fig. 19, Alt-ABP($\Delta6$)cDNA]. The other cDNA represented a fused transcript of the ABP/SHBG gene (exons 1–5) and the histidine decarboxylase (HDC) gene (Fig. 21), encoding a 98,000 M_r precursor protein (Sullivan et al., 1991). The two domains were joined at splice junctions of the ABP/SHBG and HDC genes, which were localized to rat chromosomes 10 and 3, respectively. The results indicated that the joining of the two domains was by a *trans* (donor and acceptor)-splicing mechanism. This was the first example of trans splicing in higher eukaryotes and the first described *trans* splicing that affected the encoded protein. Immunoblot experiments indicated that the protein was expressed in the fetal liver. Although the function of the fusion protein is not known, the domain structure suggests that the protein should be targeted to the rough endoplasmic reticulum (i.e., the cDNA encoded the ABP signal peptide). Furthermore, the interaction of the ABP/SHBG receptor-binding region with the protein trafficking system could direct the catalytically active HDC domain to a subcellular compartment, not accessible by normal HDC.

Alternate P_A transcripts have also been described in rat brain (Wang et al., 1990; Joseph et al., in press). The level of heterogeneous 2.3-kb mRNA varied between brain regions and individual rats (see Fig. 24) (Wang et al., 1990). RNA PCR experiments revealed that the 2.3-kb RNA is very heterogeneous and represents alternatively spliced RNAs

---

and intron 3 (I3) have been found only in brain. ABP Alt-ABP cDNA (1–6), which encodes the ABP/SHBG-like protein, is present in brain, testis and fetal liver. Alt-ABP ($\Delta6$) cDNA represents a fetal liver ABP mRNA (Sullivan et al., 1993). From Joseph et al., in press.

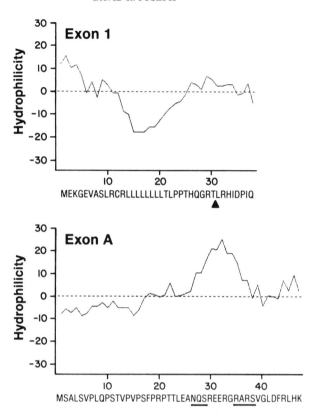

FIG. 20. Predicted hydrophilicity plot of exon 1 and exon A amino acid sequences. Hydrophilic regions are shown above the line (zero hydrophilicity) and hydrophobic regions below. The plot is according to the algorithm of Hopp and Woods (1981). The running average of the relative hydrophilicity value is given in the left margin. Amino acid residue numbers are indicated below. The arrowhead marks the signal peptide cleavage site in the ABP precursor. Potential sites of N-glycosylation and phosphorylation in the exon A sequence are underlined. Note the lack of hydrophobicity in the exon A sequence.

originating from promoter P_A (Joseph *et al.,* in press). Each RNA contained sequences of all or parts of three previously unknown exons (exons B, C, and D) located between P_A and P_1 (Fig. 16). Because neither exon has an open-reading frame, the protein encoding properties of these alternate RNAs are ambiguous. Nevertheless, with most of the characterized large RNAs, internal initiation in exon B sequence, downstream of numerous Met codons, results in a hypothetical

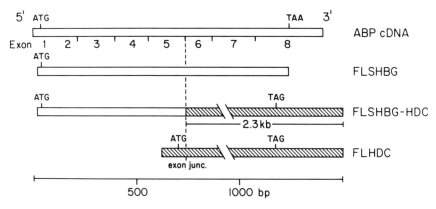

FIG. 21. Diagram of the *trans*-spliced fetal rat liver SHBG mRNA. FLSHBG cDNA represents fetal liver SHBG mRNA, which is identical to testicular ABP cDNA. FLHDC cDNA represents histidine decarboxylase (HDC) 2.5-kb mRNA, which encodes the 73,000 M_r protein (Joseph *et al.*, 1990a; Zahnow *et al.*, 1991). FLSHBG-HDC contains SHBG cDNA (exons 1–5) and HDC cDNA (all exons except number 1) sequences (Sullivan *et al.*, 1991). FLSHBG-HDC encodes a 98,000 M_r fusion protein, containing a SHBG domain, including the signal peptide and the receptor-binding region, and a catalytically active HDC domain. HDC catalyzes the conversion of histidine to histamine, an agent important in neurotransmittance, gastric acid secretion, inflammation, allergic reactions, and numerous other physiological processes.

155-residue protein with an apparent hydrophobic signal sequence at the N terminus and the plasma membrane-binding sequence of ABP/SHBG (Fig. 19, clone 5–5). Another brain cDNA encoding a much smaller (15,000 Da) ABP/SHBG-like protein (Fig. 19, clone 1–7). The functions of the alternate gene products are not obvious, but whatever their function(s) it would appear to involve the ABP/SHBG receptor; they all contain the receptor-binding sequences (Fig. 19, asterisks).

Several alternate human ABP RNAs have been described in human testis. Each RNA was characterized by two laboratories who used the same testis cDNA library, which was made from the testicular RNA of a 50-year-old man (Hammond *et al.*, 1989; Gershagen *et al.*, 1989). One isolated cDNA had the identical sequence with liver SHBG cDNA and encoded human ABP. Two other alternate transcripts had unique 5′ sequences, which replaced exon 1 sequence (Fig. 17). One transcript was missing all of the exon 7 sequence. Both alternate exon 1 sequences were located in the gene approximately 2 kb upstream of the initiating Met codon. Because these RNA forms have not been found in immature rat testis, Hammond *et al.* (1989) suggested that the alternate forms may represent differential expression during testis development. Like some of the rat brain transcripts described earlier, each RNA had ambiguous coding properties. The RNA missing the exon 7 sequence appeared to encode a truncated ABP/SHBG-related protein with a unique N-terminal amino acid sequence, but the clone lacked the full sequence with the initiating Met residue. The biological significance of these human alternate ABP/SHBG-related proteins is not known. Because the human and rat cDNAs were isolated from different tissues and their N-terminal sequences share no homology, their functions may be different. Like the rat alternate proteins, they all contain the receptor-binding region.

C. HORMONAL REGULATION

In vivo and *in vitro* studies have demonstrated that both FSH and testosterone are involved in the regulation of ABP production by Sertoli cells. In the early 1970s, it was shown that ABP activity was regulated by FSH (Hansson *et al.*, 1973; Vernon *et al.*, 1972). Following this initial work, several studies expanded the experiments (Fritz *et al.*, 1974; Hansson *et al.*, 1975a, b, 1976b; Sanborn *et al.*, 1975; Means *et al.*, 1976; Elkington *et al.*, 1977; Tindall *et al.*, 1978). Detailed reviews of these regulation studies were presented by Hansson *et al.* (1976b) and Westphal (1986). I will briefly summarize the findings. Hypophysectomy, which removes the source of FSH and the hormonal requirements (LH) for testosterone production by the Leydig cells, was shown to result in a dramatic decline in epididymal and testicular ABP. FSH treatment of the regressed rat testis reinitiates secretion of ABP into the epididymis. Interestingly, testosterone treatment immediately after hypophysectomy maintains secretion of ABP without administration of FSH. However, after regression of the seminiferous epithelium, testosterone alone does not reinitiate ABP transport to the epididymis (Hansson *et al.*, 1976a). It was concluded that after a long period of hypophysectomy, the Sertoli cell is transformed into a state of androgen insensitivity (Hansson *et al.*, 1976a).

In another study, Elkington *et al.* (1977) found that testosterone alone did reinitiate ABP synthesis in the testis, but transport to the epididymis remained blocked. The administration of cyclohexamide or actinomycin D reduced the stimulatory effect of testosterone, suggesting regulation at the transcriptional and translational levels (Means *et al.*, 1976). Studies with intact young rats demonstrated that the response to testosterone treatment was biphasic. Administration of low doses of testosterone resulted in a reduction in ABP levels, whereas at high doses there was a stimulation of ABP (Weddington *et al.*, 1976). Experiments with the Tfm rat, which lack the androgen receptor, revealed that the Tfm rat testis contained 10 times the normal level of ABP. This clearly showed that ABP is formed even in the absence of androgen stimulation (Hansson *et al.*, 1976a). The high Tfm testis ABP levels were attributed to the lack of an epididymis (i.e., transport was blocked).

A few subsequent studies have contributed to our understanding of ABP regulation. Using a GnRH antagonist to suppress testicular function, Danzo *et al.* (1990) found that FSH, hCG, or testosterone overcame the inhibitory effect of the antagonist. They concluded that androgens were the primary regulator of ABP production and transport to the epididymis. FSH effects appeared to act by a different mechanism, possibly facilitating the structural development of the seminiferous tubule. Huang *et al.* (1991) found that FSH and androgen act synergistically to maintain spermatogenesis and ABP status. On the basis of the effects of hormone administration to hypophysectomized rats, they concluded that the ABP status in the testis and epididymis is closely related to the extent of maintenance of spermatogenesis. With the development of molecular biological tools, two groups showed that FSH or testosterone administration maintains a high testicular ABP mRNA level in hypophysectomized rats, whereas the level decreases without hormone administration (Hall *et al.*, 1986; Reventos *et al.*, 1988). These observations suggest that at least part of the FSH and androgen effects on ABP are due to changes in mRNA levels.

The use of primary Sertoli cell cultures has allowed studies of ABP regulation in the absence of other cell types (Dorrington *et al.*, 1975; Steinberger *et al.*, 1975; Fritz *et al.*, 1976; Louis and Fritz, 1977, 1979; Kierszenbaum *et al.*, 1980; Wright *et al.*, 1981). In Sertoli cell cultures, FSH and dibutyryl cAMP increased ABP production and secretion, generally by two- to threefold. In culture, androgens also appeared to increase the levels of medium ABP, but the effects were minimal. It was not clear if the androgen-induced changes were due to increased ABP secretion or were caused by increasing the stability of ABP. Using DNA hybridization, Hall *et al.* (1990) studied FSH effects on ABP mRNA steady-state levels in Sertoli cell cultures. FSH did not induce ABP mRNA, but maintained the 1.7-kb ABP mRNA if the hormone was administered at plating; the level decreased in the absence of hormone. However, a minor 2.3-kb ABP mRNA was increased transiently after FSH administration. The nature of this alternate RNA is unknown.

The foregoing regulation experiments demonstrated that ABP

mRNA, ABP synthesis, and secretion are regulated by FSH and androgens. The mechanisms of regulation could be at the gene level or regulation could be indirect. Thus far, the data suggest that FSH regulation is indirect, not mediated by a cAMP response element (CRE). Furthermore, in support of this conclusion, FSH did not increase ABP mRNA levels in Sertoli cell cultures, whereas in the same experiments numerous other cAMP-regulated mRNAs, such as c-fos, c-jun, inhibin, and tissue plasminogen activator mRNAs, were dramatically increased by FSH (Hall et al., 1988, 1990). Although the initial sequence analysis of the ABP gene suggested the presence of a cyclic AMP response element (Joseph et al., 1988), further comparisons of these sequences revealed no apparent CRE.

Although androgens do have dramatic effects on ABP and ABP mRNA in vivo, they have no apparent effect on Sertoli cell ABP mRNA levels. These data suggest that androgen regulation of ABP mRNA is primarily indirect, possibly mediated by androgen-regulated paracrine factors (Skinner and Fritz, 1985). The rat ABP/SHBG gene does contain two androgen regulatory element half-sites, but it is not known if they are functional.

Hormonal effects on plasma SHBG levels have been reviewed by several authors (Anderson, 1974; Lobl, 1981; Westphal, 1986; Vermeulen, 1988; von Schoultz and Carlström, 1989; Rosner, 1990). Briefly, studies of various hormonal states have implicated androgens, estrogens, and thyroid hormone in the regulation of SHBG. For example, the high concentration of SHBG in the plasma of pregnant women has been attributed to increased estrogens. Most in vivo studies found that estrogens and thyroid hormones up-regulated SHBG, whereas androgens down-regulated the protein. Glucocorticoids also appeared to suppress plasma SHBG (Blake et al., 1988). A single study has looked at the effect of sex hormones on SHBG mRNA levels. Kottler et al. (1990) measured the effect of testosterone administration on SHBG mRNA in castrated monkeys. Even though testosterone treatment lowered the concentration of circulating SHBG, in contrast, the SHBG mRNA level increased. They concluded that mechanisms other than hormonal transcriptional regulation control the plasma SHBG concentration.

In vitro experiments with hepatocyte cultures also implicate hormones in SHBG regulation (Rosner et al., 1984; Lee et al., 1987; Plymate et al., 1988a,b; Mercier-Bodard et al., 1989, 1991; Edmunds et al., 1990; Raggatt et al., 1992). In cell culture, estrogens and androgens generally led to increased SHBG levels in hepatocyte cultures, whereas insulin and prolactin inhibit SHBG production. Mercier-Bodard et *al.* (1991) found that in hepatocyte cultures (HepG2 cells) thyroid hormone causes a dramatic increase in the steady-state SHBG mRNA levels, whereas estradiol and testosterone only slightly increase the mRNA level. Surprisingly, the estrogen antagonist tamoxifen and the antiandrogen cyproterone acetate increased SHBG mRNA levels to a greater extent than E_2 or DHT, respectively (Fig. 22). Again, these data suggest that regulation is posttranscriptional. Because HepG2 cells may synthesize SHBG, the effects could be mediated by SHBG. In a similar study, Raggatt *et al.* (1992) confirmed that thyroid hormone stimulated the level of secreted SHBG and SHBG mRNA in HepG2 cells, whereas the hormone had no effect on CBG or CBG mRNA. This thyroid hormone-mediated stimulation may be at the gene level, but no thyroid hormone regulatory element has been identified in the gene regulatory region.

Steroid hormones may exert their regulatory effects at the clearance level rather than on synthesis and secretion. Tardivel-Lacombe and Degrelle (1991) have demonstrated differences in SHBG glycan microheterogeneity in males and females (females have much higher SHBG

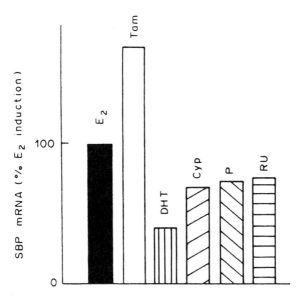

FIG. 22. Regulation of SBP mRNA in H5A cells by steroids and antagonists. H5A cells were grown in phenol red-free DCC–FCS–DMEM, divided in six groups, and treated for 60 hr as follows: E_2, 100 n*M;* tamoxifen (Tam), 100 n*M;* DHT, 100 n*M;* cyproterone acetate (Cyp), 100 n*M;* P, 100 n*M;* and RU486, 100 n*M.* Natural hormones were renewed every 12 hr. From Mercier-Bodard *et al.* (1991).

levels) based on their affinities for concanavalin-A. Also, Danzo *et al.*
(1991b) have described differences in the oligosaccharide types of male
and female rabbit SHBG. These glycosylation differences in male and
female SHBG may affect SHBG's function and metabolism. As found
with T_4-binding globulin (TBG) (Ain *et al.*, 1987), the glycosylation
state could influence the clearance rate of SHBG. Estrogens induce a
greater degree of sialylation in TBG, leading to a decreased clearance
rate. Thus, sex steroid regulation of SHBG maturation may alter its
plasma levels by altering its clearance rate. This mechanism would
explain the differences in sex hormone effects that were found in *in
vitro* and *in vivo* experiments.

Although sex steroids are considered to be important physiological
regulators of plasma SHBG in many human physiological and pathological conditions, SHBG concentrations cannot be explained only on
the basis of steroidal regulation (Toscano *et al.*, 1992). The SHBG plasma levels during the development of males and females do not correlate with the sex steroid pattern. For example, changes during puberty
and aging correlate better with insulin and IGF-1 than with steroid
hormone levels (Toscano *et al.*, 1992). In subjects with hyperandrogenic
conditions, the level of SHBG was not always inversely correlated with
androgen levels. They concluded that several other factors, such as
calorie intake, energy balance, and growth factors, may be involved in
regulation of plasma SHBG and each factor may act to a different
extent depending on the various periods of the life cycle. In a review of
the literature, von Shoultz and Carlström (1989) came to a similar
conclusion. The authors suggested that SHBG is primarily regulated
by growth hormone, IGF-I, and other growth factors; sex steroids have
an indirect, modulating influence.

Like the rat ABP/SHBG gene, there is no evidence for transcriptional regulation of the human gene by steroid hormones. Because the
human gene contains no potential androgen–glucocorticoid or estrogen regulatory elements in the characterized promoter region, it would
appear that sex steroid hormone effects are not controlled at the gene
level. A full understanding of the hormone regulatory characteristics
of the human and rat ABP/SHBG gene awaits further transcriptional
studies.

D. Expression in the Ovary, Brain, and Other Tissues

Another tissue in which the ABP/SHBG gene expression is well
documented is the rat brain. Immunoreactive ABP/SHBG was found to
be present in neuronal cell bodies throughout the brain as well as in

fibers of the hypothalamic median eminence (Wang *et al.*, 1990). Figure 23 demonstrates immunoreactive ABP/SHBG in the supraoptic and paraventricular nuclei and in fibers near the third ventricle and in the median eminence. ABP/SHBG mRNA (1.7 kb) was present in all brain regions examined, whereas a 2.3-kb species varied between brain regions (Fig. 24; see Section V, B). The rat brain does express the mRNA that encodes testicular ABP/SHBG, however, the level of the steroid-binding activity in brain is barely detectable (Wang *et al.*, 1990). An alternate mRNA codes for one form of ABP/SHBG in the brain. This mRNA, which was originally identified in the fetal liver, encodes an ABP/SHBG-like protein with an alternate N-terminal sequence (Sullivan *et al.*, 1993). As described earlier, this protein is not secreted and does not appear to bind steroid. Furthermore, experiments with an antibody specific for the ABP/SHBG-like protein N terminus also localized immunoreactive ABP/SHBG in neurons (D. R. Joseph, P. Petrusz and P. Ordronneau, unpublished results).

Another study revealed the brain cell types that synthesize ABP/SHBG (Joseph *et al.*, 1993). It was demonstrated by Northern hybridization analysis that primary neuronal and astrocyte cultures express ABP/SHBG mRNA. PCR analysis of the cDNA revealed that astroglia express RNA transcripts that encode ABP/SHBG and the ABP/SHBG-like protein and neuronal transcripts encode primarily the ABP/SHBG-like protein. Western blots of astrocyte proteins yielded immunoreactive ABP/SHBG migrating as 38,000, 34,000, and 24,000 M_r proteins. The sizes of these proteins do not correspond to the protein-encoding properties of the known ABP/SHBG mRNAs, suggesting that they are derived by proteolytic processing.

Immunoreactive ABP/SHBG has been localized to several SHBG target tissues, including prostate, endometrium, and breast (Bordin and Petra, 1980; Mercier-Bodard *et al.*, 1987; Sinnecker *et al.*, 1988, 1990; Larriva-Sahd *et al.*, 1991). In none of these experiments was the origin of the immunoreactivity experimentally addressed. However, Mercier-Bodard *et al.* (1991) have analyzed ABP/SHBG mRNA by Northern hybridization in cell lines derived from each of these tissues. Endometrial carcinoma Ishikawa cells and LNCaP prostate adenocarcinoma cells contained hybridizing mRNA, whereas there was no detectable ABP/SHBG mRNA in breast cancer MCF-7 cells or endometrial carcinoma RL95-2 cells. Obviously, more experimentation is necessary to determine the origin of immunoreactive SHBG in these tissues. The issue of whether the reactivity represents ABP/SHBG or alternate forms of the protein must also be addressed. In a recent publication, Larrea *et al.* (1993) provided evidence that the human

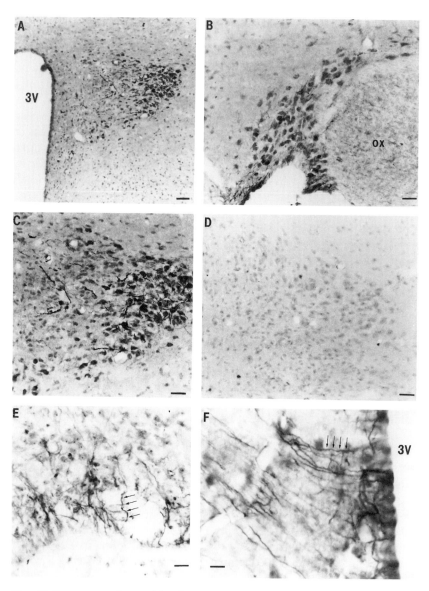

Fig. 23. Immunocytochemical localization of ABP in the hypothalamus. Coronal sections of adult virgin female rat brain were treated with ABP antiserum against peptide A, and immunoreactivity was detected with the peroxidase–antiperoxidase method. ABP-like immunoreactivity was found in neurons in all brain regions examined. The most intense staining was observed in neuronal cell bodies of the hypothalamic paraventricular (A) and supraoptic (B) nuclei. The strong staining of the magnocellular paraventricular nucleus (C) was markedly diminished in an adjacent section by preadsorption of the antiserum with peptide A (D). ABP immunoreactivity was also found in fibers of the median eminence, on the ventromedial surface of the hypothalamus just

placenta is a site of ABP/SHBG gene expression. The existence of placental ABP was demonstrated by steroid-binding assays, immunohistochemistry, and Northern blot analysis.

Evidence also exists for ABP/SHBG gene expression in the ovary. ABP/SHBG immunoreactivity was observed in the granulosa cells of rat ovarian follicles (P. Ordronneau and D. R. Joseph, unpublished results). Additional experiments demonstrated by Northern blot analysis that the rat ovary contains ABP/SHBG mRNA, which is differentially expressed during development (S. Power and D. R. Joseph, unpublished results). These data support previous findings that demonstrated large amounts of SHBG activity in human follicular fluid (Ben-Rafael *et al.*, 1986). The new data suggest that the follicular binding protein may be synthesized in the ovary and not originate totally from the blood as originally thought. Recent transgenic experiments also found ABP/SHBG gene expression in the ovary (D. R. Joseph and P. Petrusz, unpublished results) (see Section IV,D). The transgenic mouse ovary yielded the native 1.7-kb rat ABP mRNA and immunoreactive ABP in the granulosa and theca of the follicles. The ovary appears to be the source of the female transgenic plasma ABP (see Section IV,D).

E. Expression during Development

It was noted earlier how the conservation of sequences in ABP and its homologs raises the possibility that ABP/SHBG has actions in differentiation and development analogous to its homologs. Laminin, the most widely studied of these proteins, exerts diverse effects on epithelial cell development through a variety of activities, including stimulating cell adhesion, migration, proliferation, and differentiation, as well as stimulating neurite outgrowth; several of these activities are mediated by cell-surface receptors. Thus, ABP/SHBG, which also binds to a membrane receptor, may have activities in differentiation and development other than binding steroids. Other than homology, support for this idea comes from evidence that ABP/SHBG is temporally regulated and the RNA is subject to alternative splicing.

---

lateral to the median eminence (E), and around the third ventricle in the region of the median eminence (F). Arrows point to representative immunoreactive fibers. No staining was observed in the absence of primary antibody. Scale bars represent 100 μm (A), 50 μm (B–D), and 25 μm (E and F); 3V, third ventricle; OX, optic chiasm. From Wang *et al.* (1990).

ABP mRNA in Brain Regions

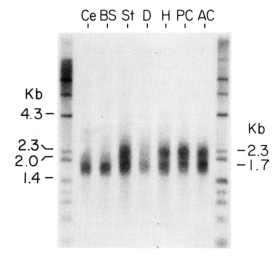

FIG. 24. Localization of ABP mRNA in rat brain by Northern blot hybridization. Brains from 40-day-old male rats were dissected, and poly(A) RNA was isolated from each region. After fractionation of the RNA (10 μg) by agarose gel electrophoresis, the ABP RNAs were characterized by hybridization blot analysis with ^{32}P-labeled ABP cDNA as probe. ^{32}P-labeled single-stranded DNA size markers flank the lanes containing brain RNA (*Hind*III digest of phage λ DNA and *Hae*III digest of phage φX174 DNA). The sizes of relevant marker DNAs are indicated on the left, and the size of each major hybridizing species relative to the markers is indicated at the right. Lane Ce, cerebellum; BS, brain stem; ST, striatum; D, diencephalon; H, hippocampus; PC, posterior cortex; AC, anterior cortex. From Wang *et al.* (1990).

In the male rat, transcription of the ABP/SHBG gene begins in the testis at 10–15 days after birth and the level of ABP mRNA reaches a maximal level around 20–25 days (Hall *et al.*, 1986; Reventos *et al.*, 1988). The level of ABP mRNA in testis remains fairly constant until at least 60 days of age (Hall *et al.*, 1986). Secreted ABP appears in the blood immediately after transcription begins and plasma ABP reaches a peak around 25 days and declines until 35–40 days of age (Gunsalus *et al.*, 1984; Carreau, 1986; Danzo and Eller, 1985). The decrease in plasma ABP coincides with the development of the blood–testis barrier and formation of the epididymal lumen (Carreau, 1986; Danzo and Eller, 1985). By 40 days of age most secreted testicular ABP is transported to the epididymis, but a small amount is secreted basally to the plasma compartment (Gunsalus *et al.*, 1980). The transient high blood concentrations of ABP from 20 to 40 days of age mimic those species that have a circulating SHBG, but the function this plasma ABP serves is unknown.

SHBG and SHBG mRNA are transiently expressed during development of the rat; SHBG mRNA (1.7 kb) is present in fetal liver from 15 to 17 days of gestation (Sullivan et al., 1991). At this time, SHBG appears in the male and female fetal blood (Gunsalus et al., 1984; Becker and Iles, 1985; Carreau, 1986). Its source is thought to be the liver, but there may be contributions from other tissues. After 17 days postconception, the plasma levels decline and are barely detectable at birth. This transient SHBG expression occurs during extremely rapid tissue growth and differentiation of the sex organs, including the Wolffian duct development into the epididymis and seminal vesicle. Becker and Iles (1985) speculated that SHBG may function in the embryo as a protection against excessive concentrations of free androgen. Our hypothesis was that the Wolffian duct was a target tissue of fetal plasma SHBG, which was functioning to modulate local levels of testosterone, which is required for Wolffian duct differentiation. However, immunocytochemistry of whole embryos revealed no immunoreactivity in the Wolffian duct or testis during this period. Interestingly, the most intensely stained tissue was not the liver, but was cardiac and striated muscle fibers. This staining occurred with three ABP antisera, one raised against purified ABP and two against ABP peptides (D. R. Joseph, P. M. Sullivan, and P. Ordronneau, unpublished results). Moreover, the identical structures reacted with an antiserum raised against a peptide specific for the N terminus of the ABP/SHBG-like protein. Thus, it would appear that the majority of the immunoreactive muscle protein is the nonsecreted ABP/SHBG-like protein. As described earlier, the function of this protein is not known.

A process that is often associated with genes that have roles in differentiation is alternative RNA splicing (Koenig et al., 1989; Yen et al., 1991; Joseph and Baker, 1992). Quite often the products of alternatively spliced mRNAs have actions that are antagonistic to the protein (Koenig et al., 1989; Yen et al., 1991). Alternatively sliced ABP/SHBG mRNAs have been described in rat fetal liver, rat brain, and human testis. One coding characteristic of all the alternate forms of ABP/SHBG is the presence of the receptor-binding region (Figs. 17 and 19). Based on expression and mutagenesis experiments, none of these alternate proteins would be expected to bind steroids. Receptor-binding properties of alternate proteins without steroid-binding properties fit the structure one would expect for an antagonist.

VI. Summary

Despite over 20 years of research, the functions of ABP and SHBG remain elusive. The major reason for this lack of knowledge has been the unavailability of natural mutants with clinical defects for study. There is strong evidence that these binding proteins do act to modulate the gene regulatory actions of nuclear sex steroid receptors by controlling the availability of androgens and estrogens. In plasma, SHBG controls the metabolic clearance rate of sex steroids. In addition there is strong evidence that they have a much broader function. The identification of plasma membrane receptors in target tissues and the finding of homologous domains in several developmental proteins support other functions. Moreover, other experiments suggest the proteins may actually be hormones or growth factors. These findings are not compatible with a model that has the proteins only regulating free steroid hormone levels. Obviously, much more experimentation will be necessary to reveal the functions of ABP and SHBG. The recent discoveries have offered several clues to their functions and open new routes for study. These experiments, coupled with newly developed techniques, such as gene knockout by homologous recombination, make one optimistic that the functions of these unique proteins will be deciphered in the near future.

Acknowledgments

I thank William Rosner and Benjamin Danzo for their evaluation of the review and helpful suggestions. I also thank Meg Hollowbush for help with typing and organizing the references and Robert Winston for his help in the library.

REFERENCES

Ain, K. B., Mori, Y., and Refetoff, S. (1987). Reduced clearance rate of thyroxine-binding globulin (TBG) with increased sialylation: A mechanism for estrogen-induced elevation of serum TBG concentration. *J. Clin. Endocrinol. Metab.* **65**, 689–696.

Amaral, L., Lin, K., Samuels, A. J., and Werthamer, S. (1974). Human liver nuclear transcortin. Its postulated role in glucocorticoid regulation of genetic activity. *Biochim. Biophys. Acta* **362**, 332–345.

Anderson, D. C. (1974). Sex hormone-binding globulin. *Clin. Endocrinol. (Oxford)* **3**, 69–96.

Anthony, C. T., Danzo, B. J., and Orgebin-Crist, M. C. (1984a). Investigations on the relationship between sperm fertilizing ability and the androgen binding protein in the restricted rat. *Endocrinology (Baltimore)* **114**, 1413–1418.

Anthony, C. T., Danzo, B. J., and Orgebin-Crist, M. C. (1984b). Investigations on the relationship between sperm fertilizing ability and the androgen binding protein in the hypophysectomized, pregnenolone-injected rat. *Endocrinology (Baltimore)* **114**, 1419–1425.

Avvakumov, G. V., Matveentseva, I. V., Akhrem, L. V., Strel'chyonok, O. A., and Akhrem, A. A. (1983). Study of the carbohydrate moiety of human serum sex hormone-binding globulin. *Biochim. Biophys. Acta* **760,** 104–110.

Avvakumov, G. V., Zhuk, N. I., and Strel'chyonok, O. A. (1986). Subcellular distribution and selectivity of the protein-binding component of the recognition system for sex-hormone-binding protein–estradiol complex in human decidual endometrium. *Biochim. Biophys. Acta* **881,** 489–498.

Avvakumov, G. V., Zhuk, N. I., and Strel'chyonok, O. A. (1988). Biological function of the carbohydrate component of the human sex steroid-binding globulin. *Biokhimiya (Moscow)* **53,** 838–841.

Baker, M. E., French, F. S., and Joseph, D. R. (1987). Vitamin K-dependent protein S is similar to rat androgen-binding protein. *Biochem. J.* **243,** 293–296.

Bardin, C. W., Musto, N., Gunsalus, G., Kotite, N., Cheng, S.-L., Larrea, F., and Becker, R. (1981). Extracellular androgen binding proteins. *Annu. Rev. Physiol.* **43,** 189–198.

Bardin, C. W., Cheng, C. Y., Musto, N. A., and Gunsalus, G. L. (1988). The Sertoli Cell. *In* "The Physiology of Reproduction" (E. Knobil and J. Neill, eds.) pp. 933–973. Raven Press, New York.

Bartke, A., Steele, R. E., Musto, N. A., and Caldwell, B. V. (1973). Fluctuations in plasma testosterone levels in adult male rats and mice. *Endocrinology (Baltimore)* **92,** 1223–1228.

Beck, K., Hunter, I., and Engel, J. (1990). Structure and function of laminin: Anatomy of a multidomain glycoprotein. *FASEB J.* **4,** 148–160.

Becker, R. R., and Iles, D. J. (1985). Developmental pattern of androgen-binding protein secretion during the critical period of sexual differentiation. *Arch. Androl.* **14,** 107–114.

Ben-Rafael, Z., Mastroianni, L., Jr., Meloni, F., Lee, M. S., and Flickinger, G. L. (1986). Total estradiol, free estradiol, sex hormone-binding globulin, and the fraction of estradiol bound to sex hormone-binding globulin in human follicular fluid. *J. Clin. Endocrinol. Metab.* **63,** 1106–1111.

Bérubé, D., Séralini, G. E., Gagné, R., and Hammond, G. L. (1990). Localization of the human sex hormone-binding globulin gene (SHBG) to the short arm of chromosome 17 (17p12-p13). *Cytogenet. Cell Genet.* **54,** 65–67.

Blake, R. E., Rajguru, S., Nolan, G. H., and Ahluwalia, B. S. (1988). Dexamethasone suppresses sex-hormone binding globulin. *Fertil. Steril.* **49,** 66–70.

Blaquier, J. A. (1971). Selective uptake and metabolism of androgens by rat epididymis. The presence of a cytoplasmic receptor. *Biochem. Biophys. Res. Commun.* **45,** 1076–1082.

Bocchinfuso, W. P., Warmels-Rodenhiser, S., and Hammond, G. L. (1991). Expression and differential glycoslylation of human sex hormone-binding globulin by mammalian cell lines. *Mol. Endocrinol.* **5,** 1723–1729.

Bocchinfuso, W. P., Warmels-Rodenhiser, S., and Hammond, G. L. (1992a). Structure/function analyses of human sex hormone-binding globulin by site-directed mutagenesis. *FEBS Lett.* **301,** 227–230.

Bocchinfuso, W. P., Ma, K.-L., Lee, W. M., Warmels-Rodenhiser, S., and Hammond, G. L. (1992b). Selective removal of glycosylation sites from sex hormone-binding globulin by site-directed mutagenesis. *Endocrinology (Baltimore)* **131,** 2331–2336.

Bordin, S., and Petra, P. H. (1980). Immunocytochemical localization of the sex steroid-binding protein of plasma in tissues of the adult monkey *Macaca nemestrina. Proc. Natl. Acad. Sci. U.S.A.* **77,** 5678–5682.

Bordin, S., Lewis, J., and Petra, P. H. (1978). Monospecific antibodies to the sex steroid-binding protein (SBP) of human and rabbit sera: Cross-reactivity with other species. *Biochem. Biophys. Res. Commun.* **85**, 391–401.

Bordin, S., Torres, R., and Petra, P. H. (1982). An enzyme-immunoassay (ELISA) for the sex steroid-binding protein (SBP) of human serum. *J. Steroid Biochem.* **17**, 453–457.

Brady, J., Radonovich, M., Vodkin, M., Natarajan, V., Thoren, M., Das, G., Janik, J., and Salzman, N. P. (1982). Site-specific base substitution and deletion mutations that enhance or suppress transcription of the SV40 major late RNA. *Cell (Cambridge, Mass.)* **31**, 625–633.

Carreau, S. (1986). L'"androgen binding protein" (ABP) chez le rat, le bélier, le taureau et l'homme: Analyse comparée. *Colloq.–Inst. Natl. Sante Rech. Med.* **149**, 293–303.

Casali, E., Petra, P. H., and Ross, J. B. A. (1990). Fluorescence investigation of the sex steroid binding protein of rabbit serum: Steroid binding and subunit dissociation. *Biochemistry* **29**, 9334–9343.

Chang, C.-F., and Lee, Y.-H. (1992). Purification of the sex steroid-binding protein from common carp (*Cyprinus carpio*) plasma. *Comp. Biochem. Physiol. B* **101B**, 587–590.

Cheng, C.-Y., Frick, J., Gunsalus, G. L., Musto, N. A., and Bardin, C. W. (1984). Human testicular androgen-binding protein shares immunodeterminants with serum testosterone-estradiol-binding globulin. *Endocrinology (Baltimore)* **114**, 1395–1401.

Cheng, C.-Y., Musto, N. A., Gunsalus, G. L., Frick, J., and Bardin, C. W. (1985). There are two forms of androgen binding proteins in the testis. Comparison of their protomeric variants with serum testosterone-estradiol binding globulin. *J. Biol. Chem.* **260**, 5631–5640.

Cheng, S.-L., and Musto, N. A. (1982). Purification and characterization of androgen binding protein from rabbit epididymis. *Biochemistry* **21**, 2400–2405.

Cheng, S.-L., Kotite, N., and Musto, N. A. (1984). Comparison of rabbit androgen binding protein with testosterone estradiol binding globulin-I. Physical and chemical properties. *J. Steroid Biochem.* **21**, 669–676.

Corvol, P., and Bardin, C. W. (1973). Species distribution of testosterone-binding globulin. *Biol. Reprod.* **8**, 277–282.

Dählback, B., Lundwall, A., and Stenflo, J. (1986). Primary structure of bovine vitamin K-dependent protein S human liver cDNA encoding a protein S precursor. *Proc. Natl. Acad. Sci. U.S.A.* **83**, 4199–4203.

Dählback, B., Hildebrand, B., and Malm, J. (1990). Characterization of functionally important domains in human vitamin K-dependent protein S using monoclonal antibodies. *J. Biol. Chem.* **265**, 8127–8135.

Damassa, D. A., and Gustafson, A. W. (1988). Effects of chronic infusions of sex steroid-binding protein on the testosterone-mediated inhibition of gonadotropin secretion and maintenance of sex accessory glands in male rats. *Endocrinology (Baltimore)* **123**, 1885–1892.

Damassa, D. A., Lin, T.-M., Sonnenschein, C., and Soto, A. M. (1991). Biological effects of sex hormone-binding globulin on androgen-induced proliferation and androgen metabolism in LNCaP prostate cells. *Endocrinology (Baltimore)* **129**, 75–84.

Danzo, B. J., and Bell, B. W. (1988). The microheterogeneity of androgen-binding protein in rat serum and epididymis is due to differences in glycosylation of their subunits. *J. Biol. Chem.* **263**, 2402–2408.

Danzo, B. J., and Black, J. H. (1990a). Structure of asparagine-linked oligosaccharides on human and rabbit testosterone-binding globulin. *Biol. Reprod.* **42**, 472–482.

Danzo, B. J., and Black, J. H. (1990b). Analysis of the oligosaccharides on rat androgen-binding protein using serial lectin chromatography. *Biol. Reprod.* **43**, 219–228.

Danzo, B. J., and Eller, B. C. (1975). Steroid-binding proteins in rabbit plasma: Separation of testosterone-binding globulin (TeBG) from corticosteroid-binding globulin (CBG), preliminary characterization of TeBG and changes in TeBG concentration during sexual maturation. *Mol. Cell. Endocrinol.* **2**, 351–368.

Danzo, B. J., and Eller, B. C. (1985). The ontogeny of biologically active androgen-binding protein in rat plasma, testis, and epididymis. *Endocrinology (Baltimore)* **117**, 1380–1388.

Danzo, B. J., and Joseph, D. R. (1994). Structure/function relationships of rat androgen-binding protein/sex hormone-binding globulin: The effect of mutagenesis on steroid-binding parameters. *Endocrinology,* in press.

Danzo, B. J., Orgebin-Crist, M.-C., and Toft, D. O. (1973). Characterization of a cytoplasmic receptor for 5α-dihydrotestosterone in caput epididymidis of intact rabbits. *Endocrinology (Baltimore)* **92**, 310–317.

Danzo, B. J., Eller, B. C., and Orgebin-Crist, M.-C. (1974). Studies on the site of origin of the androgen binding protein present in epididymal cytosol from mature intact rabbits. *Steroids* **24**, 107–122.

Danzo, B. J., Cooper, T. G., and Orgebin-Crist, M.-C. (1977). Androgen binding protein (ABP) in fluids collected from the rete testis and cauda epididymidis of sexually mature and immature rabbits and observations on morphological changes in the epididymis following ligation of the ductuli efferentes. *Biol. Reprod.* **17**, 64–77.

Danzo, B. J., Taylor, C. A., Jr., and Schmidt, W. N. (1980). Binding of the photoaffinity ligand 17β-hydroxy-4,6-androstadien-3-one to rat androgen-binding protein: Comparison with the binding of 17β-hydroxy-5α-androstan-3-one. *Endocrinology (Baltimore)* **107**, 1169–1175.

Danzo, B. J., Eller, B. C., and Bell, B. W. (1987). The apparent molecular weight of androgen-binding protein (ABP) in the blood of immature rats differs from that of ABP in the epididymis. *J. Steroid Biochem.* **28**, 411–419.

Danzo, B. J., Black, J. H., and Bell, B. W. (1989a). The microheterogeneity of rabbit testosterone-binding globulin is due to differential glycosylation of its single protomer. *Biol. Reprod.* **41**, 957–965.

Danzo, B. J., Bell, B. W., and Black, J. H. (1989b). Human testosterone-binding globulin is a dimer composed of two identical protomers that are differentially glycosylated. *Endocrinology (Baltimore)* **124**, 2809–2817.

Danzo, B. J., Pavlou, S. N., and Anthony, H. L. (1990). Hormonal regulation of androgen-binding protein in the rat. *Endocrinology (Baltimore)* **127**, 2829–2838.

Danzo, B. J., Parrott, J. A., and Skinner, M. K. (1991a). Analysis of the steroid binding domain of rat androgen-binding protein. *Endocrinology (Baltimore)* **129**, 690–696.

Danzo, B. J., Black, J. H., and Bell, B. W. (1991b). Analysis of the oligosaccharides on androgen-binding proteins: Implications concerning their role in structure/function relationships. *J. Steroid Biochem. Mol. Biol.* **40**, 821–831.

Dayhoff, M. O., Barker, W. C., and Hunt, L. T. (1983). Establishing homologies in protein sequences. *In* "Methods in Enzymology" (C. Hirs and S. Timasheff, eds.), Vol. 91, pp. 524–545. Academic Press, New York.

Degrelle, H. (1986). Dosage immunologique de la "sex steroid-binding protein" (SBP) dans le sérum humain. *Colloq.—Inst. Natl. Sante Rech. Med.* **149**, 221–226.

DePhilip, R. M., Feldman, M., Spruill, W. A., French, F. S., and Kierszenbaum, A. L. (1982). The secretion of androgen-binding protein and other proteins by rat Sertoli cells in culture: A structural and electrophoretic study. *Ann. N.Y. Acad. Sci.* **383**, 360–371.

Dorrington, J. H., Roller, N. F., and Fritz, I. B. (1975). Effects of follicle stimulating hormone on cultures of Sertoli cell preparations. *Mol. Cell. Endocrinol.* **3**, 57–70.

Edmunds, S. E. J., Stubbs, A. P., Santos, A. A., and Wilkinson, M. L. (1990). Estrogen and androgen regulation of sex hormone binding globulin secretion by a human liver cell line. *J. Steroid Biochem. Mol. Biol.* **37**, 733–739.

Ehrig, K., Leivo, I., Argraves, W. S., Ruoslahti, E., and Engvall, E. (1990). Merosin, a tissue-specific basement membrane protein, is a laminin-like protein. *Proc. Natl. Acad. Sci. U.S.A.* **87**, 3264–3268.

Ekins, R. P. (1990). Measurement of free hormones in blood. *Endocr. Rev.* **11**, 5–46.

Ekins, R. P., and Edwards, P. R. (1988). Plasma protein-mediated transport of steroid and thyroid hormones. A critique. *Ann. N.Y. Acad. Sci.* **538**, 193–203.

Elkington, J. S. H., Sanborn, B. M., Martin, M. W., Chowdhury, A. K., and Steinberger, E. (1977). Effect of testosterone propionate on ABP levels in rats hypophysectomized at different ages using individual sampling. *Mol. Cell. Endocrinol.* **6**, 203–209.

Ellison, S. A., and Pardridge, W. M. (1990). Reduction of testosterone availability to 5α-reductase by human sex hormone-binding globulin in the rat ventral prostate gland *in vivo. Prostate* **17**, 281–291.

Engel, J. (1989). EGF-like domains in extracellular matrix proteins: Localized signals for growth and differentiation? *FEBS Lett.* **251**, 1–7.

Engel, J. (1992). Laminins and other strange proteins. *Biochemistry* **31**, 10643–10651.

Englebienne, P. (1984). The serum steroid transport proteins. Biochemistry and clinical significance. *Mol. Aspects Med.* **7**, 313–396.

Englebienne, P., Van Hoorde, P., and Verheyden, R. (1987). Dimerization of SHBG by gelatin and dithiothreitol. Implications for the measurement of SHBG binding capacity in human serum. *J. Steroid Biochem.* **26**, 527–534.

Fawell, S. E., Lees, J. A., and Parker, M. G. (1989). A proposed consensus steroid-binding sequence—A reply. *Mol. Endocrinol.* **3**, 1002–1005.

Felden, F., Leheup, B., Fremont, S., Bouguerne, R., Egloff, M., Nicolas, J. P., Grignon, G., and Gueant, J. L. (1992a). The plasma membrane of epididymal epithelial cells has a specific receptor which binds to androgen-binding protein and sex steroid-binding protein. *J. Steroid Biochem. Mol. Biol.* **42**, 279–285.

Felden, F., Guéant, J. L., Ennya, G. A., Gérard, A., Frémont, S., Nicolas, J. P., and Gérard, H. (1992b). Photoaffinity labelled rat androgen-binding protein and human sex hormone steroid-binding protein bind specifically to rat germ cells. *J. Mol. Endocrinol.* **9**, 39–46.

Felden, F., Martin, M.-E., Gueant, J.-L., Benassayag, C., and Nunez, E. A. (1993). Free fatty acid-induced alterations in the steroid-binding properties of rat androgen-binding protein. *Biochem. Biophys. Res. Commun.* **190**, 602–608.

Feldman, M., Lea, O. A., Petrusz, P., Tres, L. L., Kierszenbaum, A. L., and French, F. S. (1981). Androgen-binding protein. Purification from rat epididymis, characterization and immunocytochemical localization. *J. Biol. Chem.* **256**, 5170–5175.

Feng, D., and Doolittle, R. F. (1987). Progressive sequence alignment as a prerequisite to correct phylogenetic trees. *J. Mol. Evol.* **25**, 351–360.

Fernlund, P., and Laurell, C.-B. (1981). A simple two-step procedure for the simultaneous isolation of corticosteroid binding globulin and sex hormone binding globulin from human serum by chromatography on cortisol-sepharose and phenyl-sepharose. *J. Steroid Biochem.* **14**, 545–552.

Fortunati, N., Fissore, F., Fazzari, A., Berta, L., Giudici, M., and Frairia, R. (1991). Sex steroid-binding protein interacts with a specific receptor on human premenopausal endometrium membrane, modulating effect of estradiol. *Steroids* **56**, 341–346.

Fortunati, N., Fissore, F., Fazzari, A., Berta, L., Varvello, L., and Frairia, R. (1992a). Receptor for sex steroid-binding protein of endometrium membranes: Solubilization, partial characterization and role of estradiol in steroid-binding protein-soluble receptor interaction. *Steroids* **57**, 464–470.

Fortunati, N., Frairia, R., Fissore, F., Berta, L., Fazzari, A., and Gaidano, G. (1992b). The receptor for human sex steroid binding protein (SBP) is expressed on membranes of neoplastic endometrium. *J. Steroid Biochem. Mol. Biol.* **42**, 185–191.

Fortunati, N., Fissore, F., Fazzari, A., Berta, L., Benedusi-Pagliano, E., and Frairia, R. (1993). Biological relevance of the interaction between sex steroid binding protein and its specific receptor of MCF-7 cells: Effect on the estradiol-induced cell proliferation. *J. Steroid Biochem. Mol. Biol.* **45**, 435–444.

Foucher, J.-L., and Le Gac, F. (1989). Evidence for an androgen binding protein in the testis of teleost fish (*Salmo gairdneri* R): A potential marker of Sertoli cell function. *J. Steroid Biochem.* **32**, 545–552.

Foucher, J.-L., Niu, P. D., Mourot, B., Vaillant, C., and Le Gac, F. (1991). *In vivo* and *in vitro* studies on sex steroid binding protein (SBP) regulation in rainbow trout (*Oncorhynchus mykiss*): Influence of sex steroid hormones and of factors linked to growth and metabolism. *J. Steroid Biochem.* **39**, 975–986.

Frairia, R., Fortunati, N., Berta, L., Fazzari, A., Fissore, F., and Gaidano, G. (1991). Sex steroid binding protein (SBP) receptors in estrogen sensitive tissues. *J. Steroid Biochem. Mol. Biol.* **40**, 805–812.

Frairia, R., Fortunati, N., Fissore, F., Fazzari, A., Zeppegno, P., Varvello, L., Orsello, M., and Berta, L. (1992). The membrane receptor for sex steroid binding protein is not ubiquitous. *J. Endocrinol. Invest.* **15**, 617–620.

French, F. S., and Ritzén, E. M. (1973). A high-affinity androgen-binding protein (ABP) in rat testis: Evidence for secretion into efferent duct fluid and absorption by epididymis. *Endocrinology (Baltimore)* **93**, 88–95.

Fritz, I. B. (1979). Sites of actions on androgen and follicle-stimulating hormone on cells of the seminiferous tubule. *In* "Biochemical Actions of Hormones" (G. Litwack, ed.), pp. 249–281. Academic Press, New York.

Fritz, I. B., Kopec, B., Lam, K., and Vernon, R. G. (1974). Effects of FSH on levels of androgen binding protein in the testis. *In* "Hormone Binding and Target Cell Activation in the Testis" (M. L. Dufau and A. R. Means, eds.), pp. 311–327. Plenum, New York.

Fritz, I. B., Rommerts, F. G., Louis, B. G., and Dorrington, J. H. (1976). Regulation by FSH and dibutyryl cyclic AMP of the formation of androgen-binding protein in Sertoli cell-enriched cultures. *J. Reprod. Fertil.* **46**, 17–24.

Gasic, G. P., Arenas, C. P., Gasic, T. B., and Gasic, G. J. (1992). Coagulation factors X, Xa and protein S are potent mitogens of cultured aortic smooth muscle cells. *Proc. Natl. Acad. Sci. U.S.A.* **89**, 2317–2320.

Gerard, A., Khanfri, J., Gueant, J. L., Fremont, S., Nicolas, J. P., Grignon, G., and Gerard, H. (1988). Electron microscope radioautographic evidence of *in vivo* androgen-binding protein internalization in the rat epididymis principal cells. *Endocrinology (Baltimore)* **122**, 1297–1307.

Gerard, A., Egloff, M., Gerard, H., El Harate, A., Domingo, M., Gueant, J. L., Dang, C. D., and Degrelle, H. (1990). Internalization of human sex steroid-binding protein in the monkey epididymis. *J. Mol. Endocrinol.* **5**, 239–251.

Gerard, A., En Nya, A., Egloff, M., Domingo, M., Degrelle, H., and Gerard, H. (1992). Endocytosis of human sex steroid-biding protein in monkey germ cells. *Ann. N.Y. Acad. Sci.* **637**, 258–276.

Gerard, H., Gerard, A., Nya, A. E., Felden, F., and Gueant, J. L. (1994). Spermatogenic cells do internalize Sertoli androgen-binding protein: A transmission electron microscopy autoradiographic study in rat. *Endocrinology* **134**, 1515–1527.

Gershagen, S., Henningsson, K., and Fernlund, P. (1987a). Subunits of human sex hormone binding globulin. Interindividual variation in size. *J. Biol. Chem.* **262**, 8430–8437.

Gershagen, S., Fernlund, P., and Lundwall, A. (1987b). A cDNA coding for human sex hormone binding globulin. Homology to vitamin K-dependent protein S. *FEBS Lett.* **220**, 129–135.

Gershagen, S., Lundwall, A., and Fernlund, P. (1989). Characterization of the human sex hormone-binding globulin (SHBG) gene and demonstration of two transcripts in both liver and testis. *Nucleic Acids Res.* **17**, 9245–9258.

Gershagen, S., Fernlund, P., and Edenbrandt, C.-M. (1991). The genes for SHBG/ABP and the SHBG-like region of vitamin K-dependent protein S have evolved from a common ancestral gene. *J. Steroid Biochem. Mol. Biol.* **40**, 763–769.

Grenot, C., de Montard, A., Blachère, T., Mappus, E., and Cuilleron, C.-Y. (1988). Identification d'un site de photomarquage de la protéine plasmatique de liaison de la testostérone et de l'oestradiol (SBP) par l'hydroxy-17β oxo-3 androstadiene-4,6 tritié. *C. R. Seances Acad. Sci.* **307**, 391–396.

Grenot, C., de Montard, A., Blachère T, de Ravel, M. R., Mappus, E., and Cuilleron, C. Y. (1992). Characterization of Met-139 as the photolabeled amino acid residue in the steroid binding site of sex hormone binding globulin using *delta* 6 derivatives of either testosterone or estradiol as unsubstituted photoaffinity labeling reagents. *Biochemistry* **31**, 7609–7621.

Griffin, P. R., Kumar, S., Shabanowitz, J., Charbonneau, H., Namkung, P. C., Walsh, K. A., Hunt, D. F., and Petra, P. H. (1989). The amino acid sequence of the sex steroid-binding protein of rabbit serum. *J. Biol. Chem.* **264**, 19066–19075.

Griswold, M. D., Morales, C., and Sylvester, S. R. (1988). Molecular biology of the Sertoli cell. *Oxford Rev. Reprod. Biol.* **10**, 124–136.

Gunsalus, G. L., Musto, N. A., and Bardin, C. W. (1978). Immunoassay of androgen-binding protein in blood: A new approach for the study of the seminiferous tubule. *Science* **200**, 65–66.

Gunsalus, G. L., Musto, N. A., and Bardin, C. W. (1980). Bidirectional release of a Sertoli cell product, androgen binding protein, into the blood and seminiferous tubule. *In* "Testicular Development: Structure and Function" (A. Steinberger and E. Steinberger, eds.), pp. 291–297. Raven Press, New York.

Gunsalus, G. L., Larrea, F., Musto, N. A., Becker, R. R., Mather, J. P., and Bardin, C. W. (1981). Androgen binding protein as a marker for Sertoli cell function. *J. Steroid Biochem.* **15**, 99–106.

Gunsalus, G. L., Carreau, S., Vogel, D. L., Musto, N. A., and Bardin, C. W. (1984). Use of androgen-binding protein to monitor development of the seminiferous epithelium. *In* "Sexual Differentiation: Basic and Clinical Aspects" (M. Serio, ed.), pp. 53–64. Raven Press, New York.

Gunsalus, G. L., de Besi, L., Musto, N. A., and Bardin, C. W. (1986). Measurement of rat androgen-binding protein (rABP) by steroid binding, radioimmunoassay (RIA) and enzyme-linked immunosorbent assay (ELISA). *Colloq.—Inst. Natl. Sante Rech. Med.* **149**, 227–235.

Hadley, M. A., Weeks, B. S., Kleinman, H. K., and Dym, M. (1990). Laminin promotes formation of cord-like structures by Sertoli cells *in vitro. Dev. Biol.* **140,** 318–327.

Hagen, F. S., Arguelles, C., Sui, L.-M., Zhang, W., Seidel, P. R., Conroy, S. C., and Petra, P. H. (1992). Mammalian expression of the human sex steroid-binding protein of plasma (SBP or SHBG) and testis (ABP). *FEBS Lett.* **299,** 23–27.

Hagenäs, L., Ritzen, E. M., Ploen, L., Hansson, V., French, F. S., and Nayfeh, S. N. (1975). Sertoli cell origin of testicular androgen-binding protein (ABP). *Mol. Cell. Endocrinol.* **2,** 339–350.

Hall, S. H., Joseph, D. R., Conti, M., and French, F. S. (1986). Regulation of androgen binding protein messenger RNA. *In* "Molecular and Cellular Endocrinology of the Testis" (M. Stefanini, M. Conti, R. Geremia, and E. Ziparo, eds.), pp. 139–149. Excerpta Medica, Amsterdam and New York.

Hall, S. H., Joseph, D. R., French, F. S., and Conti, M. (1988). Follicle-stimulating hormone induces transient expression of the protooncogene *c-fos* in primary Sertoli cell cultures. *Mol. Endocrinol.* **2,** 55–61.

Hall, S. H., Conti, M., French, F. S. and Joseph, D. R. (1990). Follicle-stimulating hormone regulation of androgen-binding protein messenger RNA in Sertoli cell cultures. *Mol. Endocrinol.* **4,** 349–355.

Hammond, G. L. (1986). Relative merits of steroid-binding capacity assays and immunoassays for corticosteroid-binding globulin (CBG). *Colloq.—Inst. Natl. Sante Rech. Med.* **149,** 191–198.

Hammond, G. L. (1990). Molecular properties of corticosteroid binding globulin and the sex steroid binding proteins. *Endocr. Rev.* **11,** 65–79.

Hammond, G. L., Langley, M. S., and Robinson, P. A. (1985). A liquid-phase immunoradiometric assay (IRMA) for human sex hormone binding globulin (SHBG). *J. Steroid Biochem.* **23,** 451–460.

Hammond, G. L., Robinson, P. A., Sugino, H., Ward, D. N., and Finne, J. (1986). Physicochemical characteristics of human sex hormone binding globulin: Evidence for two identical subunits. *J. Steroid Biochem.* **24,** 815–824.

Hammond, G. L., Underhill, D. A., Smith, C. L., Goping, I. S., Harley, M. J., Musto, N. A., Cheng, C.-Y., and Bardin, C. W. (1987). The cDNA-deduced primary structure of human sex hormone-binding globulin and location of its steroid-binding domain. *FEBS Lett.* **215,** 100–104.

Hammond, G. L., Underhill, D. A., Rykse, H. M., and Smith, C. L. (1989). The human sex hormone-binding globulin gene contains exons for androgen-binding protein and two other testicular messenger RNAs. *Mol. Endocrinol.* **3,** 1869–1876.

Hansson, V. (1972). Further characterization of the 5α-dihydrotestosterone binding protein in the epididymal cytosol fraction. *In vitro* studies. *Steroids* **20,** 575–596.

Hansson, V., Reusch, E., Trygstad, O., Torgersen, O., Ritzén, E. M., and French, F. S. (1973). FSH stimulation of testicular androgen binding protein. *Nature (London) New Biol.* **246,** 56–58.

Hansson, V., Weddington, S. C., Naess, O., and Attramadal, A. (1975a). Testicular androgen binding protein (ABP)—A parameter of Sertoli cell secretory function. *In* "Hormonal Regulation of Spermatogenesis" (F. S. French, V. Hansson, E. M. Ritzen, and S. N. Nayfeh, eds.), pp. 323–335. Plenum, New York.

Hansson, V., Weddington, S. C., McLean, W. S., Smith, A. A., Nayfeh, S. N., French, F. S., and Ritzén, E. M. (1975b). Regulation of seminiferous tubular function by FSH and androgen. *J. Reprod. Fertil.* **44,** 363–375.

Hansson, V., Weddington, S. C., French, F. S., McLean, W. S., Smith, A., Nayfeh, S. N.,

Ritzén, E. M., and Hagenas, L. (1976a). Secretion and role of androgen-binding proteins in the testis and epididymis. *J. Reprod. Fertil., Suppl.* **24**, 17–33.

Hansson, V., Calandra, R., Purvis, K., Ritzén, E. M., and French, F. S. (1976b). Hormonal regulation of spermatogenesis. *Vitam. Horm. (N.Y.)* **34**, 187–214.

Harrington, W. F. (1975). The effects of pressure in ultracentrifugation of interacting systems. *Fractions* **1**, 10–18.

Hessing, M. (1991). The interaction between complement component C4b-binding protein and the vitamin K-dependent protein S forms a link between coagulation and the complement system. *Biochem. J.* **277**, 581–592.

Heyns, W. (1986). Measurement of steroid-binding proteins by steroid binding. *Colloq.— Inst. Natl. Sante Rech. Med.* **149**, 181–190.

Hiramatsu, R., Dunn, J. F., and Nisula, B. C. (1986). Testosterone-binding globulin assays. *Colloq.—Inst. Natl. Sante Rech. Med.* **149**, 199–206.

Hopp, T. P., and Woods, K. R. (1981). Prediction of protein antigenic determinants from amino acid sequences. *Proc. Natl. Acad. Sci. U.S.A.* **78**, 3824–3828.

Hoskins, J., Norman, D. K., Beckman, R. J., and Long, G. L. (1987). Cloning and characterization of human liver cDNA encoding a protein S precursor. *Proc. Natl. Acad. Sci. U.S.A.* **84**, 349–353.

Hryb, D. J., Khan, M. S., and Rosner, W. (1985). Testosterone-estradiol-binding globulin binds to human prostatic cell membranes. *Biochem. Biophys. Res. Commun.* **128**, 432–440.

Hryb, D. J., Khan, M. S., Romas, N. A., and Rosner, W. (1989). The solubilization and partial characterization of the sex hormone-binding globulin receptor from human prostate. *J. Biol. Chem.* **264**, 5378–5383.

Hryb, D. J., Khan, M. S., Romas, N. A., and Rosner, W. (1990). The control of the interaction of sex hormone-binding globulin with its receptor by steroid hormones. *J. Biol. Chem.* **265**, 6048–6054.

Hsu, A.-F., and Troen, P. (1978). An androgen binding protein in the testicular cytosol of human testis. *J. Clin. Invest.* **61**, 1611–1619.

Huang, H. F. S., Pogach, L. M., Nathan, E., Giglio, W., and Seebode, J. J. (1991). Synergistic effects of follicle-stimulating hormone and testosterone on the maintenance of spermiogenesis in hypophysectomized rats: Relationship with the androgen-binding protein status. *Endocrinology (Baltimore)* **128**, 3152–3161.

Joseph, D. R., and Baker, M. (1992). Sex hormone-binding globulin, androgen-binding protein, and vitamin K-dependent protein S are homologous to laminin A, merosin, and *Drosophila crumbs* protein. *FASEB J.* **6**, 2477–2481.

Joseph, D. R., and Lawrence, W. (1993). Mutagenesis of essential functional residues of rat androgen-binding protein/sex hormone-binding globulin. *Mol. Endocrinol.* **7**, 488–496.

Joseph, D. R., Hall, S. H., and French, F. S. (1985). Identification of complementary DNA clones that encode rat androgen binding protein. *J. Androl.* **6**, 392–395.

Joseph, D. R., Hall, S. H., and French, F. S. (1986). Rat androgen binding protein: Structure of the gene, mRNA and protein. *Colloq.—Inst. Natl. Sante Rech. Med.* **149**, 123–135.

Joseph, D. R., Hall, S. H., and French, F. S. (1987). Rat androgen-binding protein: Evidence for identical subunits and amino acid sequence homology with human sex hormone-binding globulin. *Proc. Natl. Acad. Sci. U.S.A.* **84**, 339–343.

Joseph, D. R., Hall, S. H., Conti, M., and French, F. S. (1988). The gene structure of rat androgen-binding protein: Identification of potential regulatory deoxyribonucleic acid elements of a follicle-stimulating hormone-regulated protein. *Mol. Endocrinol.* **2**, 3–13.

Joseph, D. R., Sullivan, P. M., Wang, Y.-M., Kozak, C., Fenstermacher, D. A., Behrendsen, M. E., and Zahnow, C. A. (1990a). Characterization and expression of the complementary DNA encoding rat histidine decarboxylase. *Proc. Natl. Acad. Sci. U.S.A.* **87**, 733–737.

Joseph, D. R., Hall, S. H., Sullivan, P. M., Wang, Y.-M., French, F. S., Sar, M., Bayliss, D. A., Millhorn, D. E., Marschke, K. B., Wilson, E. M., Conti, M., and Zahnow, C. A. (1990b). Regulation and function of androgen-binding protein. *In* "Hormonal Communicating Events in the Testis" (A. Isidori, A. Fabbri, and M. L. Dufau, eds.), pp. 137–148. Raven Press, New York.

Joseph, D. R., Adamson, M. C., and Kozak, C. A. (1991). Genetic mapping of the gene for androgen-binding protein/sex hormone-binding globulin to mouse chromosome 11. *Cytogenet. Cell Genet.* **56**, 122–124.

Joseph, D. R., Lawrence, W., and Danzo, B. J. (1992). The role of asparagine-linked oligosaccharides in the subunit structure, steroid binding, and secretion of androgen-binding protein. *Mol. Endocrinol.* **6**, 1127–1134.

Joseph, D. R., Wang, Y.-M., and Deschepper, C. F. (1993). Demonstration of androgen-binding protein gene expression in primary neuronal and astrocyte cultures. *Mol. Cell. Neurosci.* **4**, 432–439.

Joseph, D. R., Wang, Y.-M., and Sullivan, P. S. (1994). Characterization and sex hormone regulation of multiple alternate androgen-binding protein/sex hormone-binding globulin RNA transcripts in rat brain. *Endocrinol. J.,* in press.

Kallunki, P., and Tryggvason, K. (1992). Human basement membrane heparin sulfate proteoglycan core protein: A 467-kD protein containing multiple domains resembling elements of the low density lipoprotein receptor, laminin, neural cell adhesion molecules, and epidermal growth factor. *J. Cell Biol.* **116**, 559–571.

Kato, T., and Horton, R. (1968). Studies of testosterone binding globulin. *J. Clin. Endocrinol. Metab.* **28**, 1160–1168.

Khan, M. S., and Rosner, W. (1990). Histidine 235 of human sex hormone-binding globulin is the covalent site of attachment of the nucleophilic steroid derivative, 17 *beta*-bromoacetoxydihydrotestosterone. *J. Biol. Chem.* **265**, 8431–8435.

Khan, M. S., Knowles, B. B., Aden, D. P., and Rosner, W. (1981). Secretion of testosterone-estradiol-binding globulin by a human hepatoma-derived cell line. *J. Clin. Endocrinol. Metab.* **53**, 448–449.

Khan, M. S., Ewen, E., and Rosner, W. (1982). Radioimmunoassay for human testosterone-estradiol-binding globulin. *J. Clin. Endocrinol. Metab.* **54**, 705–710.

Khan, M. S., Ehrlich, P., Birken, S., and Rosner, W. (1985). Size isomers of testosterone-estradiol-binding globulin exist in the plasma of individual men and women. *Steroids* **45**, 463–472.

Khan, M. S., Hryb, D. J., Hashim, G. A., Romas, N. A., and Rosner, W. (1990). Delineation and synthesis of the membrane receptor-binding domain of sex hormone-binding globulin. *J. Biol. Chem.* **265**, 18362–18365.

Kierszenbaum, A. L., Feldman, M., Lea, O. A., Spruill, W. A., Tres, L. L., Petrusz, P., and French, F. S. (1980). Localization of androgen-binding protein in proliferating Sertoli cells in culture. *Proc. Natl. Acad. Sci. U.S.A.* **77**, 5322–5326.

Kissinger, C., Skinner, M. K., and Griswold, M. D. (1982). Analysis of Sertoli cell-secreted proteins by two-dimensional gel electrophoresis. *Biol. Reprod.* **27**, 233–240.

Kleinman, H. K., Weeks, B. S., and Schnaper, H. W., Kibbey, M. C., Yamamura, K., and Grant, D. S. (1993). The laminins: A family of basement membrane glycoproteins important in cell differentiation and tumor metastases. *Vitam. Horm. (N.Y.)* **47**, 161–186.

Koenig, R. J., Lazar, M. A., Hodin, R. A., Brent, G. A., Larsen, P. R., Chin, W. W., and Moore, D. D. (1989). Inhibition of thyroid hormone actions by a non-hormone binding c-*erb*A protein generated by alternative mRNA splicing. *Nature (London)* **337**, 659–661.

Kotite, N. J., Nayfeh, S. N., and French, F. S. (1978). FSH and androgen regulation of Sertoli cell function in the immature rat. *Biol. Reprod.* **18**, 65–73.

Kotite, N. J., Cheng, S.-L., Musto, N. A., and Gunsalus, G. L. (1986). Comparison of rabbit epididymal androgen binding protein and serum testosterone estradiol binding globulin-II. Immunological properties. *J. Steroid Biochem.* **25**, 171–176.

Kottler, M.-L., Ribot, G., Tardivel-Lacombe, J., Counis, R., and Degrelle, H. (1988). Identification of the primary translation product of the sex steroid-binding protein from monkey liver mRNA in a cell-free system. *Biochimie* **70**, 1423–1427.

Kottler, M.-L., Counis, R., and Degrelle, H. (1989). Sex steroid-binding protein: Identification and comparison of the primary product following cell-free translation of human and monkey (*Macaca fascicularis*) liver RNA. *J. Steroid Biochem.* **33**, 201–207.

Kottler, M.-L., Dang, C. D., Salmon, R., Counis, R., and Degrelle, H. (1990). Effect of testosterone on regulation of the level of sex steroid-binding protein mRNA in monkey (*Macaca fascicularis*) liver. *J. Mol. Endocrinol.* **5**, 253–257.

Kovacs, W. J., Bell, B. W., Turney, M. K., and Danzo, B. J. (1988). Monoclonal antibodies to rat androgen-binding protein recognize both of its subunits and cross-react with rabbit and human testosterone-binding globulin. *Endocrinology (Baltimore)* **122**, 2639–2647.

Krupenko, N. I., Avvakumov, G. V., and Strel'chyonok, O. A. (1990). Binding of sex hormone-binding globulin–androgen complexes to the placental syncytiotrophoblast membrane. *Biochem. Biophys. Res. Commun.* **171**(3), 1279–1283.

Krupenko, S. A., Krupenko, N. I., and Danzo, B. D. (1994). Interaction of sex hormone-binding globulin with plasma membranes from the rat epididymis and other tissues. *J. Steroid Biochem. Mol. Biol.*, in press.

Larrea, F., Diaz, L., Carino, C., Larriva-Sahd, J., Carrillo, L., Orozco, H., and Ulloa-Aguirre, A. (1993). Evidence that human placenta is a site of sex hormone-binding globulin gene expression. *J. Steroid Biochem. Mol. Biol.* **46**, 497–505.

Larrea, F., Musto, N. A., Gunsalus, G. L., Mather, J. P., and Bardin, C. W. (1981a). Origin of heavy and light protomers of androgen-binding protein from rat testis. *J. Biol. Chem.* **256**, 12566–12573.

Larrea, F., Musto, N., Gunsalus, G., and Bardin, C. W. (1981b). The microheterogeneity of rat androgen-binding protein from the testis, rete testis fluid, and epididymis, as demonstrated by immunoelectrophoresis and photoaffinity labeling. *Endocrinology (Baltimore)* **109**, 1212–1220.

Larrea, F., Oliart, R. M., Granados, J., Mutchinick, O., Diaz-Sanchez, V., and Musto, N. A. (1990). Genetic polymorphism of the human sex hormone-binding globulin: Evidence of an isoelectric focusing variant with normal androgen-binding affinities. *J. Steroid Biochem.* **36**, 541–548.

Larriva-Sahd, J., Orozco, H., Hernandez-Pando, R., Oliart, R. M., Musto, N. A., and Larrea, F. (1991). Immunohistochemical demonstration of androgen-binding protein in the rat prostatic gland. *Biol. Reprod.* **45**, 417–423.

Lasnitzki, I., and Franklin, H. R. (1972). The influence of serum on uptake, conversion and action of testosterone in rat prostate glands in organ culture. *J. Endocrinol.* **54**, 333–342.

Lasnitzki, I., and Franklin, H. R. (1975). The influence of serum on the uptake, conversion and action of dihydrotestosterone in rat prostate glands in organ culture. *J. Endocrinol.* **64,** 289–297.

La Spada, A. R., Wilson, E. M., Lubahn, D. B., Harding, A. E., and Fischbeck, K. H. (1991). Androgen receptor gene mutations in X-linked spinal and bulbar muscular atrophy. *Nature (London)* **352,** 77–79.

Lea, O. A., and Støa, K. F. (1972). The binding of testosterone to different serum proteins: A comparative study. *J. Steroid Biochem.* **3,** 409–419.

Lee, I. R., Dawson, S. A., Wetherall, J. D., and Hahnel, R. (1987). Sex hormone-binding globulin secretion by human hepatocarcinoma cells is increased by both estrogens and androgens. *J. Clin. Endocrinol. Metab.* **64,** 825–831.

Le Gaillard, F., Han, K.-K., and Dautrevaux, M. (1975). Caractérisation et propriétés physico-chimiques de la transcortine humaine. *Biochimie* **57,** 559–568.

Lobl, T. J. (1981). Androgen transport proteins: Physical properties, hormonal regulation, and possible mechanism of TeBG and ABP action. *Arch. Androl.* **7,** 133–151.

Louis, B. G., and Fritz, I. B. (1977). Stimulation by androgens of the production of androgen binding protein by cultured Sertoli cells. *Mol. Cell. Endocrinol.* **7,** 9–16.

Louis, B. G., and Fritz, I. B. (1979). Follicle-stimulating hormone and testosterone independently increase the production of androgen binding protein by Sertoli cells in culture. *Endocrinology (Baltimore)* **104,** 454–461.

Luckock, A., and Cavalli-Sforza, L. L. (1983). Detection of genetic variation with radioactive ligands. V. Genetic variants of testosterone-binding globulin in human serum. *Am. J. Hum. Genet.* **35,** 49–57.

Mahoney, P. A., Weber, U., Onofrechuk, P., Biessmann, H., Bryant, P. J., and Goodman, C. S. (1991). The fat tumor suppressor gene in *Drosophila* encodes a novel member of the cadherin gene superfamily. *Cell* **67,** 853–868.

Maillard, C., Berruyer, M., Serre, C. M., Dechavanne, M., and Delmas, P. D. (1992). Protein-S, a vitamin K-dependent protein, is a bone matrix component synthesized and secreted by osteoblasts. *Endocrinology (Baltimore)* **130,** 1599–1604.

Mak, P., and Callard, G. V. (1987). A novel steroid-binding protein in the testis of the dogfish (*Squalus acanthias*). *Gen. Comp. Endocrinol.* **68,** 104–112.

Manfioletti, G., Brancolini, C., Avanzi, G., and Schneider, C. (1993). The protein encoded by a growth arrest-specific gene (gas6) is a new member of the vitamin K-dependent proteins related to protein S, a negative coregulator in the blood coagulation cascade. *Mol. Cell. Biol.* **13,** 4976–4985.

Martin, M.-E., Vranchkx, R., Benassayag, C., and Nunez, E. A. (1986). Modifications of the properties of human sex steroid-binding protein by nonesterified fatty acids. *J. Biol. Chem.* **261,** 2954–2959.

Mather, J. P., Gunsalus, G. L., Musto, N. A., Cheng, C. Y., Parvinen, M., Wright, W., Perez-Infante, V., Margioris, A., Liotta, A., Becker, R., Kreiger, D. T., and Bardin, C. W. (1983). The hormonal and cellular control of Sertoli cell secretion. *J. Steroid Biochem.* **19,** 41–51.

Means, A. R., Fakunding, J. L., Huckins, C., Tindall, D. J., and Vitale, R. (1976). Follicle-stimulating hormone, the Sertoli cell and spermatogenesis. *Recent Prog. Horm. Res.* **32,** 477–522.

Means, A. R., Dedman, J. R., Tash, J. S., Tindall, D. J., van Sickle, M., and Welsh, M. J. (1980). Regulation of the testis Sertoli cell by follicle stimulating hormone. *Annu. Rev. Physiol.* **42,** 59–70.

Mendel, C. M. (1989). The free hormone hypothesis: A physiologically based mathematical model. *Endocr. Rev.* **10,** 232–274.

Mendel, C. M. (1990). Rates of dissociation of sex steroid hormones from human sex hormone-binding globulin: A reassessment. *J. Steroid Biochem. Mol. Biol.* **37,** 251–255.

Mendel, C. M. (1992). The free hormone hypothesis. Distinction from the free hormone transport hypothesis. *J. Androl.* **13,** 107–116.

Mercier, C., Alfsel, A., and Baulieu, E. E. (1966). Testosterone binding globulin in human plasma. *Int. Congr. Ser.—Excerpta Med.* **101,** 212 (abstr.).

Mercier-Bodard, C., Marchut, M., Perrot, M., Picard, M.-T., Baulieu, E.-E., and Robel, P. (1976). Influence of purified plasma proteins on testosterone uptake and metabolism by normal and hyperplastic human prostate in "constant-flow organ culture". *J. Clin. Endocrinol. Metab.* **43,** 374–386.

Mercier-Bodard, C., Renoir, J.-M., and Baulieu, E.-E. (1979). Further characterization and immunological studies of human sex steroid binding plasma protein. *J. Steroid Biochem.* **11,** 253–259.

Mercier-Bodard, C., Radanyi, C., Roux, C., Groyer, M. T., Robel, P., Dadoune, J. P., Petra, P. H., Jolly, D. J., and Baulieu, E. E. (1987). Cellular distribution and hormonal regulation of h-SBP in human hepatoma cells. *J. Steroid Biochem.* **27,** 297–307.

Mercier-Bodard, C., Baville, F., Bideux, G., Binart, N., Chambraud, B., and Baulieu, E.-E. (1989). Regulation of SBP synthesis in human cancer cell lines by steroid and thyroid hormones. *J. Steroid Biochem.* **34,** 199–204.

Mercier-Bodard, C., Nivet, V., and Baulieu, E.-E. (1991). Effects of hormones on SBP mRNA levels in human cancer cells. *J. Steroid Biochem. Mol. Biol.* **40,** 777–785.

Mickelson, K. E., and Petra, P. H. (1975). Purification of the sex steroid binding protein from human serum. *Biochemistry* **14,** 957–963.

Mickelson, K. E., and Petra, P. H. (1978). Purification and characterization of the sex-steroid-binding protein of rabbit serum. *J. Biol. Chem.* **253,** 5293–5298.

Mickelson, K. E., Harding, G. B., Forsthoefel, M., and Westphal, U. (1982). Steroid–protein interactions. Human corticosteroid-binding globulin: Characterization of dimer and electrophoretic variants. *Biochemistry* **21,** 654–660.

Moore, J. W., and Bulbrook, R. D. (1988). The epidemiology and function of sex hormone-binding globulin. *Oxford Rev. Reprod. Biol.* **10,** 180–236.

Murphy, B. E. P. (1968). Binding of testosterone and estradiol in plasma. *Can. J. Biochem.* **46,** 299–302.

Musto, N. A., Gunsalus, G. L., Miljkovic, M., and Bardin, C. W. (1977). A novel affinity column for isolation of androgen binding protein from rat epididymis. *Endocr. Res. Commun.* **4,** 147–157.

Musto, N. A., Gunsalus, G. L., and Bardin, C. W. (1980). Purification and characterization of rat androgen binding protein from the rat epididymis. *Biochemistry* **19,** 2853–2860.

Nakhla, A. M., Khan, M. S., and Rosner, W. (1990). Biologically active steroids activate receptor-bound human sex hormone-binding globulin to cause LNCaP cells to accumulate adenosine 3′,5′-monophosphate. *J. Clin. Endocrinol. Metab.* **71,** 398–404.

Nakhla, A. M., Khan, M. S., Romas, N. P., Rosner, W. (1994). Estradiol causes the rapid accumulation of cAMP in human prostate. *Proc. Natl. Acad. Sci. U.S.A.,* in press.

Nakano, Y., Guerrero, I., Hidalgo, A., Taylor, A., Whittle, J. R. S., and Ingham, P. W. (1989). A protein with several possible membrane-spanning domains encoded by the *Drosophila* segment polarity gene patched. *Nature* **341,** 508–513.

Namkung, P. C., Stanczyk, F. Z., Cook, M. J., Novy, M. J., and Petra, P. H. (1989). Half-life of plasma sex steroid-binding protein (SBP) in the primate. *J. Steroid Biochem.* **32,** 675–680.

Namkung, P. C., Kumar, S., Walsh, K. A., and Petra, P. H. (1990). Identification of lysine

134 in the binding site of the sex steroid-binding protein of human plasma. *J. Biol. Chem.* **265,** 18345–18350.

Noé, G., Cheng, Y. C., Dabiké, M., and Croxatto, H. B. (1992). Tissue uptake of human sex hormone-binding globulin and its influence on ligand kinetics in the adult female rat. *Biol. Reprod.* **47,** 970–976.

Pardridge, W. M. (1988a). Selective delivery of sex steroid hormones to tissues *in vivo* by albumin and by sex hormone-binding globulin. *Ann. N.Y. Acad. Sci.* **538,** 173–192.

Pardridge, W. M. (1988b). Selective delivery of sex steroid hormones to tissues by albumin and by sex hormone-binding globulin. *Oxford Rev. Reprod. Biol.* **10,** 237–292.

Parvinen, M. (1993). Cyclic function of Sertoli cells. *In* "The Sertoli Cell" (L. D. Russell and M. D. Griswold, eds.), pp. 331–347. Cache River Press, Clearwater, FL.

Pearlman, W. H., and Crépy, O. (1967). Steroid–protein interaction with particular reference to testosterone binding by human serum. *J. Biol. Chem.* **242,** 182–189.

Pearlman, W. H., Crépy, O., and Murphy, M. (1967). Testosterone-binding levels in the serum of women during the normal menstrual cycle, pregnancy and the post-partum period. *J. Clin. Endocrinol. Metab.* **27,** 1012–1018.

Pelliniemi, L. J., Dym, M., Gunsalus, G. L., Musto, N. A., Bardin, C. W., and Fawcett, D. W. (1981). Immunocytochemical localization of androgen-binding protein in the male rat reproductive tract. *Endocrinology (Baltimore)* **108,** 925–931.

Petra, P. H. (1986). Measurement of the sex steroid-binding protein of human plasma by an enzyme-linked immunosorbent assay, ELISA. *Colloq.—Inst. Natl. Sante Rech. Med.* **149,** 215–220.

Petra, P. H. (1991). The plasma sex steroid binding protein (SBP or SHBG). A critical review of recent developments on the structure, molecular biology and function. *J. Steroid Biochem. Mol. Biol.* **40,** 735–753.

Petra, P. H., and Lewis, J. (1980). Modification in the purification of the sex steroid-binding protein of human serum by affinity chromatography. *Anal. Biochem.* **105,** 165–169.

Petra, P. H., Stanczyk, F. Z., Senear, D. F., Namkung, P. C., Novy, M. J., Ross, J. B. A., Turner, E., and Brown, J. A. (1983). Current status of the molecular structure and function of the plasma sex steroid-binding protein (SBP). *J. Steroid Biochem.* **19,** 699–706.

Petra, P. H., Stanczyk, F. Z., Namkung, P. C., Fritz, M. A., and Novy, M. J. (1985). Direct effect of sex steroid-binding protein (SBP) of plasma on the metabolic clearance rate of testosterone in the rhesus macaque. *J. Steroid Biochem.* **22,** 739–746.

Petra, P. H., Namkung, P. C., Titani, K., and Walsh, K. A. (1986a). Characterization of the plasma sex steroid-binding protein. *Colloq.—Inst. Natl. Sante Rech. Med.* **149,** 15–30.

Petra, P. H., Titani, K., Walsh, K. A., Joseph, D. R., Hall, S. H., and French, F. S. (1986b). Comparison of the amino acid sequence of the sex steroid-binding protein of human plasma (SBP) with that of the androgen-binding protein (ABP) of rat testis. *Colloq.—Inst. Natl. Sante Rech. Med.* **149,** 137–142.

Petra, P. H., Namkung, P. C., Senear, D. F., McCrae, D. A., Rousslang, K. W., Teller, D. C., and Ross, J. B. A. (1986c). Molecular characterization of the sex steroid binding protein (SBP) of plasma. Re-examination of rabbit SBP and comparison with the human, macaque and baboon proteins. *J. Steroid Biochem.* **25,** 191–200.

Petra, P. H., Que, B. G., Namkung, P. C., Ross, J. B. A., Charbonneau, H., Walsh, K. A., Griffin, P. R., Shabanowitz, J., and Hunt, D. F. (1988). Affinity labeling, molecular cloning, and comparative amino acid sequence analyses of sex steroid-binding protein of plasma. *Ann. N.Y. Acad. Sci.* **538,** 10–24.

Petra, P. H., Griffin, P. R., Yates, J. R., III, Moore, K., and Zhang, W. (1992). Complete enzymatic deglycosylation of native sex steroid-binding protein (SBP or SHBG) of human and rabbit plasma: Effect on the steroid-binding activity. *Protein Sci.* **1**, 902–909.

Picado-Leonard, J., and Miller, W. L. (1988). Homologous sequences in steroidogenic enzymes, steroid receptors and a steroid binding protein suggest a consensus steroid-binding sequence. *Mol. Endocrinol.* **2**, 1145–1150.

Plymate, S. R., Matej, L. A., Jones, R. E., and Friedl, K. E. (1988a). Inhibition of sex hormone-binding globulin production in the human hepatoma (Hep G2) cell line by insulin and prolactin. *J. Clin. Endocrinol. Metab.* **67**, 460–464.

Plymate, S. R., Matej, L. A., Jones, R. E., and Friedl, K. E. (1988b). Regulation of sex hormone binding globulin (SHBG) production in HEP G2 cells by insulin. *Steroids* **52**, 339–340.

Porto, C. S., Gunsalus, G. L., Bardin, C. W., Phillips, D. M., and Musto, N. A. (1991). Receptor-mediated endocytosis of extracellular steroid-binding protein (TeBG) in MCF-7 human breast cancer cells. *Endocrinology (Baltimore)* **129**, 436–445.

Porto, C. S., Abreu, L. C., Gunsalus, G. L., and Bardin, C. W. (1992a). Binding of sex-hormone-binding globulin (SHBG) to testicular membranes and solubilized receptors. *Mol. Cell. Endocrinol.* **89**, 33–38.

Porto, C. S., Musto, N. A., Bardin, C. W., and Gunsalus, G. L. (1992b). Binding of an extracellular steroid-binding globulin to membranes and soluble receptors from human breast cancer cells (MCF-7 cells). *Endocrinology (Baltimore)* **130**, 2931–2936.

Power, S. G. A., Boccinfuso, W. P., Pallesen, M., Warmels-Rodenhiser, S., van Baelen, H., and Hammond, G. L. (1992). Molecular analyses of a human sex hormone-binding globulin variant: Evidence for an additional carbohydrate chain. *J. Clin. Endocrinol. Metab.* **75**, 1066–1070.

Que, B. G., and Petra, P. H. (1987). Characterization of a cDNA coding for sex steroid-binding protein of human plasma. *FEBS Lett.* **219**, 405–409.

Raggatt, L. E., Blok, R. B., Hamblin, P. S., and Barlow, J. W. (1992). Effects of thyroid hormone on sex hormone-binding globulin gene expression in human cells. *J. Clin. Endocrinol. Metab.* **75**, 116–120.

Renoir, J.-M., and Mercier-Bodard, C. (1974). Purification of SBP ("sex steroid binding plasma protein") from human pregnancy plasma by affinity chromatography. *J. Steroid Biochem.* **5**, 328 (abstr.).

Renoir, J.-M., Fox, L. L., Baulieu, E.-E., and Mercier-Bodard, C. (1977). An antiserum specific for human sex steroid-binding plasma protein (SBP). *FEBS Lett.* **75**, 83–88.

Renoir, J.-M., Mercier-Bodard, C., and Baulieu, E.-E. (1980). Hormonal and immunological aspects of the phylogeny of sex steroid binding plasma protein. *Proc. Natl. Acad. Sci. U.S.A.* **77**, 4578–4582.

Reventos, J., Hammond, G. L., Crozat, A., Brooks, D. E., Gunsalus, G. L., Bardin, C. W., and Musto, N. A. (1988). Hormonal regulation of rat androgen-binding protein (ABP) messenger ribonucleic acid and homology of human testosterone-estradiol-binding globulin and ABP complementary deoxyribonucleic acids. *Mol. Endocrinol.* **2**, 125–132.

Reventos, J., Sullivan, P. M., Joseph, D. R., and Gordon, J. W. (1993). Tissue-specific expression of the rat androgen-binding protein/sex hormone-binding globulin gene in transgenic mice. *Mol. Cell. Endocrinol.* **96**, 69–73.

Ritzen, E. M., Nayfeh, S. N., French, F. S., and Dobbins, M. C. (1971). Demonstration of androgen-binding components in rat epididymis cytosol and comparison with binding components in prostate and other tissues. *Endocrinology (Baltimore)* **89**, 143–151.

Ritzén, E. M., Dobbins, M. C., Tindall, D. J., French, F. S., and Nayfeh, S. N. (1973). Characterization of an androgen binding protein (ABP) in rat testis and epididymis. *Steroids* **21**, 593–607.

Ritzén, E. M., Hansson, V., and French, F. S. (1981). The Sertoli cell. *In* "The Testis" (H. Burger and D. de Kretser, eds.), pp. 171–197. Raven Press, New York.

Ritzén, E. M., Biotani, C., Parvinen, M., French, F. S., and Feldman, M. (1982). Stage-dependent secretion of ABP by rat seminiferous tubules. *Mol. Cell. Endocrinol.* **25**, 25–33.

Roberts, K. P., and Zirkin, B. R. (1993). Androgen binding protein inhibition of androgen-dependent transcription explains the high minimal testosterone concentration required to maintain spermatogenesis in rat. *Endocr. J.* **1**, 41–47.

Rosner, W. (1986). Measurement of TeBG in biological fluids: Evolution and problems. *Colloq.—Inst. Natl. Sante Rech. Med.* **149**, 207–214.

Rosner, W. (1990). The functions of corticosteroid-binding globulin and sex hormone-binding globulin: Recent advances. *Endocr. Rev.* **11**, 80–91.

Rosner, W., and Darmstadt, R. A. (1973). Demonstration and partial characterization of a rabbit serum protein which binds testosterone and dihydrotestosterone. *Endocrinology (Baltimore)* **92**, 1700–1707.

Rosner, W., and Deakins, S. M. (1968). Testosterone-binding globulins in human plasma: Studies on sex distribution and specificity. *J. Clin. Invest.* **47**, 2109–2116.

Rosner, W., and Smith, R. N. (1975). Isolation and characterization of the testosterone-estradiol-binding globulin from human plasma. Use of a novel affinity column. *Biochemistry* **14**, 4813–4820.

Rosner (Rosenbaum), W., Christy, N. P., and Kelly, W. G. (1966). Electrophoretic evidence for the presence of an estrogen-binding β-globulin in human plasma. *J. Clin. Endocrinol. Metab.* **26**, 1399–1403.

Rosner, W., Aden, D. P., and Kahn, M. S. (1984). Hormonal influences on the secretion of steroid-binding proteins by a human hepatoma-derived cell line. *J. Clin. Endocrinol. Metab.* **59**, 806–808.

Rosner, W., Hryb, D. J., Khan, M. S., Singer, C. J., and Nakhla, A. M. (1988). Are corticosteroid-binding globulin and sex hormone-binding globulin hormones? *Ann. N.Y. Acad. Sci.* **538**, 137–145.

Rothberg, J. M., Jacobs, J. R., Goodman, C. S., and Artavanis-Tsakonas, S. (1990). *Slit:* An extracellular protein necessary for development of midline glia and commissural axon pathways contains both EGF and LRR domains. *Genes Dev.* **4**, 2169–2187.

Rupp, F., Payan, D. G., Magil-Solc, C., Cowan, D. M., and Scheller, R. H. (1991). Structure and expression of a rat agrin. *Neuron* **6**, 811–823.

Russel, L. D., and Griswold, M. D., eds. (1993). "The Sertoli Cell." Cache River Press, Clearwater, FL.

Sanborn, B. M., Elkington, J. S. H., Chowdhury, M., Tcholakian, R. K., and Steinberger, E. (1975). Hormone influences on the level of testicular androgen binding activity: Effect of FSH following hypophysectomy. *Endocrinology (Baltimore)* **96**, 304–312.

Sanborn, B. M., Steinberger, A., Tcholakian, R. K., and Steinberger, E. (1977). Direct measurement of androgen receptors in cultured Sertoli cells. *Steroids* **29**, 493–502.

Sasaki, M., Kleinman, H. K., Huber, H., Deutzmann, R., and Yamada, Y. (1988). Laminin, a multidomain protein. *J. Biol. Chem.* **263**, 16536–16544.

Scatchard, G. (1949). The attractions of proteins for small molecules and ions. *Ann. N.Y. Acad. Sci.* **51**, 660–672.

Schmidt, W. N., Taylor, C. A., Jr., and Danzo, B. J. (1981). The use of a photoaffinity ligand to compare androgen-binding protein (ABP) present in rat Sertoli cell culture media with ABP present in epididymal cytosol. *Endocrinology (Baltimore)* **108**, 786–794.

Selby, C. (1990). Sex hormone binding globulin: Origin, function and clinical significance. *Ann. Clin. Biochem.* **27**, 532–541.

Siiteri, P. K., and Simberg, N. H. (1986). Changing concepts of active androgens in blood. *Clin. Endocrinol. Metab.* **15**, 247–259.

Siiteri, P. K., Murai, J. T., Hammond, G. L., Nisker, J. A., Raymoure, W. J., and Kuhn, R. W. (1982). The serum transport of steroid hormones. *Recent Prog. Horm. Res.* **38**, 457–510.

Sinnecker, G., Hiort, O., Mitze, M., Donn, F., and Neumann, S. (1988). Immunohistochemical detection of a sex hormone binding globulin like antigen in tissue sections of normal human prostate, benign prostatic hypertrophy and normal human endometrium. *Steroids* **52**, 335–336.

Sinnecker, G., Hiort, O., Kwan, P. W. L., and DeLellis, R. A. (1990). Immunohistochemical localization of sex hormone-binding globulin in normal and neoplastic breast tissue. *Horm. Metab. Res.* **22**, 47–50.

Skinner, M. K., and Fritz, I. B. (1985). Testicular peritubular cells secret a protein under androgen control that modulates Sertoli cell functions. *Proc. Natl. Acad. Sci. U.S.A.* **82**, 114–118.

Skubitz, A. P. N., Letourneau, P. C., Wayner, E., and Furcht, L. T. (1991). Synthetic peptides from the carboxy-terminal globular domain of the A chain of laminin: Their ability to promote cell adhesion and neurite outgrowth, and interact with heparin and the $\beta1$ integrin subunit. *J. Cell Biol.* **115**, 1137–1148.

Snouwaert, J. N., Brigman, K. K., Latour, A. M., Malouf, N. N., Boucher, R. C., Smithies, O., and Koller, B. H. (1992). An animal model for cystic fibrosis made by gene targeting. *Science* **257**, 1083–1088.

Steeno, O., Heyns, W., van Baelen, H., and DeMoor, P. (1968). Testosterone binding in human plasma. *Ann. Endocrinol.* **29**, 141–148.

Steinberger, A., Heindel, J. J., Lindsey, J. N., Elkington, J. S. H., Sanborn, B. M., and Steinberger, E. (1975). Isolation and culture of FSH responsive Sertoli cells. *Endocr. Res. Commun.* **2**, 261–272.

Strel'chyonok, O. A., and Avvakumov, G. V. (1990). Specific steroid-binding glycoproteins of human blood plasma: Novel data on their structure and function. *J. Steroid Biochem.* **35**, 519–534.

Strel'chyonok, O. A., Survilo, L. I., Tsapelik, G. Z., and Sviridov, O. V. (1983). Purification and physicochemical properties of the sex steroid-binding globulin of human blood plasma. *Biokhimiya (Moscow)* **48**, 756–762.

Strel'chyonok, O. A., Avvakumov, G. V., and Survilo, L. I. (1984). A recognition system for sex-hormone-binding protein–estradiol complex in human decidual endometrium plasma membranes. *Biochim. Biophys. Acta* **802**, 459–466.

Stupnicki, R., and Bartke, A. (1976). Binding of testosterone in mouse plasma. *Endokrinologie* **68**, 150–154.

Sui, L.-M., Cheung, A. W. C., Namkung, P. C., and Petra, P. H. (1992). Localization of the steroid-binding site of human sex steroid-binding protein of plasma (SBP or SHBG) by site-directed mutagenesis. *FEBS Lett.* **310**, 115–118.

Sullivan, P. M., Petrusz, P., Szpirer, C., and Joseph, D. R. (1991). Alternative processing of androgen-binding protein RNA transcripts in fetal rat liver. Identification of a transcript formed by *trans* splicing. *J. Biol. Chem.* **266**, 143–154.

Sullivan, P. M., Wang, Y.-M., and Joseph, D. R. (1993). Identification of an alternate promoter in the rat androgen-binding protein/sex hormone-binding globulin gene that regulates synthesis of a mRNA encoding a protein with altered function. *Mol. Endocrinol.* **7**, 702–715.

Suzuki, Y., and Sinohara, H. (1984). Subunit structure of sex-steroid-binding plasma proteins from man, cattle, dog, and rabbit. *J. Biochem. (Tokyo)* **96**, 751–759.

Tardivel-Lacombe, J., and Degrelle, H. (1991). Hormone-associated variation of the glycan microheterogeneity pattern of human sex steroid-binding protein (hSBP). *J. Steroid Biochem. Mol. Biol.* **39**, 449–453.

Taylor, C. A., Jr., Smith, H. E., and Danzo, B. J. (1980). Photoaffinity labeling of rat androgen binding protein. *Proc. Natl. Acad. Sci. U.S.A.* **77**, 234–238.

Tepass, U., Theres, C., and Knust, E. (1990). *Crumbs* encodes an EGF-like protein expressed on apical membranes of *Drosophila* epithelial cells and required for organization of epithelia. *Cell (Cambridge Mass.)* **61**, 787–799.

Thomas, C. M. G., van den Berg, R. J., and Segers, M. F. G. (1987). Precision of assays for sex-hormone-binding globulin: An I-IRMA kit and two Eu-IFMA kits compared. *Clin. Chem. (Winston-Salem, N.C.)* **33**, 2120.

Tindall, D. J., Schrader, W. T., and Means, A. R. (1974a). The production of androgen binding protein by Sertoli cells. *In* "Hormone Binding and Target Cell Activation in the Testis" (M. L. Dufau and A. R. Means, eds.), pp. 167–175. Plenum, New York.

Tindall, D. J., Hansson, V., Sar, M., Stumpf, W. E., French, F. S., and Nayfeh, S. N. (1974b). Further studies on the accumulation and binding of androgen in rat epididymis. *Endocrinology (Baltimore)* **95**, 1119–1128.

Tindall, D. J., Vitale, R., and Means, A. R. (1975). Androgen binding protein as a biochemical marker of formation of the blood–testis barrier. *Endocrinology (Baltimore)* **97**, 636–648.

Tindall, D. J., Mena, C. R., and Means, A. R. (1978). Hormonal regulation of androgen-binding protein in hypophysectomized rats. *Endocrinology (Baltimore)* **103**, 589–594.

Toscano, V., Balducci, R., Bianchi, P., Guglielmi, R., Mangiantini, A., and Sciarra, F. (1992). Steroidal and non-steroidal factors in plasma sex hormone binding globulin regulation. *J. Steroid Biochem. Mol. Biol.* **43**, 431–437.

Turner, T. T., and Roddy, M. S. (1990). Intraluminal androgen binding protein alters ^{3}H-androgen uptake by rat epididymal tubules *in vitro*. *Biol. Reprod.* **43**, 414–419.

Turner, T. T., and Yamamoto, M. (1991). Different mechanisms are responsible for ^{3}H-androgen movement across the rat seminiferous and epididymal epithelia *in vivo*. *Biol. Reprod.* **45**, 358–364.

Turner, T. T., Jones, C. E., Howards, S. S., Ewing, L. L., Zegeye, B., and Gunsalus, G. L. (1984). On the androgen microenvironment of maturing spermatozoa. *Endocrinology (Baltimore)* **115**, 1925–1932.

van Baelen, H., Convents, R., Cailleau, J., and Heyns, W. (1992). Genetic variation of human sex hormone-binding globulin: Evidence for a worldwide bi-allelic gene. *J. Clin. Endocrinol. Metab.* **75**, 135–139.

Veeramachaneni, D. N. R., and Amann, R. P. (1991). Endocytosis and androgen binding protein, clusterin, and transferrin in the efferent ducts and epididymis of the ram. *J. Androl.* **12**, 288–294.

Veeramachaneni, D. N. R., Amann, R. P., Palmer, J. S., and Hinton, T. (1990). Proteins in luminal fluid of the ram excurrent ducts: Changes in composition and evidence for differential endocytosis. *J. Androl.* **11**, 140–154.

Vermeulen, A. (1988). Physiology of the testosterone-binding globulin in man. *Ann. N.Y. Acad. Sci.* **538**, 103–111.

Vermeulen, A., and Verdonck, L. (1968). Studies on the binding of testosterone to human plasma. *Steroids* **11**, 609–635.

Vernon, R. G., Dorrington, J. H., and Fritz, I. B. (1972). Testosterone binding by rat testicular seminiferous tubules. *Int. Congr. Ser. Excerpta Med.* **256**, 200 (abstr.).

Vernon, R. G., Kopec, B., and Fritz, I. B. (1974). Observations on the binding of androgens by rat testis seminiferous tubules and testis extracts. *Mol. Cell. Endocrinol.* **1**, 167–187 (abst.).

von Schoultz, B., and Carlström, K. (1989). On the regulation of sex-hormone-binding globulin—A challenge of an old dogma and outlines of alternative mechanism. *J. Steroid Biochem.* **32**, 327–334.

Walsh, K. A., Titani, K., Takio, K., Kumar, S., Hayes, R., and Petra, P. H. (1986). Amino acid sequence of the sex steroid-binding protein of human blood plasma. *Biochemistry* **25**, 7584–7590.

Wang, Y.-M., Sullivan, P. M., Petrusz, P., Yarbrough, W., and Joseph, D. R. (1989). The androgen-binding protein gene in expressed in CD1 mouse testis. *Mol. Cell Endocrinol.* **63**, 85–92.

Wang, Y.-M., Bayliss, D. A., Millhorn, D. E., Petrusz, P., and Joseph, D. R. (1990). The androgen-binding protein gene is expressed in male and female rat brain. *Endocrinology (Baltimore)* **127**, 3124–3130.

Weddington, S. C., Hansson, V., Purvis, K., Varaas, T., Verjans, H. L., Eik-nes, K. B., Ryan, W. H., French, F. S., and Ritzén, E. M. (1976). Biphasic effect of testosterone propionate on Sertoli cell secretory function. *Mol. Cell. Endocrinol.* **5**, 137–145.

Wenn, R. V., Kamberi, I. A., Keyvanjah, M., and Johannes, A. (1977). Distribution of testosterone-estradiol binding globulin (TeBG) in the higher vertebrates. *Endokrinologie* **69**, 151–156.

Westphal, U. (1986). Steroid–protein interactions. II. "Monographs on Endocrinology," pp. 198–301. Springer-Verlag, Berlin.

Wiley, S. R., Kraus, R. J., and Mertz, J. E. (1992). Functional binding of the "TATA" box binding component of transcription factor TFIID to the −30 region of TATA-less promoters. *Proc. Natl. Acad. Sci. U.S.A.* **89**, 5814–5818.

Wright, W. W., Musto, N. A., Mather, J. P., and Bardin, C. W. (1981). Sertoli cells secrete both testis-specific and serum proteins. *Proc. Natl. Acad. Sci. U.S.A.* **78**, 7565–7569.

Yen, J., Wisdom, R. M., Tratner, I., and Verma, I. M. (1991). An alternative spliced form of *Fos*B is a negative regulator of transcriptional activation and transformation by *Fos* proteins. *Proc. Natl. Acad. Sci. U.S.A.* **88**, 5077–5081.

Zahnow, C. A., Yi, H.-F., McBride, W., and Joseph, D. R. (1991). Cloning of the cDNA encoding human histidine decarboxylase from an erythroleukemia cell line and mapping of the gene locus to chromosome 15. *DNA Sequence-J. DNA Sequencing Mapping* **1**, 395–400.

Molecular Biology of Vitamin D Action

TROY K. ROSS, HISHAM M. DARWISH, AND HECTOR F. DELUCA

Department of Biochemistry
College of Agricultural and Life Sciences
University of Wisconsin–Madison
Madison, Wisconsin 53706

I. Introduction

The disease rickets (which we now know is an affliction of vitamin D deficiency) appeared in epidemic proportions around the turn of the century in northern Europe, North America, and northern Asia. This disease was first produced experimentally in dogs by feeding them a diet of oatmeal and maintaining them indoors (Mellanby, 1919). Mellanby was also able to show that the disease could be cured or prevented by the provision of cod liver oil, and concluded that the ability to cure rickets might likely be another property of the fat-soluble vitamin A found in cod liver oil by McCollum *et al.* (1916). However, McCollum and co-workers (1922) demonstrated by aeration and heating of cod liver oil that the ability to cure rickets was retained, while the vitamin A activity that supports growth and prevents xerophthalmia was destroyed. McCollum called this substance that cures rickets fat-soluble vitamin D.

On the other hand, ultraviolet light from sunlight or lamps was also found to cure rickets in children (Huldshinsky, 1919). Steenbock and Black (1924), and sometime later, Hess and Weinstock (1924) demonstrated that ultraviolet irradiation of an inert substance in the lipid fraction of skin and foods caused its conversion to vitamin D. Irradiation of foods resulted in the elimination of rickets as a major medical problem and provided the means for generation of sizable quantities of vitamin D, leading to the identification of vitamin D_2 (Askew et al., 1931). Vitamin D_3 was isolated and identified by Windaus and co-workers (1936), completing the discovery of the principal nutritional forms of vitamin D. In 1978 and 1980, two groups demonstrated that vitamin D_3 is produced naturally in skin by ultraviolet activation of 7-dehydrocholesterol (Esvelt et al., 1978; Holick et al., 1980).

During the 1930s, 1940s, and 1950s, much was learned concerning the physiology of vitamin D-mediated action, including the intestinal absorption of calcium and phosphorus (Nicolaysen and Eeg-Larsen, 1953). Carlsson (1952) showed that vitamin D, rather than acting directly on the mineralization of bone, actually brings about the mobilization of calcium from the skeleton to the plasma. By initiating intestinal calcium and phosphorus transport, mobilization of calcium from bone, and renal reabsorption of calcium, vitamin D brings about an elevation of plasma calcium and phosphorus to levels that support proper skeletal mineralization and neuromuscular function and prevent tetany, rickets, and osteomalacia (Lamm and Neuman, 1985; DeLuca, 1967) (Fig. 1). Underwood and DeLuca (1984) then showed that when plasma calcium and phosphorus are normalized in vitamin D-deficient rats, mineralization of the rachitic skeleton occurs normally. This confirmed that vitamin D does not participate in the mineralization process.

Kodicek and his colleagues completed 10 years of investigation of the metabolism of vitamin D and concluded that vitamin D was not converted to active forms before functioning (Kodicek, 1956). However, chemical synthesis of radiolabeled vitamin D with high specific activity permitted studies with physiological doses of vitamin D (Neville and DeLuca, 1966). This study demonstrated that vitamin D is inactive and must be metabolized to a functional form (Lund and DeLuca, 1966; Morii et al., 1967). At the same time, Haussler and Norman (1967) concluded that vitamin D acts directly on target tissues without alteration. The first active form of vitamin D was isolated, chemically identified as 25-hydroxyvitamin D (25-OH-D_3), and synthesized (Blunt and DeLuca, 1969; Blunt et al., 1968). However, Cousins et al. (1970), Lawson et al. (1969), and Haussler et al. (1968) provided clear evidence that this metabolite is metabolized very rapidly into more polar forms. Holick et al. (1971) isolated from the intestine a target tissue of vitamin D, the active metabolite of vitamin D, and identified its structure as 1,25-dihydroxyvitamin D_3 (1,25-$(OH)_2D_3$). This structure was confirmed by chemical synthesis as having the α-configuration at the 1-hydroxy position (Semmler et al., 1972). Fraser and Kodicek (1970)

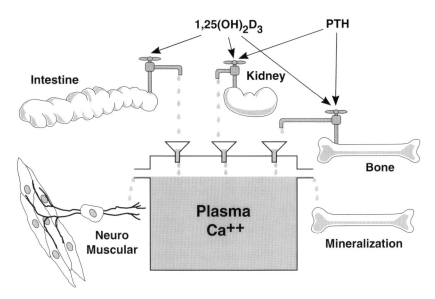

FIG. 1. A representation of the classic physiological actions of 1,25-(OH)$_2$D$_3$. The hormonal form of vitamin D functions in the intestine, bone, and the distal renal tubule to mobilize calcium. In the case of the intestine, 1,25-(OH)$_2$D$_3$ serves to mobilize phosphorus into the plasma. Saturating levels of calcium and phosphorus support mineralization in the skeleton and neuromuscular junction activity.

provided clear evidence that this metabolite is produced by the kidney, which was found to be the exclusive site of synthesis in normal animals. Although other metabolites of vitamin D were identified primarily by the DeLuca group during the 1970s, very strong evidence has been produced that the only active form of vitamin D in mammals is 1,25-(OH)$_2$D$_3$ (Brommage and DeLuca, 1985). All other metabolites result from degradation of the hormone, with the final excretory form being calcitroic acid (Esvelt et al., 1979).

With the use of fluoro derivatives, it was conclusively demonstrated that activation of the vitamin D molecule involves 25-hydroxylation in the liver and subsequent 1α-hydroxylation in the kidney to produce the final vitamin D hormone 1,25-(OH)$_2$D$_3$ (Brommage and DeLuca, 1985). It is this compound, and not its precursor, that is active on intestine and bone in anephric animals (Boyle et al., 1972; Holick et al., 1972). This form of vitamin D is also linked to the autosomal recessive disorder vitamin D-dependency rickets type I, which is likely a defect in the 1α-hydroxylase enzyme (Fraser et al., 1973).

With the finding that 1,25-(OH)$_2$D$_3$ is the active form of vitamin D, tracing it by frozen section autoradiography provided evidence that this steroid localizes in the nucleus of target tissues (Stumpf et al., 1979). Thus, the existence of a receptor for this steroid seemed almost certain, and eventually its presence was demonstrated by Brumbaugh and Haussler (1975b) and by Kream et al. (1976). The receptor has been cloned, its structure established, and much has been learned concerning the molecular biology of its action in target tissues (Baker et al., 1988; Burmester et al., 1988a). The essentiality of the receptor for vitamin D function is demonstrated by vitamin D-dependency rickets type II (Brooks et al., 1978), which is by now a well-characterized series of mutations in the vitamin D receptor (VDR) gene as described in the following. The existence of the VDR in tissues not previously thought to be target tissues has led to the discovery of new functions of vitamin D, and a pursuit of the molecular biology of the receptor has provided much new information on the molecular mechanism of action of vitamin D. This review is designed to provide an insight into the more recent advances.

II. THE VITAMIN D RECEPTOR

A. PURIFICATION AND CHARACTERIZATION

The hormonal activities of $1,25\text{-}(OH)_2D_3$ are mediated through its intracellular VDR (DeLuca and Schnoes, 1983). The VDR is a trace protein, and even within its primary source, intestine, the receptor represents only about 0.001% of the total soluble protein. Characteristic biochemical functions and structural motifs have led to the classification of the VDR as a member of the steroid/thyroid superfamily of hormone receptors (O'Malley, 1990). A study using tissue fractionation is suggestive of VDR localization in the nucleus (Nakada et al., 1984). Subsequently, fluorescence immunocytochemical techniques were used on fixed tissues from treated and nontreated cells (Barsony et al., 1990). Results from this study indicated that although some VDR is located in the nucleus, a majority of VDR exists in the cytosol. Because these reports are inconsistent, with the former representing an in vivo approach and the latter in vitro, the intracellular localization of receptor will remain controversial until confirmatory data become available.

Because of its low abundance and instability, initial attempts at VDR purification were futile. A notable advance in receptor purification occurred, however, when ligand affinity chromatography on DNA cellulose and blue dextran–Sepharose was used to obtain VDR from chicken intestinal mucosa (Pike and Haussler, 1979). At that time just 26 μg of semipure VDR was obtained from the intestines of 350 rachitic chickens. Later attempts at purification using low-salt nuclei isolation and high-salt receptor extraction from vitamin D-replete chickens provided some modifications for improvement (Simpson and DeLuca, 1982). A purification protocol reported in 1983 included a five-step scheme performed on VDR from chicken intestines that resulted in a 5800-fold purification with an 8% yield of VDR (Simpson et al., 1983). With the advent of monoclonal antibody production, VDR purification techniques would become considerably more efficient. Because hybridoma technology required only moderate amounts of semipure VDR, receptor prepared from avian intestine was used to produce antibodies that were directed against a single epitope, and were found to cross-react with mammalian receptors (Pike et al., 1983). As more abundant sources of receptor were sought, it was realized that nuclear extract from young pig intestine contains approximately twice the amount of receptor per milligram protein than that from nuclear extract of chicken intestine. A five-step purification scheme was used to obtain sufficient quantities of approximately 24% pure receptor for molecular weight determination (Dame et al., 1985) and the subsequent generation of 24 different monoclonal antibodies (Dame et al., 1986). Anti-VDR monoclonal antibodies provided the tools necessary for the development of efficient immunoaffinity purification techniques (Brown et al., 1988; Pike et al., 1987). The yield and purity of receptor were greatly enhanced using this method and thus permitted limited biochemical analyses to be performed, including the determination of two peptide sequences (Brown et al., 1988).

The production of monoclonal antibodies directed against the VDR and the purification of significant amounts of the protein were landmark contributions to the understanding of VDR structure and function. Antibody development led to their use as radiolabeled probes to screen cDNA expression libraries for VDR-producing recombinant phages. The molecular cloning of a portion of the chicken VDR cDNA was the first to be described (McDonnell et al., 1987). Shortly thereafter, the cloning of a major portion of the rat VDR coding region was reported (Burmester et al., 1988a). The entire coding sequence of the rat receptor was completely defined using a primer extension technique on rat intestinal mRNA (Burmester et al., 1988b). The deduced amino acid sequence is consistent with the peptide sequence determined by direct analysis of immunopurified porcine VDR. Using a portion of the chicken cDNA as a molecular probe, the human VDR cDNA was eventually identified (Baker *et al.*, 1988). The cloning of these complete cDNAs permitted a comparison between the deduced amino acid sequences and those amino acid sequences for other characterized steroid hormone receptors. The total size of the deduced amino acid sequence of the receptor protein is 423 for the rat (Burmester *et al.*, 1988b) and 427 for the human (Baker *et al.*, 1988). The cysteine-rich DNA binding domain at the amino terminus is 100% conserved between the human and rat species and the steroid binding domain at the carboxy terminus is 93% conserved. The existence of putative trans-activation domains, and the localization of the steroid binding domain within amino acids 114 to 425, was suggested in a study of 5' and 3' deletion mutants of the human VDR cDNA (McDonnell *et al.*, 1989a). Steroid binding activity residing in a large region at the carboxy terminus of the VDR is consistent with observations of other members of the steroid/thyroid hormone receptor family (Beato, 1989; Evans, 1988). When the cDNAs were used as probes, and hybridized with RNA populations from their respective species, it was determined that the rat receptor hybridizes with a single 4.4-kb mRNA (Burmester *et al.*, 1988b) and the human with a single 4.6-kb transcript (Baker *et al.*, 1988).

With the complete nucleotide sequence of the human receptor coding region available, it became possible to address the genetic basis for dysfunctional receptors such as those associated with the rare autosomal recessive disease rickets (reviewed in DeLuca, 1988). Unlike type I rickets, which can be treated with physiological doses of $1,25\text{-}(OH)_2D_3$, the type II form of the disease displays a target tissue resistance to the hormone (Brooks *et al.*, 1978; Liberman and Marx, 1990). The clinical characteristics of type II rickets include normal levels of $25\text{-}OH\text{-}D_3$, elevated circulating levels of $1,25\text{-}(OH)_2D_3$ and parathyroid hormone, and low serum calcium concentrations. The high levels of $1,25\text{-}(OH)_2D_3$ indicated that the target organ resistance results from a defect in the VDR (Marx *et al.*, 1984). A polymerase chain reaction (PCR) technology-based study of two families with affected children homozygous for this disorder led to the identification of two different mutations in the VDR DNA binding domain (Hughes *et al.*, 1988). Another study of the DNA from four affected children in three related families identified one mutation that produces a 292-amino-acid, truncated version of the VDR (Ritchie *et al.*, 1989). An independent investigation of three patients with type II rickets found three different single-nucleotide substitutions in the VDR cDNA sequence (Wiese *et al.*, 1993). The identified mutations in this study permitted the prediction of three truncated versions of the VDR, which would result in their having a dysfunctional ligand binding domain. Additional studies on VDR mutations associated with this disease will certainly contribute to our understanding of VDR structure and function.

B. REGULATION

An unresolved question regarding VDR-mediated responses is the control of receptor levels in target cells. There have been at least five mechanisms linked to the control of cellular VDR abundance, with the most widely studied being hormonal regulation. The glucocorticoids (Massaro *et al.*, 1983), estrogen (Walters, 1981), retinoic acid (Petkovich *et al.*, 1984), parathyroid hormone (Pols *et al.*, 1988), and $1,25$-$(OH)_2D_3$ itself have each been shown to stimulate VDR levels. Another known regulatory mechanism involves activation of the cyclic AMP signal pathway. Cyclic AMP agonists, including forskolin, have been found to induce levels of VDR mRNA in mouse NIH-3T3 (Krishnan and Feldman, 1992) and UMR 106 osteosarcoma (van Leeuwen *et al.*, 1992) cell lines. Response element nucleotide sequences for cyclic AMP have been identified in the rat osteocalcin (Lian *et al.*, 1989), somatostatin (Montminy *et al.*, 1986), and cytosolic phosphoenolpyruvate carboxykinase (Wynshaw-Boris *et al.*, 1984) genes. Because of the effects observed by cyclic AMP agonists on VDR mRNA, there may be reason to speculate that a cyclic AMP response element sequence exists in the promoter region of the VDR gene.

Another point of VDR regulation involves calcium. Although calcium reportedly increases renal VDR levels in rats maintained in a vitamin D-deficient state, a more dramatic increase was detected in vitamin D-deficient rats with normal levels of serum calcium that had been administered $1,25$-$(OH)_2D_3$ (Sandgren and DeLuca, 1990). Thus, dietary calcium is required for vitamin D to positively regulate the VDR levels in rat kidney. This response to calcium is very slow and adaptive in nature and likely involves parathyroid hormone (Uhland-Smith and DeLuca, 1993). Another group showed by indirect evidence that calcium is functionally involved in VDR regulation using calcium channel blockers and chelating agents (van Leeuwen *et al.*, 1990). A fourth means by which VDR levels have been shown to increase is by induction with serum or growth factors. Krishnan and Feldman (1991) studied the effects of several mitogenic factors, including insulin, epidermal growth factor, and insulinlike growth factor-I on the human breast cancer line, MCF-7, and the murine cell line, NIH-3T3. These agents were all found to stimulate VDR abundance and cellular proliferation.

Finally, developmental timing mechanisms have been implicated in the regulation of VDR levels. The appearance of steroid binding activity in the neonatal development of the rat was found to be between 14 and 21 days postpartum (Halloran and DeLuca, 1981a). A follow-up study using anti-VDR monoclonal antibodies to measure receptor appearance indicated no VDR present in rat intestine prior to Day 18 postpartum (Pierce and DeLuca, 1988), and this was confirmed by an analysis of VDR transcript levels in neonatal rat intestinal RNA populations (Burmester *et al.*, 1988b). Detection of VDR correlates well with the sensitivity of intestinal calcium transport to 1,25-$(OH)_2D_3$ (Halloran and DeLuca, 1981a). The appearance of VDR transcript by Day 21 in the neonatal rat was later confirmed independently by Huang *et al.* (1989). In this same report it was demonstrated that receptor induction parallels an increase in mRNA for the calbindin-D_{9k} gene.

Although it is clear that there are several control mechanisms regulating VDR expression, auto-regulation is one of the better-characterized means of modulating 1,25-$(OH)_2D_3$ receptor levels. Other steroid hormone receptors, including estrogen (Berkenstam *et al.*, 1989) and glucorticoids (Okret *et al.*, 1986), have been shown to be under control of their respective ligands at the transcriptional and postranslational levels. There have been several reports that suggest that VDR levels are under 1,25-$(OH)_2D_3$ control at the level of transcription (Costa and Feldman, 1986; McDonnell *et al.*, 1987; Strom *et al.*, 1991). However, investigations of VDR mRNA and protein levels have demonstrated clearly that increases in VDR following 1,25-$(OH)_2D_3$ treatment result from protein stabilization and not transcriptional activation. This observation has been made in mouse 3T6 fibroblasts and intestinal IEC-6 cells by Wiese *et al.* (1992) and in

---

FIG. 2. Regulation of VDR Levels in rat osteosarcoma (ROS 17/2.8) cells upon 1,25-$(OH)_2D_3$ administration. (A) Time course of induction of VDR protein abundance. ROS 17/2.8 cells were treated with hormone at a final concentration of 10 nM for the internal indicated. VDR levels were quantified and expressed in fmol receptor per mg of protein. Values are displayed as the mean $\pm$ S.E. with each point representing an average of three determinations. Statistical significance: *$P<0.01$, **$P<0.05$, ***$P<0.005$, ****$P<0.025$. (B) Stabilization of the VDR by 1,25-$(OH)_2D_3$ in cycloheximide-treated ROS 17/2.8 cells. Cycloheximide (10 μM) was administered to ROS 17/2.8 cells in the presence of ethanol of 1,25-$(OH)_2D_3$ (10 nM). At the indicated times, cells were harvested and VDR levels determined. Values are expressed in fmol VDR per mg protein versus time and are given as the mean $\pm$ S.E. Each point on the graph is representative of three independent determinations. Statistical significance: *$P<0.025$, **$P<0.10$. From Arbour *et al.* (1993b).

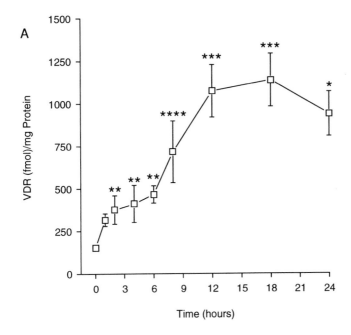

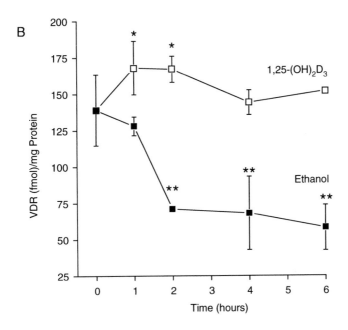

transfected COS-1 monkey kidney cells (Santiso-Mere *et al.*, 1993). Additionally, our group has confirmed that VDR auto-regulation occurs via a protein stabilization mechanism in the rat osteosarcoma cell line, ROS 17/2.8 (Arbour *et al.*, 1993b). Administration of hormone to confluent cells results in a time-dependent increase in VDR protein levels, with a maximum level of eightfold above the baseline detected by 18 hr postadministration (Fig. 2A). This stimulation is not reflected at the mRNA level, as no significant changes in VDR mRNA were observed over the 24 hr period of hormone administration. Cycloheximide treatment caused a rapid degradation of VDR protein in the absence of hormone, whereas the addition of $1,25\text{-}(OH)_2D_3$ to cycloheximide-treated cells stabilized VDR levels (Fig. 2B).

Other means of VDR regulation will likely be observed in future investigations. A recent report suggested that allelic variation in the 3′ untranslated region of the VDR gene may exert control over VDR mRNA levels (Morrison *et al.*, 1994). In this study, it was demonstrated that these allelic variants may be used to predict bone density and ultimately could lead to early intervention in those individuals at an elevated risk of osteoporotic fracture. Control of VDR at the level of transcription may be definitively addressed upon the cloning and characterization of the promoter region of the VDR receptor gene. Efforts on this front have been hampered by the existence of what are apparently very large intronic sequences in the 5′ region of the gene. Failed attempts by other groups to decipher this portion of various steroid receptor genes have been attributed to similar circumstances (Keaveney *et al.*, 1991; Kuiper *et al.*, 1989; Zong *et al.*, 1990).

C. MOLECULAR BIOLOGY OF RECEPTOR FUNCTION

An important biological control mechanism for the activity of transcription factors such as steroid hormone receptors is phosphorylation (Jackson, 1992). Many steroid receptors, including the VDR (Brown and DeLuca, 1990; Pike and Sleator, 1985), glucocorticoid (Singh and Moudgil, 1985), progesterone (Logeat *et al.*, 1985), and estrogen receptors (Auricchio *et al.*, 1987), have been reported to be phosphoproteins. Evidence indicates, for example, that phosphorylation of the chicken progesterone receptor is fundamental to the transcriptional activity mediated by this receptor (Denner *et al.*, 1990). Also, ligand-dependent phosphorylation of the estrogen receptor reportedly modulates its transcription factor activity (Ali *et al.*, 1993).

It has become increasingly apparent that phosphorylation of VDR plays a central role in the mechanism of vitamin D-influenced gene expression. The VDR phosphorylation event has been demonstrated in several cell types, including 3T6 mouse fibroblasts (Pike and Sleator, 1985), ROS 17/2.8 cells (Haussler *et al.*, 1988), and transiently transfected COS-7 cells (Jurutka *et al.*, 1993a). In most cases, the phosphorylation of steroid receptors occurs in the presence of their respective ligands (Auricchio, 1989), and the VDR is no exception. It has been demonstrated that phosphorylation of VDR takes place rapidly upon exposure to $1,25\text{-}(OH)_2D_3$ (Brown and DeLuca, 1990). Studied in a fully responsive, chicken duodenal organ culture system, VDR phosphorylation occurs within 1 hr after ligand administration. In the absence of $1,25\text{-}(OH)_2D_3$, no phosphorylation of receptor was detected in this system. It is important to note that the ligand-induced phosphorylation occurs before any other known physiological response such as the uptake of calcium or the induction of calcium binding protein. The site(s) of the $1,25\text{-}(OH)_2D_3$-induced phosphorylation was shown to lie entirely within a 23-kDa peptide of the porcine VDR (Brown and DeLuca, 1991). The segment containing the site(s) of phosphorylation was localized to a region of the VDR representing the designated hinge region and approximately one-half of the likely steroid binding domain. An independent study using an *in vitro* system indicated that phosphorylation of the human VDR occurs between the amino acids Arg[121] and Asp[232] (Jones *et al.*, 1991). In that report it was suggested that the VDR may be a substrate for casein kinase II or another related kinase. A later study from the same laboratory localized the site of VDR phosphorylation to a serine at position 208 (Jurutka *et al.*, 1993b). The position of Ser[208] is within an a consensus amino acid sequence for casein kinase II and therefore lends support for their *in vitro* phosphorylation data. It is difficult, however, to draw definitive conclusions regarding intracellular events based on this enzymatic study.

A phosphoamino acid analysis on porcine kidney cell-derived VDR indicated serine as the likely site(s) of phosphorylation (Brown and DeLuca, 1991). A report based on site-directed mutagenesis of the VDR implicated serine 51 as the site of phosphorylation on the human VDR amino acid sequence (Hsieh *et al.*, 1991). The *in vitro* study essentially consisted of the incubation of VDR with protein kinase C and demonstration that the VDR was phosphorylated. When serine 51 was mutated, phosphorylation was abolished. Additionally, the mutated VDR was incapable of gene trans-activation when co-transfected with a reporter gene construct into CV-1 cells. In a follow-up report, the same group described the site-directed mutagenesis of serine 51 to glycine and reported just a 60% reduction of basal level recombinant human VDR phosphorylation in COS-7 cells (Hsieh *et al.*, 1993). However, it is important to note that this mutation to a nonphosphorylatable amino acid did not reduce the 1,25-$(OH)_2D_3$-induced phosphorylation of the VDR. Serine [51] lies well outside of the 23-kDa peptide segment mapped as the site of porcine VDR phosphorylation and is, therefore, in contradiction with the Brown and DeLuca (1991) study. Ligand-induced VDR phosphorylation within the 23-kDa peptide provides a basis for the physiological relevance and authenticity of the event.

Once the cDNA for the VDR became available from various sources, and because the isolation of such a trace protein was tedious, an emphasis was placed on the overproduction of recombinant VDR. The ultimate objective has been to generate milligram quantities of highly purified and nearly authentic VDR for X-ray crystallographic analyses and NMR studies. The first report of VDR overproduction described the cloning of the human VDR cDNA in a high-copy yeast expression plasmid, with transcription under the control of the copper-inducible metallothionein (CUP-1) promoter (Sone et al., 1990). Transformed Saccharomyces cerevisiae cells, grown in the presence of copper and 1,25-$(OH)_2D_3$, were shown to produce VDR at a level of 0.5% of the total soluble protein. Other reports the following year described the production of recombinant rat (Ross et al., 1991) and human (Mac-Donald et al., 1991) VDR in insect cells using the baculovirus system, and recombinant human VDR in mammalian cells using an adenovirus expression system (Smith et al., 1991). A later report described the expression of human VDR in a bacterial system (Kumar et al., 1992). Although the level of rat VDR produced in insect cells was reported to be 5% of the total soluble protein (Ross et al., 1991), the bacterial source reportedly produces up to 60% of the total cellular protein (Kumar et al., 1992). The authors claimed that milligram quantities of functional VDR can be obtained from this bacterial source. In general, all of these reports provide adequate data, such as dissociation constants, DNA binding results, and the antigenic properties supporting the authenticity of the VDR produced in these various systems. One of these established overproduction systems will undoubtedly prove to be the source for VDR that will lead to the crystallization of the protein. However, to date, no one has reported an advance on this front.

Although the three-dimensional structure of the VDR has yet to be determined, overproduction of the receptor has led to the major discovery that a nuclear accessory factor (NAF) is required for VDR-specific DNA binding. The site on the chromosome at which specific VDR binding occurs is a nucleotide sequence called a vitamin D response element (DRE). It was information in the study of VDR expression in yeast cells that provided the first evidence of a required, mammalian-derived protein factor for VDR–DRE interaction (Sone *et al.*, 1990). A previous report by the same group described the recombinant VDR as capable of gene trans-activation using a β-galactosidase reporter gene system in the yeast cell background (McDonnell *et al.*, 1989b). It is interesting to note that in this first report using the overproduced VDR, reporter gene trans-activation was detected in a yeast reconstitution system; however, in the subsequent report by these investigators, the yeast-derived VDR was incapable of DRE binding without the addition of COS-1 cell extract (Sone *et al.*, 1990). This inconsistency may suggest that the reporter gene activity detected by these investigators was not VDR mediated.

A later study confirmed the NAF requirement for VDR–DRE complex formation and reported its approximate molecular size (Ross *et al.*, 1992a). The NAF was reported to be a 59- to 64-kDa protein from porcine intestinal extracts and is required for binding of a DRE by wild-type porcine receptor. This same investigation demonstrated the tissue distribution of NAF in rat tissues such as kidney, liver, and heart, but was not present in spleen and skeletal muscle. Another study provided evidence of a 55-kDa NAF in monkey kidney cells and included a list of cellular and tissue distributions of NAF (Sone *et al.*, 1991a). One of the cell lines included in this report as containing the VDR NAF is the insect cell *Spodoptera frugiperda*. Our data are not consistent with that information as we could not detect DRE binding by insect cell-produced, recombinant VDR unless mammalian nuclear extract was provided (Ross *et al.*, 1992b). In addition, the report on the overexpression of human VDR in insect cells indicated that enhanced binding of DRE occurred in the presence of extract from CV-1 cells (MacDonald *et al.*, 1991). It is unclear from this study whether the investigators detected any specific VDR–DRE binding in the absence of mammalian extract. From our interpretation of data presented in their report, it is a nonspecific binding that was detected from the recombinant insect cell extract. If that interpretation were accurate, then their conclusions would be consistent with our results, which indicate that *Spodoptera frugiperda* insect cells are incapable of providing the VDR NAF for DRE binding (Ross *et al.*, 1992a).

Studies have demonstrated that the retinoid X receptors (RXRs) may form heterodimers with the VDR and, in doing so, may serve as an accessory factor for VDR–DRE complex formation (Kliewer *et al.*, 1992; Yu *et al.*, 1991). Also, these same reports suggest that RXRs serve as universal co-regulators of receptor function as in the cases of thyroid hormone receptor and retinoic acid receptors. If the RXRs are required for the retinoic acid receptors to bind retinoic acid response elements, then the necessary RXR, or an appropriate homolog, is present in *Spodoptera frugiperda* insect cells (Ross *et al.*, 1992b). Because of this finding, combined with a comparison of the tissue distribution of NAF activity (Ross *et al.*, 1992a) and RXR transcript presence (Mangelsdorf *et al.*, 1990; Yu *et al.*, 1991), the possibility of one of the RXRs being identical to the required VDR NAF is remote. Additional data will be needed to confirm the identity of the NAF and to determine its role in the specificity of protein–protein interaction.

As previously stated, the expression controlled by $1,25\text{-}(OH)_2D_3$ is mediated by an interaction of the VDR with a target DRE. These nucleotide sequences have been, thus far, defined only in locations near the target gene's promoter. Although a region of the rat osteocalcin gene promoter had been determined to confer $1,25\text{-}(OH)_2D_3$ responsiveness to the transcription of that gene (Demay *et al.*, 1989; Kerner *et al.*, 1989; Yoon *et al.*, 1988), the localization of the site for VDR–DRE complex formation was not defined until the electrophoresis mobility shift assay was employed. The first report of this finding demonstrated the presence of a DRE in the rat osteocalcin gene promoter region encompassing nucleotides -456 to -442 (Demay *et al.*, 1990). Later studies provided the nucleotide sequences of the DREs located in the human osteocalcin (Ozono *et al.*, 1990), mouse osteopontin (or *Spp-1*) (Noda *et al.*, 1990), rat calbindin-D_{9k} (Darwish and DeLuca, 1992), human parathyroid hormone (Demay *et al.*, 1992), mouse calbindin-D_{28k} (Gill and Christakos, 1993), and rat 25-hydroxyvitamin D 24-hydroxylase genes (Zierold *et al.*, 1993). A comparison of the nucleotide sequences between these elements is shown in Table I. All of the DREs published to date mediate a $1,25\text{-}(OH)_2D_3$-induced enhancement of target gene transcription; the one exception to this is the DRE located in the human parathyroid hormone gene, which mediates its transcriptional repression (Demay *et al.*, 1992). The elemental pattern of "half-sites," and three-base-pair spacing between the "half-sites," has not been clearly defined for the calbindin-D_{28k} gene's DRE (Gill and Christakos, 1993). The DREs were reviewed in detail in a previous article from this laboratory (Darwish and DeLuca, 1993).

The role played by $1,25\text{-}(OH)_2D_3$ in the overall mechanism of controlling target gene expression has been addressed in several investigations. Included in an initial report of the NAF requirement for VDR–DRE interaction were data supporting the additional requirement of $1,25\text{-}(OH)_2D_3$ (Liao *et al.*, 1990). A later study seemed to suggest that unoccupied receptor is capable of DRE binding in the presence of mammalian nuclear extract (Sone *et al.*, 1991b). An

TABLE I
VITAMIN D RESPONSE ELEMENTS

Gene	Nucleotide sequence	Reference
Rat calbindin-D_{9k}	GGGTGT CGG AAGCCC	Darwish and DeLuca (1992)
Rat osteocalcin	GGGTGA ATG AGGACA	Demay et al. (1990)
Human osteocalcin	GGGTGA ACG GGGGCA	Ozono et al. (1990)
Mouse osteopontin	GGTTCA CGA GGTTCA	Noda et al. (1990)
Rat 25-hydroxy- vitamin D 24-hydroxylase	CGCACC CGC TGAACC	Zierold et al. (1993)
Mouse calbindin-D_{28k}	GGGGGATGTGAGGAGAAATGAGT	Gill and Christakos (1993)
Human parathyroid hormone	TGTCTGCTT TGAACCT	Demay et al. (1992)

independent investigation using insect cell-produced VDR demonstrated no detectable difference in the DRE binding behavior in the presence or absence of 1,25-$(OH)_2D_3$ (Ross et al., 1992b). These studies are not all in agreement; however, another investigation on the role of ligand compared the DRE binding behavior of intestinal VDR derived from vitamin D-deficient and vitamin D-replete rats (Ross et al., 1993). The VDR from vitamin D-deficient rats exhibited DRE binding activity in the mobility shift assay and provided the confirmation that VDR–DRE complex formation may occur in a ligand-independent fashion. In this same study, additional mobility shift assays were performed in varying salt environments and elution profiles of VDR bound to DRE-linked Sepharose were determined over a linear salt gradient. Collectively, these data provided the evidence needed to submit that 1,25-$(OH)_2D_3$ is not required for VDR–DRE binding, but rather, it increases the strength of the interaction.

Perhaps the most important result of the binding of 1,25-$(OH)_2D_3$ in the mechanism of target gene expression may be to facilitate the phosphorylation of VDR, thus imparting transcription factor activity upon its cognate receptor. The specific phosphorylation of VDR may precipitate the target gene trans-activation response as evidenced by reconstitution of a 1,25-$(OH)_2D_3$-dependent transcription system in CV-1 cells (Darwish et al., 1993). The effects of various protein kinase A modulators on this reporter gene system directly affected the phosphorylation state of the VDR. The effect of 8-bromo-cyclic AMP, an activator of protein kinase A, was the trans-activation of gene expression in a VDR-dependent fashion, but in the absence of exogenous

1,25-$(OH)_2D_3$. The addition of 1,25-$(OH)_2D_3$ to cells treated with 8-bromo-cAMP resulted in an additive effect of enhanced gene expression mediated by VDR. The effects of the phosphatase inhibitor okadaic acid included both an accumulation of phosphorylated VDR and trans-activation. The addition of 1,25-$(OH)_2D_3$ to the okadaic acid-treated cells results in twice the trans-activation effect detected with okadaic acid treatment alone. This evidence is supportive of VDR–NAF–DRE formation in the absence of ligand, as these compounds appeared to mimic the effect of 1,25-$(OH)_2D_3$-induced phosphorylation. This approach was similar to the one taken by those who investigated the phosphorylation of the progesterone receptor (Denner et al., 1990). Additional studies performed on the progesterone receptor indicated that a DNA-dependent protein kinase is responsible for its phosphorylation (Weigel et al., 1992). It remains to be determined whether a similar kinase might be important in VDR phosphorylation.

Our group has proposed a model for 1,25-$(OH)_2D_3$-influenced target gene expression based on our interpretation of available information (Ross et al., 1993). The model depicts an initial interaction of VDR with the required NAF. Evidence suggests that this initial interaction may occur in the absence of DRE-containing DNA (Sone et al., 1991b). Kinetic studies performed in this laboratory have revealed a cooperative interaction of VDR and DRE, suggesting that two or more VDR–NAF complexes bind to individual DREs (H. F. Darwish and H. F. DeLuca, unpublished observations). It has also been determined that 1,25-$(OH)_2D_3$ secures the complex of VDR–NAF–DRE (Ross et al., 1993). This may occur because of a ligand-induced conformational change in the VDR, possibly rendering the receptor DNA binding domain a more attractive binding site. We have proposed that the tightly bound, occupied VDR–NAF–DRE complex may serve as the substrate for the VDR phosphorylation event. The binding of ligand by VDR may cause a shift in the equilibrium to provide greater quantities of securely bound substrate for kinase action. The general transcription factor TFIIB has been implicated in the target gene trans-activation mechanism for several members of the steroid superfamily of receptors (Ing et al., 1992). These investigators suggested that TFIIB may serve as a stabilizing bridge between the RNA polymerase II initiation complex and activated steroid hormone receptor. Eventually, this idea may be extended to the VDR and in the case of vitamin D system, the TFIIB transcription factor may only be capable of interaction with the phosphorylated form of VDR. Once the VDR complex–RNA polymerase II initiation complex interaction occurs, the 1,25-$(OH)_2D_3$ molecule may no longer be a requirement for the target gene transcription effect. Although it has not been experimentally determined, this may result in the release of the $1,25\text{-}(OH)_2D_3$ molecule from its receptor. The realizations of many aspects of VDR biology have occurred in the past 15 years. From the discovery of $1,25\text{-}(OH)_2D_3$ binding activity to monoclonal antibody development to the cloning of VDR cDNAs, this field has steadily advanced and expanded. Dynamic areas of investigation centered around the VDR continue to be the identification of the accessory factor(s) for DRE binding, the event of phosphorylation, and the seemingly insurmountable task of determining the three-dimensional structure of the VDR. Because the VDR is central to $1,25\text{-}(OH)_2D_3$ hormonal action and because there appears to be some universality among the mechanisms of steroid action, investigations into VDR biology promise to remain at the forefront of steroid hormone research.

III. Target Tissues for Vitamin D

Several tissues have long been known to play a role in calcium homeostasis, including the intestine, kidney, and bone. Many others have been shown to contain VDR and can, therefore, be considered vitamin D target tissues (Table II). In the intestine, the presence of the specific high-affinity VDR has been clearly demonstrated (Brumbaugh and Haussler, 1974, 1975a; Kream *et al.*, 1976, 1977). Furthermore, $1,25\text{-}(OH)_2D_3$ was localized in the nucleus of the crypt and villus cells of the intestine after injecting the radiolabeled hormone *in vivo*

TABLE II
$1,25\text{-}(OH)_2D_3$ Target Cells

Proven	Putative
1. Intestinal enterocyte	1. Islet cell—pancreas
2. Osteoblast	2. Endocrine cells—stomach
3. Distal renal cells	3. Pituitary cells
4. Parathyroid cells	4. Ovarian cells
5. Keratinocytes of skin	5. Placenta
6. Promyelocytes, monocytes	6. Epididymis
7. Lymphocytes	7. Brain (hypothalamus)
8. Colon enterocytes	8. Myoblasts (developing)
9. Shell gland	9. Mammary epithelium
10. Chick chorioallantoic membrane	10. Aortic endothelial cells
	11. Skin fibroblasts

(Stumpf *et al.*, 1979). The binding of $1,25\text{-}(OH)_2D_3$ to its intestinal receptor results in stimulation of calcium absorption and phosphorus through the villus epithelial cells leading eventually to the elevation of blood calcium, which is needed for bone mineralization and neuromuscular function (DeLuca, 1978, 1982). The process of calcium transport proceeds in a biphasic fashion upon $1,25\text{-}(OH)_2D_3$ administration with an initial rapid stimulation of calcium transport observed, reaching a peak after 6 hr. The response is followed by a decline before increasing to a new, higher level by 24 hr postadministration, with that level maintained for at least 96 hr (Halloran and DeLuca, 1981b). Wasserman *et al.* (1992) showed an increase in the activity of the calcium pump in the basolateral membrane in response to $1,25\text{-}(OH)_2D_3$ treatment. Cai *et al.* (1993) demonstrated an increase in the mRNA levels encoding this calcium pump protein. It is not clear if this increase in mRNA levels is a result of transcriptional or posttranscriptional regulation.

In addition to the calcium pump, $1,25\text{-}(OH)_2D_3$ also results in a direct increase in the expression of several intestinal genes. Some of these genes include the vitamin D-dependent intestinal calcium binding protein genes, calbindin-D_{9k} (Darwish and DeLuca, 1992) and calbindin-D_{28k} (Christakos *et al.*, 1989), alkaline phosphatase (Strom *et al.*, 1991), the Ca-ATPase (Cai *et al.*, 1993), and others (Kessler *et al.*, 1986). Thus far, all the identified vitamin D-responsive genes and proteins do not seem to account for the observed vitamin D-dependent stimulation of calcium and phosphorus transport in the intestine. It was shown in some studies that $1,25\text{-}(OH)_2D_3$ causes perturbations in the brush border membranes of the rat intestine (Deliconstantinos *et al.*, 1986). However, the influence of such changes on the intestinal calcium transport is not yet clear. Other still unidentified factors are likely needed to fully characterize the exact molecular events that participate in this function of vitamin D.

Similar to the intestine, the kidney is also a major target organ for vitamin D. In addition to being the site for the final synthesis of $1,25\text{-}(OH)_2D_3$ by the action of 25-OH-D_3 1α-hydroxylase (Paulson and DeLuca, 1985; Reeve *et al.*, 1983), the kidney responds to $1,25\text{-}(OH)_2D_3$ by increasing calcium reabsorption from its filtrate (Yamamoto *et al.*, 1984). Stumpf *et al.* (1980) showed that $1,25\text{-}(OH)_2D_3$ was localized in the nuclei of the distal tubules cells of the rat kidney after a single injection of the labeled hormone. The nuclear localization of the hormone in the distal tubules was a different site from the one where it was synthesized (Stumpf *et al.*, 1980). This part of the kidney is primarily active in reabsorption of calcium in response to $1,25\text{-}(OH)_2D_3$

treatment (Yamamoto *et al.*, 1984). Two major genes expressed in the kidney were shown to be markers of responsiveness to $1,25-(OH)_2D_3$ and its receptor, that is, the calbindin-D_{28k} (Christakos *et al.*, 1989) and the 25-OH-D_3 24-hydroxylase gene (Chen *et al.*, 1993; Ohyama *et al.*, 1991). In the case of the hydroxylase gene, $1,25-(OH)_2D_3$ regulates its activity directly at the transcriptional level (Zierold *et al.*, 1993). The activity of the kidney 25-OH-D_3 1α-hydroxylase is also regulated by $1,25-(OH)_2D_3$ (Tanaka and DeLuca, 1984; Tanaka *et al.*, 1975). However, current data do not indicate the level at which regulation of this enzyme occurs.

The osteoblasts of bone represent a key target of vitamin D. It is well established that $1,25-(OH)_2D_3$ is indirectly required for normal bone mineralization during skeletal growth and remodeling (DeLuca, 1986). The mobilization of calcium from bone is a vitamin D-dependent process during early neonatal life, in contrast to intestinal calcium transport, which starts at a later time in development (Halloran and De-Luca, 1981a). The VDR is localized in osteoblasts and osteoprogenitor cells of bone, whereas no receptor for vitamin D was detected in the osteocytes, chondrocytes, and osteoclasts (Narbaitz *et al.*, 1983). The VDR was also detected in the bone marrow-derived stromal cells (Bellido *et al.*, 1993). Vitamin D causes the synthesis and secretion of a number of bone-specific proteins in the osteoblasts, such as osteocalcin (Evans *et al.*, 1988; Lian and Gundberg, 1988), osteopontin (Butler, 1989), collagen (Lichtler *et al.*, 1989; Rowe and Kream, 1982) and bone alkaline phosphatase (Kyeyune-Nyombi *et al.*, 1989). In the case of osteocalcin and osteopontin, the regulation of these proteins has been shown to be at the transcriptional level of their respective genes (Lian and Stein, 1992). Other proteins of the bone cells were also shown to be induced by $1,25-(OH)_2D_3$, like the EGF and TGF-β receptors. Furthermore, because parathyroid hormone works in conjunction with $1,25-(OH)_2D_3$ to regulate calcium mobilization from bone, this function of the parathyroid hormone could not be detected in the absence of vitamin D (Rasmussen *et al.*, 1963). Therefore, bone is a very active and sensitive tissue to vitamin D. However, even with the large amount of information about the physiological responses of bone to vitamin D, the molecular mechanism(s) of these processes is poorly understood. Part of the problem resides in the complexity of bone and the various factors that are involved in its regulation and preservation.

The clonal osteosarcoma cells, which are osteoblastic cells, have served as a model for studying a variety of vitamin D effects at the cellular level. These cells have the receptor for vitamin D (Manolagas *et al.*, 1980), are able to secrete osteocalcin into culture medium (Lajeunesse *et al.*, 1990; Price and Baukol, 1980), and possess other features of the osteoblastic phenotype such as alkaline phosphatase activity (Majeska and Rodan, 1982; Manolagas *et al.*, 1981) and parathyroid hormone responsiveness (Lormi and Marie, 1990). The numerous studies on these cells suggest that 1,25-$(OH)_2D_3$ acts directly on bone, in addition to its primary role in raising blood calcium and phosphorus levels needed for the mineralization of bone.

Apart from the intestine, kidney, and bone, which are the most notable primary target sites for vitamin D, several other tissues have VDR and, therefore, represent additional sites of 1,25-$(OH)_2D_3$ action. Stumpf *et al.* (1983) have shown that the placenta and yolk sac contain VDR. However, because placental transfer of calcium does not appear to be a vitamin D-dependent process (Brommage and DeLuca, 1984), 1,25-$(OH)_2D_3$ is likely to play a developmental role in this tissue. VDR was also detected in the β cells of the pancreas (Clark *et al.*, 1980). Several studies have linked vitamin D status to insulin secretion (Clark *et al.*, 1981; Ishida *et al.*, 1983; Tanaka *et al.*, 1984), which may be related to blood calcium levels (Tanaka *et al.*, 1986a). However, because the pancreas has a vitamin D-dependent calcium binding protein, it is possible that the observed effect of 1,25-$(OH)_2D_3$ on the secretion of insulin results from its direct effect on the pancreatic cells, primarily the β cells. VDR was also detected in certain neurons in the brain (Stumpf *et al.*, 1982). It was also shown that 1,25-$(OH)_2D_3$ increased choline acetyltransferase activity in certain brain nuclei (Sonnenberg *et al.*, 1986). Although the brain expresses calbindin-D_{28k} similar to the kidney protein, the expression of the brain protein is not regulated by 1,25-$(OH)_2D_3$ (Christakos *et al.*, 1989). The pituitary gland constitutes another site of vitamin D action, as the VDR was detected in the anterior pituitary thyrotrophes (Stumpf *et al.*, 1982). VDR was also detected in the parathyroid gland of the chicken and other species (Brumbaugh *et al.*, 1975; Hughes and Haussler, 1978). It is very well documented that 1,25-$(OH)_2D_3$ decreases the secretion of parathyroid hormone (PTH) in blood *in vivo* (Slatopolsky *et al.*, 1984). This results from a direct suppression of preproparathyroid gene expression (Okazaki *et al.*, 1991).

A number of studies have focused on the effect of vitamin D on the reproductive system. Vitamin D deficiency diminished mating success and fertility in female rats and also led to smaller litter size and slowed neonatal growth (Halloran and DeLuca, 1980). The effect of 1,25-$(OH)_2D_3$ on this system differs between males and females. Whereas the observed effect on females seems to be direct (Kwiecinski *et al.*, 1989), the effect of vitamin D on the male rat appears to be indirect and may occur through increasing calcium in the blood (Uhland *et al.*, 1992). However, VDR was detected in both the epididymis (Walters *et al.*, 1983) and the testis (Levy *et al.*, 1985; Merke *et al.*, 1983). This clearly suggests a direct action of vitamin D on these cells, which remains unidentified.

Skeletal muscle has no detectable receptor or receptor transcript (Burmester *et al.*, 1988b). VDR was detected in cultured myoblasts (Boland *et al.*, 1985), and direct action of $1,25\text{-}(OH)_2D_3$ on calcium transport in muscle was reported in cultured myoblasts and cultured skeletal muscle tissue (de Boland and Boland, 1985; Giuliani and Boland, 1984). It has also been suggested that $1,25\text{-}(OH)_2D_3$ may have a direct action on the calcium channels in the muscle membrane and on the mobilization of calcium from muscle to serum (Boland, 1986). Whether $1,25\text{-}(OH)_2D_3$ functions in skeletal muscle remains controversial.

Several other tissues were also shown to have VDR and thus comprise additional target sites for vitamin D action. These include the stomach (Stumpf *et al.*, 1979), lungs (Nguyen *et al.*, 1990), liver (Clemens *et al.*, 1988), the immune system (see the following section for details), skin (Stumpf *et al.*, 1979), mammary gland (Narbaitz *et al.*, 1981), avian shell gland (Haussler, 1986), and chorioallantoic membrane (Narbaitz *et al.*, 1980). These numerous target tissues are a testament to the diverse effects of vitamin D on a wide range of biological functions. It is presumed that the list of vitamin D target sites will continue to grow, resulting in the opening of new frontiers in vitamin D research. It should be noted that keratinocytes of skin are an important target of vitamin D, but the subject has been previously reviewed (Holick, 1990).

IV. Vitamin D and the Immune System

The relationship to the immune system represents one of the most recent arrivals to the vitamin D repertoire. It was the finding that mononuclear and other cells of the peripheral blood have receptors for $1,25\text{-}(OH)_2D_3$ (Bhalla *et al.*, 1983; Provvedini *et al.*, 1983) that prompted the notion that vitamin D could play a role in the immune system. Furthermore, $1,25\text{-}(OH)_2D_3$ causes the differentiation and maturation of malignant and normal cells of the monocytic lineage at various stages in their development (Amento, 1987).

The effect of $1,25\text{-}(OH)_2D_3$ is on both the humoral (Yang *et al.*, 1993a) and cell-mediated immune mechanisms of this system (Lemire and Adams, 1992). The major effect of 1,25-$(OH)_2D_3$ on the immune system is the result of its regulation of T cell activity (Lemire et al., 1984, 1985). It is interesting to note here that in contrast to other target tissues, the VDR is detected only in activated T cells (Provvedini et al., 1983). The expression of the VDR was found to be at equal levels in both the helper/inducer and suppressor/cytotoxic T lymphocyte subsets (Provvedini and Manolagas, 1989). Furthermore, the receptor was found to be expressed in human tonsils and the thymus (Provvedini et al., 1987). In rat thymus, the VDR is localized in the medullary cells, whereas no expression was detected in the cortical cells (Provvedini et al., 1989). At high doses, 1,25-$(OH)_2D_3$ is an immunosuppressive agent (Lemire and Archer, 1991). It suppresses IL-2 mRNA levels (Rigby et al., 1987), inhibits IL-2 secretion from T cells (Rigby et al., 1984), and down-regulates the interferon gene in T cells (Rigby et al., 1987). Besides IL-2, vitamin D also inhibits the secretion of IL-1 by monocytes (Manolagas et al., 1989). 1,25$(-OH)_2D_3$ inhibits the proliferation of activated T cells by interfering with their progression from G1a to G1b in the G phase of the cell cycle (Rigby et al., 1985). This suppressive effect is selective because it is limited to the T helper lymphocyte (Th) population (Lemire et al., 1985). Furthermore, this suppression is probably augmented by enhancing T suppressor activity, whereas the generation of both the T cytotoxic and natural killer (NK) lymphocytes is inhibited (Merino et al., 1989). 1,25-$(OH)_2D_3$ also causes a decrease in immunoglobulin production by the B lymphocytes (Iho et al., 1986). Although a direct effect of the hormone on the B cells was suggested (Iho et al., 1986), this effect seems to be indirect through the suppressive activity of 1,25-$(OH)_2D_3$ on the Th cells (Lemire et al., 1985). Th cells are known to be involved in the immunoglobulin production by B lymphocytes. The effect of vitamin D on the immune system also extends to other cell functions, which include the induction of monocytic differentiation to macrophages (Koeffler et al., 1984), increasing the antigen-presenting activity of macrophages (Amento and Cotter, 1988), and the enhancing of phagocytic activity of macrophages (Goldman, 1984).

Despite the large accumulation of in vitro data on the influence of 1,25-$(OH)_2D_3$ on the immune system, little work has been done to assess the effect of vitamin D deficiency on this system in vivo. It was reported that rachitic infants are more likely to acquire recurrent infections like bronchitis and bronchopneumonia (Stroder, 1975). These patients suffer from various deficiencies within the immune system. Some of these patients experience low levels of serum immunoglobulins (Branbe and Vaccari, 1967), and all have impaired phagocytic activity and their leukocytes show normal bacteriocidal activity (Stroder and Kasal, 1970). In addition, these patients were shown to suffer from impaired neutophil migration and have decreased ability to generate an inflammatory response (Lorente *et al.,* 1976; Stroder and Franzen, 1975). Because of the observation that 1,25-$(OH)_2D_3$ causes the differentiation of murine (Abe *et al.,* 1981) and human (Miyaura *et al.,* 1981) myeloid leukemia cells *in vitro,* it is, therefore, suspected that the monocyte/macrophage function might be impaired in vitamin D deficiency. It was shown that when mice were deprived of vitamin D, depressed phagocytic activity of their cells was observed, and there was an impaired accumulation of cells at the site of inflammation (Bar-Shavit *et al.,* 1981). These effects and others could be reversed either by the addition of 1,25-$(OH)_2D_3$ *in vitro* or by vitamin D repletion of the animals (Bar-Shavit *et al.,* 1981). The effect of vitamin D deficiency on the immune system was also studied in rats deprived of vitamin D prenatally and postnatally (Wientrob *et al.,* 1989). In these animals it was found that the response of both the thymocytes and splenocytes to a mitogen (Con A) was impaired, especially in prenatally deprived animals (Wientrob *et al.,* 1989). Interestingly, in this case, the impairment could not be restored by the addition of 1,25-$(OH)_2D_3$ *in vitro* (Wientrob *et al.,* 1989), implying a developmental defect. Depressed macrophage migration was also observed in these animals, however, no effect could be detected on the levels of either IL-1 or IL-2 as a result of deficiency (Wientrob *et al.,* 1989). Contrary to the findings in rats, macrophages from vitamin D-deficient mice produced less IL-1 than did their vitamin D-sufficient counterparts after lipopolysaccharide stimulation (Kanakova *et al.,* 1991). This discrepancy between the two findings could be due to species differences or to incomplete deficiency. In addition, it was also shown that macrophages from vitamin D-deficient mice have reduced cytolytic and cytostatic activities against tumor cells (Gavison and Bar-Shavit, 1989).

The effect of vitamin D deficiency on the cell-mediated immunity still requires detailed investigation. Some limited information is available from studies on human subjects. The relationship between serum levels of 25-OH-D_3 and the delayed hypersensitivity reaction was studied in a group of 63 elderly people (Toss and Symreng, 1983). The results of these studies showed that those individuals having decreased delayed hypersensitivity have low serum 25-OH-D_3 levels (Toss and Symreng, 1983). Interestingly, treatment of some of these individuals with vitamin D restored their response to these antigens. Furthermore, it was observed that patients with end-stage renal disease with low levels of serum 1,25-$(OH)_2D_3$ experience a decreased response to recall antigens (Lind *et al.,* 1990). Treatment with $1,25\text{-}(OH)_2D_3$ restored the response in these patients (Tabata *et al.,* 1986). The effect of vitamin D deficiency on the delayed hypersensitivity was studied in truly vitamin D-deficient mice, in which mice were produced by prolonged feeding of purified vitamin D-deficient diet and vitamin D-deficiency was assessed by severe hypocalcemia and the absence of $25\text{-}OH\text{-}D_3$ from the blood (Yang *et al.,* 1993b). When tested with contact sensitization antigens, the vitamin D-deficient mice showed a marked decrease in their delayed hypersensitive reaction (Yang *et al.,* 1993b). Similar to studies in humans, prolonged treatment of these mice with $1,25\text{-}(OH)_2D_3$ restored their normal response (Yang *et al.,* 1993b). Furthermore, when vitamin D-deficient guinea pigs were infected with mycobacterium tuberculosis, they showed decreased ability to mount delayed hypersensitivity reaction to tuberculin and also showed decreased lymphoproliferative response to purified protein derivative (Hernandez-Fontera and McMurray, 1993). It is evident from these studies that vitamin D plays a role in the cell-mediated immune response *in vivo,* a process that still awaits further investigation to determine the exact molecular changes produced by vitamin D in this system.

In addition to this effect, vitamin D was also found to play a role in the expression of the cellular antigenic markers on the cell surface (Öberg *et al.,* 1993). Vitamin D treatment seemed to extend the survival of transplanted grafts *in vivo* (Jordan *et al.,* 1988). Furthermore, treatment with excess vitamin D significantly suppressed the progression of autoimmune disorders (Lemire, 1992). In mice, high doses of $1,25\text{-}(OH)_2D_3$ and its 19-nor analog suppressed the delayed hypersensitivity response and immunoglobulin production (Yang *et al.,* 1993a). Therefore, vitamin D deficiency impairs this system and excess $1,25\text{-}(OH)_2D_3$ also acts as an immunosuppressant. The apparent conflicting findings illustrate the need for detailed studies of the immune system in vitamin D deficiency on one hand and of the role of $1,25\text{-}(OH)_2D_3$ as an immunosuppressant on the other. It also suggests that caution should be exercised in the interpretation of data obtained by the *in vitro* treatment of immune system cells with high concentrations of $1,25\text{-}(OH)_2D_3$.

V. Vitamin D and Cell Differentiation

During the process of cell differentiation, primary cells undergo a set of molecular, biochemical, and morphological changes that enables them to perform a specific set of functions. Differentiating cells also lose their ability to proliferate as compared to their parental cells. In recent years, 1,25-$(OH)_2D_3$ has joined the group of agents that cause cellular differentiation.

1,25-$(OH)_2D_3$ plays a role in controlling the growth and differentiation of several cell lines (Bar-Shavit et al., 1983; Koeffler et al., 1984; Mangelsdorf et al., 1984; Morel et al., 1986; Provvedini et al., 1983; Reitzma et al., 1983; Rigby et al., 1985; Terada et al., 1978). These include activated lymphocytes (Tsoukas et al., 1984), myeloma cells (Miyaura et al., 1981), and several leukemia cell lines (Abe et al., 1981; Tanaka et al., 1982). Furthermore, 1,25-$(OH)_2D_3$ was also shown to control the growth of tumors in vivo (Eisman et al., 1987). The human promyelocytic leukemia cell line (HL-60) represents the most widely used model of 1,25-$(OH)_2D_3$-induced differentiation (Bar-Shavit et al., 1983; Koeffler et al., 1984; Mangelsdorf et al., 1984; Ostrem and De-Luca, 1987). Normal human bone marrow cells will undergo differentiation into monocytes and macrophages after exposure to 1,25-$(OH)_2D_3$ in vitro (McCarthy, 1983; Tanaka et al., 1983). Three criteria are generally used to assess the differentiation of the HL-60 cells in response to 1,25-$(OH)_2D_3$, including phagocytosis, NBT reduction, and nonspecific esterase activity (Ostrem and DeLuca, 1987). Besides 1,25-$(OH)_2D_3$, other agents and factors also induce the differentiation of the HL-60 cells. Treatment with phorbol esters cause HL-60 cells to differentiate into monocytes (Morel et al., 1986), whereas treatment with dimethylsulfoxide (Collins et al., 1978) or retinoic acid (Breitman et al., 1980) causes them to differentiate into granulocytes. This implies that the 1,25-$(OH)_2D_3$ effect is specific and is not a result of a general differentiation pathway. The differentiation effect of 1,25-$(OH)_2D_3$ is mediated through interaction with its receptor (Mangelsdorf et al., 1984; Ostrem et al., 1987).

The binding affinity of 1,25-$(OH)_2D_3$ and its analogs to the receptor is one criterion that is used to evaluate the differentiation activity of analogs. In general, the affinity of the vitamin D analogs to the VDR parallels their ability to induce differentiation (Ostrem et al., 1987). This is not always the case as the differentiation activity of the 24-homologated analogs of vitamin D does not correlate with their binding affinity to the receptor (Perlman et al., 1990a). Furthermore, though homologation of 1,25-$(OH)_2D_3$ at the 24-position increased HL-60 differentiation, it diminished intestinal calcium transport activity and bone calcium mobilizing activity (Perlman et al., 1990a). These results suggest that perhaps more than one form of the VDR is expressed, which can account for the differential effect by these compounds. However, the cloning of the VDR from HL-60 cells showed it to be identical to the previously cloned human receptor from skin and intestine (Goto *et al.*, 1992), indicating the presence of one receptor that mediates both functions of 1,25-$(OH)_2D_3$. It should be mentioned here that these results do not completely rule out the presence of VDR subtypes but make it highly unlikely.

The mechanism of 1,25-$(OH)_2D_3$-induced differentiation is probably a result of an initial signal expressed at the level of the gene that commits the cells into the differentiation pathway. Several studies have shown that treatment of HL-60 cells with 1,25-$(OH)_2D_3$ results in a number of molecular changes that are expressed in the mature differentiated cells. One of the most notable effects is the down-regulation of c-*myc* proto-oncogene mRNA. Although this effect was reported initially to be at the transcriptional level (Simpson *et al.*, 1978), subsequent work demonstrated that 1,25-$(OH)_2D_3$ decreased the stability of the c-*myc* message (Mangasarian and Mellon, 1993). This represents an early event in the process, but it is unclear if it is a primary or secondary response. The down-regulation of the c-*myc* mRNA is a general marker of differentiation as it is shared by both the monocytic and granulocytic differentiation pathways of HL-60 cells (Einat *et al.*, 1985; Reitzma *et al.*, 1983; Westin *et al.*, 1982) and other cell types when stimulated to differentiate (Griep and DeLuca, 1986). Other responses to 1,25-$(OH)_2D_3$ treatment include the transcriptional up-regulation of protein kinase C (Obeid *et al.*, 1990), the up-regulation of the mannose receptor (Clohisy *et al.*, 1987), the induction of c-*fms* gene expression (Sariban *et al.*, 1985), and an increase in the phosphorylation of a number of nuclear proteins, including laminin B and histones (Martel *et al.*, 1992). Furthermore, 1,25-$(OH)_2D_3$ treatment induced the turnover of sphingomyelin (Okazaki *et al.*, 1989), stimulated the production of thromboxane (Honda *et al.*, 1986), and resulted in the expression of some calcium binding proteins (Roth *et al.*, 1993). What role these changes play in the differentiation process is not clear.

In addition to upregulation, 1,25-$(OH)_2D_3$ causes down-regulation of the serine protease myeloblastin. Similar to the effect on c-*myc* expression, myeloblastin down-regulation is shared by both pathways leading to the differentiation of HL-60 cells to monocytes or granulocytes (Bories *et al.*, 1989). When the expression of this enzyme was inhibited by antisense oligodeoxynucleotides, the cells stopped proliferation and were induced to differentiate (Bories *et al.*, 1989). This suggests that it is also involved in the early steps of the differentiation process. Because the major function of $1,25\text{-}(OH)_2D_3$ is to maintain calcium homeostatis, the question that presents itself is what role does calcium play in this process? Hruska *et al.* (1988) showed that HL-60 cell differentiation in response to $1,25\text{-}(OH)_2D_3$ is a calcium-independent process. However, there has been no proposed mechanism for the molecular effects of $1,25\text{-}(OH)_2D_3$-induced differentiation.

VI. ANALOGS OF $1,25\text{-}(OH)_2D_3$

The multiple functions of vitamin D described in the foregoing review, and the multiple genes affected by the hormone, through the VDR, suggest the possibility that analogs of the hormone may be constructed that might be selective for particular functions. This has been the aim of several industrial concerns and at least one investigative university laboratory.

In the 1970s with the discovery of multiple metabolites of vitamin D, much effort was expended in attempting to determine if different metabolites exert different functions from those of vitamin D (DeLuca, 1979). However, it has become abundantly clear that the only activated or hormonal form of vitamin D is $1,25\text{-}(OH)_2D_3$ and the other metabolites, some 33 of them, are either degradation products or side reaction products (Brommage and DeLuca, 1985; DeLuca and Schnoes, 1983).

The discovery by Suda and his colleagues (1984) that $1,25\text{-}(OH)_2D_3$ has a differentiative activity suggested that this action might be distinguishable from the action of the hormone on calcium. This was fueled by the idea that some form of vitamin D might be used to suppress cancerous growth and to cause differentiation of malignant cells into functional cells (Eisman, 1984). The first demonstration that the function of differentiation might be separable from the calcemic action of the vitamin came as a result of an investigation that yielded metabolites that became homologated at the 24-carbon or the 26-carbon (Tanaka *et al.*, 1986b). This led to the chemical synthesis of 26-homo-$1,25\text{-}(OH)_2D_3$ and 24-homo-$1,25\text{-}(OH)_2D_3$ (Sai *et al.*, 1986). These compounds were tested in differentiation and exhibited higher differentiation activity than the native hormone (Ostrem and DeLuca, 1987). On the other hand, the 24-homologated forms appeared to be one order of magnitude less active on calcium than that of the native hormone, or the 26-homologated analogs. The DeLuca group synthesized several homologs, demonstrating that further homologation of

1,25-$(OH)_2D_3$ at the 24-carbon either increased or did not significantly affect the action of differentiation of HL-60 cells, though it markedly decreased activity in raising blood calcium *in vivo* (Perlman *et al.*, 1990a).

Leo Pharmaceuticals synthesized a compound known as MC-903 and found it also to have very little calcemic activity while retaining activity in causing differentiation (Calverley, 1987). Chugai Pharmaceuticals synthesized 22-oxa-1,25-$(OH)_2D_3$ (Abe *et al.*, 1989), and the Hoffmann-La Roche group synthesized 16-ene-analogs and found them to have similarly reduced calcium activity while being very active in differentiation (Uskokovic *et al.*, 1991). Finally, the DeLuca group has provided 19-nor analogs (Perlman *et al.*, 1990b) and the 1α-hydroxypregnacalciferol (Ostrem and DeLuca, 1987; Ostrem *et al.*, 1987) with similar activity. There are now a large number of analogs that show this property. However, a major problem is that the differentiation activity is tested *in vitro* using a cell culture system and the calcemic activity is tested *in vivo*. *In vivo* analogs are subject to metabolic degradation, differential binding to the vitamin D plasma transport protein, transport to the receptor, binding to the receptor, and a differential response to the receptor. A problem with any one of these steps could render a compound noncalcemic, though it might still retain activity in binding to the receptor and in causing *in vitro* cellular differentiation. It is known that 22-oxa-1,25-$(OH)_2D_3$ and the Leo Pharmaceutical MC-903 have little activity *in vivo* because they have a very short *in vivo* lifetime (Abe *et al.*, 1989; Bouillon *et al.*, 1991).

One possible explanation for differential activity of analogs was that more than one receptor for 1,25-$(OH)_2D_3$ exists. Thus, one receptor may play a role in cell differentiation and another in the calcemic activities. However, Goto *et al.* (1992) have cloned the receptor from HL-60 cells that differentiate in response to 1,25-$(OH)_2D_3$ and found it to be identical to the cloned human receptor from intestinal cells that carry out calcemic activity. Further, all the 1,25-$(OH)_2D_3$ binding activity in HL-60 cells could be immunoprecipitated with two of the different epitope-specific antibodies for the VDR. It is likely, therefore, that there is a single VDR that carries out both calcemic and differentiative actions.

It is interesting that in a gene reporter system, it has been shown that 1,25-$(OH)_2D_2$ and its 24-epi compound are more effective at exerting transcriptional activity to the osteocalcin gene DRE-containing reporter construct compared with the effect of the vitamin D_3 compounds (Arbour *et al.*, 1993a). It is also of some interest that selective *in vivo* activity can be detected with analogs. The 24-epi-1,25$(OH)_2D_2$

compound appears to have little activity in mobilizing calcium from bone, while having respectable activity in intestinal calcium transport and mineralization of the skeleton (DeLuca *et al.*, 1988).

Additionally, an *in vivo* study demonstrated that the analog 24-dihomo-1,25-$(OH)_2D_3$ can stimulate calbindin-D_{9k} mRNA accumulation in the intestine, but at the same time will not stimulate intestinal calcium transport (Krisinger *et al.*, 1991). Thus, in the same tissue, this 24-homologated analog can produce one type of response and not another, indicating that discrimination also takes place at the cellular level *in vivo*.

Regardless of the underlying reason for how an analog can express selected activity *in vivo*, the compound can nevertheless be useful in the treatment of disease because of the selective activity. For example, MC-903, which is rapidly metabolized, may still be a very effective compound to be used against the disease psoriasis when topically applied (Binderup and Bramm, 1988; Kragballe *et al.*, 1991). If MC-903 should enter the circulation, it is rapidly degraded and therefore will not produce calcemic activity. Thus, for topical application, the noncalcemic analogs should be ideal.

This review is not the place to report on all of the many interesting analogs of 1,25-$(OH)_2D_3$ that have been prepared. However, several are either in use or have already been approved for use in humans. 26,27-Hexafluoro-1,25$(OH)_2D_3$ is being developed for the treatment of osteoporosis and other metabolic bone diseases in Japan (Nishizawa *et al.*, 1991). MC-903 is being used for the treatment of psoriasis in Europe and is rapidly being developed for use in the United States (Kragballe *et al.*, 1991). The analog 1α-hydroxyvitamin D_2 is nearing completion of development for osteoporosis in the United States and elsewhere (Gallagher *et al.*, 1994).

The ultimate goal of medicinal chemists is rational drug design. Rational drug design in the vitamin D area cannot occur until detailed three-dimensional molecular analyses of the receptor-ligand interaction and receptor-genome interaction are performed. Until that time, analog synthesis will be largely a matter of intuition.

ACKNOWLEDGMENTS

This work was supported in part by a Program Project Grant, No. DK-14881, from the National Institutes of Health, a fund from the National Foundation for Cancer Research, and a fund from the Wisconsin Alumni Research Foundation. T.K.R. is a postdoctoral fellow of the National Institutes of Health (DK-08424).

The authors thank Pat Mings, Nancy Arbour, Mike Bonds, Wendy Hellwig, and Tracey Baas for their assistance in the preparation of this chapter.

REFERENCES

Abe, G., Miyaura, C., Sakagami, H., Tanaka, M., Konno, K., Yamazaki, T., Yoshiki, S., and Suda, T. (1981). Differentiation of mouse myeloid leukemia cells induced by $1\alpha,25$-dihydroxyvitamin D_3. *Proc. Natl. Acad. Sci. U.S.A.* **78**, 4990–4994.

Abe, J., Takita, Y., Nakano, T., Miyaura, C., Suda, T., and Nishii, Y. (1989). A synthetic analogue of vitamin D_3, 22-oxa-$1\alpha,25$-dihydroxyvitamin D_3, is a potent modulator of *in vivo* immunoregulating activity without inducing hypercalcemia in mice. *Endocrinology (Baltimore)* **124**, 2645–2647.

Ali, S., Metzger, D., Bornert, J.-M., and Chambon, P. (1993). Modulation of transcriptional activation by ligand-dependent phosphorylation of the human oestrogen receptor A/B region. *EMBO J.* **12**, 1153–1160.

Amento, E. P. (1987). Vitamin D and the immune system. *Steroids* **49**, 55–72.

Amento, E. P., and Cotter, A. C. (1988). 1,25-Dihydroxyvitamin D_3 augments antigen presentation by murine monocyte/macrophages. *J. Bone Miner. Res.* **3**, S217.

Arbour, N. C., Darwish, H. M., and DeLuca, H. F. (1993a). 1,25-Dihydroxyvitamin D_3 analogs influence osteocalcin mRNA levels in rat osteosarcoma cells. *15th Annu. Meet. Am. Soc. Bone Miner. Res.*, Abstr. No. 398.

Arbour, N. C., Prahl, J. M., and DeLuca, H. F. (1993b). Stabilization of the vitamin D receptor in rat osteosarcoma cells through the action of 1,25-dihydroxyvitamin D_3. *Mol. Endocrinol.*, **7**, 1307–1312.

Askew, F. A., Bourdillon, R. B., Bruce, H. M., Jenkins, R. G. C., and Webster, T. A. (1931). The distillation of vitamin D. *Proc. R. Soc. London, Ser. B* **107**, 76–90.

Auricchio, F. (1989). Phosphorylation of steroid receptors. *J. Steroid Biochem.* **32**, 613–622.

Auricchio, F., Migliaccio, A., DiDomenico, M., and Nola, E. (1987). Oestradiol stimulates tyrosine phosphorylation and hormone binding activity of its own receptor in a cell-free system. *EMBO J.* **6**, 2923–2929.

Baker, A. R., McDonnell, D. P., Hughes, M., Crisp, T. M., Mangelsdorf, D. J., Haussler, M. R., Pike, J. W., Shine, J., and O'Malley, B. W. (1988). Cloning and expression of full-length cDNA encoding human vitamin D receptor. *Proc. Natl. Acad. Sci. U.S.A.* **85**, 3294–3298.

Bar-Shavit, Z., Noff, D., Edelstein, S., Meyer, M., Shibolet, S., and Goldman, R. (1981). 1,25-Dihydroxyvitamin D_3 and the regulation of macrophage function. *Calcif. Tissue Int.* **33**, 673–676.

Bar-Shavit, Z., Teitelbaum, S. L., Reitsma, P., Hall, A., Pegg, L. G., Trail, J., and Kahn, A. J. (1983). Induction of monocytic differentiation and bond readsorption by 1,25-dihydroxyvitamin D_3. *Proc. Natl. Acad. Sci. U.S.A.* **80**, 5907–5911.

Barsony, J., Pike, J. W., DeLuca, H. F., and Marx, S. J. (1990). Immunocytology with microwave-fixed fibroblasts shows $1\alpha,25$-dihydroxyvitamin D_3-dependent rapid and estrogen-dependent slow reorganization of vitamin D receptors. *J. Cell Biol.* **111**, 2385–2395.

Beato, M. (1989). Gene regulation by steroid hormones. *Cell (Cambridge, Mass.)* **56**, 335–344.

Bellido, T., Girasole, G., Passeri, G., Yu, X.-P., Mocharla, H., Jilka, R. L., Notides, A., and Manolagas, S. C. (1993). Demonstration of estrogen and vitamin D receptors in bone marrow-derived stromal cells: Up-regulation of the estrogen receptor by 1,25-dihydroxyvitamin D_3. *Endocrinology (Baltimore)* **132**, 553–562.

Berkenstam, A., Glaumann, H., Martin, M., Gustafsson, J. A., and Norstedt, G. (1989). Hormonal regulation of estrogen receptor messenger ribonucleic acid in $T_{47}D$ and MCF-7 breast cancer cells. *Mol. Endocrinol.* **3**, 22–28.

Bhalla, A. K., Amento, E. P., Clemens, T. L., Holick, M. F., and Krane, S. M. (1983). Specific high-affinity receptors for 1,25-dihydroxyvitamin D3 in human peripheral blood mononuclear cells. Presence in monocytes and induction in T lymphocytes following activation. *J. Clin. Endocrinol. Metab.* **57**, 1308–1310.

Binderup, L., and Bramm, E. (1988). Effects of a novel vitamin D analogue MC 903 on cell proliferation and differentiation *in vitro* and on calcium metabolism *in vivo*. *Biochem. Pharmacol.* **37**, 889–895.

Blunt, J. W., and DeLuca, H. F. (1969). The synthesis of 25-hydroxycholecalciferol. A biologically active metabolite of vitamin D_3. *Biochemistry* **8**, 671–675.

Blunt, J. W., DeLuca, H. F., and Schnoes, H. K. (1968). 25-Hydroxycholecalciferol. A biologically active metabolite of vitamin D_3. *Biochemistry,* **7**, 3317–3322.

Boland, R. (1986). Role of vitamin D in skeletal muscle function. *Endocr. Rev.* **7**, 434–488.

Boland, R., Norman, A., Ritz, E., and Hasselbach, W. (1965). Presence of a 1,25-dihydroxyvitamin D_3 receptor in chick skeletal muscle myoblasts. *Biochem. Biophys. Res. Commun.* **128**, 305–311.

Bories, D., Raynal, M. C., Solomon, D. H., Darzynkiewicz, L., and Cayre, Y. G. (1989). Down-regulation of a serine protease, myeloblastin, causes growth arrest and differentiation of promyelocytic leukemia cells. *Cell (Cambridge, Mass.)* **59**, 959–968.

Bouillon, R., Allewaert, K., Xiang, D. Z., Tan, B. K., and van Baelen, H. (1991). Vitamin D analogs with affinity for the vitamin D binding protein: enhanced *in vitro* and decreased *in vivo* activity. *J. Bone Miner. Res.* **6**, 1051–1057.

Boyle, I. T., Miravet, L., Gray, R. W., Holick, M. F., and DeLuca, H. F. (1972). The response of intestinal calcium transport to 25-hydroxy and 1,25-dihydroxyvitamin D in nephrectomized rats. *Endocrinology (Baltimore)* **90**, 605–608.

Branbe, M., and Vaccari, E. (1967). Alterazione dei meccanism immunitari and lattanti affetti di rachitismo ipocalcemio. *Minerva Pediatr.* **19**, 377–381.

Breitman, T. R., Selonick, S. G., and Collins, S. J. (1980). Induction of differentiation of the human promyelocytic leukemia cell line (HL-60) by retinoic acid. *Proc. Natl. Acad. Sci. U.S.A.* **77**, 2936–2940.

Brommage, R., and DeLuca, H. F. (1984). Placental transport of calcium and phosphorus is not regulated by vitamin D. *Am. J. Physiol.* **246**, E526–E529.

Brommage, R., and DeLuca, H. F. (1985). Evidence that 1,25-dihydroxyvitamin D_3 is the physiologically active metabolite of vitamin D_3. *Endocr. Rev.* **6**, 491–511.

Brooks, M. H., Bell, N. H., Love, L., Stern, P. H., Orfei, E., Queener, S. F., Hamstra, A. J., and DeLuca, H. F. (1978). Vitamin D-dependent rickets type II. Resistance of target organs to 1,25-dihydroxyvitamin D_3. *N. Engl. J. Med.* **298**, 996–999.

Brown, T. A., and DeLuca, H. F. (1990). Phosphorylation of the 1,25-dihydroxyvitamin D_3 receptor. *J. Biol. Chem.* **265**, 10025–10029.

Brown, T. A., and DeLuca, H. F. (1991). Sites of phosphorylation and photoaffinity labeling of the 1,25-dihydroxyvitamin D_3 receptor. *Arch. Biochem. Biophys.* **286**, 466–472.

Brown, T. A., Prahl, J. M., and DeLuca, H. F. (1988). Partial amino acid sequence of porcine 1,25-dihydroxyvitamin D_3 receptor isolated by immunoaffinity chromatography. *Proc. Natl. Acad. Sci. U.S.A.* **85**, 2454–2458.

Brumbaugh, P. F., and Haussler, M. R. (1974). 1α,25-Dihydroxycholecalciferol receptors in intestine. I. Association of 1α,25-dihydroxycholecalciferol with intestinal mucosa chromatin. *J. Biol. Chem.* **249**, 1251–1257.

Brumbaugh, P. F., and Haussler, M. R. (1975a). Specific binding of 1α,25-dihydroxycholecalciferol to nuclear components of chick intestine. *J. Biol. Chem.* **250**, 1588–1594.

Brumbaugh, P. F., and Haussler, M. R. (1975b). Nuclear and cytoplasmic binding components for vitamin D metabolites. *Life Sci.* **16**, 353–362.

Brumbaugh, P. F., Hughes, M. R., and Haussler, M. R. (1975). Ctyoplasmic and nuclear binding components for the 1α,25-dihydroxyvitamin D_3 in chick parathyroid glands. *Proc. Natl. Acad. Sci. U.S.A.* **72**, 4871–4876.

Burmester, J. K., Maeda, N., and DeLuca, H. F. (1988a). Isolation and expression of rat 1,25-dihydroxyvitamin D_3 receptor cDNA. *Proc. Natl. Acad. Sci. U.S.A.* **85**, 1005–1009.

Burmester, J. K., Wiese, R. J., Maeda, N., and DeLuca, H. F. (1988b). Structure and regulation of the rat 1,25-dihydroxyvitamin D_3 receptor. *Proc. Natl. Acad. Sci. U.S.A.* **85**, 9499–9502.

Butler, W. T. (1989). The nature and significance of osteopontin. *Connect. Tissue Rev.* **23**, 123–136.

Cai, Q., Chandler, J. S., Wasserman, R. H., Kumar, R., and Penniston, J. T. (1993). Vitamin D and adaptation to dietary calcium and phosphate deficiencies increase intestinal plasma membrane calcium pump gene expression. *Proc. Natl. Acad. Sci. U.S.A.* **90**, 1345–1349.

Calverley, M. J. (1987). Synthesis of MC 903, a biologically active vitamin D metabolite analogue. *Tetrahedron Lett.* **43**, 4609–4619.

Carlsson, A. (1952). Tracer experiments on the effect of vitamin D on the skeletal metabolism of calcium and phosphorus. *Acta Physiol. Scand.* **26**, 212–220.

Chen, K.-S., Prahl, J. M., and DeLuca, H. F. (1993). Isolation and expression of human 1,25-dihydroxyvitamin D_3 24-hydroxylase cDNA. *Proc. Natl. Acad. Sci. U.S.A.* **90**, 4543–4547.

Christakos, S., Gabrielides, C., and Rhoten, W. B. (1989). Vitamin D-dependent calcium binding proteins: Chemistry, distribution, functional considerations, and molecular biology. *Endocr. Rev.* **10**, 3–26.

Clark, S. A., Stumpf, W. E., Sar, M., DeLuca, H. F., and Tanaka, Y. (1980). Target cells for 1,25-dihydroxyvitamin D_3 in the pancreas. *Cell Tissue Res.* **209**, 515–520.

Clark, S. A., Stumpf, W. E., and Sar, M. (1981). Effect of 1,25-dihydroxyvitamin D_3 on insulin secretion. *Diabetes* **30**, 382–386.

Clemens, T. L., Garrett, K. P., Zhou, X. Y., Pike, J. W., Haussler, M. R., and Dempster, D. W. (1988). Immunocytochemical localization of the 1,25-dihydroxyvitamin D_3 receptor in target cells. *Endocrinology (Baltimore)* **122**, 1224–1230.

Clohisy, D. R., Bar-Shavit, Z., Chappel, J. C., and Teitelbaum, S. L. (1987). 1,25-Dihydroxyvitamin D_3 modulates bone marrow macrophage precursor proliferation and differentiation. *J. Biol. Chem.* **262**, 15922–15929.

Collins, S. J., Ruscelli, F. W., Gallagher, R. G., and Gallo, R. C. (1978). Terminal differentiation of human promyelocytic leukemia cells induced by dimethyl sulfoxide and other polar compounds. *Proc. Natl. Acad. Sci. U.S.A.* **75**, 2458–2462.

Costa, E. M., and Feldman, D. (1986). Homologous up-regulation of the 1,25-$(OH)_2$ vitamin D_3 receptor in rats. *Biochem. Biophys. Res. Commun.* **137**, 742–747.

Cousins, R. J., DeLuca, H. F., and Gray, R. W. (1970). Metabolism of 25-hydroxycholecalciferol in target and nontarget tissues. *Biochemistry* **9**, 3649–3652.

Dame, M. C., Pierce, E. A., and DeLuca, H. F. (1985). Identification of the porcine intestinal 1,25-dihydroxyvitamin D_3 receptor on sodium dodecyl sulfate/polyacrylamide gels by renaturation and immunoblotting. *Proc. Natl. Acad. Sci. U.S.A.* **82**, 7825–7829.

Dame, M. C., Pierce, E. A., Prahl, J. M., Hayes, C. E., and DeLuca, H. F. (1986). Monoclonal antibodies to the porcine intestinal receptor for 1,25-dihydroxyvitamin D_3. Interaction with distinct receptor domains. *Biochemistry* **25**, 4523–4534.

Darwish, H. M., and DeLuca, H. F. (1992). Identification of a 1,25-dihydroxyvitamin D_3 response element in the 5'-flanking region of the rat calbindin-D_{9k} gene. *Proc. Natl. Acad. Sci. U.S.A.* **89,** 603–607.

Darwish, H. M., and DeLuca, H. F. (1993). Vitamin D-regulated gene expression. *Crit. Rev. Eukaryotic Gene Express.* **3,** 89–116.

Darwish, H. M., Burmester, J. K., Moss, V. E., and DeLuca, H. F. (1993). Phosphorylation is involved in transcriptional activation by the 1,25-dihydroxyvitamin D_3 receptor. *Biochim. Biophys. Acta* **1167,** 29–36.

de Boland, A. R., and Boland, R. (1985). Suppression of 1,25-dihydroxyvitamin D_3-dependent calcium transport by protein synthesis inhibitors and changes in phospholipids in skeletal muscle. *Biochim. Biophys. Acta* **845,** 237–241.

Deliconstantinos, G., Kopeikina-Tsiboukidou, L., and Tsakiris, S. (1986). Perturbations of rat intestinal brush border membranes induced by Ca^{2+} and vitamin D_3 are detected using steady-state fluorescence polarization and alkaline phosphatase as membrane probes. *Biochem. Pharmacol.* **35,** 1633–1637.

DeLuca, H. F. (1967). Mechanism of action and metabolic fate of vitamin D. *Vitam. Horm. N.Y.* **25,** 315–367.

DeLuca, H. F. (1978). Vitamin D and calcium transport. *Ann. N.Y. Acad. Sci.* **307,** 356–376.

DeLuca, H. F. (1979). Vitamin D and its metabolites: Physiology, biochemistry and pharmacology. *Adv. Pharmacol. Ther. Proc., Int. Congr. Pharmacol., 7th, 1978,* pp. 15–24.

DeLuca, H. F. (1982). The control of calcium and phosphorus metabolism by the vitamin D endocrine system. *Ann. N.Y. Acad. Sci.* **355,** 1–15.

DeLuca, H. F. (1986). Significance of vitamin D in age-related bone diseases. *In* "Nutrition and Aging" (M. L. Hutchinson and H. N. Munroe, eds.), pp. 217–234. Academic Press, Orlando, FL.

DeLuca, H. F. (1988). The vitamin D story: A collaborative effort of basic science and clinical medicine. *FASEB J.* **2,** 224–236.

DeLuca, H. F., and Schnoes, H. K. (1983). Vitamin D: Recent advances. *Annu. Rev. Biochem.* **52,** 411–439.

DeLuca, H. F., Sicinski, R., Tanaka, Y., Stern, P. H., and Smith, C. M. (1988). The biological activity of 1,25-dihydroxyvitamin D_2 and 24-epi-1,25-dihydroxyvitamin D_2. *Am. J. Physiol.* **17,** E402–E406.

Demay, M. B., Roth, D. A., and Kronenberg, H. M. (1989). Regions of the rat osteocalcin gene which mediate the effect of 1,25-dihydroxyvitamin D_3 on gene transcription. *J. Biol. Chem.* **264,** 2279–2282.

Demay, M. B., Gerardi, J. M., DeLuca, H. F., and Kronenberg, H. M. (1990). DNA sequences in the rat osteocalcin gene that bind the 1,25-dihydroxyvitamin D_3 receptor and confer responsiveness to 1,25-dihydroxyvitamin D_3. *Proc. Natl. Acad. Sci. U.S.A.* **87,** 369–373.

Demay, M. B., Kiernan, M. S., DeLuca, H. F., and Kronenberg, H. M. (1992). Sequences in the human parathyroid hormone gene that bind 1,25-dihydroxyvitamin D_3 receptor and mediate transcriptional repression in response to 1,25-dihydroxyvitamin D_3. *Proc. Natl. Acad. Sci. U.S.A.* **89,** 8097–8101.

Denner, L. A., Weigel, N. L., Maxwell, B. L., Schrader, W. T., and O'Malley, B. W. (1990). Regulation of progesterone receptor-mediated transcription by phosphorylation. *Science* **250,** 1740–1743.

Einat, M., Resnitzky, D., and Kimchi, A. (1985). Close link between reduction of c-myc expression by interferon and G_0/g_1 arrest. *Nature (London)* **313,** 597–600.

Eisman, J. A. (1984). 1,25-Dihydroxyvitamin D_3 receptor and role of 1,25-$(OH)_2D_3$ in human cancer cells. *In* "Vitamin D" (R. Kumar, ed.), pp. 365–382. Martinus Nijhoff, Boston.

Eisman, J. A., Barkla, D. H., and Tutton, P. J. M. (1987). Suppression of *in vivo* growth of human cancer solid tumor xenografts by 1,25-dihydroxyvitamin D_3. *Cancer Res.* **47**, 21–25.

Esvelt, R. P., Schnoes, H. K., and DeLuca, H. F. (1978). Vitamin D_3 from rat skins irradiated *in vitro* with ultraviolet light. *Arch. Biochem. Biophys.* **188**, 282–286.

Esvelt, R. P., Schnoes, H. K., and DeLuca, H. F. (1979). Isolation and characterization of 1α-hydroxy-tetranor-vitamin D-23-carboxylic acid: A major metabolite of 1,25-dihydroxyvitamin D_3. *Biochemistry* **18**, 3977–3983.

Evans, R. M. (1988). The steroid and thyroid hormone receptor superfamily. *Science* **240**, 889–895.

Evans, D. B., Bunning, R. A. D., and Russell, R. G. G. (1988). Studies on the interaction between retinoic acid and $1,25(OH)_2D_3$ on human bone-derived osteoblast-like cells. *In* "Vitamin D: Molecular, Cellular and Clinical Endocrinology" (A. W. Norman, K. Schafer, H. G. Grigoleit, and E. von Herrath, eds.), p. 606. de Gruyter, New York.

Fraser, D. R., and Kodicek, E. (1970). Unique biosynthesis by kidney of a biologically active vitamin D metabolite. *Nature (London)* **228**, 764–766.

Fraser, D. R., Kooh, S. W., Kind, H. P., Holick, M. F., Tanaka, Y., and DeLuca, H. F. (1973). Pathogenesis of hereditary vitamin D dependent rickets: An inborn error of vitamin D metabolism involving defective conversion of 25-dihydroxyvitamin D to 1α,25-dihydroxyvitamin D. *N. Engl. J. Med* **289**, 817–822.

Gallagher, J. C., Bishop, C. W., Knutson, J. C., Mazess, R. B., and DeLuca, H. F. (1994). Effects of increasing doses of 1-alpha-hydroxyvitamin D_2 (1a-OH-D_2) on calcium homeostasis in postmenopausal osteoponic women. *J. Bone Miner. Res.* **9**, 607–614.

Gavison, R., and Bar-Shavit, Z. (1989). Impaired macrophage activation in vitamin D_3 deficiency: Differential *in vitro* effects of 1,25-dihydroxyvitamin D_3 on mouse peritoneal macrophage functions. *J. Immunol.* **143**, 3686–3690.

Gill, R. K., and Christakos, S. (1993). Identification of sequence elements in the mouse calbindin-D_{28k} gene that confer 1,25-dihydroxyvitamin D_3- and butyrate-inducible responses. *Proc. Natl. Acad. Sci. U.S.A.* **90**, 2984–2988.

Giuliani, D. L., and Boland, R. L. (1984). Effects of vitamin D metabolites in calcium fluxes in intact chicken skeletal muscle and myoblasts cultured *in vitro*. *Calcif. Tissue Int.* **36**, 200.

Goldman, R. (1984). Induction of a high phagocytic capability in P388D1, a macrophage-like tumor cell line, by 1α,25-dihydroxyvitamin D_3. *Cancer Res.* **44**, 11–19.

Goto, H., Chen, K. S., Prahl, J. M., and DeLuca, H. F. (1992). A single receptor identical with that from intestine/T47D cells mediates the action of 1,25-dihydroxyvitamin D_3 in HL-60 cells. *Biochim. Biophys. Acta* **1132**, 103–108.

Griep, A. G., and DeLuca, H. F. (1986). Decreased c-myc expression is an early event in retinoic acid-induced differentiation of F9 teratocarcinoma cells. *Proc. Natl. Acad. Sci. U.S.A.* **83**, 5539–5543.

Halloran, B. P., and DeLuca, H. F. (1980). Effect of vitamin D deficiency on fertility and reproductive capability in the female rat. *J. Nutr.* **110**, 1573–1580.

Halloran, B. P., and DeLuca, H. F. (1981a). Appearance of the intestinal cytosolic receptor for 1,25-dihydroxyvitamin D_3 during neonatal development in the rat. *J. Biol. Chem.* **256**, 7338–7342.

Halloran, B. P., and DeLuca, H. F. (1981b). Intestinal calcium transport: Evidence for two distinct mechanisms of action of 1,25-dihydroxyvitamin D_3. *Arch. Biochem. Biophys.* **208**, 477–488.

Haussler, M. R. (1986). Vitamin D receptors: Nature and function. *Annu. Rev. Nutr.* **6**, 527–552.

Haussler, M. R., and Norman, A. W. (1967). The subcellular distribution of physiological doses of vitamin D_3. *Arch Biochem. Biophys.* **118**, 145–153.

Haussler, M. R., Myrtle, J. F., and Norman, A. W. (1968). The association of a metabolite of vitamin D_3 with intestinal mucosa chromatin *in vivo. J. Biol. Chem.* **243**, 4055–4064.

Haussler, M. R., Terpening, C. M., Komm, B. S., Whitfield, G. K., and Haussler, C. A. (1988). Vitamin D hormone receptors: Structure, regulation, and molecular function. *In* "Vitamin D: Molecular, Cellular and Clinical Endocrinology" (A. W. Norman, K. Shaefer, H. G. Grigoleit, and D. von Herrath, eds.), pp. I205–214. de Gruyter, New York.

Hernandez-Fontera, E., and McMurray, D. N. (1993). Dietary vitamin D affects cell-mediated hypersensitivity but not resistance to experimental pulmonary tuberculosis in guinea pigs. *Infect. Immunol.* **61**, 2116–2121.

Hess, S. F., and Weinstock, M. (1924). Antiarchitic properties imparted to lettuce and to growing wheat by ultraviolet irradiation. *Proc. Soc. Exp. Biol. Med.* **22**, 5–6.

Holick, M. F. (1990). The intimate relationship between the sun, skin and vitamin D: A new perspective. *Bone* **7**, 66–69.

Holick, M. F., Schnoes, H. K., DeLuca, H. F., Suda, T., and Cousins, R. J. (1971). Isolation and identification of 1,25-dihydroxycholecalciferol. A metabolite of vitamin D active in intestine. *Biochemistry* **10**, 2799–2804.

Holick, M. F., Garabedian, M., and DeLuca, H. F. (1972). 1,25-Dihydroxycholecalciferol: Metabolite of vitamin D_3 active on bone in anephric rats. *Science* **176**, 1146–1147.

Holick, M. F., MacLaughlin, J. A., Clark, M. B., Holick, S. A., Potts, Jr., J. T., Anderson, R. R., Blank, I. H., Parrish, J. A., and Elias, P. (1980). Photosynthesis of previtamin D_3 in human skin and the physiologic consequences. *Science* **210**, 203–205.

Honda, A., Morita, I., Murota, S. I., and Mori, Y. (1986). Appearance of arachionic acid metabolic pathway in human promyelocytic leukemia (HL-60) cells during monocytic differentiation: Enhancement of thromboxane synthesis by $1\alpha,25$-dihydroxyvitamin D_3. *Bhiochim. Biophys. Acta* **877**, 423–432.

Hruska, K. Bar-Shavit, Z., Malone, J. D., and Teitelbaum, S. (1988). Ca^{2+} priming during vitamin D-induced monocyte differentiation of a human leukemia cell line. *J. Biol. Chem.* **263**, 16039–16044.

Hsieh, J.-C., Jurutka, P. W., Galligan, M. A., Terpening, C. M., Haussler, C. A., Samuels, D. S., Shimizu, Y., Shimizu, N., and Haussler, M. R. (1991). Human vitamin D receptor is selectively phosphorylated by protein kinase C on serine 51, a residue crucial to its *trans*-activation function. *Proc. Natl. Acad. Sci. U.S.A.* **88**, 9315–9319.

Hsieh, J.-C., Jurutka, P. W., Nakajima, S., Galligan, M. A., Haussler, C. A., Shimizu, Y., Shimizu, N., Whitfield, G. K., and Haussler, M. R. (1993). Phosphorylation of the human vitamin D receptor by protein kinase C. *J. Biol. Chem.* **268**, 15118–15126.

Huang, C. Y., Lee, S., Stolz, R., Gabrielides, C., Pansini-Porta, A., Bruns, M. E., Bruns, D. E., Miffin, T. E., Pike, J. W., and Christakos, S. (1989). Effect of hormones and development on the expression of the rat 1,25-dihydroxyvitamin D_3 receptor gene. *J. Biol. Chem.* **264**, 17454–17461.

Hughes, M. R., and Haussler, M. R. (1978). 1,25-Dihydroxyvitamin D_3 receptors in parathyroid glands. *J. Biol. Chem.* **253**, 1065–1073.

Hughes, M. R., Malloy, P. J., Kieback, D. G., Kesterson, R. A., Pike, J. W., Feldman, D., and O'Malley, B. W. (1988). Point mutations in the human vitamin D receptor gene associated with hypocalcemic rickets. *Science* **242**, 1702–1705.

Huldshinsky, K. (1919). Heilung von rachitis durch künstliche höhensonne. *Dtsch. Med. Wochenschr.* **45**, 712–713.

Iho, S., Takahashi, T., Kura, F., Sugiyama, H., and Hoshino, T. (1986). The effect of 1,25-dihydroxyvitamin D_3 on *in vitro* immunoglobulin production in human B cells. *J. Immunol.* **136**, 4427–4431.

Ing, N. H., Beekman, J. M., Tsai, S. Y., Tsai, M.-J., and O'Malley, B. W. (1992). Members of the steroid hormone receptor superfamily interact with TFIIB (S300-II). *J. Biol. Chem.* **267**, 17617–17623.

Ishida, H., Seino, Y., Seino, S., Tsuda, K., Takemura, J., Nisi, S., Ishizuka, S., and Imura, H. (1983). Effect of 1,25-dihydroxyvitamin D_3 on pancreatic B and D cell function. *Life Sci.* **33**, 1179.

Jackson, S. P. (1992). Regulating transcription factor activity by phosphorylation. *Trends Cell Biol.* **2**, 104–108.

Jones, B. B., Jurutka, P. W., Haussler, C. A., Haussler, M. R., and Whitfield, G. K. (1991). Vitamin D receptor phosphorylation in transfected ROS 17/2.8 cells is localized to the N-terminal region of the hormone-binding domain. *Mol. Endocrinol.* **5**, 1137–1146.

Jordan, S. C., Nigata, M., and Mullen, Y. (1988). 1,25-Dihydroxyvitamin D_3 prolongs rat cardiac allograft survival. *In* "Vitamin D: Molecular, Cellular, and Clinical Endocrinology" (A. W. Norman, K. Schafer, H. G. Grigoleit, and D. von Herrath, eds.), pp. 334–335. de Gruyter, New York.

Jurutka, P. W., Hsieh, J. C., and Haussler, M. R. (1993a). Phosphorylation of the human 1,25-dihydroxyvitamin D_3 receptor by cAMP dependent protein kinase, *in vitro,* and in transfected COS-7 cells. *Biochem. Biophys. Res. Commun.* **191**, 1089–1096.

Jurutka, P. W., Hsieh, J.-C., MacDonald, P. N., Terpening, C. M., Haussler, C. A., Haussler, M. R., and Whitfield, G. K. (1993b). Phosphorylation of serine 208 in the human vitamin D receptor. *J. Biol. Chem.* **268**, 6791–6799.

Kanakova, M., Lulini, W., Pedrazzoni, M., Riganti, F., Sironi, M., Bottazzi, B., Mantovani, A., and Vecchi, A. (1991). Impairment of cytokine production in mice fed a vitamin D3-deficient diet. *Immunology* **73**, 466–471.

Keaveney, M., Klug, J., Dawson, M. T., Nestor, P. V., Neilan, J. G., Forde, R. C., and Gannon, F. (1991). Evidence for a previously unidentified upstream exon in the human oestrogen receptor gene. *J. Mol. Endocrinol.* **6**, 111–115.

Kerner, S. A., Scott, R. A., and Pike, J. W. (1989). Sequence elements in the human osteocalcin gene confer basal activation and inducible response to hormonal vitamin D_3. *Proc. Natl. Acad. Sci. U.S.A.* **86**, 4455–4459.

Kessler, M. A., Lamm, L., Jarnagin, K., and DeLuca, H. F. (1986). 1,25-Dihydroxyvitamin D_3-stimulated mRNA in rat small intestine. *Arch. Biochem. Biophys.* **251**, 403–412.

Kliewer, S. A., Umesono, K., Mangelsdorf, D. J., and Evans, R. M. (1992). Retinoid X receptor interacts with nuclear receptors in retinoic acid, thyroid hormone and vitamin D signalling. *Nature (London)* **355**, 446–449.

Kodicek, E. (1956). Metabolic studies on vitamin D. *Bone Struct. Metab., Ciba Found. Symp., 1955,* pp. 161–174.

Koeffler, H. P., Amatruda, T., Ikekawa, N., Kobayashi, Y., and DeLuca, H. F. (1984). Induction of macrophage differentiation of human normal and leukemic myeloid stem cells by 1,25-dihydroxyvitamin D_3 and its fluorinated analogues. *Cancer Res.* **44**, 5624–5628.

Kragballe, K., Gjertsen, B. T., DeHoop, D., Karlsmark, T., Van de Kerkhof, P. C. M., Larko, O., Nieboer, C., Roed-Petersen, J., Strand, A., and Tikjob, G. (1991). Double-blind, right/left comparison of calcipotriol and betamethasone valerate in treatment of *psoriasis vulgaris. Lancet* **337**, 193–196.

Kream, B. G., Reynolds, R. D., Knutson, J. C., Eisman, J. A., and DeLuca, H. F. (1976). Intestinal cytosol binders of 1,25-dihydroxyvitamin D_3 and 25-dihydroxyvitamin D_3. *Arch. Biochem. Biophys.* **176**, 779–787.

Kream, B. G., Yamada, Y., Schnoes, H. K., and DeLuca, H. F. (1977). Specific cytosol-binding protein for 1,25-dihydroxyvitamin D_3 in rat intestine. *J. Biol. Chem.* **254**, 9488–9491.

Krishnan, A. V., and Feldman, D. (1991). Stimulation of 1,25-dihydroxyvitamin D_3 receptor gene expression in cultured cells by serum and growth factors. *J. Bone Miner. Res.* **6**, 1099–1107.

Krishnan, A. V., and Feldman, D. (1992). Cyclic adenosine $3',5'$-monophosphate up-regulates 1,25-dihydroxyvitamin D_3 receptor gene expression and enhances hormone action. *Mol. Endocrinol.* **6**, 198–206.

Krisinger, K., Strom, M., Darwish, H. M., Perlman, K., Smith, C., and DeLuca, H. F. (1991). Induction of calbindin-D 9k mRNA but not calcium transport in rat intestine by 1,25-dihydroxyvitamin D_3 24-homologues. *J. Biol. Chem.* **266**, 1910–1913.

Kuiper, G. G. J. M., Faber, P. W., van Rooij, H. C. J., van der Korput, J. A. G. M., Ris-Stalpers, C., Klaassen, P., Trapman, J., and Brinkmann, A. O. (1989). Structural organization of the human androgen receptor gene. *J. Mol. Endocrinol.* **2**, R1–R4.

Kumar, R., Schaefer, J., and Wieben, E. (1992). The expression of milligram amounts of functional human 1,25-dihydroxyvitamin D_3 receptor in a bacterial system. *Biochem. Biophys. Res. Commun.* **189**, 1417–1423.

Kwiecinski, G. G., Petrie, G. I., and DeLuca, H. F. (1989). 1,25-Dihydroxyvitamin D_3 restores fertility on vitamin D-deficient female rats. *Am. J. Physiol.* **256**, E483–E487.

Kyeyune-Nyombi, G., Lau, K.-H. W., Baylink, D. J., and Stong, D. D. (1989). Stimulation of cellular alkaline phosphatase activity and its messenger RNA level in a human osteosarcoma cell line by 1,25-dihydroxyvitamin D_3. *Arch. Biochem. Biophys.* **275**, 363.

Lajeunesse, D., Frondoza, C., Schofield, B., and Stacktor, B. (1990). Osteocalcin secretion by the human osteosarcoma cell line MG-63. *J. Bone Miner. Res.* **5**, 915–922.

Lamm, M., and Neuman, W. F. (1958). On the role of vitamin D calcification. *Arch. Pathol.* **66**, 204–209.

Lawson, D. E. M., Wilson, P. W., and Kodicek, E. (1969). Metabolism of vitamin D: A new cholecalciferol metabolite, involving the loss of hydrogen at C-1, in chick intestinal nuclei. *Biochemistry* **115**, 269–277.

Lemire, J. M. (1992). Immunomodulatory role of 1,25-dihydroxyvitamin D_3. *J. Cell. Biol.* **49**, 26–31.

Lemire, J. M., and Adams, J. S. (1992). 1,25-Dihydroxyvitamin D_3 inhibits the passive transfer of cellular immunity by a myelin basic protein-specific T cell clone. *J. Bone Miner. Res.* **7**, 171–177.

Lemire, J. M., and Archer, D. C. (1991). 1,25-Dihydroxyvitamin D_3 prevents *in vivo* induction of murine experimental autoimmune encephalomyelitis. *J. Clin. Invest.* **87**, 1103–1107.

Lemire, J. M., Adams, J. S., Sakai, R., and Jordan, S. C. (1984). 1α, 25-Dihydroxyvitamin D_3 suppresses proliferation and immunoglobulin production by normal human peripheral blood mononuclear cells. *J. Clin. Invest.* **74**, 657–661.

Lemire, J. M., Adams, J. S., Kermani-Arab, V., Sakai, R., and Jordan, S. C. (1985). 1, 25-Dihydroxyvitamin D_3 suppresses human T-helper/inducer lymphocyte activity *in vitro*. *J. Immunol.* **134**, 3032–3035.

Levy, F. O., Eikvar, L., Jutte, N. H. P. M., Cervenka, J., Yogana-than, T., and Hansson, V. (1985). Appearance of the rat testicular receptor for calcitriol (1,25-dihydroxyvitamin D_3) during development. *J. Steroid Biochem.* **23,** 51–56.

Lian, J. B., and Gundberg, C. M. (1988). Osteocalcin: Biochemical considerations and clinical applications. *Clin. Orthop. Relat. Res.* **226,** 267–291.

Lian, J. B., and Stein, G. S. (1992). Transcriptional control of vitamin D-regulated proteins. *J. Cell. Biochem.* **43,** 37–45.

Lian, J. B., Stewart, C., Puchacz, E., Mackowiak, S., Shalhoub, V., Collart, D., Zambetti, G., and Stein, G. (1989). Structure of the rat osteocalcin gene and regulation of vitamin D-dependent expression. *Proc. Natl. Acad. Sci. U.S.A.* **86,** 1143–1147.

Liao, J., Ozono, K., Sone, T., McDonnell, D. P., and Pike, J. W. (1990). Vitamin D receptor interaction with specific DNA requires a nuclear protein and 1,25-dihydroxy-vitamin D_3. *Proc. Natl. Acad. Sci. U.S.A.* **87,** 9751–9755.

Liberman, U. A., and Marx, S. J. (1990). Vitamin D dependent rickets. *In* "Primer on the Metabolic Bone Diseases and Disorders of Mineral Metabolism" (M. J. Favus, ed.), pp. 274–279. Am. Soc. Bone Miner. Res., Kelseyville, CA.

Lichtler, A., Stover, M. L., Angilly, J., Kream, B., and Rowe, D. W. (1989). Isolation and characterization of the rat α1(l) collagen promoter. *J. Biol. Chem.* **264,** 3072–3077.

Lind, L., Wengle, B., Sorenson, O. H., and Ljunghall, S. (1990). Delayed hypersensitivity in primary and secondary hyperparathyroidism. Treatment with active vitamin D. *Exp. Clin. Endocrinol.* **95,** 271–274.

Logeat, F., Le Cunff, M., Pamphile, R., and Milgrom, E. (1985). The nuclear-bound form of the progesterone receptor is generated through a hormone-dependent phospho-rylation. *Biochem. Biophys. Res. Commun.* **131,** 421–427.

Lorente, F., Fontan, G., Jara, P., Casas, C., Garcia-Rodriguez, M. C., and Ojeda, J. A. (1976). Defective neutrophil motility in hypo-vitaminosis D rickets. *Acta Paediatr. Scand.* **65,** 695–699.

Lormi, A., and Marie, P. J. (1990). Changes in cytoskeletal proteins in response to parathyroid hormone and 1,25-dihydroxyvitamin D in human osteoblastic cells. *Bone Miner.* **10,** 1–12.

Lund, J., and DeLuca, H. F. (1966). Biologically active metabolite of vitamin D_3 from bone, liver, and blood serum. *J. Lipid Res.* **7,** 739–744.

MacDonald, P. N., Haussler, C. A., Terpening, C. M., Galligan, M. A., Reeder, M. C., Whitfield, G. K., and Haussler, M. R. (1991). Baculovirus-mediated expression of the human vitamin D receptor. *J. Biol. Chem.* **266,** 18808–18813.

Majeska, R. J., and Rodan, G. A. (1982). The effect of 1,25($OH)_2D_3$ on alkaline phospha-tase in osteoblastic osteosarcoma cells. *J. Biol. Chem.* **257,** 3362–3365.

Mangasarian, K., and Mellon, W. S. (1993). 1,25-Dihydroxyvitamin D_3 destabilizes c-*myc* in HL-60 leukemia cells. *Biochim. Biophys. Acta* **1172,** 55–63.

Mangelsdorf, D. J., Koeffler, H. P., Donaldson, C. A., Pike, J. W., and Haussler, M. R. (1984). 1,25-Dihydroxyvitamin D_3-induced differentiation in a human promyelocy-tic leukemia cell line (HL-60): Receptor mediated maturation to macrophage like cells. *J. Cell Biol.* **98,** 239–398.

Mangelsdorf, D. J., Onz, E. S., Dyck, J. A., and Evans, R. M. (1990). Nuclear receptor that identifies a novel retinoic acid response pathway. *Nature (London)* **345,** 224–229.

Manolagas, S. C., Haussler, M. R., and Deftos, L. J. (1980). 1,25-Dihydroxyvitamin D_3 receptor-like macromolecule in rat osteogenic sarcoma cell lines. *J. Biol. Chem.* **255,** 4414–4417.

Manolagas, S. C., Burton, D. W., and Deftos, L. J. (1981). 1,25-Dihydroxyvitamin D_3

stimulates the alkaline phosphatase activity of osteoblast-like cells. *J. Biol. Chem.* **256,** 7115–7117.

Manolagas, S. C., Hustmyer, F. G., and Yu, X.-P. (1989). 1,25-Dihydroxyvitamin D_3 and the immune system. *J. Soc. Exp. Biol. Med.* **191,** 238–245.

Martel, R. G., Strahler, J. R., and Simpson, R. U. (1992). Identification of lamin B and histones as 1,25-dihydroxyvitamin D_3-regulated nuclear phosphoproteins in HL-60 cells. *J. Biol. Chem.* **267,** 7511–7519.

Marx, S. J., Liberman, U. A., Eli, C., Gamblin, G. T., DeGrange, D. A., and Balsan, S. (1984). Hereditary resistance to 1,25-dihydroxyvitamin D. *Recent Prog. Horm. Res.* **40,** 589–620.

Massaro, E. R., Simpson, R. U., and DeLuca, H. F. (1983). Glucocorticoids and appearance of 1,25-$(OH)_2D_3$ receptor in rat intestine. *Am. J. Physiol.* **244,** E230–E235.

McCarthy, D. M., San Miguel, J. F., Freake, H. C., Green, P. M., Zola, H., Catovsky, D., and Goldman, J. M. (1983). 1,25-Dihydroxyvitamin D_3 inhibits proliferation of human promyelocytic leukaemia (HL-60) cells and induces monocyte-macrophage differentiation in HL-60 and normal human bone marrow cells. *Leukemia Res.* **7,** 51–55.

McCollum, E. V., Simmonds, N., and Pitz, W. (1916). The relation of the unidentified dietary factors, the fat-soluble A, and water-soluble B, of the diet to the growth-promoting properties of milk. *J. Biol. Chem.* **27,** 33–43.

McCollum, E. V., Simonds, N., Becker, J. E., and Shipley, P. G. (1922). Studies of experimental rickets. XXI. An experimental demonstration of the existence of a vitamin which promotes calcium deposition. *Bull. Johns Hopkins Hosp.* **33,** 229–230.

McDonnell, D. P., Mangelsdorf, D. J., Pike, J. W., Haussler, M. R., and O'Malley, B. W. (1987). Molecular cloning of complementary DNA encoding the avian receptor for vitamin D. *Science* **235,** 1214–1217.

McDonnell, D. P., Scott, R. A., Kerner, S. A., O'Malley, B. W., and Pike, J. W. (1989a). Functional domains of the human vitamin D_3 receptor regulate osteocalcin gene expression. *Mol. Endocrinol.* **3,** 635–644.

McDonnell, D. P., Pike, J. W., Drutz, D. J., Butt, T. R., and O'Malley, B. W. (1989b). Reconstitution of the vitamin D-responsive osteocalcin transcription unit in *Saccharomyces cerevisiae. Mol. Cell. Biol.* **9,** 3517–3523.

Mellanby, E. (1919). An experimental investigation on rickets. *Lancet* **1,** 407–412.

Merino, F., Alvearez-Mon, M., De La Hera, A., Ales, J. E., Bonilla, F., and Durantez, A. (1989). Regulation of natural killer cytotoxicity by 1,25-dihydroxyvitamin D_3. *Cell. Immunol.* **118,** 328–336.

Merke, J., Kreusser, W., Bier, B., and Ritz, E. (1983). Demonstration and characterization of a testicular receptor for 1,25-dihydroxycholecalciferol in the rat. *Eur. J. Biochem.* **130,** 303–308.

Miyaura, C., Abe, G., Kuribayuashi, T., Tanaka, H., Konno, K., Nishii, Y., and Suda, T. (1981). 1α,25-Dihydroxyvitamin D_3 induces differentiation of human myeloid leukemia cells. *Biochem. Biophys. Res. Commun.* **102,** 937–943.

Montminy, M. R., Sevarino, K. A., Wagner, J. A., Mandel, G., and Goodman, R. H. (1986). Identification of a cyclic-AMP-responsive element within the rat somatostatin gene. *Proc. Natl. Acad. Sci. U.S.A.* **83,** 6682–6686.

Morel, P. A., Manolagas, S. C., Provvedini, D. M., Wegman, D. R., and Chiller, J. M. (1986). Interferon-γ-induced IA expression in wehi-3 cells is enhanced by the presence of 1,25-dihydroxyvitamin D_3. *J. Immunol.* **136,** 2181–2186.

Morii, H., Lund, J., Neville, P. F., and DeLuca, H. F. (1967). Biological activity of a vitamin D metabolite. *Arch. Biochem. Biophys.* **120,** 508–512.

Morrison, N. A., Qi, J. C., Tokita, A., Kelley, P. J., Crofts, L., Nguyen, T. V., Sambrook, P. N., and Eisman, J. R. (1994). Prediction of bone density from vitamin D receptor alleles. *Nature* **367**, 284–287.

Nakada, M., Simpson, R. U., and DeLuca, H. F. (1984). Subcellular distribution of DNA-binding and non-DNA-binding 1,25-dihydroxyvitamin D_3 receptors in chicken intestine. *Proc. Natl. Acad. Sci. U.S.A.* **81**, 6711–6713.

Narbaitz, R., Stumpf, W. E., Sar, M., DeLuca, H. F., and Tanaka, Y. (1980). Autoradiographic demonstration of target cells for 1,25-dihydroxycholecalciferol in the chick embryo chorioallantoic membrane, duodenum, and parathyroid glands. *J. Clin. Invest.* **81**, 270–273.

Narbaitz, R., Sar, M., Stumpf, W. E., Haung, S., and DeLuca, H. F. (1981). 1,25-Dihydroxyvitamin D3 target cells in rat mammary gland. *Horm. Res.* **15**, 263–269.

Narbaitz, R., Stumpf, W. E., Sar, M., Haung, S., and DeLuca, H. F. (1983). Autoradiographic localization of target cells for $1\alpha,25$-dihydroxyvitamin D_3 in bones from fetal rats. *Calcif. Tissue Int.* **35**, 177–182.

Neville, P. F., and DeLuca, H. F. (1966). The synthesis of $[1,2\text{-}^3H]$ vitamin D_3 and the tissue localization of a 0.25 µg (10 IU) dose per rat. *Biochemistry* **5**, 2201–2207.

Nguyen, T. M., Guillozo, H., Marin, L., Dufour, M. E., Tordet, C., Pike, J. W., and Garabedian, M. (1990). 1,25-Dihydroxyvitamin D_3 receptors in rat lung during the perinatal period. Regulation and immunohistochemical localization. *Endocrinology (Baltimore)* **127**, 1755–1762.

Nicolaysen, R., and Eeg-Larsen, N. (1953). The biochemistry and physiology of vitamin D. *Vitam. Horm. (N.Y.)* **11**, 29–60.

Nishizawa, Y., Morii, H., Ogura, Y., and DeLuca, H. F. (1991). Clinical trial of 26,26,26,27,27,27-hexafluoro-1,25-dihydroxyvitamin D_3 in uremic patients on hemodialysis: Preliminary report. *Contrib. Nephrol.* **90**, 196–203.

Noda, M., Vogel, R. L., Craig, A. M., Prahl, J., DeLuca, H. F., and Denhardt, D. T. (1990). Identification of a DNA sequence responsible for binding of the 1,25-dihydroxyvitamin D_3 receptor and 1,25-dihydroxyvitamin D_3 enhancement of mouse secreted phosphoprotein 1 (*Spp-1* or osteopontin) gene expression. *Proc. Natl. Acad. Sci. U.S.A.* **87**, 9995–9999.

Obeid, L. M., Okazaki, T., Karolak, L. A., and Hannuu, Y. A. (1990). Transcriptional regulation of protein kinase C by 1,25-dihydroxyvitamin D_3 in HL-60 cells. *J. Biol. Chem.* **265**, 2370–2374.

Öberg, F., Botling, J., and Nilsson, K. (1993). Functional antagonism between vitamin D_3 and retinoic acid in the regulation of CD14 and CD23 expression during monocytic differentiation of U-937 cells. *J. Immunol.* **150**, 3487–3495.

Ohyama, Y., Noshiro, M., and Okuda, K. (1991). Cloning and expressing of cDNA encoding 25-hydroxyvitamin D_3 24-hydroxylase. *FEBS Lett.* **278**, 195–198.

Okazaki, T., Bell, R. M., and Hannun, Y. A. (1989). Sphingomyelin turnover induced by vitamin D_3 in HL-60 cells. *J. Biol. Chem.* **264**, 19076–19080.

Okazaki, T., Zajac, J. D., Igarashi, T., Ogata, E., and Kronenberg, H. M. (1991). Negative regulatory elements in the human parathyroid hormone gene. *J. Biol. Chem.* **266**, 21903–21910.

Okret, S., Poellinger, L., Dong, Y., and Gustafsson, J. A. (1986). Down-regulation of glucocorticoid receptor mRNA by glucocorticoid hormones and recognition by the receptor of a specific binding sequence within a receptor cDNA clone. *Proc. Natl. Acad. Sci. U.S.A.* **83**, 5899–5903.

O'Malley, B. W. (1990). The steroid receptor superfamily: More excitement predicted for the future. *Mol. Endocrinol.* **4**, 363–69.

Ostrem, V. K., and DeLuca, H. F. (1987). The vitamin D-induced differentiation of HL-60 cells: Structural requirements. *Steroids* **496**, 73–102.

Ostrem, V. K., Lau, W. F., Lee, S. H., Perlman, K., Prahl, J., Schnoes, H. K., and DeLuca, H. F. (1987). Induction of monocytic differentiation of HL-60 cells by 1,25-dihydroxyvitamin D analogs. *J. Biol. Chem.* **262**, 14164–41471.

Ozono, K., Liao, J., Kerner, S. A., Scott, R. A., and Pike, J. W. (1990). The vitamin D-responsive element in the human osteocalcin gene. *J. Biol. Chem.* **265**, 21881–21888.

Paulson, S. K., and DeLuca, H. F. (1985). Subcellular location and properties of rat renal 25-hydroxyvitamin D_3-1α-hydroxylase. *J. Biol. Chem.* **260**, 11488–11492.

Perlman, K. L., Kutner, A., Prahl, J., Smith, C., Inaba, M., Schnoes, H. K., and DeLuca, H. F. (1990a). 24-Homologated 1,25-dihydroxyvitamin D_3 compounds: Separation of calcium and cell differentiation activities. *Biochemistry* **29**, 190–196.

Perlman, K. L., Sicinski, R. R., Schnoes, H. K., and DeLuca, H. F. (1990b). 1α,25-Dihydroxy-19-nor-vitamin D_3, a novel vitamin D-related compound with potential therapeutic activity. *Tetrahedron Lett.* **31**, 1823–1824.

Petkovich, P. M., Heersche, J. N. M., Tinker, D. O., and Jones, G. (1984). Retinoic acid stimulates 1,25-dihydroxyvitamin D3 binding in rat osteosarcoma cells. *J. Biol. Chem.* **259**, 8274–8290.

Pierce, E. A., and DeLuca, H. F. (1988). Regulation of the intestinal 1,25-dihydroxyvitamin D_3 receptor during neonatal development in the rat. *Arch. Biochem. Biophys.* **261**, 241–249.

Pike, J. W. (1991). Vitamin D_3 receptors: Structure and function in transcription. *Annu. Rev. Nutr.* **11**, 189–216.

Pike, J. W., and Haussler, M. R. (1979). Purification of chicken intestinal receptor for 1α,25-dihydroxyvitamin D. *Proc. Natl. Acad. Sci. U.S.A.* **76**, 5485–5489.

Pike, J. W., and Sleator, N. M. (1985). Hormone-dependent phosphorylation of the 1,25-dihydroxyvitamin D_3 receptor in mouse fibroblasts. *Biochem. Biophys. Res. Commun.* **131**, 378–385.

Pike, J. W., Marion, S. L., Donald, C. A., and Haussler, M. R. (1983). Serum and monoclonal antibodies against the chicken intestinal receptor for 1,25-dihydroxyvitamin D_3. Generation by a preparation enriched in a 64,000-dalton protein. *J. Biol. Chem.* **258**, 1289–1296.

Pike, J. W., Sleator, N. M., and Haussler, M. R. (1987). Chicken intestinal receptor for 1,25-dihydroxyvitamin D_3. *J. Biol. Chem.* **262**, 1305–1311.

Pols, H. A. P., van Leeuwen, J. P. T. M., Schilte, J. P., Visser, T. J., and Birkenhager, J. C. (1988). Heterologous up-regulation of the 1,25-dihydroxyvitamin D_3 receptor by parathyroid hormone and PTH-like peptide in osteoblast-like cells. *Biochem. Biophys. Res. Commun.* **156**, 588–594.

Price, P. A., and Baukol, S. A. (1980). 1,25-Dihydroxyvitamin D_3 increases synthesis of the vitamin K-dependent bone protein by osteosarcoma cells. *J. Biol. Chem.* **255**, 11660–11663.

Provvedini, D. M., and Manolagas, S. C. (1989). 1,25-Dihydroxyvitamin D3 receptor distribution and effects on subpopulations of normal human T lymphocytes. *J. Immunol.* **68**, 774–779.

Provvedini, D. M., Tsoukas, C. D., Deftos, L. J., and Manolagas, S. C. (1983). 1,25-Dihydroxyvitamin D_3 receptors in human leukocytes. *Science* **221**, 1181–1183.

Provvedini, D. M., Tsoukas, C. D., Deftos, L. J., and Manolagas, S. C. (1987). 1a,25-Dihydroxyvitamin D3 receptors in human thymic and tonsillar lymphocytes. *J. Bone Miner. Res.* **2**, 239–247.

Provvedini, D. M., Sakagami, Y., and Manolagas, S. C. (1989). Distinct target cells and effects of $1\alpha,25$-dihydroxyvitamin D_3 and glucocorticoids in the rat thymus gland. *Endocrinology (Baltimore)* **124**, 1532–1538.

Rasmussen, H., DeLuca, H. F., Arnaud, C., Hawker, C., and von Stedingk, M. (1963). The relationship between vitamin D and parathyroid hormone. *J. Clin. Invest.* **42**, 1940–1946.

Reeve, L., Tanaka, Y., and DeLuca, H. F. (1983). Studies on the site of 1,25-dihydroxyvitamin D_3 synthesis *in vivo*. *J. Biol Chem.* **258**, 3615–3617.

Reitzma, P. H., Rothberg, P. G., Astrin, S. M., Trial, J., Bar-Shavit, Z., Hall, A., Teitelbaum, S., and Kahn, A. J. (1983). Regulation of *myc* gene expression HL-60 leukemia cells by a vitamin D metabolite. *Nature (London)* **306**, 492–494.

Rigby, W. F. C., Stacy, T., and Fanger, M. W. (1984). Inhibition of T lymphocyte mitogenesis by 1,25-dihydroxyvitamin D_3 (calcitriol). *J. Clin. Invest.* **74**, 1451–1455.

Rigby, W. F. C., Noelle, R. J., Kranse, K., and Franger, M. W. (1985). Effects of 1,25-dihydroxyvitamin D_3 on human T lymphocyte activation and proliferation: A cell cycle analysis. *J. Immunol.* **135**, 2279–2286.

Rigby, W. F. C., Denome, S., and Fanger, M. W. (1987). Regulation of lymphokine production and human T lymphocyte activation by 1,25-dihydroxyvitamin D_3. *J. Clin. Invest.* **79**, 1659–1664.

Ritchie, H. H., Hughes, M. R., Thompson, E. T., Malloy, P. J., Hochberg, Z., Feldman, D., Pike, J. W., and O'Malley, B. W. (1989). An ochre mutation in the vitamin D receptor gene causes hereditary 1,25-dihydroxyvitamin D_3-resistant rickets in three families. *Proc. Natl. Acad. Sci. U.S.A.* **86**, 9783–9787.

Ross, T. K., Prahl, J. M., and DeLuca, H. F. (1991). Overproduction of rat 1,25-dihydroxyvitamin D_3 receptor in insect cells using the baculovirus expression system. *Proc. Natl. Acad. Sci. U.S.A.* **88**, 6555–6559.

Ross, T. K., Moss, V. E., Prahl, J. M., and DeLuca, H. F. (1992a). A nuclear protein essential for binding of rat 1,25-dihydroxyvitamin D_3 receptor to its response elements. *Proc. Natl. Acad. Sci. U.S.A.* **89**, 256–260.

Ross, T. K., Prahl, J. M., Herzberg, I. M., and DeLuca, H. F. (1992b). Baculovirus-mediated expression of retinoic acid receptor type γ in cultured insect cells reveals a difference in specific DNA binding behavior with the 1,25-dihydroxyvitamin D_3 receptor. *Proc. Natl. Acad. Sci. U.S.A.* **89**, 10282–10286.

Ross, T. K., Darwish, H. M., Moss, V. E., and DeLuca, H. F. (1993). Vitamin D-influenced gene expression via a ligand-independent, receptor–DNA complex intermediate. *Proc. Natl. Acad. Sci. U.S.A.* **90**, 9257–9260.

Roth, J., Goebeler, M., Bos, C. V. D., and Sorg, C. (1993). Expression of calcium-binding proteins MRP8 and MRP14 is associated with distinct monocytic differentiation pathways in HL-60 cells. *Biochem. Biophys. Res. Commun.* **191**, 565–570.

Rowe, D. W., and Kream, B. G. (1982). Regulation of collagen synthesis in fetal rat calavaria by 1,25-dihydroxyvitamin D_3. *J. Biol. Chem.* **257**, 8009–8015.

Sai, H., Takatsuo, S., Ikekawa, N., Tanaka, Y., and DeLuca, H. F. (1986). Synthesis of some side chain homologues of $1\alpha,25$-dihydroxyvitamin D_3 and investigation of their biological activities. *Chem. Pharm. Bull.* **34**, 4508–4515.

Sandgren, M. E., and DeLuca, H. F. (1990). Serum calcium and vitamin D regulate 1,25-dihydroxyvitamin D_3 receptor concentration in rat kidney *in vivo*. *Proc. Natl. Acad. Sci. U.S.A.* **87**, 4312–4314.

Santiso-Mere, D., Sone, T., Hilliard, G. M., IV, Pike, J. W., and McDonnell, D. P. (1993). Positive regulation of the vitamin D receptor by its cognate ligand in heterologous expression systems. *Mol. Endocrinol.* **7**, 833–839.

Sariban, G., Mitchell, T., and Kufe, D. (1985). Expression of the *c-fms* protooncogene during human monocytic differentiation. *Nature (London)* **316**, 64–66.

Semmler, E. J., Holick, M. F., Schnoes, H. K., and DeLuca, H. F. (1972). The synthesis of 1α,25-dihydroxycholecalciferol—A metabolically active form of vitamin D_3. *Tetrahedron Lett.* **40**, 4147–4150.

Simpson, R. U., and DeLuca, H. F. (1982). Purification of chicken intestinal receptor for 1α,25-dihydroxyvitamin D_3 to apparent homogeneity. *Proc. Natl. Acad. Sci. U.S.A.* **79**, 16–20.

Simpson, R. U., Hsu, T., Begley, D. A., Mitchell, B. S., and Alizaheh, B. N. (1978). Transcriptional regulation of the c-*myc* protooncogene by 1,25-dihydroxyvitamin D_3 in HL-60 promyelocytic leukemia cells. *J. Biol. Chem.* **262**, 4104–4108.

Simpson, R. U., Hamstra, A., Kendrick, N. C., and DeLuca, H. F. (1983). Purification of the receptor for 1α,25-dihydroxyvitamin D_3 from chicken intestine. *Biochemistry* **22**, 2586–2594.

Singh, V. B., and Moudgil, V. K. (1985). Phosphorylation of rat liver glucocorticoid receptor. *J. Biol. Chem.* **260**, 3684–3690.

Slatopolsky, E., Weerts, C., Thieland, J., Horst, R., Harter, H., and Martin, K. (1984). Marked suppression of secondary hyperparathyroidism by i.v. administration of 1,25-dihydroxycholecalciferol in uremic patients. *J. Clin. Invest.* **74**, 2136.

Smith, C. L., Hager, G. L., Pike, J. W., and Marx, S. J. (1981). Overexpression of the human vitamin D_3 receptor in mammalian cells using recombinant adenovirus vectors. *Mol. Endocrinol.* **5**, 867–878.

Sone, T., McDonnell, D. P., O'Malley, B. W., and Pike, J. W. (1990). Expression of human vitamin D receptor in *Saccharomyces cervisiae*. *J. Biol. Chem.* **265**, 21997–22003.

Sone, T., Ozono, K., and Pike, J. W. (1991a). A 55-kilodalton accessory factor facilitates vitamin D receptor DNA binding. *Mol. Endocrinol.* **5**, 1578–1586.

Sone, T., Kerner, S., and Pike, J. W. (1991b). Vitamin D receptor interaction with specific DNA. *J. Biol. Chem.* **266**, 23296–23305.

Sonnenberg, J., Luine, V. N., Krey, L. C., and Christakos, S. (1986). 1,25-Dihydroxyvitamin D_3 treatment results in increased choline acetyltransferase activity in specific brain nuclei. *Endocrinology (Baltimore)* **118**, 1433–1439.

Steenbock, H., and Black, A. (1924). Fat-soluble vitamins. XVII. The induction of growth-promoting and calcifying properties in a ration by exposure to ultraviolet light. *J. Biol. Chem.* **61**, 405–422.

Stroder, J. (1975). Immunity in vitamin D deficient rickets. *In* "Vitamin D and Problems of Uremic Bone Disease" (A. W. Norman, K. Schaefer, H. G. Grigoleit, D. von Herrath, and E. Ritz, eds.), p. 675. de Gruyter, Berlin.

Stroder, J., and Franzen, C. H. (1975). Die unpezifische entzundungsreaktion bei vitamin-D-mangel rachitis. *Klin. Paediatr.* **187**, 461–464.

Stroder, J., and Kasal, P. (1970). Evaluation of phagocytosis in rickets. *Acta Paediatr. Scand.* **59**, 288–292.

Strom, M., Krisinger, J., and DeLuca, H. F. (1991). Isolation of an mRNA that encodes a putative intestinal alkaline phosphatase regulated by 1,25-dihydroxyvitamin D_3. *Biochim. Biophys. Acta* **1090**, 299–304.

Stumpf, W. E., Sar, M., Reid, F. A., Tanaka, Y., and DeLuca, H. F. (1979). Target cells for 1,25-dihydroxyvitamin D_3 in intestinal tract, stomach, kidney, skin, pituitary gland and parathyroid. *Science* **206**, 1188–1190.

Stumpf, W. E., Sar, M., Narbaitz, R., Reid, F. A., DeLuca, H. F., and Tanaka, Y. (1980). Cellular and subcellular localization of 1,25-$(OH)_2$-vitamin D_3 in rat kidney: Comparison with localization of parathyroid hormone and estradiol. *Proc. Natl. Acad. Sci. U.S.A.* **77**, 1149–1153.

Stumpf, W. E., Sar, M., Clark, S. A., and DeLuca, H. F. (1982). Brain target sites for 1,25-dihydroxyvitamin D_3. *Science* **215**, 1403–1405.

Stumpf, W. E., Narbaitz, R., Haung, S., and DeLuca, H. F. (1983). Autoradiographic localization of 1,25-dihydroxyvitamin D_3 in rat placenta and yolk sac. *Horm. Res.* **18**, 215–220.

Suda, T., Abe, E., Miyaura, C., Tanaka, H., Shiina, Y., and Kuribayashi, T. (1984). Vitamin D in the differentiation of myeloid leukemia cells. *In* "Vitamin D" (R. Kumar, ed.), pp. 343–363. Marinus Nijhoff, Boston.

Tabata, T., Suzuki, R., Kikunami, K., Matsushita, Y., Inoue, T., Inoue, T., Okamato, T., Miki, T., Nishizawa, Y., and Morii, H. (1986). The effect of 1α-hydroxyvitamin D_3 on cell-mediated immunity in hemodialyzed patients. *J. Clin. Endocrinol. Metab.* **63**, 1218–1221.

Tanaka, H., Abe, G., Miaura, C., Kuribayashi, T., Konno, K., Nishii, Y., and Suda, T. (1982). 1α,25-Dihydroxyvitamin D_3 and a human myeloid leukemia cell line (HL-60). The presence of a ctyosol receptor and induction of differentiation. *J. Biochem. (Tokyo)* **204**, 713–719.

Tanaka, H., Abe, M., Miyaura, C., Shiina, Y., and Suda, T. (1983). 1α,25-Dihydroxyvitamin D_3 induces differentiation of human promyelocytic leukemia cells (HL-60) into monocyte-macrophages, but not into granulocytes. *Biochem. Biophys. Res. Commun.* **117**, 86–92.

Tanaka, Y., and DeLuca, H. F. (1984). Rat renal 25-hydroxyvitamin D_3 1- and 24-hydroxylases: Their *in vivo* regulation. *Am. J. Physiol.* **246**, E168–E173.

Tanaka, Y., Lorenc, R. S., and DeLuca, H. F. (1975). The role of 1,25-dihydroxyvitamin D_3 and parathyroid hormone in the regulation of chick renal 25-hydroxyvitamin D_3-24-hydroxylase. *Arch. Biochem. Biophys.* **171**, 521–526.

Tanaka, Y., Seino, Y., Ishida, M., Yamaoka, K., Yabuuchi, H., Ishida, H., Seino, S., Seino, Y., and Imura, H. (1984). Effect of vitamin D_3 on the pancreatic secretion of insulin and somatostatin. *Acta Endocrinol. (Copenhagen)* **105**, 528–533.

Tanaka, Y., Seino, Y., Ishida, M., Yamakoa, K., Satomura, K., Yabuuchi, H., Seino, Y., and H, I. (1968a). Effect of 1,25-dihydroxyvitamin D_3 on insulin secretion: Direct or mediated? *Endocrinology (Baltimore)* **118**, 1971–1978.

Tanaka, Y., Sicinski, R. R., DeLuca, H. F., Sai, H., and Ikekawa, N. (1986b). Unique rearrangement of ergocalciferol side-chain *in vitro:* Production of biologically highly active homolog of 1,25-dihydroxyvitamin D_3. *Biochemistry* **25**, 5512–5518.

Terada, M., Nudel, M., Fibach, G., Rifkind, R. A., and Marks, P. A. (1978). Changes in DNA associated with induction of erythroid differentiation by dimethyl sulfoxide in murine rythroleukemia cells. *Cancer Res.* **38**, 835–840.

Toss, G., and Symreng, T. (1983). Delayed hypersensitivity response and vitamin D deficiency. *Int. J. Vitam. Nutr. Res.* **53**, 27–31.

Tsoukas, C. D., Provvedini, D. M., and Manolagas, S. C. (1984). 1,25-Dihydroxyvitamin D3: A novel immunoregulatory hormone. *Science* **224**, 1438–1440.

Uhland, A. M., Kwiecinski, G. G., and DeLuca, H. F. (1992). Normalization of serum calcium restores fertility in vitamin D-deficient male rats. *J. Nutr.* **122**, 1338–1344.

Uhland-Smith, A., and DeLuca, H. F. (1993). The necessity for calcium for increased renal vitamin D receptor in response to 1,25-dihydroxyvitamin D. *Biochim. Biophys. Acta* **1176**, 321–326.

Underwood, J. L., and DeLuca, H. F. (1984). Vitamin D is not directly necessary for bone growth and mineralization. *Am. J. Physiol.* **246**, E493–E498.

Uskokovic, M. R., Baggiolini, E., Shiuey, S.-J., Lacobelli, J., Hennessy, B., Kriegiel, J., Daniewski, A. R., Pizzolato, G., Courtney, L. F., and Horst, R. L. (1991). The 16-ene analogs of 1,25-dihydroxycholecalciferol. Synthesis and biological activity. *In* "Vitamin D: Gene Regulation, Structure–Function Analysis and Clinical Application" (A. W. Norman, R. Bouillon, and M. Thomasset, eds.), pp. 139. de Gruyter, New York.

van Leeuwen, J. P. T. M., Birkenhager, J. C., Burman, C. J., Schilte, J. P., and Pols, H. A. P. (1990). Functional involvement of calcium in the homologous up-regulation of the 1,25-dihydroxyvitamin D_3 receptor in osteoblast-like cells. *FEBS Lett.* **270**, 165–167.

van Leeuwen, J. P. T. M., Birkenhager, J. C., Wijngaarden, T. V. V., van den Bemd, G. J. C. M., and Pols, H. A. P. (1992). Regulation of 1,25-dihydroxyvitamin D_3 receptor gene expression by parathyroid hormone and cAMP-agonists. *Biochem. Biophys. Res. Commun.* **185**, 881–886.

Walters, M. R. (1981). an estrogen-stimulated 1,25-$(OH)_2$ vitamin D_3 receptor in rat uterus. *Biochem. Biophys. Res. Commun.* **103**, 721–726.

Walters, M. R., Cuneo, D. L., and Jamison, A. P. (1983). Possible significance of new target tissues for 1,25-dihydroxyvitamin D_3. *J. Steroid Biochem.* **19**, 913–920.

Wasserman, R. H., Smith, C. A., Brindak, M. G., de Talamoni, N., Fullmer, C. S., Penniston, J. T., and Kumar, R. (1992). Vitamin D and mineral deficiencies increase the plasma membrane calcium pump of chicken intestine. *Gastroenterology* **102**, 886–894.

Weigel, N. L., Carter, T. H., Schrader, W. T., and O'Malley, B. W. (1992). Chicken progesterone receptor is phosphorylated by a DNA-dependent protein kinase during *in vitro* transcriptional assays. *Mol. Endocrinol.* **6**, 8–14.

Westin, G. H., Wong-Staal, F., Gelmann, G. P., Favera, R. D., Papas, T. S., Lautenberger, J. A. R., Eva, A., Reddy, E. P., Tronick, S. R., Aaronson, S. A., and Gallo, R. C. (1982). Expression of cellular homologues of retroviral *onc* genes in human hematopoietic cells. *Proc. Natl. Acad. Sci. U.S.A.* **79**, 2490–2494.

Wientrob, S., Winter, C. C., Wahl, S. M., and Wahl, L. M. (1989). Effect of vitamin D deficiency on macrophage and lymphocyte function in the rat. *Calcif. Tissue Int.* **44**, 125–130.

Wiese, R. J. A., U.-S., Ross, T. K., Prahl, J. M., and DeLuca, H. F. (1992). Up-regulation of the vitamin D receptor in response to 1,25-dihydroxyvitamin D_3 results from ligand-induced stabilization. *J. Biol. Chem.* **267**, 20082–20086.

Wiese, R. J., Goto, H., Prahl, J. M., Marx, S. J., Thomas, M., Al-Aqeel, A., and DeLuca, H. F. (1993). Vitamin D-dependency rickets type II: Truncated vitamin D receptor in three kindreds. *Mol. Cell. Endocrinol.* **90**, 197–201.

Windaus, A., Schenck, F., and von Werder, F. (1936). Über das antirachitisch wirksame Bestrahlungs-produkt aus 7-dehydro-cholesterin. *Hoppe-Seyler's Z. Physiol. Chem.* **241**, 100–103.

Wynshaw-Boris, A., Lugo, T. G., Short, J. M., Fournier, R. E. K., and Hanson, R. W. (1984). Identification of a cAMP regulatory region in the gene for rat cytosolic phosphoenolpyruvate carboxykinase (GTP). *J. Biol. Chem.* **259**, 12161–12169.

Yamamoto, M., Kawanobe, Y., Takahashi, H., Shimazawa, E., Kimura, S., and Ogata, E. (1984). Vitamin D deficiency and renal calcium transport in the rat. *J. Clin. Invest.* **74**, 507–513.

Yang, S., Smith, C., and DeLuca, H. F. (1993a). 1α,25-dihydroxyvitamin D_3 and 19-nor-1α,25-dihydroxyvitamin D_2 suppress immunoglobin production and thymic lymphocyte proliferation *in vivo*. *Biochim. Biophys. Acta.* **1158**, 279–286.

Yang, S., Smith, C., Prahl, J. M., Lou, X., and DeLuca, H. F. (1993b). Vitamin D deficiency suppresses cell-mediated immunity *in vivo*. *Arch. Biochem. Biophys.* **303**, 98–106.

Yoon, K., Rutledge, S. J. C., Buenaga, R. F., and Rodan, G. A. (1988). Characterization of the rat osteocalcin gene: Stimulation of promoter activity by 1,25-dihydroxyvitamin D_3. *Biochemistry* **27**, 8521–8526.

Yu, V. C., Delsert, C., Andersen, B., Holloway, J. M., Devary, O. V., Näär, A. M., Kim, S. Y., Boutin, J.-M., Glass, C. K., and Rosenfeld, M. G. (1991). RXRβ: A coregulator that enhances binding of retinoic acid, thyroid hormone, and vitamin D receptors to their cognate response elements. *Cell (Cambridge, Mass.)* **67,** 1251–1266.

Zierold, C., Darwish, H. M., and DeLuca, H. F. (1993). Identification of a vitamin D response element in the rat 25-hydroxyvitamin D 24-hydroxylase gene. *Proc. Natl. Acad. Sci. U.S.A.* **91,** 900–902.

Zong, J., Ashraf, J., and Thompson, E. B. (1990). The promoter and first, untranslated exon of the human glucocorticoid receptor gene are GC rich but lack consensus glucocorticoid receptor element sites. *Mol. Cell. Biol.* **10,** 5580–5585.

Nuclear Retinoid Receptors and Their Mechanism of Action

MAGNUS PFAHL, RAINER APFEL, IGOR BENDIK, ANDREA FANJUL, GERHART GRAUPNER, MI-OCK LEE, NATHALIE LA-VISTA, XIAN-PING LU, JAVIER PIEDRAFITA, MARIA ANTONIA ORTIZ, GILLES SALBERT, AND XIAO-KUN ZHANG

La Jolla Cancer Research Foundation
La Jolla, California 92037

I. INTRODUCTION

Early in this century the important biological roles of vitamins and hormones were recognized. Since then, many different hormones and vitamins have been purified and characterized, but so far only limited knowledge of their molecular mechanism of action has been achieved.

Vitamin A and its active derivative retinoic acid (RA) regulate a large number of biological processes, including development, differentiation, morphogenesis, growth, metabolism, and homeostasis (1–3). Clinically RA and its synthetic derivatives (retinoids) are used for the treatment of several skin diseases (4) but have also shown promise for the treatment and prevention of cancers (5).

An understanding of the signaling pathways that underlie the large diversity of responses to vitamin A-derived hormones has been greatly advanced by the cloning of specific receptors. These receptors belong to one of the largest family of transcription factors known today, the steroid/thyroid hormone receptor superfamily (6,7). At least six genes encode retinoid receptors in mammals. The receptors fall into two subfamilies, the classic retinoic acid receptors (RARs) α, β, and γ (8–13) and the retinoid X receptors (RXRs) α, β, and γ (14–18). From the individual receptor genes, different isoforms can be generated by differential splicing and/or different promoter usage (19,20, and references therein). RARs and RXRs belong to a subgroup of the nuclear receptor superfamily that has similar DNA binding domains (6,7,21) and that is able to recognize and function from overlapping and related response elements. This subfamily of receptors includes the thyroid hormone receptors (TR) (22), the vitamin D3 receptor (VDR), the peroxisome proliferator-activated receptor (PPAR) (23), and several orphan receptors (receptors for which specific ligands are not known). In analogy with the steroid hormone receptors, retinoid receptors function by binding to specific DNA regions (called response elements) (24–27, and references therein). Recently, major advances in our knowledge of how RARs and RXRs function have been made. It is now clear that RARs require heterodimerization with RXRs for efficient DNA binding and action. RXRs are also coreceptors for TRs, VDRs, and other receptors (17,18,28–32). In addition, RXRs can also function on their own in the presence of specific ligands and form homodimers (33,34). Furthermore, the retinoid receptors can regulate transcription without binding to specific DNA sequences, but by interaction with other transcription factors (35–37). In the following sections we will review the mechanisms of retinoid receptor action, their interaction with their ligands, their relationship to other receptors, and retinoid receptor involvement in diseases, as well as advances in the development of retinoids as therapeutics.

II. RETINOID RECEPTORS AND THE SUPERFAMILY

The molecular cloning of steroid hormone receptors stimulated an increasing interest in the molecular mechanisms of hormone and vita-

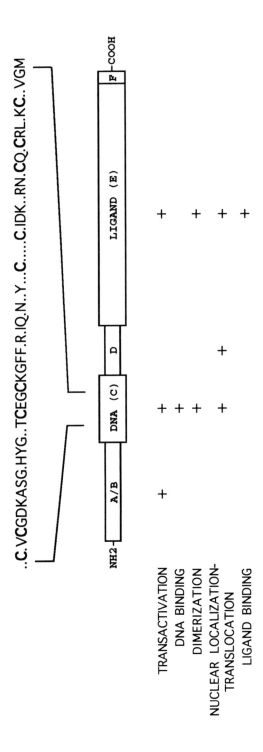

Fig. 1. General features of nuclear receptors. The basic structure of a nuclear receptor is shown and the different functions assigned to specific domains. The sequence marked between brackets is the consensus sequence for the DNA binding domain of 41 nuclear receptors. The nine cisteine residues are highlighted.

min action. During the past decade this led to the discovery of a large number of nuclear receptors. These receptors have a general building plan depicted in Fig. 1 (6,7,38). Although originally six subdomains (A–F) were proposed for these receptors (6), only some of these domains were subsequently found to be present in all receptors and serve a particular function. The two major domains are the DNA binding domain (DBD) and the ligand binding domain (LBD) (Fig. 1), each of which is composed of subdomains.

Computer-aided alignments of the amino acid sequences of the more conserved domains of the receptors, the DBDs and LBDs, allows one to carefully analyze the homologies shared by these receptors and to construct evolutionary trees (39), which can further our understanding of the relationship among the different members of this family and their interactions. A number of interesting results arise from such studies, examples of which are shown in Table I and Fig. 2. A minimum of four different groups of receptors can be distinguished sharing the same P-box (see the following), a short sequence located in the first zinc finger and responsible for the DNA binding specificities of the receptors (21,40,91). The first group contains the glucocorticoidlike receptors glucocorticoid (GR), mineralocorticoid (MR), progesterone (PR), and androgen receptors (AR), whereas the second group comprises the estrogen receptor (ER) and the estrogenlike receptors ERR1 and ERR2. A third group is composed of TRs, RXRs, PPARs, and VDR and also contains many orphan receptors. Most of the nuclear receptors from *Drosophila* belong to a fourth group together with v-erbA and COUP. In addition, at least three orphan receptors (TLL, HNF4, and FTZ-F1) remain unclassified, each one with a different sequence for the P-box. Thus these receptors could represent members of new subfamilies (Table I). Interestingly, many of the receptors that have identical P-box sequences were shown to bind the same DNA sequences. For instance, all receptors from the glucocorticoid group bind the steroid response element (38), whereas the retinoid receptors bind the same consensus sequence as TRs, VDRs, and PPARs (42). On the other hand, COUP and EAR2 belong to a separate P-box group but also bind retinoid/thyroid receptor response elements (see the following).

An evolutionary tree based on the sequence identities of the complex DNA binding domain reveals a more complex picture of how the various receptors relate to each other (Fig. 2A). The two retinoid receptor groups RAR and RXR belong to different subbranches and the RXRs are clearly the evolutionarily older receptors, consistent with the notion that an RXR homolog (USP) is found already in *Drosophila* (43) whereas RARs are not present in insects. The evolutionary tree of the

TABLE I
Nuclear Receptor Classification
According to the P-Box Sequence

P-BOX	NUCLEAR RECEPTOR	D-BOX	SUBFAMILY
CGSCKV	GR MR PR AR	AGRND AGRND AGRND ASRND	GLUCOCORTICOID
CEGCKA	ER ERR1 ERR2	PATNQ PASNE PATNE	ESTROGEN-RELATED
CEGCKG	E75 EAR1 THRα THRβ NGF1B VDR ECR RXRα RXRβ RXRγ USP TR2 RARα RARβ RARγ PPAR	TKNQQ LKNEN KYDSC KYEGK LANKD PFNGD KFGRA RDNKD RDNKD RDNKD RENRN RGSKD HRDKN HRDKN HRDKN --DRS	THYROID/RETINOID
CEGCKS	KNL KNRL EGON SVP COUP EAR2	KNEGK KNNGE KHNGD RGSRN RANRN RSNRD	DROSOPHILA-LIKE
CESCKG	FT2-F1	VAERS	ORPHAN I
CDGCKG	HNF4	RFSRQ	ORPHAN II
CDGCAG	TLL	KSQKQ	ORPHAN III

Note. A classification in several groups into subfamilies can be made, based on the homology of the P-Box sequence.

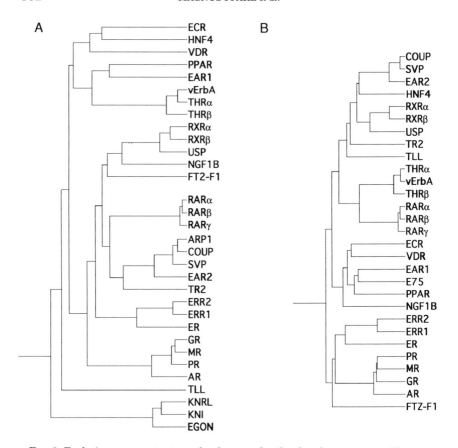

FIG. 2. Evolutionary receptor trees for the superfamily of nuclear receptors. These are UPGMA (Unweighted Pair Group Method with Arithmetic Mean) trees, which show the calculated evolutionary relationships of previously aligned sequences. The length of the horizontal lines connecting one sequence with another is proportional to the estimated genetic distance between the sequences. (A) The relationships are based on homologies in the DBD or (B) LBD. For alignments, essentially the method of Laudet *et al.* (39) was followed.

LBDs (Fig. 2B) looks quite different, however, and again RARs and RXRs are on different subbranches, and RARs are much more recent events.

Overall, the relative high conservation of the DNA and ligand binding domains in this superfamily of nuclear receptors points to a common ancestor for all receptor genes that, by duplication processes, transformed a single original sequence into a complicated tree of divergent genes. However, clearly DBD and LBD appear to have developed separately. The result is one of the largest families of transcription factors known today, a superfamily of proteins with common structural characteristics but with a wide diversity of functions (39).

III. Specific Domains

A. The DBD, a Complex Network of "Boxes"

The DBD turned out to be significantly more complex than originally anticipated and encodes several distinct functions essential for efficient DNA binding. The specific regions called "boxes" of the DBD have been defined by different approaches such as mutagenesis (21,40,41,44), nuclear magnetic resonance (NMR) (45,46), and crystallography (47). The specificity of DNA sequence recognition by the receptors is directed by essentially three amino acids of the P-box (Table I and Fig. 3). For instance, it has been possible to convert GR into a receptor that binds an estrogen response element (ERE) simply by replacing the GR P-box with the ER P-box (21,40). ER could be converted into a GR response element (GRE) binding protein (41) by interchanging the P-box in ER with that of GR. It has also been shown that the side chain of the Val 462 residue, the last residue of the GR P-box, establishes van der Waals contacts with the methyl group of the T at position 4 in the 3' half-site of the GRE (see the following for response elements) (47), a base that is highly specific for GREs. As

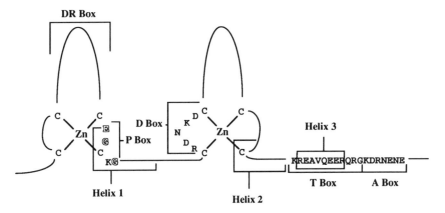

FIG. 3. Schematic structure of the DNA binding domain of RXRα (the three amino acids of the P-box that are essential for DNA binding specificity are outlined).

shown in Table I, all three RXRs (α, β, and γ), all three RARs, and TRs and VDR share the same P-box EGCKG and can bind to the same recognition half-site. However, in the natural DNA recognition sites of these receptors the half-sites are distinctly arranged as will be discussed later. Structural studies of one receptor (47) also revealed the importance of several residues in the second zinc finger of the DBD for DNA interaction. However, the second zinc finger serves indirectly for DNA binding in that it allows receptor dimerization. These residues have been named D-box and differ considerably between RARs, RXRs, and other members of the retinoid/TR subfamily (see Table I), suggesting different dimerization properties. An additional region, called the DR-box, has been defined by Perlmann *et al.* (48). The DR-box is located within the loop of the first zinc finger of RARs and TRs, and has been proposed to interact with the RXR D-box, allowing for asymmetric complex formation of heterodimers that interact with direct repeat response elements.

A unique feature of the RXRs is a third helix in their DBD (Figs. 3 and 4). In previous X-ray crystallographic analyses of the GR (47), or NMR studies of the GR (45), ER (49), and RARβ (50), only the first two helices (see Fig. 3) were observed in the DBDs. The third helix, part of a region named the T-box (44), was revealed by NMR studies of RXRα and is necessary for RXRα (46) and RXRβ (44) homodimerization. Furthermore, Lee *et al.* (46) suggest that the helix 3 is not only necessary for protein–protein interactions but also for efficient DNA–protein interaction. These observations in fact extend the DBD into the so-called "hinge region" (also called the D-domain).

Finally, an additional box has been defined at the carboxy-terminal end of the DBD, the A-box, first reported for the orphan receptor

	T BOX	A BOX
RXRα	KREAVQEERQRG	KDRNENE
RXRβ, γ	KREAVQEERQRG	KDKDGDG
RARα	S KESVRNDRNKK	KKEVKEE
RARβ	S KESVRNDRNKK	KKETSKQ
RARγ	S KEAVRNDRNKK	KKEVKEE

FIG. 4. Comparison of the T- and A-boxes of retinoid receptors.

NGFI-B (which presumably binds DNA as a monomer). The A-box has been proposed to be involved in the recognition of bases upstream of the NGFI-B recognition sequence (44). Until recently it was unknown whether the A-box could also play any role in the specificity of DNA recognition by RXRs and RARs, or if it was only a characteristic of monomeric receptors such as NGFI-B and SF1 (44, 51). Sequence comparisons, however, revealed that residues forming the putative A-box of RXRs and RARs were very well conserved within these receptors (Fig. 4). Although a divergence between RAR and RXR A-boxes is noticeable, both contained approximately 60% charged residues. The functional importance of the A-box was demonstrated by Kurokawa *et al.* (52), who showed that mutations in the A-box of TRβ or RXRα can dramatically affect their DNA recognition. For instance, point mutations in the TR A-box are sufficient to abolish binding of TR homodimers to the DR4 element and an inverted palindrome with 5-bp spacer (52). The same TR point mutations also affected the affinity of TR/RXR heterodimers for DR4 and TREp, but to a lesser extent (52). Similarly, mutations of the RXRα A-box can affect the behavior of the TR/RXR heterodimer. RXR mutations did not appear to modify the stability of the protein–DNA interactions, but decreased the selectivity of binding (when complexed in a dimer with TR) to the 5′ half-site of direct repeat elements (52). Wild-type RXR usually occupies the 5′ half-site when heterodimers interact with direct repeat sequences. (52). These experiments thus allow the conclusion that the basic residues in the TR and RXR A-boxes are important for stabilizing DNA binding, and for defining the receptor position in heterodimers, probably by their ability to contact the DNA minor groove upstream of the half-site (52).

In conclusion, receptors binding to the same half-site (AGGTCA) via their common P-box are apparently able to discriminate between different spacing and orientation of the two half-sites of various response elements by using a combination of DBD subdomains (boxes) so that appropriate protein–protein and protein–DNA interactions can occur.

B. Transcriptional Activation Domains

Activation functions (AFs) of nuclear receptors have been assigned to two different domains (53): the so-called AF1, located in the N-terminal region (or A/B domain), which does not require ligand for its activity, and the AF2, which is the ligand-dependent activation domain located in the LBD (Fig. 5). These two different activation functions

A

A/B domain	Percentage of S and P residues
hRXRα	35 %
mRXRβ	49 %
mRXRγ	31 %
hRARα1	38 %
hRARβ2	28 %
hRARγ1	33 %

B

RXRs	I D T F L M E M L E A P H Q	419
RARα, β	M P P L I Q E M L E N S E G	419, 412
RARγ	M P P L I R E M L E N P E M	421
TRα	F P P L F L E V F E D Q E V	410
TRβ	L P P L F L E V F E D	456
ER	L Y D L L L E M L D A H R L	549

FIG. 5. (A) Percentage of serine (S) and proline (P) residues in the A/B domains of retinoid receptors. (B) Alignment of the conserved residues of the putative AF2 of several nuclear receptors.

have been described for several types of nuclear receptors, such as the ER (53), the PR (54), the GR (55), and also the RARs and RXRs (56–58).

1. *AF1s in the Retinoid Receptor*

AF1s have now been reported for all three RARs (α, β, and γ), but for only two of the RXRs (α and γ) (57). These AF1s show weak activation properties by themselves, but can strongly synergize with their corresponding AF2 (57). This synergistic effect of the two AFs was also previously demonstrated for ER (53). One particularity of the AF1s is that they do not show only promoter-context specificity (56,59) but also cell-type specificity (53,58). The cell-type specificity of AF1s point to the possibility that they do not interact directly with the basic transcription machinery, but rather provide intermediary contacts with cell-type-specific proteins. Interestingly, retinoid receptor AFs appear to be more similar to the ER AF1s than to the GR AF1s. GR appears to have multiple AF1s, consistent with its more complex amino-terminal region (60,61). One interesting observation is that RXRβ does not seem to have any AF1 activity. On the other hand, sequence analyses do not reveal any striking similarity between RXRα and RXRγ, or any striking dissimilarity between RXRα and RXRβ or RXRγ and RXRβ N-terminal domains. On the contrary, these domains of RXRα, γ and especially β (as well as those of the RARs) contain a high percentage of proline (P) and serine (S) residues (Fig. 5A). Proline-rich activation domains are a common feature of several transcription factors. The constitutive factor CTF/NF-1 (62) possesses an activation domain rich in proline (20 to 30%), and similar stretches of proline are seen in AP-2 (63), Oct-2 (64), and the serum response factor (65). These proline residues, which are α-helix breakers, may contribute to a specific structure for transcriptional activation. In this respect, a secondary structure prediction analysis of RAR and RXR N-terminal domains reveals a coil conformation throughout these domains. Alternatively, however, the multiple serine residues in the amino-terminal domain could be targets for phosphorylation (58), a process that has been shown to enhance the transcriptional activity of several nuclear receptors (66–68). Phosphorylation of the N-terminal domain is a common feature of PR (69), GR (70), TRα (71), and RARγ (72).

2. AF2, a Possible Acidic Activation Domain

A ligand-dependent activation function, AF2, has been assigned to the LBD of all nuclear receptors for which a specific ligand has been identified. RXR AF2s are unmasked or activated in the presence of 9-*cis* RA, and likewise RAR AF2 are activated in the presence of 9-*cis* RA or all-*trans* RA (57). Thus binding of the hormone induces a conformational change that allows the appearance of an activation domain at the surface of the protein. Other data that favor this ligand-induced conformational switch were also obtained by protease resistance analyses of the LBD in the presence or absence of ligand (73,74). Disruption of the ligand-dependent activation properties of nuclear receptors can be achieved by deletions or point mutations of the very C-terminal end of the LBD. This is not only true for the ERs (75,76) and the TRs (77,78), but also for RARs (58,77,79,80) and RXRs (57,82). Mutations in this region do not abolish DNA binding or ligand binding functions, although receptors mutated in this region show lower affinity than the wild-type receptors for their respective ligands (75,77,78). Interestingly, some of these mutants are of dominant negative phenotype (77,78).

Alignment of the C-terminal region of several nuclear receptors (Fig. 5B) reveals highly conserved acidic residues surrounded by hydrophobic residues. A secondary structure analysis prediction shows that this region could form a short amphiphatic α-helix bringing the acidic residues in contact with the solvent. Moreover, it has been shown that modification of either the hydrophobic residues or the acidic residues can abolish ligand-dependent transcriptional activation in TR (77) and ER (75). Thus these findings raise the possibility that AF2s are actually acidic activation domains (AADs), such as the AADs of Gal4 (83), GCN4 (84), and VP16 (85,86). AADs are thought to form α-helical or unstructured regions that interact with the basic amphiphatic α-helices found in the general transcription factors TFIID and TFIIB. In this respect it is noteworthy that ER and PR have been demonstrated to directly interact with TFIIB (87) via their ligand binding domains (containing AF2), whereas the N-terminal domains (containing AF1) are not able to interact with TFIIB. Nonetheless, to prove that the C-terminal domains of the nuclear receptors can account by themselves for AF2, it would be necessary to fuse these peptides to the Gal4 DNA binding domain, for instance, and determine if the fusion protein shows any activation property. On the other hand, the whole concept of acidic activation domain is questionable in light of the demonstration that the Gal4 activation domain does not form an α-helical or unstructured region but rather a β-sheet, where the acidic residues are not necessary for transcriptional activation (88,89). Finally, an adenovirus E1A-like protein has been proposed to be responsible for transactivating cooperativity of RARβ and TFIID in certain cell types (90,91).

Considering that AF1s usually show cell-type specificity and do not interact directly with TFIIB, it is tempting to speculate that an E1A-like protein could bridge AF1 domains with the general transcription factor TFIID, whereas TFIIB directly contacts AF2 as shown in the model in Fig. 6. In this model, the nuclear receptors bind as homo- or heterodimer to an HRE stabilizing TFIIB association with TFIID by the synergistic activities of the two AF2s and one AF1. These multiple interactions would greatly stabilize the formation of the preinitiation complex, the rate-limiting step of gene transcription (note that in this model a stable interaction between TFIID and one of the nuclear receptors requires both AF1 and AF2, whereas a stable interaction between TFIIB and the other receptor requires only AF2).

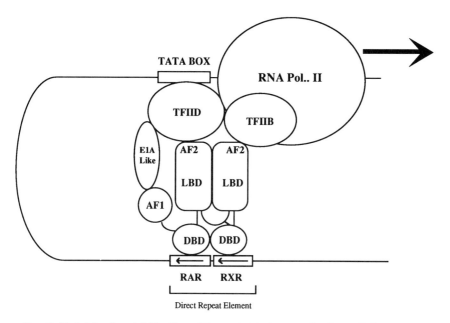

FIG. 6. Model for the stabilization of the preinitiation complex by an RXR/RAR heterodimer.

C. THE DIMERIZATION INTERFACE AND THE ZIPPERLIKE STRUCTURE OF THE LBD

Dimerization of receptor monomers is a prerequisite for transcriptional activation and efficient DNA binding for most of the nuclear receptors known to date. Even if the DBDs by themselves show a natural propensity to dimerize (as discussed in the following), the stability of these dimers is much weaker than that formed by full-length receptors. Strong interactions between monomers are accomplished through the LBDs of the receptors, as described also for ER (92) and PR (93). The dimerization domain of the LBD has been proposed to form a helical structure presenting a hydrophobic surface formed by the repetition of nine heptad motifs containing hydrophobic residues at positions 1 and 8, and hydrophobic or charged amino acids with hydrophobic side chains at position 5 (94). This succession of hydrophobic sequences would generate a leucine-zipper-like structure (95) or a helix-turn-zipper motif (96) known to be necessary for the dimerization of other transcription factors. So far, only the ninth heptad repeat has been extensively studied and it is still not known if, or to what extent, the first eight repeats contribute to dimerization.

From deletion analyses (18,32,82), it appears that the ninth repeat of RXRs and RARs is required for heterodimerization of these receptors as well as TR/RXR heterodimerization. Figure 7 shows sequence alignments of the ninth repeat of several receptors. One study (97) indicated that residues 1 and 8 of the last repeat of RAR and TR are necessary for heterodimerization of these two receptors with RXR but, surprisingly, not for the formation of TR or RAR homodimers. Furthermore, if the cognate ligand of RAR or TR mutated at position 1 is added, heterodimerization can occur again (97). These findings provide evidence that, first, homodimers and heterodimers probably do not use the same dimerization interface and, second, liganded heterodimers differ in their interactions from unliganded heterodimers. For homodimerization, RXRs are likely to use an additional dimerization interface located between the last HR and AF2 [(28,32,82); see also Section IV], a region also required for stable homodimers of the orphan receptor Coup-TF (our unpublished observations) and defined as the ligand-2 domain by Forman and Samuels (94). The fact that mutations of the τ_1 domain [defined in (94), and see following discussion] of RXRβ do not modify its dimerization properties (98) emphasizes the importance of the C-terminal dimerization domain of RXR. The τ_1 domain is a highly conserved region (Fig. 8) located between the ligand-1 domain (94) and the first heptad repeat. Involvement of the τ_1 domain in dimerization has been revealed by several studies focusing on TR and RAR (98–101). It was shown that deletion or point mutations of this domain in TR and RAR (but not in RXRβ) can abolish TR homodimerization or TR/RXR and RAR/RXR heterodimerization. As revealed by

```
RXRα      412PGRFAK LLLRLPAI RSIGLKCLEHLFFFKLI442
RXRβ      476QGRFAK LLLRLPAI RSIGLKCLEHLFFFKLI506
RXRγ      413PGRFAK LLLRLPAI RSIGLKCLEHLFFFKLI443
RARα      366PHMFPK MLMKITDL RSISAKGAERVITLKME396
RARβ      364PHMFPK ILMKITDL RSISAKGAERVITLKME394
RARγ      373PYMFPR MLMKITDL RGISTKGAERAITLKME403
TRα       360PHFWPK LLMKVTDL RMIGACHASRFLHMKVE390
TRβ       410THFWPK LLMKVTDL RMIGACHASRFLHMKVE440
VDR       375GSHLLY AKMIQKLA DLRSLNEEHSKQYRCLS405
ER        501QQRLAQ LLLILSHI RHMINKGMEHLYNMKCK531
COUP-TF   466PSRFGK LLLRLPSL RTVSSSVIEQLFFVRLV496
```

FIG. 7. Conserved amino acids of the ninth heptad repeat of retinoid receptors and related receptors.

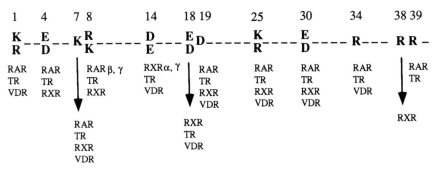

FIG. 8. Conserved charged residues in the $\tau 1$ domain of RARs, RXRs, TRs, and VDR.

the sequence alignment in Fig. 8, one striking feature of the τ_1 domain is the almost perfect conservation of the charged residues in RARs, RXRs, TRs, and VDR. It is noteworthy that the acidic residue at position 14 (Fig. 8) is absent in RXRb and all RARs but present in RXRα and γ. Moreover, mutation of residue 19 disrupts dimerization of RAR and TR without influencing the hormone binding properties of TR (98–101). Thus, these charged residues are most likely involved in dimerization functions of at least TRs and RARs. Analysis of TRβ deletion mutants (99) further emphasized the importance of τ_1, as it showed that deletion of most of the heptad repeats (including the ninth repeat) does not abolish TRβ homodimer activity, whereas a single point mutation in τ_1 (D300A) is able to abolish dimerization.

Taken together, from these studies it appears that structures beyond the zipperlike structure are necessary for dimerization of nuclear receptors. One possibility is that a helix-turn-zipper motif [as recently proposed by Litwack's group (96)] is part of an α/β barrel structure that can accommodate the ligand (102). However, this putative structure is only one component of the multiple dimerization interfaces that allow very diverse protein–protein interaction.

D. OTHER MUTATIONS IN THE RETINOID RECEPTORS

1. Separating RXR Hetero- and Homodimerization

As discussed earlier, our understanding of the structure and function of retinoid receptors is largely built on mutational studies. In addition, a number of naturally occurring mutants have been isolated from various cell types. The characterization of these naturally existing receptor mutants can significantly contribute to our knowledge of the structural features required for various retinoid receptor activities, and modulation of these structures can significantly influence or alter receptor-mediated retinoid signaling, resulting even in the development of cancer.

Since the demonstration that RAR/RXR heterodimers and RXR homodimers are the active receptor complexes that mediate retinoid action (103), an impressive and fascinating body of studies has been carried out to delineate structural features required for retinoid X receptor dimerization. Unlike other nuclear receptors, RXRs heterodimerize with a number of nuclear receptors but can also form homodimers in the presence of the ligand 9-*cis* RA (103). To investigate the unique dimerization capacity of RXRs, Zhang *et al.* (82,104,105) employed both deletions and point mutations to localize regions important for RXR dimerization, DNA binding, and transcriptional regulation activity. A very short region in the extreme C-terminal end of RXRα was revealed to be critical for dimerization and transcriptional activation activities. Deletion of 20 amino acids (amino acids 443–413) abolished DNA binding and transactivation activities of both RXR heterodimers and RXR homodimers. These 20 amino acids included the ninth heptad repeat (Fig. 6), suggesting the importance of this repeat for the RXR dimerization function. Interestingly, deletion of 29 amino acids from the RXR C-terminal end resulted in a mutant that retained RXR heterodimer activity but was impaired in RXR homodimer activity. This indicates that distinct structural features are involved in RXR homodimerization. Alternatively, the region immediately downstream of the ninth heptad repeat could be involved in RXR ligand binding because RXR homodimer formation is dependent on ligand binding (33). By single amino acid substitutions in this region, several amino acids were shown to be critical for RXR dimerization (82). One of these amino acids, Leu 422, located in the fifth position of the ninth heptad repeat (Fig. 7), appears to play a role in the ligand-induced switch that leads to RXR homodimer DNA binding. Substitution of this amino acid with Gln resulted in a receptor that bound with high affinity as homodimer to DNA even in the absence of 9-*cis* RA. Surprisingly, addition of 9-*cis* RA inhibited homodimer binding of this mutant.

Another amino acid that is critical for RXR homodimer activity is Leu 430, which is located five amino acids downstream of the ninth heptad repeat. Mutation of this amino acid to Phe impaired the ligand sensitivity of the RXR homodimer DNA binding. This mutant receptor bound strongly to DNA as a homodimer and its DNA binding was not influenced by addition of 9-*cis* RA (82). Thus, it is likely that this amino acid is part of the ligand binding pocket. The fact that this mutant receptor does bind DNA with high affinity as homodimer also suggests that it may be involved in RXR homodimer formation.

The effect of these deletions or amino acid substitutions on RXR homodimer DNA binding characteristics was observed on several response elements, including palindromic or direct repeat RAREs, suggesting that the same dimerization interface in the LBD is utilized for efficient RXR homodimer DNA binding independent of the arrangement and orientation of the half-sites. Interestingly, although the single amino acid substitution mutants bind DNA with high affinity in the absence of 9-*cis* RA, they did not show any constitutive transactivation activities when tested in transfection assays *in vitro*. Thus, RXR homodimerization can be separated from the transactivation activity. Importantly, despite impaired homodimer transactivation function and ligand responsiveness, these mutants still retain their heterodimer activities in that they enhance the DNA binding and transactivation function of RAR or TR. Thus, RXR homodimer and heterodimer activities can be separated (82). These mutants could provide useful tools for the dissection of RXR-mediated activities and the study of the role of the RXR ligand, 9-*cis* RA, in various biological processes.

2. Dominant Negative Mutants

To study the distinct effects of RA in various cells, one possible approach is to use dominant negative receptor derivatives that suppress their endogenous receptor activities. As mentioned earlier, RXRα can be converted into a potent negative regulator by either short deletions or amino acid substitutions in the RXR ligand binding domain (82). A truncation mutant of RXRα, terminating after amino acid 413, does not respond to 9-*cis* RA but acts as a potent inhibitor of cotransfected wild-type RXR receptor. In addition, the point mutant that replaces Leu 430 with Phe also functions as a dominant negative regulator. Although these mutants strongly repress RXR homodimer activities, they only partially inhibit 9-*cis* RA-induced RAR/RXR heterodimer activity and not all-*trans* RA-induced activity. Similarly, by deletion and amino acid substitution, Damm *et al.* (106) created RARα receptor derivatives that act as potent inhibitors of cotransfected wild-type RARα in CV-1 cells.

An example of how dominant negative receptor derivatives can be used as a tool to study receptor-mediated retinoid activities in various biological processes was provided by Espeseth *et al.* (107). These investigators stably transfected a truncated RARα (lacking the E/F domain) into F-9 cells, they observed a repression of the retinoid response, and the cells revealed an RA-resistant phenotype, that is, they were blocked in differentiation. Another study demonstrated that expression of an RARα mutant could inhibit RA-induced HL60 cell differentiation and alter the lineage development of a multipotent hematopoietic cell line (108). Indeed, a point mutation in the RARα ligand binding domain that results in frameshift and the truncation of 52 amino acids at the C-terminal end was detected in a mutant subclone of HL60 that showed markedly reduced sensitivity to RA. This truncated RARα exhibited dominant negative activity when assayed *in vitro* and is therefore likely responsible for the RA resistance in HL60 cells. Similarly, in P19 cells selected for RA nonresponsiveness, a rearrangement that resulted in truncation of 70 amino acids from the C-terminal end of RARα was detected (109). The truncated RARα was again found to act as a dominant negative regulator.

3. *Mutations Defining a Possible Silencing Domain*

Transcriptional repression may involve several mechanisms, such as the formation of inactive heterodimers and competition for DNA biding sites. Another mechanism may involve the direct inhibition of gene transcription, that is, the silencing. Thyroid hormone receptor and v-erbA have been shown to repress transcription by silencing (110, 111) in which the DNA binding of these receptors in the absence of ligand inhibits gene transcription independently of position and orientation. Baniahmad *et al.* (111) reported that RARα also exhibits silencer activity. However, the repression effect is cell type dependent. It has also been shown that TR silencing activity can be mediated by TR homodimers (112). Unlike TRs and v-erbA, which display repressor effects in both CV-1 and L cells, RARα shows repressing activity only in L cells, but acts as a weak activator in CV-1 cells. To localize the repression function in the C-terminal part of the RAR, a number of the C-terminal deletion mutants were tested. Deletion of 59 amino acids from the C-terminal end of RARα did not affect its repression activity. However, further deletion of 42 amino acids, which includes the ninth heptad repeat, completely abolished the repression function of the RAR. Thus, the ninth heptad repeat may also play a role in negative transcriptional regulation by retinoid receptors.

IV. Mechanisms of Receptor Action

A. Hetero- and Homodimers: Identification of an RAR Coreceptor

The large diversity of hormone response pathways require a very high specificity for a given signal, but at the same time a certain degree of flexibility generating different responses under different physiological conditions. Recent progress allows for a much better understanding of how the nuclear receptors accomplish this function (103). The existence of different receptor isoforms and the demonstration of protein–protein interactions among members of the receptor superfamily now delineate a complex molecular network that generates the diversity of the hormonal responses.

Over the years, several lines of evidence emerged to point at fundamental mechanistic differences between steroid hormone receptors and the retinoid/thyroid hormone receptor subclass. ER and GR form homodimers on their cognate response elements, which are palindromic hexanucleotides separated by a 3-bp spacer (38). When DNA binding strength and transcriptional activation by ER and TR (113) were compared, the relatively low affinity of TR binding to response elements did not correlate well with the strong activation observed in transient transfection assays. Remarkably, the *in vitro* binding strength of both TR and RAR could be significantly increased when nuclear extracts from a variety of cell lines were added to the receptor proteins, resulting in a larger presumably heteromeric complex (113–117).

The unknown protein, first called TRAP (for TR auxiliary protein),

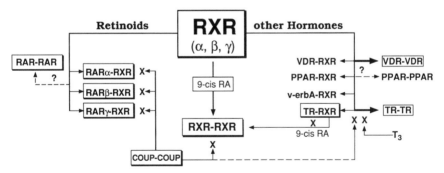

FIG. 9. Interactions among the retinoid/thyroid hormone receptor subfamily. RXRs form heterodimers with RARs, which are activated by retinoids. RXR also forms heterodimers with TRs (including v-erbA), VDR, PPAR, and probably other receptors yet to be determined. In the presence of RXR-specific retinoids like 9-*cis* RA, RXR homodimers that recognize a specific subset of RAREs are formed. The presence of ligands that induce the formation of RXR homodimers can inhibit the formation of certain heterodimers and, for instance, lead to a reduction of T3-responsive genes, whereas RAR-containing heterodimers do not appear to be negatively affected by RXR-specific ligands. COUP receptors form homodimers that bind with high affinity to several RAREs and can repress RAR/RXR heterodimer as well as RXR homodimer activity. Thus, COUP receptors can restrict retinoid responses to a subset of retinoid-responsive genes. COUPs may also antagonize other receptors. VDR homodimers bind to a subset of VDREs. TR homodimers bind only in the absence of T3 and can function as ligand-responsive repressors. Other homodimers (RAR, PPAR) may also form on specific response elements.

did not enhance TR binding to a single TRE/RARE half-site AGGTCA (116), thus excluding formation of a heterodimeric complex in which only one binding partner is in contact with DNA. The salient requirements for the auxiliary protein turned out to be TREpal half-site binding and receptor interaction through a dimerization domain, features characteristic for members of the TR/RAR receptor subfamily. Using different experimental approaches, isoforms of the retinoid X receptors were identified as the auxiliary proteins for both RARs and TRs (17,18,28–32). RAR/RXR heterodimers were active on virtually all known response elements (103). Importantly, however, the heterodimers with RXR bound specifically only to cognate response elements, that is, RAR–RXR did not bind well to TREs, and TR–RXR did not bind to RAREs (31,118). Thus, the specificity of hormone response is largely contributed to by differences in the composition of the heterodimeric complex.

RXR has subsequently been shown to form heterodimers with a multitude of nuclear receptors: VDR and PPAR (29–32,98,119), the v-erbA oncogene (120), and several orphan receptors (our and others' unpublished results). Figure 9 depicts a scheme for the receptor interactions (see the following section).

B. RAR Subtype and Isoform Function

Initially, three different RAR genes were cloned (labeled α, β, γ and generally related to as subtypes) that showed tissue-specific expression patterns during embryogenesis and in adult life (8–13). Closer analysis of the 5' upstream region led to the discovery of additional, functional transcriptional start sites that defined N-terminal isoforms of each individual receptor. So far, three different isoforms of RARα, four different isoforms of RARβ, and seven different isoforms of RARγ have been described (122). These isoforms (identified by numbers behind the Greek index) appear to be regulated by tissue-specific expression as well (123,124). Data from several laboratories now also indicate that isoforms of the three different retinoid X receptor types (RXRα, β, γ) exist (105).

In situ hybridization studies of mouse embryos were conducted to correlate the multitude of receptor isoforms with possible distinct functions in development; they demonstrated specific temporospatial patterns for each individual mouse RAR gene (122,125–127) and for each individual RXR gene (16). This suggested that each individual RAR type and isoform may have a distinct set of target genes within the spatial and temporal framework of eukaryotic development (128).

Further insight into possible selectivity rules for target genes was gained by employing different combinatorial permutations of RAR and RXR heterodimers. In *in vitro* binding and transactivation studies, RAR/RXR heterodimer formation relaxes the spacing rules that assign exclusive preference to DR+5 motifs for RAR (28,48,57,129,130), response element sequences modify affinities mostly dependent on spacing, and RAR isoform transactivationβ levels from the same response element may vary. In addition, surrounding promoter sequences may profoundly affect selectivity and transactivation levels elicited by RAR or RXR subtypes (56). On synthetic response elements displaying variable spacing of the half-site AGGTCA or AGTTCA, transcriptional activation and cooperative binding on DR+2 elements and DR+4 elements could be demonstrated, whereas only weak binding with little transactivation was observed with DR+1 and especially DR+3 elements. Interestingly, RAR heterodimers showed different selectivity for the actual binding site sequence, with a preference for AGTTCA over AGGTCA under those spacing conditions that allowed the strongest interaction (i.e., DR+5 and DR+2). The influence of different RAR isoforms on transactivation responses obtained with natural response elements and promoters was rather subtle, whereas several examples of rather striking differences between subtypes could be demonstrated. For instance, the RARγ subtype, which has previously been credited with inhibitory functions on a natural response element (131), proved to be a strong activator from the CRAB-PII element. The RARα subtype activated the RARβ promoter to higher levels than the isolated response element in front of the TK promoter, whereas the RARβ subtype, including its isoforms, showed the opposite activation preference. Furthermore, all three RXR subtypes could activate only from the RARβ promoter, yet they failed to elicit any response from the isolated response element. Not unexpectedly, both selectivity and redundancy in activation are generated through the combinatorial permutations of RAR subtypes and isoforms in heterodimerization with RXR subtypes. A more rigorous way to define isoforms and subtype roles has been employed through selective RAR isoform and/or subtype knockouts (132,133). Interestingly, knockouts of RAR isoforms yielded no phenotype, indicating an overlap in receptor isoform and/or subtype functions. On the other hand, a tissue-specific physiological role for the RARα subtype could be identified in spermatogenesis (133), whereas RARγ, consistent with earlier *in situ* hybridization studies, could be linked to major functions in the skin (132). More systematic studies with RAR and RXR knockouts have now been carried out and will undoubtedly be the major experimental pathway to reveal important principles of RAR and RXR function in development and homeostasis of living tissues.

C. ALTERNATIVE DIMERIZATION PARTNERS FOR RARs?

Since it has been found that RAR and RXR can form heterodimers in solution in the absence of DNA (28), whereas TR with RAR cannot (94), reports in the literature on heterodimer formation between RAR and TR need to be reassessed with regard to the experimental conditions under which this phenomenon was observed. Setting out from the discovery of palindromic response elements for steroid hormone receptors and, by default of knowledge about natural response element sequences for RARs and TRs, early studies were restricted to the TRE-pal, which does not display the spacing characteristics essential for selective receptor binding and, thus, allows more ambiguous results and interpretations. On direct repeat response elements, RAR/TR heterodimers have not been observed. Moreover, in the presence of both TR and RAR, competition for binding to endogenous liver protein (RXR) has been documented (134). As numerous investigators have observed that rabbit reticulocyte lysate contains a protein similar to RXR, previously observed complexes of *in vitro* synthesized receptors interpreted as TR/RAR heterodimers may in fact contain significant amounts of TR/RXR and RAR/RXR heterodimers, which copurify in ABCD assays and comigrate in the gel shift assay. Similarly, TR/RAR heterodimers observed with high amounts of protein expressed in bacteria or in Sf9 cells may in fact be heterodimeric complexes between receptor and host-cell proteins (134). According to a newly proposed model on RXR–RAR interactions, cooperative heterodimeric receptor interactions with correct spacing selection require the interaction of the DR-box on RAR with the D-box on RXR (48). It is only on direct repeat elements that those two boxes are thought to be juxtaposed, in a configuration favorable for interaction. On a palindromic element like TREpal, two D-boxes of the receptors would be juxtaposed, thus allowing weak, spacing-nonselective or receptor-unspecific protein–protein interactions that require high receptor concentrations to occur. These are in fact the rather artificial experimental conditions under which strong dimeric complexes between TR and RAR have been demonstrated. Similarly, the ability of C-terminal half-molecules of TR and RAR to repress DNA binding and transcriptional activation of full-length TR or RAR (97,135,136) is likely to not result from direct interactions between TR and RAR, but is rather due to the ability of the truncated receptors to interact with RXR (137) present in essentially all cells. However, it should also be mentioned that heterodimeric complexes among receptors could possibly be catalyzed by specific DNA sequences (still to be defined), as observed for TR homodimers (112).

D. RXRs MEDIATE DISTINCT RESPONSES BY HOMODIMER FORMATION

From the aforementioned observations, it became apparent that RXRs have a very central role. They regulate RAR DNA binding and transcriptional activation, as well as the activities of several other hormone receptors that respond to structurally unrelated ligands (Fig. 9). Therefore, it was of interest to investigate what effects RXR-specific ligands would have on RXR. As pointed out before, RXR alone does not bind to DNA effectively and requires heterodimerization with other receptors. When we investigated RXR–DNA interaction, we made the surprising observation that in the presence of 9-*cis* RA, RXR no longer required heterodimerization for DNA binding. 9-*cis* RA, previously shown to bind with high affinity to RXRs (138,139), induced the formation of RXR homodimers that bound with high affinity to DNA (33). In contrast, RAR DNA binding or homodimer formation was not induced by 9-*cis* RA that can also bind RARs (138,139). RXR homodimers were found to bind to and activate specific response elements such as the ApoAI-RARE and CRBPII-RARE, but not the bRARE and the rat CRBPI-RARE (for response elements see Fig. 10) (33). Although the ApoAI-RARE is also specifically bound and activated by RAR/RXR heterodimers, the CRABPII-RARE is activated only by RXR homodimers (33,34). These date indicated that RXR-specific ligands, like 9-*cis* RA, lead to the formation of RXR homodimers that can bind and activate a subset of RAREs and thereby induce a distinct retinoid response pathway. Even though 9-*cis* RA is not selective for RXR because it also binds to and activates RARs, we have been able to define retinoids that are RXR selective, in that they primarily activate RXR homodimers but not RXR-containing heterodimers (34). Such retinoids can serve to regulate selectivity RXR homodimer responsive genes and may, as such, represent valuable tools for examining the roles of RXR in development and disease.

Thus RXR can mediate two distinct retinoid signaling pathways triggered by all-*trans* RA or 9-*cis* RA through heterodimer or homodimer formation as illustrated in Fig. 10. VDR was also found to mediate two signaling pathways via heterodimers and homodimers (119).

The fact that RXR-specific ligands can induce RXR homodimer formation raises the possibility for competition between RXR homodimer and heterodimer formation. When Lehmann *et al.* investigated the

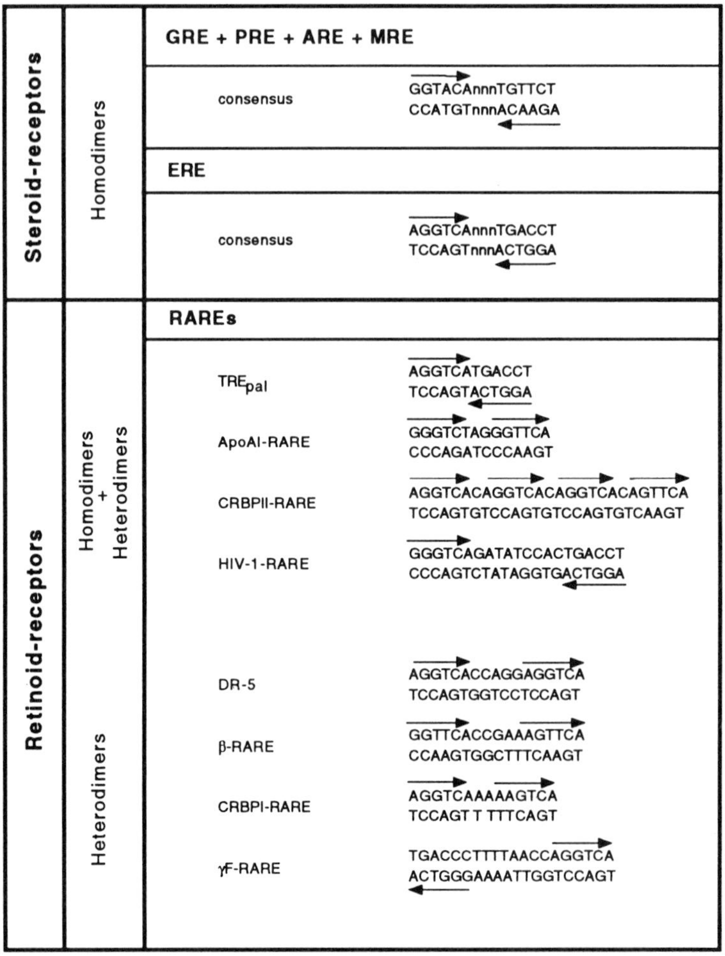

		GRE + PRE + ARE + MRE	
Steroid-receptors	**Homodimers**	consensus	GGTACAnnnTGTTCT CCATGTnnnACAAGA
		ERE	
		consensus	AGGTCAnnnTGACCT TCCAGTnnnACTGGA
Retinoid-receptors	**Homodimers + Heterodimers**	**RAREs**	
		TRE$_{pal}$	AGGTCATGACCT TCCAGTACTGGA
		ApoAI-RARE	GGGTCTAGGGTTCA CCCAGATCCCAAGT
		CRBPII-RARE	AGGTCACAGGTCACAGGTCACAGTTCA TCCAGTGTCCAGTGTCCAGTGTCAAGT
		HIV-1-RARE	GGGTCAGATATCCACTGACCT CCCAGTCTATAGGTGACTGGA
	Heterodimers	DR-5	AGGTCACCAGGAGGTCA TCCAGTGGTCCTCCAGT
		β-RARE	GGTTCACCGAAAGTTCA CCAAGTGGCTTTCAAGT
		CRBPI-RARE	AGGTCAAAAAGTCA TCCAGT T TTTCAGT
		γF-RARE	TGACCCTTTTAACCAGGTCA ACTGGGAAAATTGGTCCAGT

FIG. 10. Response elements. Retinoid receptors interact as homo- or heterodimers to a large variety of response elements. A selection of retinoic acid-responsive elements (RAREs) is compared to the typical glucocorticoid (GRE)- and estrogen (ERE)-responsive elements.

effect of 9-*cis* RA on TR/RXR activity (137) it indeed was observed that T3 induction of TRE-containing genes could be strongly repressed by 9-*cis* RA. In addition, repression was observed with RXR-selective retinoids, supporting a mechanism in which sequestering of RXR molecules from TR/RXR heterodimers into RXR homodimers leads to repression of the T3 response. Similarly, the vitamin D3 response can be inhibited by 9-*cis* RA (140). Thus, a ligand-induced squelching mechanism exists that can control the availability of RXR molecules for heterodimerization and thereby allow hormonal cross-talk (see Fig. 9). However, not all RXR-containing heterodimers are negatively affected by RXR-specific retinoids. As mentioned earlier, RAR/RXR heterodimers are not inhibited by 9-*cis* RA, but in fact are (in most cases) superactivated. The activity of PPAR/RXR heterodimers also appears to be increased when both PPAR and RXR-specific ligands are present (30). Thus, 9-*cis* RA and other retinoids that lead to RXR homodimer formation inhibit some heterodimers and enhance others.

E. RXR-INDEPENDENT PATHWAYS THROUGH RAR HOMODIMERS?

RXR-independent pathways for activation (119) and repression (112) have been demonstrated, but an RXR-independent RAR function is another important mechanistic principle yet to be proven. Because RAR expression and RXR expression overlap only partially (as judged from *in situ* hybridization studies), certain tissues seem to have an excess expression of RAR, for example, the developing limb (16,125, 141). It is in those tissues that RAR homodimers may have biological functions, in particular RARβ, which is highly inducible for RA (27). However, no structural elements seem to exist in the DNA binding domain of RAR that favor cooperative RAR homodimer binding (48, 130). In addition, currently no response element has been identified that shows high-affinity binding to RAR homodimers (130). However, as with TRs (112), special DNA sequences may exist that catalyze efficient RAR homodimer binding. Alternatively, RAR concentrations may be sufficiently high in some cell types to allow effective RAR homodimer interaction with DR-5-type RAREs.

F. RETINOID RECEPTOR INTERACTION WITH COUP

Although the various heterodimers and ligand-induced RXR homodimers allow for a large degree of diversity and specificity of the retinoid responses, the mechanism by which some hormonal responses can be restricted in certain cell types is still unclear. One of the orphan receptors, COUP-TF, belongs to the TR/RAR subfamily (see Fig. 2) (142,143), and together with its relatives EAR-2 and ARP-1 (144,145), it binds to a variety of response elements composed of direct repeats or palindromes with different-length spacers that initially have been identified as response elements for retinoid, thyroid hormone, vitamin D3, estrogen, and orphan receptors (146–151). In addition, COUP-TF also binds to the upstream region of chicken ovalbumin gene (146, 152), rat insulin II gene (153), and the human immunodeficiency virus (HIV) long terminal repeat (154).

Several laboratories (146,148,150) observed that COUP receptors could dramatically inhibit retinoid receptor activities on several response elements that are activated by RAR/RXR heterodimers or RXR homodimers. When they investigated COUP DNA binding it was observed that COUP receptors bound strongly as homodimers to these response elements, but did not form heterodimers with RXR (146). Elements that could be bound effectively included the palindromic TRE and the CRBPI-RARE, as well as the RXR homodimer binding ApoAI- and CRBPII-RAREs. The RARβ2-RARE was only bound weakly. All response elements that bound COUP homodimers effectively were also strongly inhibited in transient transfection assays by COUP (146). These data and similar results reported by others (146,148,149) suggest that the COUP receptors are repressors that can restrict RA signaling to certain genes (see Fig. 9 for a general scheme of receptor interactions). COUP receptors appear to function as receptors by binding response elements as homodimers (146,150); whether the weak heterodimer formed between COUP and RXR reported by others (148, 149) has biological significance still needs to be demonstrated. Interestingly, the COUP-TF specific binding site in the ovalbumin promoter also functions as an RARE (146). Thus, COUP orphan receptors may play an important role in the regulation of cell or tissue-specific restriction of RA-sensitive programs during development and in the adult, a role consistent with that of the *Drosophila* COUP homolog *seven-up* (43).

G. Retinoid Receptor Interaction with AP-1

The products of the two proto-oncogenes c-Fos and c-Jun and several related proteins, induced by extracellular stimuli, form a complex in the nucleus called activator protein 1 (AP-1) that binds specifically to a sequence motif referred to as AP-1 binding site, or TPA-responsive element (TRE). The c-fos proto-oncogene is the cellular homolog of the transforming gene of FBJ osteosarcoma virus (155,156).

Regulation of the expression of the c-fos proto-oncogene has for a long time been a subject of investigations, and a wide variety of agents have been discovered that influence cell growth and differentiation by inducing expression of c-Fos (157). The c-Fos/c-Jun heterodimeric complex interacts with the AP-1 binding site (158), whereas c-Fos alone does not bind to this sequence. c-Jun, however, can bind to this site as a homodimer with lower affinity. The presence of c-Fos enhances the binding of the c-Fos/c-Jun complex, primarily by stabilizing the protein/DNA interaction (159). Dimerization of c-Jun and c-Fos occurs through the leucine zipper motif (160,161).

The AP-1 binding site was first identified in the enhancer regions of simian virus 40, human metallothionein IIA (hMTIIA), and several phorbol ester-responsive genes (162,163). Similar nucleotide sequence motifs are present in negative and positive regulatory regions of several genes, such as the overlapping AP-1/glucocorticoid response element site in the mouse proliferin promoter gene (164), the AP-1 site embedded within retinoic acid/vitamin D response element of the human osteocalcin gene, (165), or the promoter region of the collagenase gene (162).

The collagenase promoter AP-1 site has served as a major model system to delineate nuclear receptor interaction with AP-1 (35–37). The collagenase promoter is induced by the inflammatory mediators interleukin-1 (IL-1) and tumor necrosis factor-a (TNFα) (162). Collagenase and stromelysin are members of a larger family of metalloproteases that are produced and secreted by synoviocytes and other tissues and that may play an important role in joint destruction in rheumatoid arthritis (166). Retinoic acid and other retinoids (167) are considered antiarthritic agents (168) because they can inhibit the production of collagenase by synovial cells. On the other hand, retinoids are also antineoplastic agents (5) and their anticancer activity could be a result of their ability to inhibit AP-1-induced gene activation. The mechanisms by which RA inhibits the AP-1-mediated induction of collagenase was therefore of general interest.

It turned out that RA inhibited the collagenase promoter through the AP-1 site (35,37,169). However, RARs and/or RXRs do not bind the AP-1. In addition, AP-1, when activated by TPA, can also inhibit RAR function (35). This mutual antagonism appears to result from direct protein–protein interaction between RAR and AP-1 components (35,37). This regulatory mechanism, by which members of two different families of transcription factors can control each other's activity, is likely to represent a major signaling switch (170). c-Jun and c-Fos induce cell proliferation, whereas nuclear receptors induce cell differentiation. Because c-Jun and c-Fos mediate the signals of peptide hormones, oncogenes, and tumor promoters, the retinoid receptor-mediated anti-AP-1 activities may represent a major pathway by which retinoids exert their anticancer activities.

V. Response Elements

A. RAREs

Nuclear retinoid receptors act as dimeric transcription factors to modulate expression of target genes by binding to specific DNA sequences known as hormone response elements (HREs) or, more specifically, retinoic acid or retinoid response elements (RARE). So far most of the RAREs identified in the promoter of natural RA target genes consist of an imperfect 6-bp direct repeat (DR) (related to a consensus motif) spaced by either 1, 2, or 5 bp (DR2 or DR5) (24,25). Genes containing such RAREs include those for RARα, RARβ, RARγ, CRBPI, CRABPII, the alcohol dehydrogenase 3, and the complement factor H genes (26,27,131,172–174). Other RAREs have been identified by *in vitro* binding and transient transfection assays in the promoter of the phosphoenol pyruvate carboxykinase, oxytocin, apolipoprotein A1, major histocompatability, osteocalcin, and CRBPII genes (165,175,176), and in the LTR of the HIV (1977).

Comparison of these naturally occurring RA response elements, combined with results obtained with synthetic RAREs, indicates that these cis-active sequences can be considered as a repetition of the consensus sequence 5′-PuG(GT)CA-3′, called a "half-site." However, the spacing, orientation, and precise sequence of these half-sites vary considerably among the described RAREs (Fig. 10). In addition to these specific RAREs, RARs can bind and activate some estrogen, vitamine D3, and T3-responsive elements (TREs); the rat growth hormone gene responds to both T3 and RA via a common HRE. A synthetic response element derived from this promoter, the TREpal, mediates transactivation by T2, RA, and estrogen nuclear receptors (178–180). Additional natural RAREs have been identified that reveal novel symmetries of the half-sites, for instance, the RARE from the g crystalline F gene (181) is an inverted palindrome with an 8-bp spacer, and an RARE found in the long terminal repeat region of HIV-1 is a palindrome with a 9-bp spacer (177) (Fig. 10).

Furthermore, the finding that RXR interacts with TR, VDR, and RAR increases (at least theoretically) dramatically the number and diversity of potential retinoid response elements. The fact that T3, retinoic acid, estrogen, and vitamin D3 activate overlapping sets of genes via closely related response elements raises the problem of the molecular basis of specificity of the hormonal responses.

It has been shown that the binding specificity can be determined through the number of nucleotides separating the half-sites. Direct repeat of AGGTCA-like half-sites spaced by either three, four, or five nucleotides would function as optimal HRE for the VDR, TRs, and RARs, respectively. Mutations that alter the distance between these hexamers can switch receptor specificity (24).

The relative orientation of the half-sites within HRE is also critical in the recognition specificity. Direct repeat, palindromic, and inverted palindromic organization of the TCAGGTCA half-site spaced by 3 bp confers specific response to retinoic acid, estrogen, and T3 receptors, respectively (25).

These observations provide a model involving both orientation and spacing of the half-sites in the determination of the selective but overlapping effects of T3, RA, vitamin D3, and estrogen receptors on natural target genes. Nevertheless, many genes exhibit some permissive properties that do follow this model. The human osteocalcin gene is activated by vitamin D3 and RA through the same 3-bp-spaced inverted repeat (165). The palindromic thyroid hormone response element TREpal, γF-RARE, and the HIV-RARE are examples for different classes of response elements that are not covered by the simple 3, 4, 5 rule.

B. RXR-RESPONSIVE ELEMENTS

Gel mobility-shift assays performed with bacterially expressed RXRα and synthetic DRs linked to a reporter gene first indicated that half-sites spaced by one nucleotide (DR1) were preferentially recognized by RXR homodimers or higher-order RXR complexes (33,34, 102,176). This property may be restricted to RXRα and β, as the RXRγ was shown to be unable to transactivate any of the DR1 tested in *S. cerevisiae*. However, a naturally occurring DR1 has been identified in the promoter of the rat CRBPII gene (Fig. 10). This complex binding site consists of four direct repeats spaced by one cytosine residue. It is also active with heterologous promoter and shows selective responsiveness *in vivo* to the RXRα but not to the RARs (33,34,176). However, it is not clear whether RA responsiveness occurs in its natural promoter context. The ApoAI response element is also recognized by RXR homodimers, but this DR2 is also recognized by RARα/RXRα heterodimers and is in addition sensitive to a negative regulation by COUP proteins (146,149). A response element that only binds RXR homodimers has not yet been described. RXR homodimer binding—at least in the well-studied case of RXRα—is always dependent on 9-*cis* RA. RXR-selective ligands like those described by Lehmann *et al.* (34) will be useful to study which genes are regulated *in vivo* by RXR homodimers.

In conclusion, studies concerning retinoid response elements have led to the identification of a large set of RAREs that overlaps with response elements regulated by other nuclear receptors. However, so far most of the known RAREs have been characterized by their ability to confer retinoid responses to a heterologous promoter. Very few studies have been carried in the native promoter context. Thus, it is still unclear whether all characterized RAREs behave similarly *in vivo*. Furthermore, the influence of the RXR or RAR isoform (α, β, or γ) on heterodimer formation and DNA binding specificity remains to be determined.

The most striking feature of the RAREs is their diversity (Fig. 10). In fact a striking difference between the steroid hormone response elements and retinoid response elements becomes apparent. One possibility to explain this difference is that the receptors that interact with them function differently. For instance, the distinct, large, amino-terminal domains of GR, PR, AR, and MR may allow differential recognition of the same binding sites in the context of specific promoters, or may allow differential activation of various promoters containing the same response element. On the other hand, in the case of the retinoid receptors a large variety of recognition sequences are used that favor activation by various heterodimers composed of various receptor subtypes, or homodimers (42) allowing for a broad diversity of biological responses.

VI. SELECTIVE RETINOIDS

Although retinoids are promising therapeutic agents for a variety of cancers, including acute myelocytic leukemia and squamous-cell carcinoma of head and neck, and represent some of the most effective drugs for the treatment of proliferative skin diseases (4) and severe acne, their therapeutic potential has been limited by their toxicity. The most common negative effects are angular cheilitis, pruritic rash, headache, dizziness and facial dermatitis (182), and congenital malformations. Hence an important area for retinoid research has been the search for effective, but less toxic, retinoids. Identification of retinoids with selective biological activities and the understanding of their mechanism of action would help to develop such retinoids. The pleiotropic effects of retinoids are mediated through the RARs and RXRs. Although these receptors are differentially expressed in different cell types and developmental stages, different sets of genes respond to different receptors, and different receptors are therefore likely to modulate different biological responses. The development of synthetic retinoids that are selective for retinoid receptors would thus help not only to develop clinically effective retinoids with higher therapeutic indices, but also further elucidate the roles of individual receptors in retinoid-dependent biological functions.

A. RAR-SELECTIVE LIGANDS

The possibility of the development of receptor-selective retinoids that activate transcription via one of the RAR subtypes has been indicated by several studies (183–185). Graupner et al. (184) and Lehmann et al. (185) first developed an assay system in which retinoids were evaluated for their transcriptional activation capabilities for the different RAR subtypes. In their studies, conformationally restricted retinoids, in which selected bonds of the polyolefinic chain of the retinoid skeleton were incorporated into aromatic ring systems, were used. Because these retinoids have fewer degrees of freedom they can provide more useful information about the geometry of the ligand binding site than the conformationally more flexible all-trans-RA (Fig. 11). In general, RARα appeared to be far more sensitive to variations in the structural geometry of the retinoids in the 4-substituted benzoic acid and 6-substituted 2-naphthalenecarboxylic acid classes than RARβ and RARγ. Modification at the 4-position of the retinoid skeleton appeared to contribute to this selectivity. Several of the retinoids analyzed, R15, R19, R20, and R22, showed striking differences in their activation capacity of individual RARs, in that they were efficient activators of RARβ and RARγ but poor activators or nonactivators of the RARα (Fig. 1). The RAR β/γ selectivity of these retinoids may be related to the higher degree of sequence similarity between the ligand binding domain of RARβ and RARγ compared to the ligand binding domain of RARα (Fig. 2) (8–13). Consistent with this observation, R20, 2-naphthalenecarboxylic acid, has been reported to bind more efficiently to RARβ than to RARα (185). Graupner and colleagues (184) analyzed the activities of structurally closely related 6'-substituted naphthalene-2-carboxylic acid derivatives in detail and showed that this class of 13-cis RA analogs also exhibited RAR subtype selectivity. Three of these retinoids (Nos. 1, 2, and 3 in Fig. 12) induced strong transcriptional activation of RARγ, whereas no activation of RARα was observed. The selectivity of RARβ/γ over RARα of the retinoid 1 has also been noted in in vitro binding assays (188). In contrast, two retinobenzoic acid derivatives,

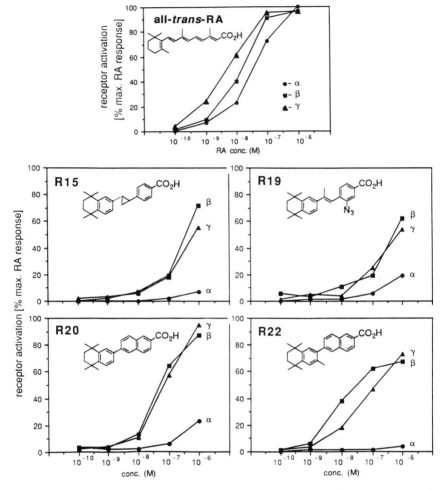

Fig. 11. Structure and transcriptional activation characteristics of R15, R19, R20, and R22. The receptor activation patterns were evaluated with the ER–RAR hybrid receptors (α, β, γ). Receptor activation of 100% represents the activity measured in the presence of 10^{-6} M all-*trans* RA.

Am80 and Am580, representatives of a distinct class of all-*trans* RA analogs (184) (Fig. 12), showed completely different receptor subtype selectivity in that they preferentially bind to RARα and also induce transcriptional activation of RARα (184,187,188).

Bernard and colleagues reported retinoids that are highly selective for either RARβ or RARγ (186). Surprisingly, retinoids exhibiting

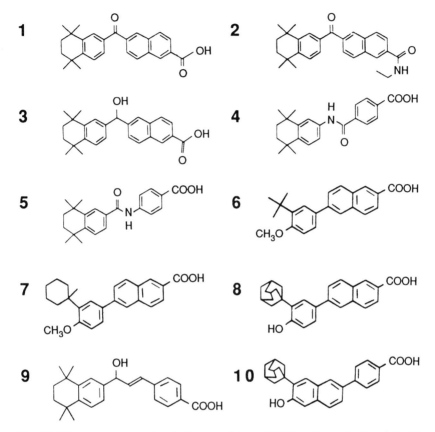

FIG. 12. The structure and chemical nomenclature of RAR subtype-selective retinoids.
1. 6-[(5,5,8,8-tetramethyl-5,6,7,8-tetrahydro-2-naphthyl)carbonyl]2-naphthalene carboxylic acid. **2.** N-ethyl-6-[(5,5,8,8-tetramethyl-5,6,7,8-tetrahydro-2-naphthyl)carbonyl]2-naphthalene carboxamide. **3.** 6-[(5,5,8,8-tetramethyl-5,6,7,8-tetrahydro-2-naphthyl)hydroxymethyl]2-naphthalene carboxylic acid. **4.** 4-(5,6,7,8-tetrahydro-5,5,8,8-tetramethyl-2-naphthalenylcarbamoyl)benzoic acid (Am80). **5.** 4-(5,6,7,8-tetrahydro-5,5,8,8-tetramethyl-2-naphthamino)benzoic acid (Am 580). **6.** 6-(3-tert-butyl-4-methoxyphenyl)-2-naphthoic acid. **7.** 6-(3-(1-methoxycyclohexyl)-4-methoxyphenyl)-2-naphthoic acid. **8.** 6-(3-(1-adamantyl)-4-hydroxyphenyl)-2-naphthoic acid. **9.** 4-(6-hydroxy-7-(1-adamantyl)-2-naphthyl)benzoic acid. **10.** (E)-4-(1-hydroxy-7-(5,6,7,8-tetrahydro-5,5,8,8-tetramethyl)-2-propenyl)benzoic acid.

RARβ selectivity and RARγ selectivity were derived from a common parental compounds, indicating that a simple chemical modification can be sufficient to differentiate between the ligand binding pockets of these two RAR subtypes. Taken together, the accumulated experimental results demonstrate that retinoids with unique profiles of RAR subtype selectivity can be defined. These retinoids now need to be tested for their impact on cellular programs and for therapeutic applications.

B. RXR-SELECTIVE LIGANDS

The central role for RXR in regulating several distinct hormonal response pathways has been described earlier (see Fig. 9). RXR homodimers have different response element specificities than the RAR/RXR heterodimers (33). Thus, compounds that activate only one of the RXR pathways are useful in separating the complexities of the retinoid response. A series of retinoids was reported that selectively activate RXR homodimers but not RAR/RXR heterodimers (34). Conformational analyses indicated that the spatial orientations of the lipophilic head and carboxyl termini of these retinoids were similar to those of 9-*cis* RA and thus activity could be related to the length and volume of the substituent group, linking the tetrahydronaphthalene and phenyl ring systems (Fig. 13). Selectivity of these retinoids for RXR homodimers was shown using cotransfection assays with several different reporter constructs. Similar to 9-*cis* RA, the RXR-selective retinoids SR11217 and SR11237 were strong activators of the CRBPII-RARE (for response elements see Fig. 10), which is activated only by RXR homodimers. However, in contrast to 9-*cis* RA, they did not induce the rat CRBPI-RARE, which is activated only by the RAR/RXR heterodimers (34). The ApoAI-RARE was effectively activated by both RAR/RXR heterodimers and RXR homodimers in the presence of 9-*cis* RA, however, the response element was activated only by homodimers when SR11217 and SR11237 were present. These results demonstrate

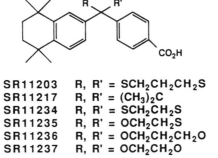

SR11203	R, R' = $SCH_2CH_2CH_2S$
SR11217	R, R' = $(CH_3)_2C$
SR11234	R, R' = SCH_2CH_2S
SR11235	R, R' = OCH_2CH_2S
SR11236	R, R' = $OCH_2CH_2CH_2O$
SR11237	R, R' = OCH_2CH_2O

FIG. 13. Structure of RXR homodimer-selective retinoids.

that specific retinoids can allow the selective activation of RXR homodimer-responsive pathways without inducing RAR-dependent response pathways. Thus this new class of RXR-specific retinoids may provide a more restricted spectrum of physiological responses than previously available retinoids.

VII. RETINOID ANTAGONISTS

Besides their importance for analyzing the physiological roles of retinoids, retinoid antagonists may have significant practical potential because RA has been shown to enhance replication of several viruses such as human immunodeficiency virus type 1 (HIV-1) (190,191) and human cytomegalovirus (CMV) (192). The retinoid activation of such viruses appears to be controlled by the long terminal repeat region (LTR) of the viral genome (177). Indeed, retinoic acid response elements that allow transcriptional activation by retinoid receptors have been identified in the LTR of HIV-1 (177), CMV (192), and human hepatitis B virus. In the case of the HIV-1-RARE, both RXRα homodimer and RARα/RXRα heterodimer are good activators of the response elements. Therefore, retinoid antagonists could provide a way of repressing viral response elements and thereby inhibit the RA-dependent transcription and replication of the viruses induced *in vivo*.

In general, the search for retinoid antagonists has been of very limited success, thus little is known about what type of retinoid modifications can result in antagonist activity. Lee and collaborators conducted a structure–activity analysis of retinoids showing low transcriptional activation activity to establish skeletal features that could be modified to enhance receptor antagonism (177). Analyses of such compounds led to an antagonist that proved to be very effective at inhibiting the ability of all-*trans* RA to induce the HIV-1-RARE. Clear inhibition by the antagonist was even observed when all-*trans* RA was used at 10^{-7} M. SR11335 also functioned as a potent inhibitor of 9-*cis* RA, however, inhibition of the heterodimers was more efficient than inhibition of RXR homodimers (177).

Three other structurally unrelated retinoid antagonists have also been reported. The retinoid antagonist activity of 4-[3-(4-dimantyl)-4-methoxybenzoamido]benzoic acid and 4-[3-(4-dimantyl)-4-methoxybenzoxy]benzoic acid has been tested by measuring inhibition of retinoid-induced differentiation of human promyelocytic leukemia HL-60 cells (193). Interestingly, another retinoid antagonist, Ro

41-5253, showed selective RARα antagonist activity (194). This antagonist counteracted the all-*trans* RA effects on HL-60 cell differentiation and on B-lymphocyte polyclonal activation, indicating that RARα is involved in these biological programs. Further development of antagonists with receptor selectivity would be useful to dissect the diverse retinoid-induced effects on biological systems.

VIII. RARs and RXRs and Their Relation to Disease and Therapy

A. Cancer and Chemoprevention

Despite the advances made in the last few decades in the treatment of certain forms of cancer, total cancer mortality has not declined but has actually increased. These findings emphasize the need for increased research efforts in the area of cancer causation and prevention. Retinoids, although recognized for more than 50 years as a family of molecules that have profound impact on many biological functions, have only more recently been applied in clinical oncology. Michael Sporn (195) originally introduced the term cancer prevention with reference to studies utilizing retinoids. Now several retinoids and their precursors are known to inhibit the carcinogenic process, and interestingly many of them are natural products present in various foods.

Direct morphological evidence connecting vitamin A with carcinogenesis came from the observation that identical histological changes occurred in prostate tissue after exposure to the carcinogen 3-methylcholanthrene or vitamin A depletion of the culture medium (196). Later it could been shown that vitamin A has a preventive effect on the induction of experimentally induced precancerous conditions like benign epithelial tumors and both metaplasia and carcinoma in animals (197–201). The central role of vitamin A and its derivatives in the growth and differentiation of normal epithelial tissue and its preventive effects further raised interest in the natural role of this vitamin and so studies were undertaken. But the serious side effects, known as hypervitaminosis A syndrome (202), led to the systematic search for synthetic, less toxic derivatives. Applications of such compounds in precancerous stages have mainly been focused on the bladder, mammary gland, and skin. Regarding the latter, Hong *et al.* reported an encouraging positive result using 13-*cis* retinoic acid to inhibit the development of secondary primary tumors of the head and the neck (203). Also of considerable interest are ongoing or planned

TABLE II
CANCER CHEMOPREVENTION: RECENT OR PROPOSED CLINICAL TRIALS[a]

Site	Agents	End point
Skin	Beta-carotene, retinoids	Cancer
Oral	Beta-carotene, retinoids	Dysplasia, micronuclei
Head and neck	13-*Cis* retinoic acid	Second primary
Esophagus	Vitamins, minerals	Dysplasia, micronuclei
Lung	Beta-carotene, retinoids, other vitamins	Dysplasia, cancer
Breast	Retinoid, tamoxifen, low fat	Cancer
Colon	Beta-carotene, vitamins, bran, low fat, Ca^{2+}	Polyps
Cervix	Retinoids, folic acid	Dysplasia
All sites	Beta-carotene	Cancer incidence

[a]From Tallman and Wiernik (198).

studies to use retinoids for prevention of breast cancer, preferentially with the retinoid 4-HPR [*N*-(4-hydroxyphenyl) retinamide] in combination with tamoxifen (204). The same retinoid has already proven its value for inhibition of prostatic carcinogenesis in animal models. The most significant current directions for clinical trials of retinoids are listed in Table II (198).

B. RETINOIDS IN THERAPY

For several decades retinoids have been used in the treatment of a wide array of dermatological disorders and neoplasms with considerable success (2,4,5). Although the mechanistic principles of pharmacological action still remain unknown, several concepts to distinguish retinoid therapy from general cytostatic therapy have been developed and tested in preclinical studies. They include direct induction of cell differentiation (1) with or without other differentiating agents (e.g., cytokines, growth factors) or induction of apoptosis (differentiation to programmed cell death) (5,208). However, clinical investigations have dampened high expectations considerably; so far the best results of retinoid therapy were achieved only in a regimen combining retinoids with other differentiating or cytotoxic agents. Besides retinal and retinoic acid, isotretinoin (13-*cis* retinoic acid, Ro 4-3780, Accutane) and etretinate (Ro 10-9359, Tigason) have been used. More recently, the retinoids 9-*cis* retinoic acid and *N*-(4-hydroxyphenyl)retinamide were added to this list. Although these latter compounds are quite promising agents in preclinical studies, both are less well documented in clinical trials.

The severe toxic side effects that have been observed under systemic retinoid treatment mimic the hypervitaminosis A syndrome. The main symptoms are cheilitis and CNS symptoms (e.g., headache, lethargy, visual disturbances, dizziness). Furthermore, ocular effects (particularly dryness of the eyes and blepharoconjunctivitis) have been encountered. All of these side effects, in particular the CNS toxicity, are retinoid dose dependent. More importantly, retinoids are very powerful teratogens and present unacceptable risks during pregnancy.

Therefore the most convincing results with retinoids have been documented in the field of dermatological disorders, where topical application can circumvent the toxic side effects seen with systemic applications. As mentioned earlier, the therapeutic effect of vitamin A for a variety of dermatoses generated interest 50 years ago. Follicular keratoses, as seen in vitamin A deficiency, provided a conceptual link to previously treatment-resistant skin disorders like keratosis follicularis and pityriasis rubra pilaris. Oral vitamin A addition was a common therapeutical practice. Only since the last decade have synthetic retinoids like Accutane and Tigason been introduced into clinical practice and now show remarkable effects. In fact, in benign dermatoses like psoriasis, cystic acne, or cutaneous disorders of keratinization, synthetic retinoids are considered the most effective treatment. Etretinate alone or in combination with other therapies for psoriasis may lead to complete remission, but continuous therapy is required. Unlike etretinate in psoriasis, isotretinoin in severe cystic acne can normally be discontinued after 4–5 months of successful therapy. Also in Darier's disease (206) and pityriasis rubra pilaris (207), both cutaneous disorders of keratinization, the synthetic retinoids must be considered as the most effective treatments.

Besides the dermatological disorders, retinoids have important impacts in specific cancer therapies. The most dramatic antitumor effects of retinoids are observed with acute promyelocytic leukemia (APL). In APL cancer, which is a rare form of acute myeloid leukemia (AML), retinoid therapy is associated with a high complete remission rate (208,209). However, the observed sensitivity of leukemic cells to retinoid treatment seems to be unique to APL cells. AML cells, excluding APL, respond only partially to retinoid treatment in combination with cytokines and other differentiating agents (214). Retinoic acid inhibits growth of the human tumor cells listed in Table III (212–230).

In other hematopoietic malignancies like myelodysplastic syndrome (230), juvenile chronic myelogenous leukemia (231), or Sézary syn-

TABLE III
GROWTH INHIBITION OF HUMAN TUMOR CELLS
BY RETINOIC ACID[a]

Tumor cell	References
APL (acute promyelocytic leukemia)	(211)
AML (acute myeloid leukemia)	(212–214)
Breast carcinoma	(215–219)
Chondrosarcoma	(220)
Germ cells	(221, 222)
Lung cancer (non-small-cell)	(223)
Lung cancer (variant small-cell)	(224)
Melanoma	(217)
Neuroblastoma	(225, 226)
Osteosarcoma	(220)
Peripheral neuroectodermal tumor	(226)
Prostate	(227)
Squamous-cell carcinoma	(228, 229)

[a]Modified after Smith et al. (5).

drome (232), retinoid therapy resulted only in modest improvement. Retinoid cancer therapy of squamous-cell carcinomas (SCC) of the upper aerodigestive tract epithelium is encouraging. The antiproliferative and differentiative effects of retinoids on SCC cell lines from human head and neck carcinomas could be demonstrated (228). Hong et al. showed a clear decrease compared with therapy with placebo in second epithelial tumors of the neck and head when using isotretinoin (203). Non-small-cell lung cancer (NSCLC) cell lines responded with growth inhibition to retinoid treatment, whereas small-cell lung cancer (SCLC) cell lines showed only minimal inhibition of clonal proliferation (223). However, clinical studies of lung cancer treatment with retinoids have so far been disappointing (233). In a clinical trial of isotretinoin in combination with interferon alpha-2a on skin squamous carcinoma, a high overall response rate was observed (234). Retinoids induced inhibition of proliferation in malignant human melanoma cell lines (217) and also inhibition of melanoma metastasis in athymic mice (235). A similar situation was observed for breast carcinomas. The growth of selected breast cancer cell lines is inhibited be retinoids (218), and etretinate has been shown to be able to prevent growth of xenotransplanted breast carcinoma cells in athymic mice (219). The effectiveness of these preclinical observations still needs to be confirmed by clinical trials. Retinoid precursors like beta-carotene and other carotenes appear to act on more than one step in the carcinogenic process; as antioxidants and by altering gene expression.

Thus, today retinoic acid and other retinoids hold great promise for "rational therapy" based on the identification of a molecular target. However, the newest generation of receptor- and response pathway-selective retinoids has not yet been introduced into the clinic. Particularly promising examples are AP-1-selective retinoids (236), as they can interfere with the chronic inflammation process through their antagonism to AP-1-dependent collagenase and stromelysin gene activation and antagonize cell proliferation. Other examples are the RXR-selective compounds, which can be expected to have fewer side effects (34). The natural retinoids, all-*trans* and 9-*cis* RA (but also 13-*cis* RA), can be expected to have the highest levels of undesirable side effects. Therefore they are likely to yield good results with acceptable side effects only in a very limited number of situations.

C. Acute Promyelocytic Leukemia: An Example

In the case of APL, a direct link between a chromosomal transloca-tion, t(15;17)(q22;q11-22), and its unique response to retinoic acid *in vitro* and *in vivo* has been established (237). The cloning of the trans-location breakpoint demonstrated the underlying molecular defect: a disruption of the RARα gene on chromosome 17 and a fusion with another previously unknown gene (PML) on chromosome 15 (238). This translocation disrupts intron 2 of the RARα gene and either in-tron 3 or intron 6 of the PML gene. In all cases the PML/RARα fusion protein contains the functional domains of both proteins, the DNA binding and RA binding domain of RARα and the DNA binding and dimerization domain of PML.

PML/RAR fusion protein may act by blocking differentiation, di-rectly or indirectly, or by interfering in a trans-dominant manner with the normal function of PML, of RARα, or of both. At the molecular level, several mechanisms could explain a dominant effect over the normal RARs and PML proteins. Unlike RARα, which requires RXRs for dimerization and function, PML/RAR can easily form a homodimer that binds strongly to RAREs. PML/RAR also appears to heterodimer-ize with normal RXR and PML, potentially disrupting the equilibrium of available RXR receptors. Furthermore, PML/RAR could act domi-nantly to delocalize PML from its natural nuclear position, and this process may be reversed by retinoic acid. Complete remission of APL is achieved by all-*trans* RA within a few weeks. RA treatment of APL may induce differentiation of the abnormal promyelocytes to form mature granulocytes by mechanisms other than those outlined here. For example, overexpression of normal RAR or an alternative retinoid receptor in response to retinoid treatment is a possibility. In addition, remission may be achieved by blocking some abnormal activities of the PML/APL fusion protein that prevents differentiation.

Remissions induced by treatment of APL with t-RA alone are unfortunately short-lived and maintenance therapy with RA is ineffective at preventing relapse (239). RA resistance may result from the selection of subclones of the leukemic pool or may arise as a consequence of the induction of a new metabolic response by RA. Serum levels of RA usually decrease after an initial achievable level of about 10^{-7} M. Clinical resistance may reflect this inability to keep the serum level of RA at this effective concentration (240). Possible mechanisms for this decrease are either induction of P450-like enzyme systems or the induction of the cellular RA binding protein CRABP1, which is normally not expressed in normal or leukemic myeloid cells, but appears to be induced in patients with APL after prolonged treatment with all-*trans* RA (241). Even though the specific mutation is not shared by other cancers, the documented cytodifferentiating action of retinoids and usefulness in cancer chemoprevention appear to be unrelated to a specific genetic background in cancers.

Other genetic diseases may involve mutations in specific retinoid receptor genes. In this context it is of particular interest to note that all three RXR genes are located in chromosomal regions that have been linked to a number of diseases (242).

REFERENCES

1. Lotan, R. (1981). Effects of vitamin A and its analogs (retinoids) on normal and neoplastic cells. *Biochim. Biophys. Acta* **605**, 33–91.
2. Roberts, A. B., and Sporn, M. B. (1984). Cellular biology and biochemistry of the retinoids. *In* "The Retinoids" (M. B. Sporn, A. B. Roberts, and D. S. Goodman, eds.), pp. 209–286. Academic Press, Orlando, FL.
3. Morris-Kay, G., ed. (1992). "Retinoids in Normal Development and Teratogenesis." Oxford Sci. Publ., Oxford.
4. Bollag, W., and Holdener, E. E. (1992). Retinoids in cancer prevention and therapy. *Ann. Oncol.* **3**, 513–526.
5. Smith, M. A., Parkinson, D. R., Cheson, B. D., and Friedman, M. A. (1992). Retinoids in cancer therapy. *J. Clin. Oncol.* **10**, 839–864.
6. Green, S., and Chambon, P. (1988). Nuclear receptors enhance our understanding of transcription regulation. *Trends Genet.* **4**, 309–314.
7. Evans, R. M. (1988). The steroid and thyroid hormone receptor family. *Science* **240**, 889–895.
8. Petkovich, M., Brand, N. J., Krust, A., and Chambon, P. (1987). Human retinoic acid receptor belongs to the family of nuclear receptors *Nature (London)* **330**, 444–450.

9. Giguere, V., Ong, E. S., Seigi, P., and Evans, R. M. (1987). Identification of a receptor for the morphogen retinoic acid. *Nature (London)* **330**, 624–629.

10. Benbrook, D., Lernhardt, W., and Pfahl, M. (1988). A new retinoic acid receptor identified from a hepatocellular carcinoma. *Nature (London)* **333**, 669–672.

11. Brand, N., Petkovich, M., Krust, A., de-Thé, H., Marchio, A., Tiollais, P., and Dejean, D. (1988). Identification of a second human retinoic acid receptor. *Nature (London)* **332**, 850–853.

12. Krust, A., Kastner, P. H., Petkovich, M., Zelent, A., and Chambon, P. (1989). A third human retinoic acid receptor, hRAR-γ. *Proc. Natl. Acad. Sci. U.S.A.* **86**, 5310–5314.

13. Giguere, V., Shago, M., Zirngibl, R., Tate, P., Rossant, J., and Varmuza, S. (1990). Identification of a novel isoform of the retinoic acid receptor γ expressed in the mouse embryo. *Mol. Cell. Biol.* **10**, 2335–2340.

14. Hamada, K., Gleason, S. L., Levi, B.-Z., Hirschfeld, S., Apella, E., and Ozato, K. (1989). H-2RIIBP, a member of the nuclear hormone receptor superfamily that binds to both the regulatory element of major histocompatability class I genes and the estrogen response element. *Proc. Natl. Acad. Sci. U.S.A.* **86**, 8289–8293.

15. Mangelsdorf, D. J., Ong, E. S., Dyck, J. A., and Evans, R. M. (1990). Nuclear receptor that identifies a novel retinoic acid pathway. *Nature (London)* **345**, 224–229.

16. Mangelsdorf, D. J., Borgmeyer, U., Heyman, R. A., Zhan, J. Y., Ong, E. S., Oro, A. E., Kakizuka, A., and Evans, R. M. (1992). Characterization of three RXR genes that mediate the action of 9-*cis* RA retinoic acid. *Genes. Dev.* **6**, 329–344.

17. Yu, V. C., Delsert, C., Andersen, B., Holloway, J. M., Devary, O. V., Näar, A. M., Kim, S. Y., Boutin, J.-M., Glass, C. K., and Rosenfeld, M. G. (1991). RXRb, a coregulator that enhances binding of retinoic acid, thyroid hormone and vitamin D receptors to their cognate response elements. *Cell (Cambridge, Mass.)* **67**, 1251–1266.

18. Leid, M., Kastner, P., Lyons, R., Nakshatri, H., Saunders, M., Zacharewski, T., Chen, J.-Y., Staub, A., Garnier, J.-M., Mader, S., and Chambon, P. (1992). Purification, cloning, and RXR identity of the HeLa cell factor with which RAR or TR heterodimerizes to bind target sequences efficiently. *Cell (Cambridge, Mass.)* **68**, 377–395.

19. Lehmann, J. M., Hoffmann, B., and Pfahl, M. (1991). Genomic organization of the retinoic acid receptor g gene. *Nucleic Acids Res.* **19**, 573–578.

20. Leroy, P., Krust, A., Kastner, P., Mendelsohn, C., Zelent, A., and Chambon, P. (1992). Retinoic acid receptors. *In* "Retinoids in Normal Development and Teratogenesis" (G. Morriss-Kay, ed.), pp. 2–25. Oxford Univ. Press, New York.

21. Umesono, K., and Evans, R. M. (1989). Determinants of target gene specificity for steroid/thyroid hormone receptors. *Cell (Cambridge, Mass.)* **57**, 1139–1146.

22. Glass, C. K., and Holloway, J. M. (1990). Regulation of gene expression by the thyroid hormone receptor. *Biochim. Biophys. Acta* **1032**, 157–176.

23. Issemann, I., and Green, S. (1990). Activation of a member of the steroid hormone receptor superfamily by peroxisome proliferators. *Nature (London)* **347**, 645–649.

24. Umesono, K., Murakami, K. K., Thompson, C. C., and Evans, R. M. (1991). Direct repeats as selective response elements for the thyroid hormone, retinoic acid and vitamin D3 receptors. *Cell (Cambridge, Mass.)* **65**, 1255–1266.

25. Näar, A. M., Boutin, J.-M., Lipkin, S. M., Yu, V. C., Holloway, J. M., Glass, C. K., and Rosenfeld, M. G. (1991). The orientation and spacing of core DNA-binding motifs dictate selective transcriptional responses to three nuclear receptors. *Cell (Cambridge, Mass.)* **65**, 1267–1279.

26. Lehmann, J. M., Zhang, X.-K., and Pfahl, M. (1992). RARγ2 expression is regulated through a retinoic acid response element embedded in Sp1 sites. *Mol. Cell. Biol.* **12**, 2976–2985.

27. Hoffmann, B., Lehmann, J. M., Zhang, X-K., Hermann, T., Graupner, G., and Pfahl, M. (1990). A retinoic acid receptor specific element controls the retinoic acid receptor-β promoter. *J. Mol. Endocrinol.* **4**, 1734–1743.

28. Zhang, X-K., Hoffmann, B., Tran, P., Graupner, G., and Pfahl, M. (1992). Retinoid X receptor is an auxiliary protein for thyroid hormone and retinoic acid receptors. *Nature (London)* **355**, 441–446.

29. Kliewer, S. A., Umesono, K., Mangelsdorf, D. J., and Evans, R. M. (1992). Retinoid X receptor interacts with nuclear receptors in retinoic acid, thyroid hormone and vitamin D3 signaling. *Nature (London)* **355**, 446–449.

30. Kliewer, S. A., Umesono, K., Noonan, D. J., Heyman, R. A., and Evans, R. M. (1992). Convergence of 9-*cis* retinoic acid and peroxisome proliferator signaling pathways through heterodimer formation of their receptors. *Nature (London)* **358**, 771–774.

31. Bugge, T. H., Pohl, J., Lonnoy, O., and Stunnenberg, H. G. (1992). RXRα, a promiscuous partner of retinoic acid and thyroid hormone receptors. *EMBO J.* **11**, 1409–1418.

32. Marks, M. S., Hallenbeck, P. L., Nagata, T., Segars, J. H., Appella, E., Nikodem, V. M., and Ozato, K. (1992). H-2RIIBP (RXRβ) heterodimerization provides a mechanism for combinatorial diversity in the regulation of retinoic acid and thyroid hormone responsive genes. *EMBO J.* **11**, 1419–1435.

33. Zhang, X.-K., Lehmann, J. M., Hoffmann, B., Dawson, M. I., Cameron, J., Graupner, G., Hermann, T., and Pfahl, M. (1992). Homodimer formation of retinoid X receptor induced by 9-*cis* retinoic acid. *Nature (London)* **358**, 587–591.

34. Lehmann, J. M., Jong, L., Fanjul, A., Cameron, J. F., Lu, X. P., Haefner, P., Dawson, M. I., and Pfahl, M. (1992). Retinoids selective for retinoid X receptor response pathways. *Science* **258**, 1944–1946.

35. Yang Yen, H.-F., Zhang, X-K., Graupner, G., Tzukerman, M., Sakamoto, B., Karin, M., and Pfahl, M. (1991). Antagonism between retinoic acid receptors and AP-1: Implication for tumor promotion and inflammation. *New Biol.* **3**, 1206–1219.

36. Zhang, X.-K., Wills, K. N., Husmann, M., Hermann, T., and Pfahl, M. (1991). Novel pathway for thyroid hormone receptor action through interaction with jun and fos oncogene activities. *Mol. Cell. Biol.* **11**, 6016–6025.

37. Schüle, R., Rangarajan, P., Yang, N., Kliewer, S., Ransone, L. J., Bolado, J., Verma, I. M., and Evans, R. M. (1991). Retinoic acid is a negative regulator of AP-1-responsive genes. *Proc. Natl. Acad. Sci. U.S.A.* **88**, 6092–6096.

38. Beato, M. (1989). Gene regulation by steroid hormone. *Cell (Cambridge, Mass.)* **56**, 335–344.

39. Laudet, V., Hänni, C., Coll, J., Catzeflie, F., and Stéhelin, D. (1992). Evolution of the nuclear receptor gene superfamily. *EMBO J.* **11**, 1003–1013.

40. Danielson, M., Hinck, L., and Ringold, G. M. (1989). Two amino acids within the knuckle of the first zinc finger specify DNA response element activation by the glucocorticoid receptor. *Cell (Cambridge, Mass.)* **57**, 1131–1138.

41. Mader, S., Kumar, V., de Verneuil, H. and Chambon, P. (1989). Three amino acids of the oestrogen receptor are essential to its ability to distinguish an oestrogen from a glucocorticoid-responsive element. *Nature (London)* **338**, 271–274.

42. Pfahl, M. (1994). Vertebrate receptors: Molecular biology, dimerization and response elements. *In* "Seminars in Cell Biology" (G. Richards, ed.), Vol. 5, pp. 201.1–9. Academic Press, London.

43. Mlodzik, M., Hiromi, Y., Weber, U., Goodman, C. S., and Rubin, G. M. (1990). The drosophila seven-up gene, a member of the steroid receptor gene superfamily, controls photoreceptor cell fates. *Cell (Cambridge, Mass.)* **60**, 211–224.

44. Wilson, T. E., Paulsen, R. E., Padgett, K. A., and Milbrandt, J. (1992). Participation of non-zinc finger residues in DNA binding by two nuclear orphan receptors. *Science* **256,** 107–110.
45. Härd, T., Kellenbach, E., Boelens, R., Maler, B. A., Dahlman, K., Freedman, L. P., Carlstedt-Duke, V., Yamamoto, K. R., Gustafsson, J.-A., and Kaptein, R. (1990). Solution structure of the glucocorticoid receptor DNA-binding domain. *Science* **249,** 157–160.
46. Lee, M. S., Kliewer, S. A., Provencal, J., Wright, P. E., and Evans, R. M. (1993). Structure of the retinoic X receptor a DNA binding domain: A helix required for homodimeric DNA binding. *Science* **260,** 1117–1121.
47. Luisi, B. F., Xu, W. X., Otwinowshi, Z., Freedman, L. P., Yamamoto, K. R., and Sigler, P. B. (1991). Crystallographic analysis of the interaction of the glucocorticoid receptor with DNA. *Nature (London)* **352,** 497–505.
48. Perlmann, T., Rangarajan, P. N., Umesono, K., and Evans, R. M. (1993). Determinants for selective RAR and TR recognition of direct repeat HREs. *Genes Dev.* **7,** 1411–1422.
49. Schwabe, J. W. R., Neuhaus, D., and Rhodes, D. (1990). Solution structure of the DNA-binding domain of the oestrogen receptor. *Nature (London)* **348,** 458–461.
50. Katahira, M., Knegtel, M. A., Boelens, R., Eib, D., Schilthuis, J. G., van der Saag, P. T., and Kaptein, R. (1992). Homo- and heteronuclear NMR studies of the human retinoic acid receptor b DNA-binding domain: Sequential assignments and identification of secondary structure elements. *Biochemistry* **31,** 6474–6480.
51. Wilson, T. E., Fahrner, T. J., and Milbrandt, J. (1993). The orphan receptors NGFI-B and steroidogenic factor 1 establish monomer binding as a third paradigm of nuclear receptor–DNA interaction. *Mol. Cell. Biol.* **13,** 5794–5804.
52. Kurokawa, R., Yu, V. C., Naar, A., Kyakumoto, S., Han, Z., Silverman, S., Rosenfeld, M. G., and Glass, C. K. (1993). Differential orientations of the DNA-binding domain and carboxy-terminal dimerization interface regulate binding site selection by nuclear receptor heterodimers. *Genes Dev.* **7,** 1423–1435.
53. Tora, L., White, J., Brou, C., Tasset, D., Webster, N., Scheer, E., and Chambon, P. (1989). The human estrogen receptor has two independent nonacidic transcriptional activation functions. *Cell (Cambridge, Mass.)* **59,** 477–487.
54. Meyer, M.-E., Pornon, A., Ji, J., Bocquel, M.-T., Chambon, P., and Gronemeyer, H. (1990). Agonistic and antagonistic activities of RU486 on the functions of the human progesterone receptor. *EMBO J.* **12,** 3923–3932.
55. Webster, N. J. G., Green, S., Jin, J. R., and Chambon, P. (1988). The hormone-binding domains of the estrogen and glucocorticoid receptors contain an inducible transcription activation function. *Cell (Cambridge, Mass.)* **54,** 199–207.
56. Nagpal, S., Saunders, M., Kastner, P., Durand, B., Nakshatri, H., and Chambon, P. (1992). Promoter context- and response element-dependent specificity of the transcriptional activation and modulating functions of retinoic acid receptors. *Cell (Cambridge, Mass.)* **70,** 1007–1019.
57. Nagpal, S., Friant, S., Nakshatri, H., and Chambon, P. (1993). RARs and RXRs: Evidence for two autonomous transactivation functions (AF-1 and AF-2) and heterodimerization *in vivo. EMBO J.* **12,** 2349–2360.
58. Folkers, G. E., van der Leede, B.-J.M., and van der Saag, P. T. (1993). The retinoic acid receptor-b2 contains two separate cell-specific transactivation domains, at the N-terminus and in the ligand-binding domain. *Mol. Endocrinol.* **7,** 616–627.
59. Kumar, V., Green, S., Stack, G., Berry, M., Jin, J.-R., and Chambon, P. (1987). Functional domains of the human estrogen receptor. *Cell (Cambridge, Mass.)* **51,** 941–951.

60. Tasset, D., Tora, L., Fromental, C., Scheer, E., and Chambon, P. (1990). Distinct classes of transcriptional activating domains function by different mechanisms. *Cell (Cambridge, Mass.)* **62,** 1177–1187.

61. Holenberg, S. M., and Evans, R. M. (1988). Multiple and cooperative trans-activation domains of the human glucocorticoid receptor *Cell (Cambridge, Mass.)* **55,** 899–906.

62. Mermod, N., O'Neill, E. A., Kelly, T. J., and Tjian, R. (1989). The proline-rich transcriptional activator of CTF/NF-I is distinct from the replication and DNA binding domain. *Cell (Cambridge, Mass.)* **58,** 741–753.

63. Williams, T., Admon, A., Luscher, B., and Tjian, R. (1988). Cloning and expression of AP-1, a cell-type-specific transcription factor that activates inducible enhancer elements. *Genes Dev.* **2,** 1557–1569.

64. Ko, H.-S., Fast, P., McBride, W., and Staudt, L. M. (1988). A human protein specific for the immunoglobulin octamer DNA motif contains a functional homeobox domain. *Cell (Cambridge, Mass.)* **55,** 135–144.

65. Norman, C., Runswick, M., Pollock, R., and Treisman, R. (1988). Isolation and properties of cDNA clones encoding SRF, a transcription factor that binds to the c-fos serum response element. *Cell (Cambridge, Mass.)* **55,** 989–1003.

66. Power, R. F., Lydon, J. P., Conneely, O. M., and O'Malley, B. W. (1991). Dopamine activation of an orphan of the steroid receptor superfamily. *Science* **252,** 1546–1548.

67. Power, R. F., Mani, S. K., Codina, J., Conneely, O. M., and O'Malley, B. W. (1991). Dopaminergic and ligand-independent activation of steroid hormone receptors. *Science* **254,** 1636–1639.

68. Huggenvik, J. I., Collard, M. W., Kim, Y.-W., and Sharma, R. P. (1993). Modification of the retinoic acid signaling pathway by the catalytic subunit of protein kinase-A. *Mol. Endocrinol.* **7,** 543–550.

69. Sullivan, W. P., Madden, B. J., McCormick, D. J., and Toft, D. O. (1988). Hormone-dependent phosphorylation of the avian progesterone receptor. *J. Biol. Chem.* **263,** 14717–14723.

70. Hoeck, W., and Groner, B. (1990). Hormone-dependent phosphorylation of the glucorticoid receptor occurs mainly in the amino-terminal transactivation domain. *J. Biol. Chem.* **265,** 5403–5408.

71. Glineur, C., Bailly, M., and Ghysdael, J. (1989). The c-erbA a-encoded thyroid hormone receptor is phosyphorylated in its amino terminal domain by casein kinase II. *Oncogene* **4,** 1247–1254.

72. Rochette-Egly, C., Lutz, Y., Saunders, M., Scheuer, I., Gaub, M.-P., and Chambon, P. (1991). Retinoic acid receptor g: Specific immunodetection and phosyphorylation. *J. Cell Biol.* **115,** 535–545.

73. Bhat, M. K., Parkison, C., McPhie, P., Liang, C.-M., and Cheng, S.-Y. (1993). Conformational changes of human b1 thyroid hormone receptor induced by binding of 3,3′,5-TRIIODO-L-thyronine. *Biochem. Biophys. Res. Commun.* **195,** 385–392.

74. Allan, G. F., Leng, X., Tsai, S. Y., Weigel, N. L., Edwards, D. P., Tsai, M.-J., and O'Malley, B. W. (1992). Hormone and antihormone induce distinct conformational changes which are central to steroid receptor activation. *J. Biol. Chem.* **267,** 19513–19520.

75. Danielian, P. S., White, R., Lees, J. A., and Parker, M. G. (1992). Identification of a conserved region required for hormone dependent transcriptional activation by steroid hormone receptors. *EMBO J.* **11,** 1025–1033.

76. Ince, B. A., Zhuang, Y., Wrenn, C. K., Shapiro, D. J., and Katzenellenbogen, B. S. (1993). Powerful dominant negative mutants of the human estrogen receptor. *J. Biol. Chem.* **268,** 14026–14032.

77. Saatcioglu, F., Bartunek, P., Deng, T., Zenke, M., and Karin, M. (1993). A conserved C-terminal sequence that is deleted in v-erb A is essential for the biological activities of c-erbA (the thyroid hormone receptor). *Mol. Cell. Biol.* **13,** 3675–3685.

78. Zenke, M., Munoz, A., Sap, J., Vennström, B., and Beug, H. (1990). v-erbA oncogene activation entails the loss of hormone-dependent regulator activity of c-erbA. *Cell (Cambridge, Mass.)* **61,** 1035–1049.

79. Damm, D., Heyman, R. A., Umesono, K., and Evans, R. M. (1993). Functional inhibition of retinoic acid response by dominant negative retinoic acid receptor mutants. *Proc. Natl. Acad. Sci. U.S.A.* **90,** 2989–2993.

80. Tsai, S., Bartelmez, S., Heyman, R., Damm, K., Evans, R. M., and Collins, S. J. (1992). A mutated retinoic acid receptor-a exhibiting dominant-negative activity alters the lineage development of a multipotent hematopoietic cell line. *Genes Dev.* **6,** 2258–2269.

81. Durand, B., Saunders, M., Leroy, P., Leid, M., and Chambon, P. (1992). All-*trans* and 9-*cis* retinoic acid induction of CRABPII transcription is mediated by RAR–RXR heterodimers bound to DR1 and DR2 repeated motifs. *Cell (Cambridge, Mass.)* **71,** 73–85.

82. Zhang, X.-K., Salbert, G., Lee, M.-O., and Pfahl, M. (1994). Mutations that alter ligand induced switches and dimerization activities in the retinoid X receptor. *Mol. Cell Biol.* **14,** 4311–4323.

83. Gill, G., and Ptashne, M. (1987). Mutants of GAL4 protein altered in an activation function. *Cell (Cambridge, Mass.)* **51,** 121–126.

84. Hope, I. A., Mahadevan, S., and Struhl, K. (1988). Structural and functional characterization of the short acidic transcriptional activation region of yeast GCN4 protein. *Nature (London)* **333,** 635–640.

85. Sadowski, I., Ma, J., Triezenberg, S., and Ptashne, M. (1988). GAL4-VP16 is an unusually potent transcriptional activator. *Nature (London)* **335,** 563–564.

86. Trienzenberg, S. J., Kingsbury, R. C., and McKnight, S. L. (1988). Functional dissection of VP16, the trans-activator of herpes simplex virus immediate early gene expression. *Genes Dev.* **2,** 718–729.

87. Ing, N. H., Beekman, J. M., Tsai, S. Y., Tsai, M.-J., and O'Malley, B. W. (1992). Members of the steroid hormone receptor superfamily interact with TFIIB (S300-II). *J. Biol. Chem.* **267,** 17617–17623.

88. Leuther, K. K., Saimeron, J. M., and Johnston, S. A. (1993). Genetic evidence that an activation domain of GAL4 does not require acidity and may form a β sheet. *Cell (Cambridge, Mass.)* **72,** 575–585.

89. Van Hoy, M., Leuther, K. K., Kodadek, T., and Johnston, S. A. (1993). The acidic activation domains of the GCN4 and GAL4 proteins are not a helical but form β sheets. *Cell (Cambridge, Mass.)* **72,** 587–594.

90. Berkenstam, A., Ruiz, M.-D.-M.V., Barettino, D., Horikoshi, M., and Stunnenberg, H. G. (1992). Cooperativity in transactivation between retinoic acid receptor and TFIID requires an activity analogous to E1A. *Cell (Cambridge, Mass.)* **69,** 401–412.

91. Kruyt, F. A. E., Folkers, G. E., Walhout, A. J. M., van der Leede, B.-J.M., and van der Saag, P. T. (1993). E1A functions as a coactivator of retinoic acid-dependent retinoic acid receptor-β2 promoter activation. *Mol. Endocrinol.* **7,** 604–615.

92. Kumar, V., and Chambon, P. (1988). The estrogen receptor binds tightly to its responsive element as a ligand-induced homodimer. *Cell (Cambridge, Mass.)* **55,** 145–156.

93. Guiochon-Mantel, A., Loosfelt, H., Ragot, T., Bailly, A., Atger, M., Misrahi, M., Perricaudet, M., and Milgrom, E. (1988). Receptors bound to antiprogestin form abortive complexes with hormone response elements. *Nature (London)* **336,** 695–698.

94. Forman, B. M., and Samuels, H. H. (1990). Interactions among a subfamily of nuclear hormone receptors: The regulatory zipper hypothesis. *Mol. Endocrinol.* **4,** 1293–1302.

95. Abel, T., and Maniatis, T. (1989). Action of leucine zippers. *Nature (London)* **341,** 24–25.

96. Maksymowych, A. B., Hsu, T.-C., and Litwack, G. (1993). A novel, highly conserved structural motif is present in all members of the steroid receptor superfamily. *Receptor* **2,** 225–239.

97. Au-Fliegner, M., Helmer, H., Casanova, J., Raaka, B. M., and Samuels, H. H. (1993). The conserved ninth C-terminal heptad in thyroid hormone and retinoic acid receptors mediates diverse responses by affecting heterodimer but not homodimer formation. *Mol. Cell. Biol.* **13,** 5725–5737.

98. Rosen, E. D., Beninghof, E. G., and Koenig, R. J. (1993). Dimerization interface of thyroid hormone, retinoic acid, vitamin D and retinoid X receptors. *J. Biol. Chem.* **268,** 11534–11541.

99. Lee, J. W., Gulick, T., and Moore, D. D. (1992). Thyroid hormone receptor dimerization function maps to a conserved subregion of the ligand binding domain. *Mol. Endocrinol.* **6,** 1867–1873.

100. O'Donnell, A. L., and Koenig, R. J. (1990). Mutational analysis identifies a new functional domain of the thyroid hormone receptor. *Mol. Endocrinol.* **4,** 715–720.

101. O'Donnell, A. L., Rosen, E. D., Darling, D. S., and Koenig, R. J. (1991). Thyroid hormone receptor mutations that interfere with transcriptional activation also interfere with receptor interaction with a nuclear protein. *Mol. Endocrinol.* **5,** 94–99.

102. McPhie, P., Parkison, C., Lee, B. K., and Cheng, S.-Y. (1993). Structure of the hormone binding domain of human B1 thyroid hormone nuclear receptor: Is it an α/β barrel? *Biochemistry* **32,** 7460–7465.

103. Zhang, X.-K., and Pfahl, M. (1993). Regulation of retinoid and thyroid action through homo- and heterodimeric receptors. *Trends Endocrinol. Metab.* **4,** 156–162.

104. Zhang, X.-K., Hoffmann, B., Trans, P., Graupner, G., and Pfahl, M. (1992). Retinoid X receptor is an auxiliary protein for thyroid hormone and retinoic acid receptors. *Nature (London)* **355,** 441–446.

105. Fleischhauer, K., Park, J. H., DiSanto, J. P., Marks, M., Ozato, K., and Yang, S. Y. (1992). Isolation of a full-length cDNA clone encoding an N-terminally variant form of the human retinoid X receptor β. *Nucleic Acids Res.* **20,** 1801.

106. Damm, K., Heymann, R. A., Umesono, K., and Evans, R. M. (1993). Functional inhibition of retinoic acid response by dominant negative retinoic acid receptor mutants. *Proc. Natl. Acad. Sci. U.S.A.* **90,** 2989–2993.

107. Espeseth, A. S., Murphy, S. P., and Linney, E. (1989). Retinoic acid receptor expression vector inhibits differentiation of F9 embryonal carcinoma cells. *Genes Dev.* **3,** 1647–1656.

108. Kent, A., Emami, B., and Collins, S. J. (1992). Retinoic acid-resistant HL-60R cells harbor a point mutation in the retinoic acid receptor ligand-binding domain that confers dominant negative activity. *Blood* **80,** 1885–1889.

109. Pratt, M. A. C., Kralova, J., and McBurney, M. W. (1990). A dominant negative mutation of the alpha retinoic acid receptor gene in a retinoic acid-nonresponsive embryonal carcinoma cell. *Mol. Cell. Biol.* **10,** 6445–6453.

110. Baniahmad, A., Steiner, C. H., Kohne, A. C., and Renkawitz, R. (1990). Modular structure of a chicken lysozyme silencer; Involvement of an unusual thyroid hormone receptor binding site. *Cell (Cambridge, Mass.)* **61,** 505–514.

111. Baniahmad, A., Kohne, A. C., and Renkawitz, R. (1992). A transferable silencing

domain is present in the thyroid hormone receptor, the v-erb A oncogene product and in the retinoic acid receptor. *EMBO J.* **11**, 1015–1023.

112. Piedrafita, J. F., Bendik, I., Ortiz, M. A., and Pfahl, M. (1994). A novel T3 response mechanism: Thyroid hormone receptor homodimers function like prokaryotic repressors. In press.

113. Murray, M. B., and Towle, H. C. (1989). Identification of nuclear factors that enhance binding of the thyroid hormone receptor to a thyroid hormone response element. *Mol. Endocrinol.* **3**, 1434–1442.

114. Burnside, J., Darling, D. S., and Chin, W. W. (1990). A nuclear factor that enhances binding of thyroid hormone receptors to thyroid hormone response elements. *J. Biol. Chem.* **265**, 2500–2504.

115. Lazar, M. A., and Berrodin, T. J. (1990). Thyroid hormone receptors form distinct nuclear proteins—Dependent and independent complexes with a thyroid hormone response element. *Mol. Endocrinol.* **4**, 1627–1635.

116. Zhang, X.-K., Tran, P., and Pfahl, M. (1991). DNA binding and dimerization determinants for TRa and its interaction with a nuclear protein. *Mol. Endocrinol.* **5**, 1909–1920.

117. Rosen, E. D., O'Donnell, A. L., and Koenig, R. J. (1991). Protein–protein interactions involving v-erbA superfamily receptors: Through the TRAPdoor. *Mol. Cell. Endocrinol.* **78**, C83–C88.

118. Hermann, T., Hoffmann, B., Zhang, X-K., Tran, P., and Pfahl, M. (1992). Heterodimeric receptor complexes determine 3,5,3'-triiodothyronine and retinoid signaling specificities. *Mol. Endocrinol.* **6**, 1153–1162.

119. Carlberg, C., Bendik, I., Wyss, A., Meier, E., Sturzenbecker, L., Grippo, J., and Hunziker, W. (1993). Two nuclear signaling pathways for vitamin D. *Nature (London)* **361**, 657–660.

120. Hermann, T., Hoffmann, B., Piedrafita, J.F., Zhang, Z.-K., and Pfahl, M. (1993). V-erbA requires auxiliary proteins for dominant negative activity. *Oncogene* **8**, 55–65.

121. Leroy, P., Krust, A., Kastner, P., Mendelsohn, C., Zelent, A., and Chambon, P. (1992). Retinoic acid receptors. *In* "Retinoids in Normal Development and Teratogenesis" (G. Morriss-Kay, ed.), pp. 2–25. Oxford Univ. Press, New York.

122. Dolle, P., Ruberte, E., Leroy, P., Morriss-Kay, G., and Chambon, P. (1990). Retinoic acid receptors and cellular retinoid binding proteins. I. A systematic study of their differential pattern of transcription during mouse organogenesis. *Development (Cambridge, UK)* **110**, 1133–1151.

123. Zelent, A., Mendelsohn, C., Kastner, P., Krust, A., Garnier, J.-M., Ruffenach, F., Leroy, P., and Chambon, P. (1991). Differentially expressed isoforms of the mouse retinoic acid receptor β are generated by usage of two promoters and alternative splicing. *EMBO J.* **10**, 71–81.

124. Leroy, P., Krust, A., Zelent, A., Mendelsohn, C., Garnier, J.-M., Kastner, P., Dierich, A., and Chambon, P. (1991). Multiple isoforms of the mouse retinoic acid receptor α are generated by alternative splicing and differential induction by retinoic acid. *EMBO J.* **10**, 59–69.

125. Dolle, P., Ruberte, E., Kastner, P., Petkovich, M., Stonber, C. M., Gudas, L., and Chambon, P. (1989). Differential expression of the genes encoding the retinoic acid receptors α, β, and γ and CRABP in the developing limbs of the mouse. *Nature (London)* **342**, 702–705.

126. Ruberte, E., Dolle, P., Krust, A., Zelent, A., Morriss-Kay, G., and Chambon, P. (1990). Specific spatial and temporal distribution of retinoic acid receptor gamma transcripts during mouse embryogenesis. *Development (Cambridge, UK)* **108**, 213–222.

127. Ruberte, E., Dolle, P., Chambon, P., and Morriss-Kay, G. (1991). Retinoic acid receptors and cellular retinoid binding proteins. II. Their differential pattern of transcription during early morphogenesis in mouse embryos. *Development (Cambridge, UK)* **111**, 45–60.

128. Kastner, P., Krust, A., Mendelsohn, C., Garnier, J. M., Zelent, A., Leroy, P., Staub, A., and Chambon, P. (1990). Murine isoforms of retinoic acid receptor-γ with specific patterns of expression. *Proc. Natl. Acad. Sci. U.S.A.* **87**, 2700–2704.

129. Mader, S., Leroy, P., Chen, J.-Y., and Chambon, P. (1993). Multiple parameters control the selectivity of nuclear receptors for their response elements. *J. Biol. Chem.* **268**, 591–600.

130. Mader, S., Chen, J.-Y., Chen, Z., White, J., Chambon, P., and Gronemeyer, H. (1993). The patterns of binding of RAR, RXR, and TR homo- and heterodimers to direct repeats are dictated by the binding specificities of the DNA binding domains. *EMBO J.* **12**, 5029–5041.

131. Husmann, M., Lehmann, J., Hoffmann, B., Hermann, T., Tzukerman, M., and Pfahl, M. (1991). Antagonism between retinoic acid receptors. *Mol. Cell. Biol.* **11**, 4097–4103.

132. Lohnes, D., Kastner, P., Dierich, A., Mark, M., LeMeur, M., and Chambon, P. (1993). Function of retinoic acid receptor γ in the mouse. *Cell (Cambridge, Mass.)* **73**, 643–658.

133. Lufkin, T., Lohnes, D., Mark, M., Dierich, A., Gorry, P., Gaub, M.-P., LeMeur, M., and Chambon, P. (1993). High postnatal lethality and testis degeneration in retinoic acid receptor α mutant mice. *Proc. Natl. Acad. Sci. U.S.A.* **90**, 7225–7229.

134. Berrodin, T. J., Marks, M. S., Ozato, K., Linney, E., and Lazar, M. A. (1992). Heterodimerization among thyroid hormone receptor, retinoic acid receptor, retinoic X receptor, chicken ovalbumin upstream promoter transcription factor, and an endogenous liver protein. *Mol. Endocrinol.* **6**, 1468.

135. Forman, B. M., Yang, C.-R., Au, M., Casanova, J., Ghysdael, J., and Samuels, H. H. (1989). A domain containing leucine-zipper-like motifs mediate novel *in vivo* interactions between thyroid hormone and retinoic acid receptors. *Mol. Endocrinol.* **3**, 1610–1626.

136. Graupner, G., Zhang, X.-K., Tzukerman, M., Wills, K., Hermann, T., and Pfahl, M. (1991). Thyroid hormone receptors repress estrogen receptor activation of a TRE. *Mol. Endocrinol.* **5**, 365–372.

137. Lehmann, J. M., Zhang, X.-K., Graupner, G., Lee, M.-O., Hermann, T., Hoffmann, B., and Pfahl, M. (1993). Formation of RXR homodimers leads to repression of T3 response: Hormonal cross-talk by ligand induced squelching. *Mol. Cell. Biol.* **13**, 7698–7707.

138. Heyman, R. A., Mangelsdorf, D. J., Dyck, J. A., Stein, R. B., Eichele, G., Evans, R. M., and Thaller, C. (1992). 9-*cis* retinoic acid is a high affinity ligand for the retinoid X receptor. *Cell (Cambridge, Mass.)* **68**, 397–406.

139. Levin, A. A., Sturzenbecker, L. J., Kazmer, S., Bosakowski, T., Huselton, C., Allenby, G., Speck, J., Kratzeisen, C., Rosenberger, M., Lovey, A., and Grippo, J. F. (1992). 9-*cis* retinoic acid steroisomer binds and activates the nuclear receptor RXRa. *Nature (London)* **355**, 359–361.

140. MacDonald, P. N., Dowd, D. R., Nakajima, S., Galligan, M. A., Reeder, M. C., Haussler, C. A., Ozato, K., and Haussler, M. (1993). Retinoid X receptors stimulate and 9-*cis* retinoic acid inhibits 1,25-dihydroxyvitamin D3-activated expression of the rat osteocalcin gene. *Mol. Cell. Biol.* **13**, 5907–5917.

141. Mendelsohn, C., Ruberte, E., and Chambon, P. (1992). Retinoid receptors in vertebrate limb development. *Dev. Biol.* **152**, 50–61.

142. Tsai, S. Y., Sagami, I., Wanh, H., Tsai, M. J., and O'Malley, B. W. (1987). Interactions between a DNA-binding transcription factor (Coup) and a non-DNA binding factor (S300-II). *Cell (Cambridge, Mass.)* **50**, 701–709.

143. Wang, L. H., Tsai, S. Y., Cook, R. G., Beattie, W. G., Tsai, M. J., and O'Malley, B. W. (1989). Coup transcription factor is a member of the steroid receptor superfamily. *Nature (London)* **340**, 163–166.

144. Ladias, J. A. A., and Karathanasis, S. K. (1991). Regulation of the apolipoprotein AI gene by ARp-1, a novel member of the steroid receptor superfamily. *Science* **251**, 561–565.

145. Miyajima, N., Kadowaki, Y., Fukushige, S. I., Semba, S. K., Yamanashi, Y., Matsubara, K. I., Toyoshima, K., and Yamamoto, K. R. (1988). Identification of two novel members of erbA superfamily by molecular cloning: The gene products of the two are highly related to each other. *Nucleic Acids Res.* **16**, 11057–11074.

146. Tran, P., Zhang, X. K., Salbert, G., Hermann, T., Lehmann, J., and Pfahl, M. (1992). Coup orphan receptors are negative regulators of retinoic acid response pathways. *Mol. Cell. Biol.* **12**, 4666–4676.

147. Mietus-Snyder, M., Sladek, F. M., Ginsburg, G. S., Kuo, F., Ladias, J. A. A., Darnell, J. E., Jr., and Karathanasis, S. K. Antagonism between apolipoprotein AI regulatory protein 1, Ear3/Coup-TF, and hepatocyte nuclear factor 4 modulates apolipoprotein CIII gene expression in liver and intestinal cells. *Mol. Cell. Biol.* **12**, 1708–1718.

148. Kliewer, S., Umesono, K., Heyman, R. A., Mangelsdorf, D. J., Dyck, J. A., and Evans, R. M. (1992). Retinoid X receptor–Coup-TF interactions modulate retinoic acid signaling. *Proc. Natl. Acad. Sci. U.S.A.* **89**, 1448–1452.

149. Windom, R. L., Rhee, M., and Karathanasis, S. K. (1992). Repression by ARP-1 sensitizes apolipoprotein AI gene responsiveness to RXRa and retinoic acid. *Mol. Cell. Biol.* **12**, 3380–3389.

150. Cooney, A. J., Tsai, S. Y., O'Malley, B. W., and Tsai, M. J. (1992). Chicken ovalbumin upstream promoter transcription factor (Coup-TF) dimers bind to different GGTCA response elements, allowing Coup-TF to repress hormonal induction of the vitamin D3, thyroid hormone, and retinoic acid receptors. *Mol. Cell. Biol.* **12**, 4153–4163.

151. Teng, C. T., Liu, Y., Yang, N., Walmer, D., and Panella, T. (1992). Differential molecular mechanism of the estrogen action that regulates lactoferrin gene in human and mouse. *Mol. Endocrinol.* **6**, 1969–1981.

152. Wang, L. H., Tsai, S. Y., Sagami, I., Tsai, M. J., and O'Malley, B. W. (1987). Purification and characterization of chicken ovalbumin upstream promoter transcription factor from Hela cells. *J. Biol. Chem.* **262**, 16080–16086.

153. Hwung, Y. P., Crowe, D. T., Wang, L. H., Tsai, S. Y., and Tsai, M. J. (1988). The coup transcription factor binds to an upstream promoter element of rat insulin II gene. *Mol. Cell. Biol.* **8**, 2070–2077.

154. Cooney, A. J., Tsai, S. Y., O'Malley, B. W., and Tsai, M. J. (1991). Chicken ovalbumin upstream promoter transcription factor binds to a negative regulatory region in the human immunodeficiency virus 1 long terminal repeat. *J. Virol.* **65**, 2853–2860.

155. Curran, T., Peter, S. G., van Beveren, C., Teich, N. M., and Verma, I. M. (1982). FBJ murine osteosarcoma virus: Identification and molecular cloning of biologically active proviral DNA. *J. Virol.* **44**, 674–682.

156. Finkel, M. P., Biskis, B. O., and Jinkis, P. B. (1966). Virus induction of osteosarcoma in mice. *Science* **151**, 698–701.

157. Curran, T., Franza, B. R., Jr. (1988). Fos. and Jun: The AP-1 connection. *Cell* **55**, 395–397

158. Bohmann, D., Bos, T. J., Admon, A., Nishimura, T., Vogt, P. K., and Tjian, R. (1987). Human proto-oncogene c-jun encodes a DNA binding protein with structural and functional properties of transcription factor AP-1. *Science* **238,** 1386–1392.

159. Abate, C., Luk, D., Gentz, R., Rauscher, F. J., III, and Curran, T. (1990). Expression and purification of the leucine zipper and DNA-binding domains of Fos and Jun: Both Fos and Jun contact DNA directly. *Proc. Natl. Acad. Sci. U.S.A.* **87,** 1032–1036.

160. Schuermann, M., Neuberg, M., Hunter, J. B., Jenuwein, T., Ryseck, R.-P., Bravo, R., and Müller, R. (1989). The leucine repeat motif in Fos protein mediates complex formation with Jun/Ap-1 and is required for transformation. *Cell (Cambridge, Mass.)* **56,** 507–516.

161. Karin, M. (1990). The AP-1 complex and its role in transcriptional control by protein kinase C. *In* "Molecular Aspects of Cellular Regulation" (P. Cohen and C. Foulkes, eds.), Vol. 6, pp. 143–161. Elsevier, Amsterdam.

162. Angel, P., Imagawa, M., Chiu, R., Stein, B., Imbra, R. J., Rahmsdorf, H. J., Jonat, C., Herrlich, P., and Karin, M. (1987). Phorbol ester-inducible genes contain a common cis element recognized by a TPA modulated trans-acting factor. *Cell (Cambridge, Mass.)* **49,** 729–739.

163. Lee, W., Mitchell, P. J., and Tjian, R. (1987). Purified transcription factor AP-1 interacts with TPA inducible enhancer elements. *Cell* **49,** 741–752.

164. Diamond, M. I., Miner, J. N., Yoshinaga, S. K., and Yamamoto, K. R. (1990). Transcription factor interactions: Selectors of positive or negative regulation from a single DNA element. *Science* **249,** 1266–1272.

165. Schüle, R., Umesono, K., Mangelsdorf, D. J., Bolado, J., Pike, J. W., and Evans, R. M. (1990). Jun-Fos and receptors for vitamins A and D recognize a common response element in the human osteocalcin gene. *Cell (Cambridge, Mass.)* **61,** 497–504.

166. Harris, E. D., Jr., DiBona, D. R., and Krane, S. M. (1969). Collagenases in human synovial fluid. *J. Clin. Invest.* **48,** 2104–2113.

167. Dawson, M. I., Chao, W. R., Hobbs, P. D., and Delair, T. (1990). Chemistry and biology of synthetic retinoids. *In* "The Inhibitory Effects of Retinoids on the Induction of Ornithine Decarboxylase and the Promotion of Tumors in Mouse Epidermis," (M. I. Dawson, and Okamura, eds.), pp. 385–466. CRC Press, Boca Raton, FL.

168. Brinckerhoff, C. E., Sheldon, L. A., Benoit, M. C., Burgess, D. R., and Wilder, R. L. (1985). Effect of retinoids on rheumatoid arthritis, a proliferative and invasive non-malignant disease. *In* "Retinoids, Differentiation and Disease," (J. Nugent and S. Clark, eds.), pp. 191–211 Pitman, London.

169. Lafyatis, R., Kim, S.-J., Angel, P., Roberts, A. B., Sporn, M. B., Karin, M., and Wilder, R. L. (1990). Interleukin-1 stimulates and all-*trans*-retinoic acid inhibits collagenase gene expression through its 5' activator protein-1 binding site. *Mol. Endocrinol.* **4,** 973–980.

170. Pfahl, M. (1993). Nuclear receptor/AP-1 interaction. *Endocr. Rev.* **14,** 651–658.

171. de Luca, L. M. (1991). Retinoids and their receptors in differentiation, embryogenesis, and neoplasia. *FASEB J.* **5,** 2924–2933.

172. Leroy, P., Nakshatri, H., and Chambon, P. (1991). Mouse retinoic acid receptor α isoform is transcribed from a promoter that contains a retinoic acid response element. *Proc. Natl. Acad. Sci. U.S.A.* **88,** 10138–10142.

173. Duester, G., Shean, M. L., McBride, M. S., and Stewart, M. J. (1991). Retinoic acid response element in the human alcohol dehydrogenase gene ADH3: Implications for regulation of retinoic acid synthesis. *Mol. Cell. Biol.* **11,** 1638–1646.

174. Munoz-Canoves, P., Vik, D. P., and Tack, B. F. (1990). Mapping of a retinoic acid

responsive element in the promoter region of the complement factor H gene. *J. Biol. Chem.* **265**, 20065–20068.

175. Rottman, J. N., Widom, R. L., Nadal-Ginard, B., Mahdavi, V., and Karathanasis, S. K. (1991). A retinoic acid-responsive element in the apolipoprotein A1 gene distinguishes between two different retinoic acid response pathways. *Mol. Cell. Biol.* **11**, 3814–3820.

176. Mangelsdorf, D. J., Umesono, K., Kliewer, S., Borgmeyer, U., Ong, E. S., and Evans, R. M. (1991). A direct repeat in the cellular retinol-binding protein type II gene confers differential regulation by RXR and RAR. *Cell* **66**, 555–556.

177. Lee, M.-O., Hobbs, P. D., Zhang, X.-K., Dawson, M. I., and Pfahl, M. (1994). A new retinoid antagonist inhibits the HIV-1 promoter. *Proc. Natl. Acad. Sci. U.S.A.*, **91**, 5632–5636.

178. Graupner, G., Wills, K. N., Tzukerman, M., Zhang, X.-K., and Pfahl, M. (1989). Dual regulatory role for thyroid-hormone receptors allows control of retinoic-acid receptor activity. *Nature (London)* **340**, 653–656.

179. Glass, C. K., Holloway, J. M., Devary, O. V., and Rosenfeld, M. G. (1988). The thyroid hormone receptor binds with opposite transcriptional effects to a common sequence motif in thyroid hormone and estrogen response elements. *Cell (Cambridge, Mass.)* **54**, 313–323.

180. Brent, G. A., Harney, J. W., Chen, Y., Warne, R. L., Moore, D. D., and Larsen, P. R. (1989). Mutations in the rat growth hormone promoter which increase and decrease response to thyroid hormone define a consensus thyroid hormone response element. *Mol. Endocrinol.* **3**, 1996–2004.

181. Tini, M., Otulakowski, G., Breitman, M. L., Tsui, L.-C., and Giguère, V. (1993). An everted repeat mediates retinoic acid induction of the γF-crystallin gene: Evidence of a direct role for retinoids in lens development. *Genes Dev.* **7**, 295–307.

182. Orfanos, C. E., Ehlert, R., and Gollnick, H. (1987). The retinoids: A review of their clinical pharmacology and therapeutic use. *Drugs* **24**, 459–503.

183. Astrom, A., Pettersson, U., Krust, A., Chambon, P., and Voorhees, J. J. (1990). Retinoic acid and synthetic analogs differentially activate retinoic acid receptor dependent transcription. *Biochem. Biophys. Res. Commun.* **173**, 339–345.

184. Graupner, G., Malle, G., Maignan, J., Lang, G., Prunieras, M., and Pfahl, M. (1991). 6'-Substituted naphthalene-2-carboxylic acid analogs, a new class of retinoic acid receptor subtype-specific ligand. *Biochem. Biophys. Res. Commun.* **179**, 1554–1561.

185. Lehmann, J. M., Dawson, M. I., Hobbs, P. D., Husmann, M., and Pfahl, M. (1991). Identification of retinoids with nuclear receptor subtype-selective activities. *Cancer Res.* **51**, 4804–4809.

186. Bernard, B. A., Bernardon, J.-M., Delescluse, C., Martin, B., Lenoir, M.-C., Maignan, J., Charpentier, B., Pilgrim, W. R., Reichart, U., and Shroot, B. (1992). Identification of synthetic retinoids with selectivity for human nuclear retinoic acid receptor g. *Biochem. Biophys. Res. Commun.* **186**, 977–983.

187. Crettaz, M., Baron, A., Siegenthaler, G., and Hunziker, W. (1990). Ligand specificities of recombinant retinoic acid receptors RARα and RARβ. *Biochem. J.* **272**, 391–397.

188. Delescluse, C., Cavey, M. T., Martin, B., Bernard, B. A., Reichert, U., Maignan, J., Darmon, M., and Shroot, B. (1991). Selective high affinity retinoic acid receptor β-γ ligands. *Mol. Pharmacol.* **40**, 556–562.

189. Hashimoto, Y. (1991). Retinobenzoic acid and nuclear retinoic acid receptors. *Cell Struc. Funct.* **16**, 113–123.

190. Turpin, J. A., Vargo, M., and Meltzer, M. S. (1992). Retinoid effects mediated through mechanisms related to cell differentiation and to a direct transcriptional action on viral gene expression *J. Immunol.* **148**, 2539–2546.

191. Maio, J., and Brown, F., (1988). Regulation of expression driven by human immunodeficiency virus type 1 and human T cell leukemia virus type 1 long terminal repeats in pluripotential human embryonic cells. *J. Virol.* **62**, 1398–1407.

192. Ghazal, P., DeMattei, C., Giulietti, E., Kliewer, S. A., Umesono, K., and Evans, R. M. (1992). Retinoic acid receptors initiate induction of the cytomegalovirus enhancer in embryonal cells. *Proc. Natl. Acad. Sci. U.S.A.* **89**, 7630–7634.

193. Kaneko, S., Kagechika, H., Kawachi, E., Hashimoto, Y., and Shudo, K. (1991). Retinoid antagonist. *Med. Chem. Res.* **1**, 220–225.

194. Apfel, C., Bauer, F. M., Crettaz, M., Forni, L., Kamber, M., Kaufmann, F., LeMotte, P., Pirson, W., and Klaus, M. (1992). A retinoic acid receptor α antagonist selectively counteracts retinoic acid effects. *Proc. Natl. Acad. Sci. U.S.A.* **89**, 7129–7133.

195. Sporn, M. B., and Newton, D. C. (1979). Chemoprevention of cancer with retinoids. *Fed. Proc., Fed. Am. Soc. Exp. Biol.* **38**, 2528–2534.

196. Latznitzki, I. (1955). The influence of a hypervitaminosis on the effect of 20-methylcholanthrene on mouse prostate glands grown *in vitro*. *Br. J. Cancer* **9**, 434–441.

197. Row, N. A., and Gorlin, R. J. (1959). The effect of vitamin A deficiency upon experimental oral carcinogenesis. *J. Dent. Res.* **38**, 72–83.

198. Tallman, M. S., and Wiernik, P. H. (1992). Retinoids in cancer treatment. *J. Clin. Pharmacol.* **32**, 868–888.

199. Chu, E. W., and Malgren, R. A. (1965). An inhibitory effect of vitamin A on the induction of tumors of forestomach and cervix in the Syrian hamster by carcionogenic polycyclic hydrocarbons. *Cancer Res.* **25**, 884–895.

200. Davies, R. E. (1967). Effect of vitamin A on 7,12-dimethylbenz(a)anthracene-induced papillomas in Rhino mouse skin. *Cancer Res.* **27**, 237–241.

201. Saffiotti, V., Montesano, R., Sellakumar, A. R., and Borg, S. A. (1967). Experimental cancer of the lung. Inhibition by vitamin A of the induction of tracheobronchial squamous metaplasia and squamous cell tumors. *Cancer (Philadelphia)* **20**, 857–864.

202. Stimson, W. H. (1961). Vitamin A intoxication in adults. *N. Engl. J. Med.* **265**, 369–373.

203. Hong, W. K., Lippman, S. M., Itri, L. M., Karp, D. D., Lee, J. S., Byers, R. M., Schants, S. P., Kramer, A. M., Lotan, R., Peters, I. J., Dimery, I. W., Brown, B. W., and Goepfert, H. (1991). Prevention of second primary tumors with isotretinoin in squamous-cell carcinoma of the head and neck. *N. Engl. J. Med.* **323**, 795–801.

204. Cobleigh, M. A., Dowlat, K., Minn, F., Benson, A. B., Rademaker, A. W., Ashenhurst, J. B., and Wade, J. L. (1991). Phase I trial of tamoxifen (TAM) with or without fenretinide (4HPR), an analogue of vitamin A, in patients with metastatic breast cancer. *14th Annu. San Antonio Breast Cancer Symp.*, Abstr. 65.

205. Waxman, S., Huang, Y., Scher, B. M., and Scher, M. (1992). Enhancement of differentiation and cytotoxicity of leukemia cells by combination of fluorinated pyrimidines and differentiation inducers: Development of DNA double strand breaks. *Biomed. Pharmacother.* **46**, 183–192.

206. Burge, S. M., and Wilkinson, J. D. (1992). Darier–White disease: A review of the clinical features in 163 patients. *J. Am. Acad. Dermatol.* **27**, 40–50.

207. Borok, M., and Lowe, N. J. (1990). Pityriasis rubra pilaris. *J. Am. Acad. Dermatol.* **22**, 792–795.

208. Chen, Z.-X., Xue, Y.-Q., Zhang, R., Tao, R. F., Xia, X.-M., Li, C., Wange, W., Zu, W.-Y., Yao, X.-Z., and Ling, B. J. (1991). A clinical and experimental study on all-*trans* retinoic acid-treated acute promyelocytic leukemia patients. *Blood* **78**, 1413.

209. Warrell, R. P., Frankel, S. R., Miller, W. H., Jr., Scheinberg, D. A., Itri, L. M., Hittelman, W. N., Vyas, R., Andreeff, M., Tafuri, A., Jakubowski, A., Gabrilove, J., Gordon, M. S., and Dmitrovsky, E. (1991). Differentiation therapy of acute promyelocytic leukemia with tretinoin (all-*trans*-retinoic acid). *N. Engl. J. Med.* **324**, 1385–1393.

210. Bell, B., Findley, H., Kirscher, J., *et al.* (1991). Phase II study of 13-*cis*-retinoic acid in pediatric patients with acute nonlymphocytic leukemia–A Pediatric Oncology Group study. *J. Immunother.* **10**, 77–83.

211. Grignani, F., Ferrucci, P. F., Testa, U., Talamo, G., Fagioloi, M., Alcalay, M., Mencarelli, A., Grignani, F., Peschle, C., Nicoletti, I., and Pelicci, P. G. (1993). The acute promyelocytic leukemia-specific PML-RAR fusion protein inhibits differentiation and promotes survival of myeloid precursor cell. *Cell* **74**, 423–431.

212. Lawrence, H., Conner, K., Kelly, M., Haussler, M. R., Wallace, P., and Bagby, G. C., Jr. (1987). Cis-retinoic acid stimulates the clonal growth of some myeloid leukeamia cells in vitro. *Blood* **69**, 302–307.

213. Findley, H., Steuber, C., Ruymann, F., McKolanis, J. R., Williams, D. L., and Ragab, A. H. (1986). Effect of retinoic acid on myeloid antigen expression and clonal growth of leukemic cells from children with acute nonmlymphocytic leukeamia—A Pediatric Oncology Group study. *Leuk. Res.* **10**, 43–50.

214. Gallagher, R., Lurie, K., Leavitt, R., and Wiernik, P. H. (1987). Effects of interferon and retinoic acid on the growth and differentiation of clonogenic leukeamia cells. *Leuk. Res.* **7**, 609–619.

215. Marth, C., Daxenbichler, G., and Dapunt, O. (1986). Synergistic antiproliferative effect of human recombinant interferons and retinoic acid in cultured breast cancer cells. *JNCI, J. Natl. Cancer Inst.* **77**, 1197–1202.

216. Fontana, J. (1987). Interaction of retinoids and tamoxifen on the inhibition of human mammary carcinoma cell proliferation. *Exp. Cell Biol.* **55**, 136–144.

217. Lotan, R. (1979). Different susceptibilities of human melanoma and breast carcinoma cell lines to retinoic acid-induced growth inhibition. *Cancer Res.* **39**, 1014–1019.

218. La Croix, A., and Lippman, M. (1980). Binding to retinoids to human breast cancer cell lines and their effects on cell growth. *J. Clin. Invest.* **65**, 586–591.

219. Halter, S., Fraker, L., Adcock, D., and Vick, S. (1988). Effect of retinoids on xenotransplanted human mammary carcinoma cells in athymic mice. *Cancer Res.* **48**, 3733–3736.

220. Thein, R., and Lotan, R. (1982). Sensitivity of cultured human osteosarcoma and chondrosarcoma cells to retinoic acid. *Cancer Res.* **42**, 4771–4775.

221. Dmitrovsky, E., Moy, D., Miller, W., Li, A., and Masui, H. (1990). Retinoic acid causes a decline in TGF-a expression, cloning efficiency, and tumorigenicity in a human embryoonal cancer cell line. *Oncogene Res.* **5**, 233–239.

222. Andrews, P. (1984). Retinoic acid induces neuronal differentiation of a cloned human embryonal carcinoma cell line *in vitro*. *Dev. Biol.* **103**, 285–293.

223. Munker, M., Munker, R., Saxton, R., and Koeffler, H. P. (1987). Effect of recombinant monokines, lymphokines and other agents on clonal proliferation of human lung cancer cell lines. *Cancer Res.* **47**, 4081–4085.

224. Doyle, L., Giangiulo, D., Hussain, A., Park, H.-J., Chin Yen, R.-W., and Borges, M.

(1989). Differentiation of human variant small cell cancer cell lines to a classic morphology by retinoic acid. *Cancer Res.* **49,** 6745–6751.

225. Sidell, N., and Horn, R. (1985). Properties of human neuroblastoma cells following induction by retinoic acid. In "Advances in Neuroblastoma Research" (A. Evans, G. D'Angio, and R. Seeger, eds.), pp. 39–53. Liss, New York.

226. Reynolds, C., Kane, D., Einhorn, P., Matthay, K. K., Crouse, V. L., Wilbur, J. R., Shunn, S. B., and Seeger, R. C. (1991). Response of neuroblastoma to retinoic acid *in vitro* and *in vivo*. In "Advances in Neuroblastoma Research" (A. Evans, G. D'Angio, A. Knudson, *et al.*, eds.), pp. 203–211. Liss, New York.

227. Peehl, D., Wong, S., and Stamey, T. (1991). Cytostatic effects of suramin on prostrate cancer cells cultured from primary tumors. *J. Urol.* **145,** 624–630.

228. Jetten, A., Kim, J., Sacks, P., Rearick, J., Lotan, D., Hong, W., and Lotan, R. (1990). Inhibition of growth and squamous-cell differentiation markers in cultured human head and neck squamous carcinoma cells by b-all-*trans* retinoic acid. *Int. J. Cancer* **45,** 195–202.

229. Sacks, P., Oke, V., Amos, B., Vasey, T., and Lotan, R. (1989). Modulation of growth, differentiation, and glycoprotein synthesis by b-*all*-trans retinoic acid in a multicellular tumor spheroid model for squamous carcinoma of the head and neck. *Int. J. Cancer* **44,** 926–933.

230. Koeffler, H., Heitjan, D., Mertelsmann, R., Kolitz, J., Schulman, P., Itri, L., Gunter, P., and Besa, E. (1988). Randomized study of 13-*cis* retinoic acid versus placebo in the myelodysplastic disorders. *Blood* **71,** 703–708.

231. Castleberry, R., Emanuel, P., Gualtieri, R., *et al.* (1991). Preliminary experience with 13-*cis*-retinoic acid (cRA) in the treatment of juvenile chronic myelogenous leukemia (JCML). *Blood* **78,** 170.

232. Wieselthier, J. S., and Koh, H. K. (1990). Sézary syndrome: Diagnosis, prognosis, and critical review of treatment options. *J. Am. Acad. Dermatol.* **22,** 381–401.

233. Grunberg, S., and Itri, L. (1987). Phase II study of isotretinoin in treatment of advanced nonsmall cell lung cancer. *Cancer Treat Rep.* **71,** 1097–1098.

234. Lippmann, S., Parkinson, D., Itri, L., *et al.* (1992). 13-*cis*-retinoic acid and interferon alpha-2a: Effective combination therapy for advanced squamous cell carcinoma of the skin. *J. Natl. Cancer Inst.* **84,** 235–241.

235. Schleicher, R., Moon, R., Patel, M., and Beattie, C. (1988). Influence of retinoids on growth and metastasis of hamster melanoma in athymic mice. *Cancer Res.* **48,** 1465–1469.

236. Fanjul, A., Hobbs, P., Graupner, G., Zhang, X.-K., Dawson, M. I., and Pfahl, M. (1994). A novel class of retinoids with selective anti-oncogene activity and their mechanism of action. Submitted for publication.

237. Stone, R. M., and Mayer, R. J. (1990). The unique aspects of acute promyelocytic leukemia. *J. Clin. Oncol.* **8,** 1913–1921.

238. Kastner, P., Perez, A., Lutz, Y., Rochette-Egly, C., Gaub, M. P., Durand, B., Lanotte, M., Berger, R., and Chambon, P. (1992). Structure, localization and transcriptional properties of two classes of retinoic acid receptors alpha fusion protein in APL: Structural similarities with a new family of oncoproteins. *EMBO J.* **11,** 629–642.

239. Miller, W. H., Kakizuko, A., Frankel, S. R., Warrell, R. P., Deblasio, A., Levine, K., Evans, R. M., and Dmitrovsky, E. (1992). RT–PCR for the rearranged retinoic receptor alpha clarifies diagnosis and detects minimal residual disease in APL. *Proc. Natl. Acad. Sci. U.S.A.* **89,** 2694–2698.

240. Muindi, J., Frankel, S. R., Huselton, C., Degrazia, F., Garland, W. A., Young, C. W.,

and Warrel, R. P. (1992). Clinical pharmacology of oral all-*trans* retinoic acid in patients with acute promyelocytic leukeamia. *Cancer Res.* **52,** 2138–2142.

241. Cornic, M., Delva, L., Guidez, F., Balitrand, N., Degos, L., and Chomienne, C. (1992). Induction of CRABP in normal and malignant myeloid cells by retinoic acid in APL patients. *Cancer Res.* **52,** 3329–3334.

242. Almasan, A., Mangelsdorf, D. J., Ong, E. S., Wahl, G. M., and Evans, R. M. (1994). Chromosomal localization of the human retinoid X receptors. In press.

VITAMINS AND HORMONES, VOL. 49

Molecular Mechanisms of Androgen Action

JONATHAN LINDZEY,* M. VIJAY KUMAR,*
MIKE GROSSMAN,† CHARLES YOUNG,*
AND DONALD J. TINDALL*,†

*Departments of *Urology and †Biochemistry/Molecular Biology*
Mayo Foundation,
Rochester, Minnesota 55905

I. Introduction

Androgens are pervasive in their effects on the development, maintenance, and regulation of male phenotype and reproductive physiology in adult males. In addition, androgens are implicated in a number of human pathologies. Despite the wide variety of target tissues, the basic sequence of events leading to androgen effects on gene transcription do not vary considerably from tissue to tissue. Testosterone (T), the dominant testicular androgen, diffuses passively into the cell and either binds directly to the androgen receptor (AR), undergoes enzymatic reduction to 5α-dihydrotestosterone (DHT), or undergoes aromatization to estrogens. Once T or DHT binds to the AR, the protein undergoes conformational changes, chaperone proteins such as heat shock proteins dissociate from the receptor, and the activated receptor can then bind DNA (Fig. 1) (for reviews, see Johnson *et al.*, 1988; Chan *et al.*, 1989). The androgen-activated AR binds to specific DNA enhancer sequences called androgen-responsive elements (AREs). Once anchored to the ARE, the AR is able to regulate transcriptional activity in either a positive or negative fashion. This review focuses on the mechanisms by which testosterone and its androgenic metabolites such as DHT act directly upon target genes through the AR.

The review will provide an overview of (1) the normal physiological actions of androgens, (2) involvement of androgens in pathologies, (3) structure–function relationships of the AR protein, (4) regulation of

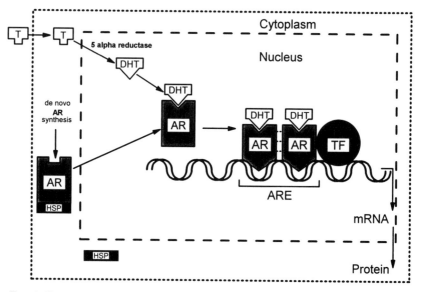

FIG. 1. Steps involved in mechanisms of androgen action. Circulating testosterone (T) enters the cell, undergoes reduction to 5α-dihydrotestosterone (DHT), and binds to the androgen receptor (AR). The activated DHT–AR complex binds to specific androgen response elements (AREs), interacts with other transcription factors (TF) and elements of the basal transcriptional machinery, and modulates transcriptional activity.

the androgen receptor gene, and (5) molecular mechanisms by which androgens regulate androgen-responsive genes. Our intent is to provide a review of the current understanding of the molecular mechanisms by which androgens act and suggest directions for future research. Thus, the first two sections are designed to provide a brief background for the latter sections that deal with our current understanding of the molecular mechanisms of androgen action.

II. Physiological Actions of Androgens

ARs are expressed in a wide variety of tissues—neural, muscular, skeletal, and endocrine tissues and cell types. Indeed it is difficult to find a tissue that fails to express some level of AR with the notable exception of the spleen. Many androgen target tissues display sexual dimorphisms in morphology and function. Early observations that testicular androgens masculinize accessory sex structures and sex behaviors led to the organizational hypothesis that early exposure to androgens permanently alters the structure and, hence, function of organs in the adult animals. The same testicular androgens then act in adulthood to activate various behavioral and physiological events. The organizational–activational hypothesis has played a key role in guiding research on the effects of androgens. In the following, we briefly discuss the role of androgens in development and regulation of the male reproductive tract, external genitalia, sebaceous glands, muscle, and brain and behavior.

A. Internal Reproductive Structures and External Genitalia

Sex determination occurs through either genetic or temperature-dependent mechanisms (Crews, 1993) that dictate development of testes or ovaries from the urogenital ridge. The gonadal sex (testis or ovary) in turn dictates the endocrine milieu to which the anlagen of the internal sex structures (Wolffian and Müllerian derivatives) and external genitalia are exposed and, thus, the path of differentiation.

Early studies by Jost (1953, 1971) demonstrated that the testes are responsible for determining the differentiation of the genital ducts in mammals. In the testis, maturing Leydig cells secrete T and Sertoli cells secrete Müllerian inhibiting substance (MIS) (Josso et al., 1977). MIS suppresses development of Müllerian derivatives (uterus and fallopian tubes) and circulating T causes the mesonephric/Wolffian ducts to differentiate into the epididymis, vas deferens, and seminal vesicles. The effects of T on these structures appear to be direct as 5α-reductase is not expressed during the critical periods of development (Siiteri and Wilson, 1974). Furthermore, aromatization clearly does not account for T effects on Wolffian derivatives because fetal ovaries do not promote differentiation and growth and antiandrogens such as cyproterone acetate inhibit the effects of T on Wolffian derivatives (Neumann *et al.*, 1970). Interestingly, this effect appears to be local in that unilateral castration inhibits Wolffian duct differentiation on that side. These data suggest that very high levels of T are needed for differentiation, a finding consistent with reports that the affinity of AR for T is less than that for DHT.

Differentiation of external genitalia from the genital tubercle, urethral folds, labioscrotal folds, and urogenital sinus is mediated by DHT. The genital tubercle gives rise to the glans penis and corpus cavernosum whereas the urethral folds and labioscrotal swellings give rise to portions of the penis and scrotum. The urogenital sinus forms accessory glands such as the prostate and bulbourethral glands. Complete differentiation of the external genitalia and urogenital sinus occurs only if androgens are available during the critical fetal periods (8–12 weeks for humans) and subsequently during puberty. During this period, 5α-reductase is expressed in the anlagen of the external genitalia. Indeed, in 5α-reductase-deficient individuals the Wolffian structures develop normally under the influences of T whereas the external genitalia and prostate are underdeveloped or absent (Grumbach and Conte, 1992). The effects of DHT appear to be on the mesenchyme, which then mediates differentiation of epithelial components of the tissues (Cunha *et al.*, 1980). At puberty the hypothalamic–pituitary–gonadal axis is reactivated and increasing circulating concentrations of T complete differentiation of secondary sex characteristics.

In the adult male, continuing exposure to T is required to maintain the morphological differentiation of a number of structures and to activate various physiological processes. For instance, castration leads to involution of the prostate and seminal vesicles, and exogenous T or DHT restores the secretory epithelia to normal morphology and function. In addition, testicular T plays a critical role in regulating spermatogenesis, steroidogenesis, and feedback on the hypothalamus and pituitary.

B. SEBACEOUS GLANDS

In some mammals aggregates of sebaceous glands are organized into glands that are utilized for olfactory communication. These glands are

typically androgen regulated and have been used as models for androgen regulation of sebaceous glands. Some of these are the hamster
flank organs (Frost and Gomez, 1972), ventral glands of gerbil
(Thiessen *et al.*, 1969), supracaudal glands of the guinea pig (Martin,
1962), and side glands of the musk shrew (*Suncus murinus*) (Dryden
and Conaway, 1967). Androgen receptors have been localized in
sebaceous glands in the human scalp, temple, forehead, chin, neck,
genitalia, and leg (Choudhry *et al.*, 1992) and side glands (Komada *et
al.*, 1989).

Sebaceous glands are functional in both human males and females
with hyperplasia occurring in both male and female adolescents (Pochi
and Strauss, 1974). Human sebaceous glands are stimulated by adrenal glands in newborns (Agache *et al.*, 1980), preadolescent children
(Pochi *et al.*, 1979), and adult females (Lookingbill *et al.*, 1985). Among
the animal models, flank organs of the male hamster are larger than
those of the female (Hamilton and Montagna, 1950) whereas the side
glands of the *Suncus murinus* are well developed in both the sexes. As
with humans, adrenal androgens are important for the maintenance of
female sebaceous gland activity (Sonada *et al.*, 1991).

Castration or treatment of intact males with the antiandrogens cyproterone acetate (Cunliffe *et al.*, 1969; Burkick and Hill, 1970; Pye *et
al.*, 1976; Vermorken *et al.*, 1986; Neumann and Kalmus, 1991), flutamide (Lutsky *et al.*, 1975; Lyons and Shuster, 1986) and spironolactone (Weismann *et al.*, 1985; Walton *et al.*, 1986) results in atrophy
of sebaceous glands in humans and animals (for review, see Thody
and Shuster, 1989). Administration of T or DHT increases the size of
the sebaceous glands, its mitotic activity, and lipogenesis (Pochi and
Strauss, 1974; Pochi *et al.*, 1979; Thody and Shuster, 1989). Furthermore, adult subjects with complete androgen insensitivity syndrome
have undetectable sebum production that is similar to that of pre-
adrenarchal children (Imperato-McGinley *et al.*, 1993). In contrast to
androgens, estrogens inhibit the activity of the sebaceous glands in
human and animal models (Thody and Shuster, 1989; Ebling and Skinner, 1983).

In sebaceous glands, T is converted into DHT by 5α-reductase (Takayasu and Adachi, 1972; Bonne and Raynaud, 1977; Komada *et al.*,
1989). However, conflicting results question the role of DHT as the
primary androgen regulating sebaceous gland function. In one study,
5α-reductase inhibitors reduced sebaceous gland function in hamsters
(Takayasu and Adachi, 1970). However, topical treatment of hamster
flank organs with the androgen antagonist 17α-propylmesterolone
and 5α-reductase inhibitors, such as 4-androsten-3-one-17β-carboxylic

acid and 17β-*N*,*N*-diethylcarbamoyl-4-methyl-4-aza-5α-androstan-3-one, suggests that DHT has no regulatory function in the growth and maintenance of sebaceous glands (Schroder *et al.*, 1989). Also, treatment of men with finasteride, a 5α-reductase inhibitor, did not decrease sebum production (Imperato-McGinley *et al.*, 1993), and male pseudohermaphrodites with 5α-reductase deficiency have normal sebum production. Thus, 5α-reductase may not play as important a role in sebaceous gland function as in other androgen-dependent targets.

C. Muscle

The observation that skeletal muscle is more developed in the male than in the female led to the supposition that androgens might be responsible for myotropic or anabolic action. Androgen receptors have been characterized in normal skeletal muscle (Michel and Baulieu, 1976; Celotti and Negri-Cesi, 1992; Mooradian *et al.*, 1987), though sexually dimorphic targets such as the levator ani muscle and frog laryngeal muscles contain higher levels of androgen receptor compared to other skeletal muscles (Max *et al.*, 1981).

Sex differences in skeletal muscle characteristics and metabolic profiles include higher aerobic and strength performance capacity in male humans (Komi and Karlsson, 1978) and increased enzymatic activities of lysosomal hydrolases and the mitochondrial enzyme cytochrome c oxidase in the gastrocnemius and soleus of male mice. Administration of T increases enzyme activities, muscle hypertrophy (Koenig *et al.*, 1980), ^{14}C-labeled tyrosine incorporation, and the rate of muscle protein synthesis in rats (Fernandez-Gonzalez *et al.*, 1989). Administration of T (Mooradian *et al.*, 1987; Fernandez-Gonzalez *et al.*, 1989; Hayden *et al.*, 1992) or synthetic anabolic steroids (Michel and Baulieu, 1976; Choo *et al.*, 1991; Lukas, 1993) alters body composition, leading to an accumulation of muscle mass.

Sexually dimorphic muscles provide excellent model systems for studying androgen effects. The muscles of the larynx are developmentally influenced by androgens and sexually dimorphic in the adult African clawed frog. At metamorphosis the larynx is similar in both sexes—in size, fiber number, and content of the androgen receptor. However, by 3 months postmetamorphosis autoradiography demonstrates sexually dimorphic DHT binding levels (Kelley *et al.*, 1989). Laryngeal muscles of adult male frogs have three- to fourfold higher androgen receptor content, greater numbers of fibers (32,000 muscle fibers versus 4000 in the female) (Sassoon and Kelley, 1986), and different fiber types compared to females (for review, see Kelley, 1986).

Androgens also regulate the smooth muscle in the ductus deferens, prostate, seminal vesicle, and penis. Androgen receptors are present in the smooth muscle cells of the prostate (Ricciardelli *et al.*, 1989), ductus deferens (Syms *et al.*, 1987; Harris *et al.*, 1989), seminal vesicle (West *et al.*, 1990), and penis (Gonzalez-Cadavid *et al.*, 1991, 1993; Lin *et al.*, 1993). Also, a smooth muscle tumor cell line derived from ductus deferens (DDT1MF-2) has been shown to proliferate in response to androgens (Syms *et al.*, 1987; Harris *et al.*, 1989).

In rats, the levator ani/bulbocavernosus muscle complex contains AR (Dube *et al.*, 1976) and the fibers undergo an androgen-dependent increase in size (Venable, 1966). These androgen-sensitive muscle cells are innervated by an androgen-sensitive group of neurons originating in the spinal nucleus of the bulbocavernosus (see following section).

D. Brain and Behavior

In many vertebrates, testicular androgens masculinize the developing nervous system in terms of morphology (at both micro- and macroscopic levels) and function. Sex differences in brain structure include larger preoptic area (POA) nuclei in male rats and humans, larger spinal nuclei that innervate penile musculature, and larger song nuclei in male song birds (for reviews, see Fishman and Breedlove, 1988; Arnold and Gorski, 1984; Gorski, 1984). Functional correlates of these anatomical dimorphisms include dimorphic sex behaviors and gonadotropin-releasing hormone (GnRH) secretion patterns, dimorphic perineal musculature, and dimorphic song repertoire.

The mechanisms by which testicular androgens masculinize the nervous system vary from species to species and from site to site. In rats, aromatization of T to 17β-estradiol (E_2) appears to be responsible for increased size of the sexually dimorphic nucleus (SND) of the POA. The role of aromatization may not be as important in humans (Breedlove, 1992). The masculinizing effect of E_2 in some vertebrates would appear to pose a dilemma for female fetuses in which circulating E_2 is elevated. However, masculinization of female fetuses is prevented because high levels of α-fetoprotein sequester circulating E_2 in female fetuses but not circulating T in male fetuses.

In adult males, circulating levels of T serve to regulate both behaviors and secretion of hormones (i.e., GnRH, FSH, LH) involved in reproduction. Castration of most male vertebrates leads to cessation of male-typical reproductive behaviors and exogenous T restores the behaviors (Baum, 1992; Moore and Lindzey, 1992). In addition, T feedback on nuclei of the POA and gonadotropes regulates secretion of GnRH and gonadotropins. In a number of species, the behavioral effects of T are due to both direct interaction of T or DHT with AR and indirect actions via aromatization into E_2 (Baum, 1992).

In humans the link between testicular androgens and sex behaviors is more tenuous. Castration often reduces sex drive and activity but this varies between individuals (Carter, 1992). Furthermore, it is not known to what extent E_2 is involved in either differentiation of neural structures or activation of sex behaviors.

III. Pathological Actions of Androgens

Androgens have been implicated in pathologies ranging from seriously debilitating diseases such as prostate cancer to the benign manifestations of male-pattern baldness. In this section we will discuss the link that androgens have to several pathologies and how research into the molecular mechanisms of androgen action may shed light on the etiology of these pathologies.

A. Androgen Insensitivity

Androgen insensitivity syndrome (AIS) is an X-linked, heritable trait that arises from a mutation(s) within the AR gene. AIS manifests itself with variable severity, ranging from complete AIS (CAIS) in which a genetic male develops as a phenotypic female to partial AIS (PAIS) in which external genitalia are masculinized to varying degrees. Some researchers have drawn further distinctions between PAIS, minimal AIS, and Reifenstein syndrome: PAIS individuals have some degree of ambiguity in external phenotype but minimal AIS and Reifenstein individuals are unambiguously male with less obvious deficits such as infertility. We choose to view minimal AIS and Reifenstein as points on a continuum of PAIS rather than separate entities.

CAIS is characterized by an XY genotype coupled with female external genitalia, blunt-ended vagina, absent uterus, and abdominal or inguinal testes (see Grumbach and Conte, 1992). The testes contain functional Leydig cells that produce normal or supranormal levels of T in adults, although because of androgen insensitivity Wolffian derivatives are underdeveloped. Despite reasonably normal levels of T, germ cells and spermatogenesis are absent from the seminiferous tubules.

The Sertoli cells also secrete MIS, resulting in regression of Müllerian derivatives (Brown *et al.*, 1993). Many of the CAIS features are similar to the phenotypes of 5α-reductase-deficient individuals with the exception that Müllerian and Wolffian derivatives are absent in CAIS. PAIS individuals are characterized by an XY genotype and varying phenotypes that may include microphallus, hypospadia, gynecomastia, and infertility (see Grumbach and Conte, 1992).

In theory the underlying causes of AIS can be due to any mutation that alters one of the critical functions of the AR, such as steroid binding or DNA binding. Characterizations of human AIS and the murine equivalents, referred to as testicular feminized male (Tfm), have revealed an array of different types of mutations that lead to a deficient AR and the resulting feminization. These include (1) point mutations that introduce amino acid substitutions within the steroid binding or DNA binding domains, (2) point mutations that introduce a stop codon causing premature termination, (3) point mutations that introduce aberrant splicing, or (4) deletion of parts or, in extreme cases, the entire gene (a partial listing of AIS mutations is presented in Table I).

Mutations that abolish steroid and DNA binding both result in feminization but the characteristics of the dysfunctional AR differ. This leads to different diagnostic features when examining AR function in genital skin fibroblasts (GSF) of patients. Patients with an altered DNA binding domain (DBD) typically express normal quantities and sizes of AR mRNA and protein according to Northern analysis, receptor binding studies, and immunodetection. These patients are referred to as receptor-positive AIS based on the ability of the AR to bind androgens. Mutations within the steroid binding domain (SBD) that prevent AR detection by ligand binding assays are referred to as receptor negative. Point mutations that lead to premature stop signals, aberrant splicing, or deletion of portions of exons can lead to either receptor-positive or receptor-negative AIS depending on the affected domain(s).

There appears to be a fairly good correlation between the severity of AIS and the extent to which a mutated AR is impaired in various *in vitro* assays such as ligand binding, DNA binding, and transactivation of reporter constructs. For instance, in six PAIS cases where molecular mechanisms and receptor characteristics were quantified (see Table I), normal amounts of receptor were expressed in 100% of the cases. There were, however, less obvious defects such as increased K_d, rapid dissociation, and thermolability associated with the mutant ARs. However, in CAIS approximately 50% of the cases listed in Table I are characterized by negligible steroid binding and 40% are receptor posi-

TABLE I
Characteristics of Mutations Leading to Androgen Insensitivity[a]

Degree of AIS	Position	WT	Mutant	Ligand binding capacity	K_d	Transactivation	Reference
CAIS	114	Gln	Stop	Reduced	Normal	→	Zoppi et al. (1993)
CAIS	588	Lys	Stop	−	ND	−	Marcelli et al. (1990a)
CAIS	614	Arg	His	+, some decrease	Normal	No[b]	Mowszowicz et al. (1993)
CAIS	615	Arg	Pro	+		No	Marcelli et al. (1991b); Zoppi et al. (1992)
CAIS	686	Asp	His	→Normal	ND	→→	Ris-Stalpers et al. (1991); Brinkmann et al. (1991)
CAIS	686	Asp	Asn	Normal	↑, rapid dissociation	→	
CAIS	717	Trp	Stop	→	ND	ND	Sai et al. (1990)
CAIS	749	Met	Val	ND	ND	ND	Jakubiczka et al. (1992)
CAIS	754	Phe	Val	−	ND	ND	Lobaccaro et al. (1993)
CAIS	765	Ala	Thr	ND	ND	ND	Sweet et al. (1992)
CAIS	772	Arg	Cys	− GSF + COS	←	→	Marcelli et al. (1991a)
CAIS	773	Arg	Cys	−	ND	ND	Prior et al. (1992)
CAIS	773	Arg	His	Normal	→ Thermolabile	ND	
CAIS	774	Arg	Cys	−	ND	− →	Brown et al. (1990)
CAIS	774	Arg	His	+	→ ↓ ←	→	DeBellis et al. (1992)
CAIS	786	Met	Val	− GSF + COS	←	ND	Nakao et al. (1992)
CAIS	794	Trp	Stop	−	−	ND	Marcelli et al. (1990b)
CAIS	820	Gly	Ala	+	Normal but thermolabile	ND	Kasumi et al. (1993)
CAIS	831	Arg	Gln	−	ND	−	Brown et al. (1990)
CAIS	864	Asp	Gly	−	ND	ND	DeBellis et al. (1992)

Phenotype	Location/codon	Wild type	Mutant	Binding	Binding properties	Transactivation	Reference
CAIS	865	Val	Met	Normal	Normal	↓	Kazemi-Esfarjam et al. (1993)
CAIS	866	Val	Met	Normal	↓ ←	↓	Lubahn et al. (1989b)
CAIS	866	Val	Met	+	←		Brown et al. (1990)
CAIS	Aberrant splicing			−	ND	↓ ↓	Ris-Stalpers et al. (1990, 1992)
CAIS	Exon 5	Deletions		↓	ND	ND	MacLean et al. (1993)
CAIS	Exons 6, 7	Deletions		−	ND	ND	Quigley et al. (1992a)
CAIS	Complete deletion of AR gene				ND	No[b]	Quigley et al. (1992a)
CAIS	Exon 3			↑	Normal	ND	Quigley et al. (1992b)
PAIS	607	Arg	Gln	ND	ND	ND	Wooster et al. (1992)
PAIS	607	Arg	Lys	ND	ND	↓	Saunders et al. (1992)
PAIS	743	Gly	Val	Normal	Normal, thermalobile	Normal	Nakao et al. (1993)
PAIS	761	Tyr	Cys	+	Normal, rapid dissociation, thermolabile[c]	Reduced[c]	McPhaul et al. (1991)
PAIS	772	Ala	Thr	+	Normal	↓	Klocker et al. (1992)
PAIS	855	Arg	His	Normal, thermolabile	↑	↓	Batch et al. (1993)
PAIS	869	Ile	Met	Normal, thermolabile	↑	↑	Saunders et al. (1992)
PAIS	866	Val	Leu	ND	ND	Normal	Saunders et al. (1992)
LNCaP	868	Thr	Ala	Normal	Normal, but altered specificity	Normal	Veldscholte et al. (1990b)
Tfm rat	734	Arg	Gln	↓	Normal	Reduced	Yabrough et al. (1990)
Tfm mouse	412	Val	Stop	↓	Normal	↓	He et al. (1991)

[a] Note: Numbering systems used by different authors can vary. ND = not done; + = detectable; − = nondetectable or reduced; ↑/↓ = increase or decrease.

[b] Decreased AR–DNA interactions correlate with decreased transactivation.

[c] Mutation of 761 Tyr→Cys exhibits increased thermolability and reduced transactivation capacity only when combined with partial deletion of a polyglutamine stretch in the NH₂.

tive. Another 10% show some decreases in steroid binding activity. Thus, severe impairment of steroid binding is almost invariably correlated with CAIS though other more subtle mutations can also lead to CAIS.

In addition to amino acid substitutions, point mutations can lead to alternative splicing, as in one case of AR-negative AIS. A mutation was found in the donor site of intron 4 that eliminates normal splicing at exon 4–intron 4 and deletes 123 nucleotides from the mRNA. This mutation eliminated steroid binding in GSF cells and overexpressed mutant AR (mutAR). Not surprisingly, the mutAR was unable to transactivate a reporter construct. In another family, two AIS siblings lack exon 5 whereas their aunt is missing exons 6 and 7. Interestingly, both deletions may originate from the same area of intron 5 (MacLean *et al.*, 1993). The most extreme example of a deletion mutation is found in a family with CAIS individuals (receptor negative) in which the entire AR gene has been deleted (Quigley *et al.*, 1992a).

The Tfm mouse presents a receptor-negative case of AIS in an animal model. The phenotypic anomalies found in a Tfm mouse parallel those found in human CAIS. Early experiments indicated that the development and physiological resistance to androgens correlated with very low levels of steroid binding and a truncated protein (~66 kDa) compared to the wild type (~110 kDa) (Attardi and Ohno, 1974, 1978; Young *et al.*, 1989). Following cloning of the AR gene, additional experiments demonstrated that although mRNA size was normal the quantities detected by Northern analysis were very low and exhibited decreased stability relative to the wild type (Gaspar *et al.*, 1990, 1991; Charest *et al.*, 1991). Comparison of the Tfm AR gene to the wt gene revealed a single-base-pair deletion (within a hexacytidine stretch) that shifts the reading frame and creates a stop codon at amino acid 412 (Charest *et al.*, 1991; Gaspar *et al.*, 1991; He *et al.*, 1991). This results in production of a truncated amino-terminal protein (52 kDa) that lacks steroid and DNA binding domains. Work in this and other laboratories demonstrated that reinitiation from an internal ATG (Met) produces a carboxy-terminal protein of 45 kDa (Gaspar *et al.*, 1991; He *et al.*, 1994). This carboxy product accounts for the residual steroid binding activity and smaller form of AR detected in Tfm tissues. Overexpression of the Tfm AR in a cell line results in a reduced number of receptors with normal affinity compared to the wild-type AR.

The Tfm rat exhibits very reduced levels of AR binding activity but normal-sized protein. Sequencing of the gene revealed an Arg734Gln mutation within the SBD (Yarbrough *et al.*, 1990). This highly conserved Arg appears to be critical for normal steroid binding as *in vitro*-expressed mutAR has only 10–15% of the binding capacity of wt AR. As with human AIS, the decreased steroid binding activity is associated with decreased transactivation.

B. 5α-Reductase Deficiency

Although T and DHT both interact with the androgen receptor, DHT has a twofold higher affinity for binding and dissociates at one-fifth the rate of T (Grino *et al.*, 1990). Thus, failure of steroid 5α-reductase to convert T into DHT results in pseudohermaphroditism in which genotypic males (46 X,Y) have masculinized internal urogenital tracts but external genitalia that are either ambiguous or feminized (Griffin and Wilson, 1989). The levels of T present during embryogenesis are able to induce the virilization of the Wolffian ducts into the epididymis, vas deferens, and seminal vesicles but the external genitalia require DHT or higher levels of T. In many individuals the increased levels of T present at puberty cause some masculinization of secondary sex characteristics (enlargement of the clitorallike penis, appearance of facial hair, and significant muscular development) (Imperato-McGinley *et al.*, 1991).

The reason that some tissues undergo virulization while others do not is unclear but may be linked to the fact that two isozymes of the 5α-reductase gene are differentially expressed (Jenkins *et al.*, 1991, 1992; Thigpen *et al.*, 1993). They have been designated 5α-reductase 1 and 2 (Andersson *et al.*, 1991). The isoenzyme 5α-reductase-2 is responsible for the differentiation of the male external genitalia and prostate, whereas the function of 5α-reductase 1 is still unknown. Deletion or mutation of the 5α-reductase 2 isozyme has been reported in 23 families that contained individuals who had 5α-reductase deficiency (Thigpen *et al.*, 1992). No patients with pseudohermaphroditism have been shown to contain any mutations in the 5α-reductase 1 isozyme (Jenkins *et al.*, 1992). Thus, deficiencies in the 5α-reductase type 2 isozyme appear to be responsible for some forms of male pseudohermaphroditism.

C. X-Linked Spinal Bulbar Muscular Atrophy

X-linked spinal bulbar muscular atrophy (Kennedy's disease or SBMA) is a motor neuron disease characterized by adult onset of symptoms dominated by progressive muscle weakness and atrophy (Harding *et al.*, 1982; Choi *et al.*, 1993). Interestingly this malady was often associated with signs of PAIS such as gynecomastia and decreased fertility (La Spada *et al.*, 1991; Choi *et al.*, 1993). Linkage analysis indicated that the "SBMA gene" was located near the AR gene (Xq11-12). Subsequent cloning and sequencing of the AR gene from SBMA patients revealed heterogeneity in a CAG repeat near the 5′ end of the coding region (Fig. 2). Affected individuals possessed approximately 75 bp of additional CAG repeats coding for the amino acid Gln (La Spada *et al.*, 1991; Matsuura *et al.*, 1992), leading to the hypothesis that the longer polyglutamine tract causes SBMA.

Amato *et al.* (1993) report no correlation between the severity of the symptoms and the length of the repeats. However, Igarashi *et al.* (1992) report that the length of the repeat correlates well with the age of onset. Another study demonstrated a tendency for increases in the repeats with successive generations (Biancalana *et al.*, 1992). The functional effects of increased polyglutamine tracts are unknown but two hypotheses seem viable. First, longer polyglutamine tracts may alter the transactivating capacity of the AR. Second, the increased polyglutamine tract may facilitate cross-linking of the AR to other proteins by transglutaminase activity (Green, 1993). Either hypothesis involves some deficiency of androgen responsiveness, though the lack of SBMA symptoms in AIS patients argues for a more complex etiology than a simple deficit in androgen responsiveness. The reason for late adult onset is unclear but may relate to low rate of physiological changes owing to mutations and a cumulative effect on the viability of the spinal neurons.

D. Benign Prostate Hyperplasia and Prostate Cancer

Androgens play a key role in the growth and function of the prostate and are therefore implicated in the etiology of benign prostate hyperplasia (BPH) and prostate cancer. Surgical removal of fetal testis or chemical abrogation of androgenic activity with antiandrogen or estro-

Fig. 2. Schematic of the functional domains of the androgen receptor. The steroid binding and DNA binding domains (SBD and DBD, respectively) are located in the carboxy (COOH) end of the androgen receptor protein. Transactivating functions are located predominantly in the amino end of the protein, including a polyglycine repeat (Gln_n). The nuclear localization signal (NLS) resides in the hinge region between the SBD and DBD.

gen inhibits development of prostate (Cunha *et al.*, 1987). Maintenance of the structural and functional activity of the adult prostate is also dependent on the continued presence of androgens (Cunha *et al.*, 1987; Klein *et al.*, 1991). Androgens regulate the growth, differentiation, and morphogenesis of the prostate epithelium indirectly via the surrounding stroma (Cunha *et al.*, 1987). The mechanism by which the stroma influence the epithelium is unclear, although components of extracellular matrix and growth factors have been implicated.

BPH is a nonmalignant neoplasm of the prostatic stroma and epithelial tissue resulting in the enlargement of the prostate. BPH is a very common disease affecting 80% of all men over 60 years of age (Berry *et al.*, 1984). As normal prostatic growth is regulated by androgens, it has been suspected that BPH is under androgen control (Coffey and Walsh, 1990). One study demonstrated higher DHT content (three- to fourfold) in BPH tissue (Siiteri and Wilson, 1970; Krieg *et al.*, 1979; Trachtenberg *et al.*, 1980). However, other studies indicate that DHT content of normal and BPH tissue is very similar (Walsh *et al.*, 1983; Coffey and Walsh, 1990).

Androgen receptors have been detected in normal, BPH, and cancerous prostate tissue (Ekman *et al.*, 1979; Fentie *et al.*, 1986; Rennie *et al.*, 1988). One study reported that nuclear AR levels were increased in BPH (Barrack *et al.*, 1983). However, using immunohistochemical analysis, it was demonstrated that there was no statistically significant change in the intensity of nuclear AR in BPH compared to in normal prostate (Chodak *et al.*, 1993; Miyamoto *et al.*, 1993). Thus, it is difficult to implicate androgens in BPH by such simple measures as content of DHT and/or AR. With recent advances in molecular biology it may now be possible to detect differences in the actions of androgens in normal and BPH tissues.

Prostate cancer is the second leading cause of cancer-related death in American men. In men over 50 years of age, 10 to 30% have latent microscopic prostate cancer, which resides asymptomatically in the prostate gland (Coffey, 1993). Androgen ablation therapy (surgical or chemical castration) is a common treatment for prostate cancer and produces a dramatic reduction in cancer mass and clinical remissions in 80% of patients with advanced metastatic cancer (for review, see Geller, 1993). In addition to orchiectomy, the available treatment options include estrogens, luteinizing hormone releasing hormone analogues, and antiandrogens, either individually or in combination (Coffey, 1993; Daneshgari and Crawford, 1993; Denis, 1993; Geller, 1993; Labrie *et al.*, 1993; McLeod, 1993; Soloway and Matzkin, 1993).

Unfortunately, a marked heterogeneity in response to castration has been noticed between different patients and different tumors in the same patient. Two mechanisms have been suggested to explain such variation in endocrine control of prostate cancer: continued androgenic stimulation due to adrenal androgens and presence of androgen-independent cells in heterogeneous carcinomas (Klein *et al.*, 1991). Positive clinical responses increase when antiandrogens are added to the foregoing treatment (Labrie *et al.*, 1993), suggesting the presence of androgen-hypersensitive prostate cells in some of the tumors. Alternatively, tumor progression after androgen deprivation could be due to the emergence of androgen-insensitive clones of prostate cancer cells (Isaacs and Kyprianou, 1987) or the influence of growth factors (Steiner, 1993).

IV. Structure and Function of the Androgen Receptor

To understand how androgens regulate normal physiology and pathological states, it is critical to gain an understanding of the structure of the receptor and how the structure relates to the functions of the receptor. These efforts have been facilitated by the cloning of AR cDNAs from human (Trapman *et al.*, 1988; Tilley *et al.*, 1989; Faber *et al.*, 1989; Lubahn *et al.*, 1988, 1989a,b; Brinkmann *et al.*, 1989; Chang *et al.*, 1988a,b), rat (Chang *et al.*, 1988a,b), and mouse (He *et al.*, 1990). Studies have demonstrated that the androgen receptor is coded for by eight exor (Faber *et al.*, 1989). The first exon codes for the amino terminus, which contains transactivating functions and a number of phosphorylation sites. Exons 2 and 3 code for sequences containing the first and second zinc fingers of the DNA binding domain (DBD, Fig. 2). Part of exon 4 codes for the hinge region, which contains a nuclear localization signal (NLS, Fig. 2). Exons 4 through 8 code for the steroid binding domain (SBD, Fig. 2). Though the AR gene and protein appear to be quite modular in nature, the modular functions of the AR act in a coordinated fashion such that disruption of one activity disrupts the normal actions of the AR.

Historically, many of the data concerning structure–function relationships have arisen through studies focusing on mechanisms underlying androgen insensitivity in humans, rats, and mice. More recently, artificially introduced mutations have provided insight into the critical regions and amino acids. A discussion of the data concerning these four activities follows.

A. Steroid Binding Domain

The SBD consists of approximately 250 amino acids (653–910) at the carboxy terminus of the AR. This region forms a hydrophobic pocket that exhibits high affinity and specificity for androgens and thus provides the key to specificity of steroid–AR interactions and regulation of physiological events by androgens. The interaction between the SBD and steroid is relatively diffuse in that mutations in a number of amino acids in different areas result in altered binding characteristics (see Table I), which makes it difficult to draw generalizations concerning the structure–function relationship of the SBD. However, this in itself argues for a complex interaction of different amino acids to achieve high-affinity, specific steroid binding.

Despite the lack of obvious patterns, a number of interesting observations and conclusions can be made based on cases of naturally occurring mutants. First, mutations tend to cluster in a few stretches of the SBD, indicating that these areas are critical for normal function. Second, mutants in the SBD are influenced by mutations in the amino terminus, indicating an interaction between amino and carboxy termini. Third, different substitutions at the same position have demonstrably different effects on steroid binding and human phenotype. Fourth, some mutations have significant effects on specificity of steroid binding.

The locations of AR mutations from a number of unrelated AIS patients have been reviewed (McPhaul *et al.*, 1992). It was found that the mutations clustered in three areas: one portion of the DBD and two regions of the SBD (726 to 772 and 826 to 864). Interestingly, the 726 to 772 stretch is highly homologous to a hot spot for mutations in the ligand binding domain of the thyroid receptor from patients with thyroid hormone resistance (McPhaul *et al.*, 1993). This region is also highly homologous to regions of the progesterone (PR), glucocorticoid (GR), and mineralocorticoid receptors (MR). Such homology suggests that this region is critical for receptor–ligand interactions in the steroid receptor superfamily. Also, the importance of this region to ligand binding in receptors with very different specificity implies that the basis for steroid specificity either (1) resides in other regions, (2) derives from the interactions of multiple amino acids that contribute in a cooperative fashion to specificity, or (3) resides in a few nonconserved nucleotides within this region. Further mutational analysis of these regions should shed light on mechanisms dictating steroid specificity.

Another case of AIS provides evidence of a complex interplay be-

tween the amino acid and carboxy ends of the receptor. It was determined that a mutation in the carboxy terminus (Tyr761Cys) causes
rapid dissociation of androgen, whereas a polyglutamine deletion in
the amino terminus interacted with the Tyr761Cys mutant to affect
thermolability (McPhaul et al., 1991). These data indicate that the
carboxy and amino termini interact to affect the key functions of the
SBD such as stable, high-affinity binding. The mechanisms underlying these interactions have yet to be determined but may include physical interactions between amino and carboxy termini that affect conformation, and/or interaction of amino termini amino acids with
accessory factors that contribute to stable AR–androgen interactions.

Two AIS-related mutations demonstrate that different mutations at
the same codon can lead to significantly different function. An
Arg773Cys mutation resulted in little or no steroid binding activity
whereas the Arg773His mutation resulted in normal binding capacity
(Prior et al., 1992). Similarly, Asp686His resulted in no binding and
Asp686Asn caused more subtle changes in the binding characteristics
(Brinkmann et al., 1991). In both instances these mutations fall within
a stretch of amino acids that are highly conserved between PR, GR,
and mineralocorticoid receptors. These mutations tell us that the type
of amino acid substitution affects the interaction between SBD and
steroid and that manipulation of amino acids with different sizes and
charges may yield insight into the nature of ligand binding.

Some of the most interesting mutations are those that change specificity of steroid binding. In a prostate cell line (LNCaP), a Thr868Ala
mutation results in altered specificity that allows this mutant AR to be
partially activated by progestins (Veldscholte et al., 1990a). Additional
studies demonstrate that substitution of Asp, Lys, or Tyr at 868 abolishes androgen binding and transactivating activity. However, substitution of Ser or Ala further relaxed the specificity of steroid binding
(Ris-Stalpers et al., 1993). These data indicate that codon 868 of the
human AR is critical for influencing steroid specificity of human AR.

Artificially induced mutations in the SBD of AR have not been pur-
sued extensively. However, truncation of the last 12 amino acids of the
SBD resulted in decreased steroid binding and transactivational capacity of the human AR (Jenster et al., 1991). Additional experiments
(Jenster et al., 1991) provide more insight into the nature of the interaction between SBD and transactivation domains. Deletions of the
SBD resulted in an AR that exhibited constitutive activity equaling
approximately 40–90% of that of the steroid-activated wild-type AR.
These data indicate that in the absence of bound androgen the SBD
actually inhibits the transactivation domain. It will be of interest to

determine if phosphorylation of the AR can supplant the need for androgen to induce transactivation.

B. DNA Binding Domain

The DBD contains two zinc fingers that provide the basis for protein–DNA interactions in the regulatory sequences of target genes. AR and other steroid receptors contain nine conserved cysteine residues of which eight interact in a coordinated fashion to form two separate tetrahedral metal binding complexes or "fingers" (Berg, 1990). The four cysteine residues in each finger bind Zn^{2+}, which permits interaction with DNA. Removal of zinc abolishes, whereas addition of Zn^{2+} reinstates, functional activity.

Based on analysis of GR and estrogen receptor (ER), specificity of interaction with enhancer elements appears to reside in two amino acids located between the two cysteine residues in the carboxy end of the first finger. Manipulation of these two amino acids in the "knuckle" of the first zinc finger alters specificity of interactions with glucocorticoid response elements (GREs) or estrogen response elements (EREs) (Danielson et al., 1989). It is likely that these two amino acids are also critical for specific AR–DNA interactions.

Some patients with receptor-positive AIS have provided documentation of the importance of the DBD. Studies have detected mutations at a highly conserved Arg within the second zinc finger of the AR gene: Arg615Pro (Zoppi et al., 1992; Marcelli et al., 1991b), Arg614His (Mowszowicz et al., 1993), or Arg 607Lys (Saunders et al., 1992) (Note: The numbering systems of the authors vary.) These mutations have potent effects on the ability of the mutant AR to transactivate a reporter construct. Furthermore, these types of mutant receptors have normal specificity of steroid binding, binding capacities, and affinities (receptor positive) but reduced DNA binding ability (Mowszowicz et al., 1993).

In another case of receptor-positive AIS, analysis of AR in GSF cells revealed supranormal steroid binding capacity but reduced DNA binding affinity. Cloning and sequencing of the cDNA indicate an in-frame deletion of exon 3 that contains the second zinc finger (Quigley et al., 1992b). Thus, the in vitro effects of these missense or deletion mutations are consistent with the receptor-positive AIS exhibited by the patients that express these mutants. In fact, AIS patients who exhibit normal AR concentrations, binding characteristics, and androgen concentrations will most likely possess mutations within the first three exons of the gene.

Studies using recombinant AR or truncated AR containing only

DBD or both DBD and SBD reveal the modular nature of the DBD. Bandshift and footprint studies indicate that DBD fusion proteins can bind to an androgen response element (ARE) (De Vos *et al.*, 1991, 1993a). Another study using a DBD–SBD fusion protein indicated that androgen was not necessary to permit interaction of this truncated protein with an ARE (Young *et al.*, 1990). These data demonstrate that the DBD can function at least minimally (i.e., bind DNA) as an independent moiety. However, other data indicate that other domains of the full-length AR influence the DBD of AR (Wong *et al.*, 1993). Unlike the truncated AR proteins, full-length recombinant AR must bind androgen before binding to an ARE. Also, the specificity or level of transactivation induced by different steroids parallels the abilities of these steroids to promote DNA binding activity. Deletion of the amino terminus eliminates the steroid binding requirement, suggesting that the amino terminus exerts inhibitory effects on the DBD (Wong *et al.*, 1993).

C. TRANSACTIVATION DOMAIN

Transactivation functions reside primarily in the amino terminus. Unfortunately, very little is known about the specific nucleotides or "domains" that contribute to transactivation functions because AIS rarely arises from point mutations within the amino terminus of the AR gene. In addition, the lack of homology between the amino termini of different steroid receptors has prevented extrapolation from more thoroughly characterized steroid receptors. There are, however, several factors that emerge from different studies: multiple elements may contribute to the transactivation function(s), residual transactivation activity resides in the carboxy terminus, and the transactivation domain appears to be suppressed by the SBD until bound by ligand.

There are several polymeric stretches within the amino terminus: polyglutamine, polyproline, and polyglutamic. In one case of AIS, a decrease in length from 20 to 12 Gln contributed to increased thermolability and decreased transactivation potential of the mutant AR (McPhaul *et al.*, 1991). Also, an increased polyglutamine tract is correlated with SBMA. These two lines of evidence suggest that the polyglutamine repeat contributes to transactivation.

Unfortunately, very few artificial mutations have been introduced into the amino terminus. Gross deletions of amino acids 1–141 of the human AR (hAR) had little effect on transactivation whereas further truncations abolished chloramphenicol acetyl transferase (CAT) activity in transient transfections (Simental *et al.*, 1991). Another study demonstrated that deletions of amino acids 51–211 or 244–360 of the hAR eliminated most transactivation (Jenster *et al.*, 1991). These data suggest that large stretches of amino acids may be involved with transactivation. However, such large-scale deletions may simply provide information concerning amino acids that are crucial for proper protein conformation rather than specific amino acids that mediate transactivation.

Studies of the Tfm mouse demonstrate that overexpression of a carboxy fragment containing the DBD and SBD inhibits transactivation by wild-type mouse AR (mAR) and results in residual androgen-dependent transactivation in transient transfections (He *et al.*, 1994). These data suggest that as in the GR, the AR may contain a weak transactivation domain within the carboxy portion of the protein. As discussed earlier, mutant hAR in which the SBD domain has been truncated exhibit a high degree of constitutive activity in transfection studies (Simental *et al.*, 1991). These data indicate that the SBD inhibits transcription activation domains until steroid binds to the AR.

D. NUCLEAR LOCALIZATION SIGNALS

Early characterizations of the AR indicated that the protein was distributed in both the cytoplasm and nucleus and that binding of androgen resulted in activation and translocation of a predominantly cytosolic receptor to the nucleus (see Johnson *et al.*, 1988; Chan *et al.*, 1989). Indeed, based on more recent data, it seems likely that an equilibrium of distribution exists and that this distribution varies with cell or tissue type and the availability of androgens (Simental *et al.*, 1991; Jenster *et al.*, 1993). Factors that may influence cellular distribution of AR include availability of androgens and presence of nuclear localization signals (NLSs).

Transfection of AR into COS cells and immunohistochemical localization indicate that in the absence of androgens the wild-type AR is located predominantly in the cytoplasm, whereas addition of androgens shifts the equilibrium to predominantly nuclear AR (Simental *et al.*, 1991; Jenster *et al.*, 1993). These findings agree with earlier biochemical characterizations indicating that castration of animals resulted in a predominantly cytoplasmic localization of AR (see Johnson *et al.*, 1988; Chan *et al.*, 1989). Although these data demonstrate that availability of androgen is one factor that regulates distribution of AR, the importance of this factor is questionable because many animals experience levels of circulating androgen that, while variable, are probably sufficient to saturate most AR populations. Thus other factors may play a more pivotal role in determining AR cellular distribution.

In the case of the human androgen receptor, putative NLSs have been identified within a stretch of basic amino acids in the hinge region (exon 4). One putative NLS shares homology with the putNLS of the SV40 large T-antigen, ratGR and rabbit PR NLS (Simental *et al.*, 1991). Deletion of the stretch of amino acids (628–657) containing the putNLS results in decreased nuclear localization in the presence of androgen but does not totally abolish nuclear localization. The residual nuclear localization is sufficient to mediate androgen-dependent induction of a mouse mammary tumor virus reporter construct (MMTV-CAT) (Simental *et al.*, 1991), suggesting that normal levels of nuclear AR may be suprathreshold for some androgen responses.

Another report has localized an NLS to amino acid residues 608–625 of the human AR [Note: Numbering systems of Simental *et al.* (1991) and Jenster *et al.* (1993) differ.] This signal overlaps the putNLS described earlier and is structurally and functionally homologous to the nucleoplasmin NLS. This NLS consists of two basic regions separated by a 10-amino-acid spacer (Jenster *et al.*, 1993). Apparently mutations within both basic regions are required to eliminate androgen-induced nuclear accumulation of AR.

In addition to specific NLS, the presence of other domains of the AR are important for proper localization. Large-scale deletions of either the NH_2 or carboxy terminus result in nuclear localization of AR even in the absence of androgens (Simental *et al.*, 1991; Jenster *et al.*, 1993). However, as noted by the authors, this may be a nonspecific artifact due to the reduced size of the protein allowing free diffusion through the nuclear membrane.

E. PHOSPHORYLATION OF THE ANDROGEN RECEPTOR

Steroid receptors are phosphoproteins and, in general, phosphorylation plays an important role in the activation of steroid receptors (Moudgil, 1990). Androgen receptors from the rat ventral prostate (Goueli *et al.*, 1984), LNCaP cells (van Laar *et al.*, 1991), and monkey kidney cells (COS) (Kemppainen *et al.*, 1992) are phosphorylated. Androgen treatments increase phosphorylation of AR 1.8-fold and 2- to 4-fold in LNCaP and COS cells, respectively. In COS cells, hormone-induced phosphorylation was specific to androgens as treatment of the cells with estradiol, progesterone, or antiandrogens did not increase AR phosphorylation (Kemppainen *et al.*, 1992).

Hormone-dependent phosphorylation of AR may occur from alterations in levels of casein kinase II, protein kinase A (pkA), phosphatases, receptor abundance, or conformational changes in the receptor.

Endogenous phosphatase activity in cytosolic and nuclear fractions of ventral prostate is inhibited by DHT (Golsteyn *et al.*, 1990). In addition, substantial data suggest that androgens increase AR protein stability and thus increase the presence of pkA substrates. Finally, androgens are known to induce conformational changes that alter the binding of proteins such as heat shock proteins and, thus, might also expose additional phosphorylation sites.

Phosphorylation occurs mainly in the N-terminal transactivation region of the AR with no phosphorylation in the DNA and steroid binding domains. In the first 300 amino acids of the N-terminal region a total of 12 potential phosphorylation sites have been located, suggesting a role of phosphorylation of the androgen receptor in transcription regulation (Kuiper *et al.*, 1993). It remains to be determined whether ligand-independent phosphorylation of the AR results in transactivation of androgen target genes as demonstrated with the PR and progesterone-regulated genes (Denner *et al.*, 1990).

V. Regulation of the Androgen Receptor Gene

To fully understand androgen regulation of genes it is necessary to consider how the regulation of the AR gene is coordinated with androgen regulation of target genes. The 5′ flanking regions of the human, rat, and mouse AR gene have been cloned (Tilley *et al.*, 1990; Baarends *et al.*, 1990; Faber *et al.*, 1991a,b; Kumar *et al.*, 1992), allowing analysis of molecular mechanisms that regulate expression of the AR gene. In this section, we will consider data concerning the promoters of the AR gene, different endocrine factors, and signaling pathways (i.e., pkA and pkC) that act to regulate the AR gene, and possible enhancer elements found in the 5′ flanking region of the AR gene.

A. Promoter

Transcriptional promoters of eukaryotic genes are diverse but can be divided into two main groups, that is, those that possess the consensus sequence TATAAAA (TATA box) and those that lack such sequences. Promoters that lack TATA boxes are usually GC rich, do not appear to bind TFIID directly, and often have multiple sites of transcription initiation. An additional element that is found in both TATA-containing and TATA-less promoters is the Initiator (consensus PyPyCAPyPyPyPyPy), which surrounds the transcription start site. This element is poorly defined but can function as a minimal promoter for TATA-less genes and appears to be a transcription control element because mutations in this region can result in the use of alternative start sites and/or a reduction in promoter strength (Smale and Baltimore, 1989). It has been demonstrated that SP1 (GC box-binding protein), Initiator, TFIID, and undetermined coactivators are required for transcription from some TATA-less promoters (Pugh and Tjian, 1990, 1991).

Genomic sequences from the 5' flanking regions of the human, rat, and mouse androgen receptors have been cloned and shown to contain promoters by functional assays such as transient transfection of reporter constructs. Sequence analysis of the promoter regions revealed no TATA or CAAT boxes (Tilley *et al.*, 1990; Baarends *et al.*, 1990; Faber *et al.*, 1991a,b; Kumar *et al.*, 1992). The AR promoter does contain an Initiator (CTTTCCACCTCCA) preceding the +1 site of transcription and a perfect GC box (consensus GGCGGG) in a GC-rich region approximately 45 bp upstream. In addition, the mAR promoter has a homopurine region that is 54 bp long and contains six GGGGA repeats. A similar region in the osteonectin gene increases promoter activity in a bone cell line but not in a kidney cell line (Ibaraki *et al.*, 1993).

Primer extension and S1 nuclease experiments have localized the transcription start sites (Faber *et al.*, 1991b; Baarends *et al.*, 1990). The mouse AR promoter contains at least two main sites of transcription initiation that are 13 bp apart. The +13 transcription start sites have been shown to be regulated by the GC box when assayed by S1-nuclease protection (Faber *et al.*, 1993). However, both of the sites have been shown to be sensitive to the synthetic androgen mibolerone (Wolf *et al.*, 1993).

B. AUTOREGULATION OF AR EXPRESSION

The androgen receptor is autoregulated in certain tissues (Quarmby *et al.*, 1990) although the exact mechanism(s) underlying this autoregulation has not been defined fully. The autoregulation is cell and tissue specific and appears to involve transcription initiation as well as altered rates of translation and/or protein stabilization.

Several studies have shown that steady-state levels of AR mRNA are down-regulated by androgens. For example, androgen treatment decreases steady-state AR mRNA in the human prostate carcinoma cell line LNCaP but increases the half-life of AR mRNA from 3.8 to 6.4 hr. Nuclear run-on analysis demonstrated that androgen treatments caused a 75% reduction in transcription initiation (Wolf *et al.*, 1993;

Blok *et al.*, 1992b). These data indicate that the autologous down-regulation of the AR mRNA in LNCaP cells is due to a decrease in transcription initiation.

In the rat, castration causes an increase in steady-state levels of AR mRNA (2- to 10-fold) in several tissues: ventral prostate, coagulating gland, epididymis, seminal vesicle, kidney, and brain. Treatment of castrated rats with T propionate reduced steady-state levels of AR mRNA from the ventral prostate to control levels whereas injection of estradiol had no effect (Quarmby *et al.*, 1990; Tan *et al.*, 1988). In addition, castration of the Tfm rat (Quarmby *et al.*, 1990) and androgen treatments of cultured GSF from CAIS patients fail to regulate AR mRNA levels. Therefore it appears that autoregulation of the AR mRNA is due partly to a decrease in rate of transcription initiation and requires a functional AR.

Androgen regulation of translation or AR protein stabilization is thought to occur because changes in levels of AR mRNA often are not mirrored by changes in the amount of AR protein. Studies of the rat ventral prostate and seminal vesicles indicate that mRNA levels increased 9- to 11-fold 48 hr after castration whereas the protein levels increased to a much lower degree as determined by immunoblotting (Shan *et al.*, 1990). Further, studies using LNCaP cells have indicated that the protein levels are either unchanged (Wolf *et al.*, 1993) or may actually be higher in cells containing androgens as shown by immunoblots and ligand binding studies (Krongrad *et al.*, 1991). AR in the ductus deferens smooth muscle tumor cell line (DDT1MF-2) possess 3.1- and 6.6-hr half-lives in the absence or presence of 1 nM R1881, respectively. In addition, rate of receptor synthesis increased from 1.35 to 2.23 fmol/µg DNA/hr when 1nM R1881 was added. This resulted in an overall increase of androgen receptors from 6.0 to 12.2 fmol/µg DNA (Syms *et al.*, 1987). These data suggest that androgens stabilize the AR protein and, in some tissues, increase the rate of translation.

One potential mechanism of androgen-induced posttranslational modification involves an actual change in conformation of the receptor. Immunohistochemical evaluation of rat prostate using antibodies to the C-terminal or N-terminal regions of the AR indicates that 4 hr following castration there is a decrease in the amount of nuclear AR detected using the C-terminal antibodies but not using the N-terminal antibodies. This indicates that conformational change of the AR may occur in the absence of androgens (Husmann *et al.*, 1990).

In conclusion, the autologous regulation of the AR by androgens occurs at the transcriptional, translational, and posttranslational levels. A functional AR is necessary for the regulation of transcription, indicating that AR protein is directly involved. Finally, the level of translation can be affected by androgens independently from the level of mRNA but this seems to be a tissue-specific event.

C. Second Messenger Regulation of AR Expression

Androgen-regulated genes are subject to nonsteroidal regulatory signals, including peptide/glycoprotein hormones, growth factors, and neurotransmitters. These different signals are transduced at the cell membrane into second messengers and two main signaling pathways are activated: cAMP-pkA and DAG/Ca^{2+}-pkC. Recent data demonstrate that these pathways also regulate the AR gene.

Several studies using rat Sertoli cells indicate that stimulation of the pkA pathway via FSH or dbcAMP results in a transient down-regulation of mRNA followed by an up-regulation of mRNA and protein (Blok *et al.*, 1989, 1992a,b; Verhoeven and Cailleau, 1988). Verhoeven and Cailleau demonstrated that FSH and androgens up-regulate expression (two- or three-fold) of the AR protein in Sertoli cells and that a combination of FSH and DHT was additive. Conversely, stimulation of pkC pathways by a tumor promoting agent results in suppression of AR mRNA in LNCaP cells (Henttu and Vihko, 1993). These data suggest that transcriptional activity of the AR gene is regulated by second messenger pathways possibly via CREB, AP2, or AP1 transcription factors.

The mouse 5' flanking region has been cloned in this laboratory and based on sequence homology it contains consensus AP1, putative CRE, and putative AP2 sequences (Table II) (Kumar *et al.*, 1992; Lindzey *et al.*, 1993). Thus, this region contains sequences that may mediate the effects of both pkA and pkC pathways. Transfection experiments into mouse pituitary cells demonstrate that a 1.5-kb fragment of the 5' flanking region contains sequences that mediate an 8- to 10-fold increase in CAT activity when challenged with 20 μM forskolin (an activator of adenylate cyclase) (Lindzey *et al.*, 1993). Similar results were obtained in a mouse hypothalamus and quail fibrosarcoma cell line. A sequence containing the putative CRE is able to confer increased cAMP inducibility on a heterologous promoter. Furthermore, this element exhibits specific binding to nuclear proteins from a mouse gonadotropin cell line. Transfection experiments also indicate that regions within the rat 5' flanking region are cAMP inducible (2-fold) (Blok *et al.*, 1992b).

TABLE II
PUTATIVE RESPONSE ELEMENTS IN THE 5′ FLANKING REGION OF THE MOUSE

Putative element	Consensus sequence	mAR sequence	Location
TATA box	TATAA	Not found	
CAAT box	CCAAT	Not found	
Sp1	GGGCGG	GGGCGG	−39 to −44
	CCGCCC	CCGCCC	+200 to +205
		CCGCCC	+557 to +562
Ap1	TGAC/GTCA	TGAGTCA	−193 to −199
Ap2	CCCCAGGC	CCCTGGGGGG	−433 to −442
		GCGGGGGCGG	−39 to −48
		CCCAGCC	+151 to +158
		CCCCACCC	+864 to +871
CRE	TGACGTCA	TCCCGTCA	+733 to +740
ARE/GRE/PRE	TGTTCT	TGTTCT	−483 to −488
half-site		TGTCCT	−449 to −454
		TGTCCT	+299 to +304
		TGTTCT	+762 to +767
Pu box	GAGGAA	TTCCTC	−375 to −380
		TTCCTC	+218 to +223
		GAGGAA	+258 to +263
OCT	ATTTGCAT	ATTTGCAC	+675 to +682

VI. ANDROGEN REGULATION OF GENES

A large number of genes are regulated by testicular androgens through either DHT or aromatization to estrogens. These range from genes coding for hormones, receptors, enzymes, and structural proteins to housekeeping genes. Furthermore, almost every tissue or organ possesses an androgen-regulated gene (Mooradian et al., 1987). Despite the number of androgen-regulated genes, only a few genes have been well characterized in terms of the mechanisms of androgen regulation. Table III lists some of the genes in which substantial progress has been made in defining the mechanisms by which androgens regulate these genes (i.e., sex-limited protein, probasin, prostatein, prostate-specific antigen, and human glandular kallikrein). In the following sections we discuss the biological functions of these genes, data demonstrating androgen regulation of these genes, and the regulatory sequences that mediate the androgen regulation.

TABLE III
ANDROGEN-REGULATED GENES AND THE ANDROGEN RESPONSE ELEMENTS (ARE)
THAT MEDIATE THE EFFECTS OF ANDROGENS ON THE TARGET GENE

Gene	Sequence	Location	References
C(3)	5'-AGTACGtgaTGTTCT-3'	1st intron	Tan *et al.* (1992)
Probasin	ARE1: 5'-ATCTTGTTCTTAGT-3'	−236 to −223	Rennie *et al.* (1993)
	ARE2: 5'-GTAAAGTACTCCA AGAACCTATT-3'	−140 to −117	
Slp	HRE1: 5'-GTAATTatcTGTTCT-3'	−126 to −112	Adler *et al.* (1991)
	HRE2: 5'-TGGTCAgccAGTTCT-3'	−143 to −128	
	HRE3: 5'-AGAACAggcTGTTTC-3'	−158 to −144	
PSA	5'-AGAACAgcaAGTGCT-3'	−170 to −156	Murtha *et al.* (1993); Riegman *et al.* (1991)
hKLK2	5'-GGAACAgcaAGTGCT-3'	−170 to −156	Murtha *et al.* (1993); Riegman *et al.* (1991)

A. TRANSCRIPTIONAL REGULATION BY ANDROGENS

The prostate gland has been an important target tissue for studying androgen regulation of gene expression (Mooradian *et al.*, 1987) owing to its clinical significance and androgen dependence. Several prostate-specific genes have been identified and characterized as androgen-regulated genes, including prostatic steroid binding protein (PSBP) also known as prostatein (Allison *et al.*, 1989; Tan *et al.*, 1992), pro-basin (Rennie *et al.*, 1993), and a number of kallikrein genes (Riegman *et al.*, 1991; Murtha *et al.*, 1993; Clements, 1989).

PSBP is the principal secretory protein of the ventral prostate of rats (Page and Parker, 1982). It is a tetramer consisting of two sub-units, one containing the polypeptides C1 and C3 and the other containing the polypeptides C2 and C3. Although C1 and C2 are coded for by a single-copy gene (Allison *et al.*, 1989), C3 is produced by two nonallelic genes: C3(1) and C3(2) (Allison *et al.*, 1989). All three genes are located at or near position q31 on chromosome 5 (Zhang *et al.*, 1988). PSBP is up-regulated by androgen at both transcriptional and posttranscriptional levels. The steady-state levels of PSBP mRNA for all three subunits decrease about 100-fold 3 days following castration (Page and Parker, 1982). Nuclear run-on assays of rat prostate cell nuclei demonstrated that the transcription rates for the three genes decrease 2- to 3-fold upon castration and are reversed by T treatment of castrated rats (Page and Parker, 1982). Because the androgen-induced changes in transcription rate cannot solely account for the androgen-induced changes in steady-state mRNA level, it has been suggested that androgen may also exert effects on the mRNA stability for the PSBP gene (Page and Parker, 1982).

The C3(1) gene contains an androgen-responsive element in the first intron (Claessens *et al.*, 1989, 1990a,b, 1993a,b; Tan *et al.*, 1992). This element binds androgen receptor *in vitro* (Claessens *et al.*, 1990b) and confers androgen responsiveness of C3(1) first intron–CAT reporter constructs in transient transfection assays (Tan *et al.*, 1992). An ARE-like sequence has also been identified in 5' flanking sequences of both the C3(1) (Claessens *et al.*, 1990a,b) and the C2 gene (Delaey *et al.*, 1987), though it is not clear that these putative AREs have *in vivo* function.

In contrast to PSBP, probasin, a single-polypeptide protein, is mainly expressed in rat dorsolateral prostate (Sweetland *et al.*, 1988). Interestingly, the probasin protein is synthesized in both nuclear and secreted forms that are derived from the same bifunctional messenger RNA (Spence *et al.*, 1989). Amino acid sequence analysis suggests that probasin may be a ligand-carrier protein (Spence *et al.*, 1989).

The expression of probasin mRNA is regulated by androgen and zinc (Matusik *et al.*, 1986). The levels of probasin mRNA increase significantly in the dorsolateral prostate after sexual maturity, whereas the low levels of probasin mRNA in ventral prostate are further down-regulated postpubertally (Sweetland *et al.*, 1988). Although the probasin mRNA and protein decrease following castration of rats, the mRNA "rebounds" to its precastration level 12 days after castration. It has been suggested that this rebound phenomenon may be mediated in part by adrenal steroids. It has been demonstrated that androgen regulation of probasin gene is mediated via a direct interaction of the androgen receptor and two AREs in the 5' flanking region of the probasin gene (Rennie *et al.*, 1993).

Human glandular kallikrein-1 (hKLK2), prostate-specific antigen (PSA), and human pancreatic/renal kallikrein are three members of a subgroup of serine proteases that are potentially involved in the activation of specific polypeptides through posttranslational processing (Clements, 1989). The hKLK2 and PSA mRNAs exhibit a high degree of sequence homology and are expressed exclusively in the prostate. Both hKLK2 and PSA mRNAs have been localized in human prostatic epithelia by *in situ* hybridization (Young *et al.*, 1992). The same studies revealed that the level of hKLK2 mRNA in human benign prostatic hyperplasia tissues is approximately half that of PSA mRNA.

PSA protein is well characterized and is an important marker for prostate cancer. The hKLK2 protein has yet to be isolated and its function is uncertain. Deduced amino acid sequences indicate that hKLK2 may be a trypsinlike serine protease (Morris, 1989), whereas PSA is a chymotrypsinlike serine protease (Lilja, 1985), therefore these two genes may have different physiological functions. Interestingly, the hKLK2 gene is located about 12 kb downstream from the PSA gene in a head-to-tail fashion on chromosome 19 (Riegman *et al.*, 1989). The similarities of gene structure and deduced amino acid sequences of these human kallikreins suggest that their evolution may involve the same ancestral gene.

Both hKLK2 and PSA mRNAs are under androgenic regulation in the human prostatic cancer cell line, LNCaP cells (Young *et al.*, 1991, 1992). The induction of mRNAs for both genes by androgens (i.e., T, DHT, and mibolerone) occurs rapidly (2–4 hr) and reaches a maximal level 24 hr after androgen treatment. The androgen regulation of hKLK2 and PSA genes is in part mediated at a transcriptional level and functional AREs have been characterized in the 5′ flanking regions of the two genes (Riegman *et al.*, 1991; Murtha *et al.*, 1993).

Sex-limited protein (Slp), a defective isoform of the mouse fourth component of complement (C4) gene (Miyagoe *et al.*, 1993), is a serum protein regulated by T (Cox and Robins, 1988). Both C4 and Slp genes are tandemly located in the S region of the H-2 major histocompatibility complex. Based on homology and proximity of location it has been suggested that Slp is a duplicated form of the C4 gene. Though Slp mRNA and protein are synthesized mainly in hepatocytes and/or macrophages, the Slp mRNA can be detected in several other tissues by RNase mapping assays (Cox and Robins, 1988). In fact, the expression of Slp mRNA is very complex and exhibits strikingly different tissue-specific as well as mouse strain-specific regulation (Cox and Robins, 1988).

It appears that the Slp gene became androgen dependent by a proviral insertion in the 5′ flanking region of the gene (Stavenhagen and Robins, 1986). A 120-bp stretch containing multiple enhancer sequences including an ARE was found within the proviral insertion of the Slp gene (Adler *et al.*, 1991, 1992). In an *in vitro* transfection assay, both the ARE and auxiliary elements in the 120-bp DNA fragment are required for gene induction by androgen receptor but not by glucocorticoid receptor (Adler *et al.*, 1992). However, the relevance of this *in vitro* data *to vivo* expression of Slp has been questioned because hypophysectomy of male mice dramatically lowers Slp mRNA to female levels, and T does not increase Slp mRNA in hypophysectomized females (Miyagoe *et al.*, 1993; Georgatsou *et al.*, 1993). Furthermore, pituitary factors such as growth hormone (GH) rapidly restore Slp mRNA level in hypophysectomized male and normal female mice up to 80% of the normal male level. There is also a similar response of Slp mRNA increase in testicular feminization mutation (Tfm/Y) mice by GH injection. The authors (Georgatsou *et al.*, 1993) suggested that androgens may act on a temporally remote step by inducing a male-specific GH secretory pattern to regulate the expression of Slp in male animals.

B. Post-transcriptional Regulation by Androgens

The guinea pig seminal vesicle produces four major androgen-regulated secretory products that derive from two larger precursor proteins. SVP2 derives from an abundant 581-nucleotide mRNA whereas SVP1, 3, and 4 derive from a 1368-nucleotide mRNA (Hagström *et al.*, 1989). Levels of mRNA and protein increase rapidly between 5 and 10 days postpartum and castration of neonates at Day 5 eliminated normal accumulation of protein but not mRNA. Treatment of castrates with DHT or T restored protein levels but estrogens were ineffective. Interestingly, manipulations of androgen levels altered translational activity to a far greater extent than transcriptional activity, suggesting a posttranscriptional regulation (Hagström *et al.*, 1992; Norvitch *et al.*, 1989, 1991). Similar data suggest that the AR gene product is regulated posttranscriptionally (see Section V,B).

C. Simple Androgen Response Elements

AREs are composed of specific sequences of DNA that are bound by the DBD of the AR and thus bring the AR into a position where its transactivating domains can interact with and influence the activity of transcription complexes. Until recently, the only known AREs also functioned as GREs or progesterone response elements (PREs) (i.e., AREs found in MMTV and TAT genes). This is not surprising because of the extensive homology between AR, GR, and PR DBD. Indeed many researchers were skeptical that a distinct and separate ARE existed. However, recent work has provided examples of AREs that exhibit preference for AR (Slp and probasin genes).

AREs typically consist of an imperfect palindrome sequence with a 3-bp spacer between the two half-sites (see Table III). They may occur in either the 5′ flanking regions, intronic sequences, or 3′ untranslated sequences. Consensus sequences (5′-GGA/TACAnnnTGTTCT-3′) have been developed based on sequence homologies between known

AREs and experimental manipulations of known or consensus AREs (Beato, 1989; Roche *et al.*, 1992). Characterization of binding affinity for AR and transactivating activity demonstrated that this consensus was optimal for AR binding in an *in vitro* system (Roche *et al.*, 1992).

Data also indicate that the first nucleotide of the spacer (G or A favored) and the 3′ flanking sequence (A and T favored) influence both AR binding and androgen-induced CAT activity in transfection studies. Furthermore, alterations in the AR binding and transactivating activity paralleled each other. Thus it appears that the efficiency of a consensus simple ARE is influenced by nucleotides within the core, spacer, and flanking sequences. Indeed, the importance of flanking sequences becomes more apparent in the case of complex AREs.

Physical interactions between AR proteins and AREs have been demonstrated with bandshift and footprint analyses. In general, mutations that abolish functional activity also abolish ARE–protein interactions. Experiments using recombinant DBD of AR suggest that the AR acts as a dimer with near-perfect AREs but weakly as a monomer with imperfect AREs or ARE half-sites (De Vos *et al.*, 1991, 1993a). In addition, a DBD fusion protein interacts with an ARE in the absence of androgen. However, recombinant full-length AR requires the presence of androgen for binding to AREs (Wong *et al.*, 1993). Removal of the NH_2 terminus of the AR eliminates the requirement for androgen binding, suggesting that the NH_2 terminus interacts with the rest of the protein to block binding to an ARE until the AR is activated by androgen binding.

Current data suggest that the hKLK2 and PSA genes are regulated by simple AREs that differ by only one nucleotide (Table III). Compared with the consensus glucocorticoid-responsive element (GGTACA nnn TGTTCT) (Beato, 1989), hKLK2 is less homologous and a less perfect palindrome than the PSA ARE. However, by *in vitro* transient transfection experiments (Murtha *et al.*, 1993), the hKLK2 ARE is more potent than the PSA ARE.

The C(3)1 gene is also regulated by a simple ARE located in the first intron (Table III) (Claessens *et al.*, 1990a,b; Rushmere *et al.*, 1987, 1990; Parker *et al.*, 1987). Two ARE-like sequences are located in this region, and both bind AR but only the downstream element acts as a functional ARE (Table III). As is the case with many simple AREs, PR or GR also modulate activity through this element.

D. COMPLEX ANDROGEN RESPONSE ELEMENTS

It is important to recognize that the distinction between simple and complex AREs will not necessarily be clear. Indeed, as our sophistication and knowledge increase, simple AREs such as those in PSA, hKLK2, and C3(1) genes may turn out to be part of a more complicated hormone-responsive unit or complex ARE. However, a convenient working definition is that a complex ARE requires multiple elements and the binding of multiple proteins to these elements for full androgen-induced activity. Thus, complex AREs contain a core ARE that is necessary but not sufficient for normal androgen-induced levels of activity. The flanking sequences contain overlapping or closely juxtaposed elements that bind other transcription or accessory factors that interact with AR bound to the core ARE sequence.

The mouse Slp and rat probasin genes present examples of complex AREs that exhibit specificity for AR. The Slp gene contains a complex ARE composed of a core ARE sequence (see Table III) and adjacent enhancers within a 120- to 160-bp stretch. Contained within this stretch are two degenerate hormone response elements (HREs) that do not function independently but synergize with the core ARE sequence (Adler *et al.*, 1991). Both GR and AR can bind to the core ARE sequence but GR cannot transactivate through this complex ARE. Thus the Slp ARE complex provides an example of an androgen-specific ARE that may derive its specificity via interactions between AR and other transcription factors/accessory factors (Adler *et al.*, 1992).

The probasin gene contains two AREs that interact to regulate androgen expression of this gene. Based on footprint analysis, ARE1 spans -236 to -223 and ARE2 spans -140 to -117. ARE1 contains a canonical ARE half-site (TGTTCT) on the coding strand whereas ARE2 contains the sequence GTTCT on the noncoding strand (Table III). Mutations in either ARE1 or ARE2 led to a drastic reduction in androgen effect. Interestingly, when transfected into PC3 cells (a human prostate carcinoma cell line) these AREs show a clear preference for androgen over progesterone or glucocorticoids compared to MMTV or TAT AREs. Furthermore, this effect appears to be tissue specific in that this steroid specificity is not as pronounced in Hela cells (Rennie *et al.*, 1993).

E. ACCESSORY FACTORS

An emerging area of research involves the importance of accessory factors in androgen regulation of gene expression. An accessory factor can be defined as a protein that facilitates the effects of another transcription factor but does not exert an effect on its own. In the previous section we discussed a complex ARE (Slp gene) in which AR effects are enhanced by additional nonreceptor proteins. Furthermore, recent data suggest that highly purified recombinant AR can be deficient in

DNA binding activity until nuclear extracts are added to the purified AR. Both lines of evidence suggest that accessory factors are involved in AR-regulated events.

It appears that at least two general types of accessory factors exist: a DNA binding protein and a protein that interacts directly with the AR. De Vos *et al.* (1993b) observed that in gel shift assays nuclear extracts of different cell lines and tissues (FTO-2B, COS, prostate, liver, kidney) synergized with *E. coli* expressed DBD of AR and GR. Furthermore, the nuclear extracts alone shifted the ARE at a different mobility than AR alone. This indicates that the factor is widespread and possesses DNA binding activity. Whether this DNA binding activity is specific for certain DNA sequences remains to be determined.

Another accessory factor has been found in reticulocyte lysate. This factor, called a receptor accessory factor, is thermolabile, trypsin sensitive, and has a mass of 130 kDa. In bandshift assays, this factor increases interaction of recombinant AR (25-fold) and GR (6-fold) but does not bind DNA and appears to interact directly with the amino terminus of the AR (Kupfer *et al.*, 1993). These data indicate that multiple types of accessory factors exist and raise the possibility that AR effects may be facilitated by accessory factors in a gene-specific and tissue-specific fashion.

Accessory factors may act by (1) anchoring the AR to the ARE or transcription complex, (2) modifying the transcriptional activity of the AR, or (3) excluding transcriptional activity of GR or PR and, hence, providing steroid specificity. These possibilities raise the questions of whether there are general or specific accessory factors that enhance different classes of receptors.

VII. Summary

Androgens directly regulate a vast number of physiological events. These direct androgen effects are mediated by a nuclear receptor that exhibits four major functions or activities: steroid binding, DNA binding, transactivation, and nuclear localization. The SBD consists of a hydrophobic pocket of amino acids that exhibits high-affinity, androgen-specific binding. Based on studies of mutant AR, it appears that a number of different amino acids contribute to the steroid binding characteristics of the AR. The DNA binding domain confers sequence-specific binding to structures called androgen-responsive elements. The specificity of steroid binding and DNA binding provides a crucial basis for androgen-specific regulation of target genes. The nuclear localization signal shares homology with known nuclear localization signals and, coupled with the presence of androgens, is responsible for localizing the AR to the nucleus. The transactivation functions reside mostly in the NH_2 terminus but the responsible domains are as yet poorly defined. Though the different domains can act as independent moieties, one domain can clearly alter the behavior of another domain. For instance, the SBD appears to inhibit the transactivating functions until steroid is bound and the amino terminus prevents DNA binding activity until steroid is bound. The relative ease of introducing mutations with polymerase chain reaction technology will facilitate further delineation of critical amino acids and domains responsible for the various activities of the AR.

The recent cloning and characterization of AR promoters revealed that the AR genes are driven by a TATA-less promoter characteristic of housekeeping genes. Analysis of transcription rates, mRNA levels, and protein levels indicates that androgens and pkA and pkC pathways modulate expression of AR mRNA and protein. This indicates that the same signal pathways that interact to regulate androgen target genes also regulate the levels of AR in the target tissues.

Surprisingly few androgen-regulated genes have been well characterized for the mechanisms by which androgen regulates the gene. The C(3), Slp, probasin, PSA, and hKLK2 genes have provided examples where androgens regulate transcription. Posttranscriptional regulation by androgens has been demonstrated for the SVP1, 2, 3, and 4 and AR genes. The mechanisms underlying posttranscriptional regulation are poorly defined but substantial progress has been made in defining the critical elements that mediate transcriptional effects of androgens. Transcriptional effects are mediated through binding of androgen–AR complexes to specific DNA sequences called AREs. Simple AREs such as those found in C(3) and kallikrein genes tend to be permissive in that GR and PR can also act through the same element. However, the more complex AREs appear to be capable of some degree of AR-specific interaction probably because of the presence of multiple enhancer elements, and interactions with other transcription factors or accessory factors.

Historically, the AR gene was the last of the major steroid receptors to be cloned and our knowledge of its molecular mechanisms has therefore lagged behind that of other classes of steroid receptors. However, rapid progress is now possible with recent applications of technology, such as polymerase chain reaction to *in vitro* mutagenesis and detection of new androgen-regulated genes by differential display polymerase chain reaction.

ACKNOWLEDGMENT

We thank Ms. Kelli J. Ambroson for assistance with literature searches and typing of the manuscript.

REFERENCES

Adler, A. J., Scheller, A., Hoffman, Y., and Robins, D. M. (1991). Multiple components of a complex androgen-dependent enhancer. *Mol. Endocrinol.* **5**, 1587–1596.

Adler, A. J., Danielson, M., and Robins, D. M. (1992). Androgen-specific gene activation via a consensus glucocorticoid response element is determined by interaction with nonreceptor factors. *Proc. Natl. Acad. Sci. U.S.A.* **89**, 11660–11663.

Agache, P., Blac, D., Barrand, C., and Laurent, R. (1980). Sebum levels during the first year of life. *Br. J. Dermatol.* **103**, 643–649.

Allison, J., Zhang, Y.-L., and Parker, M. G. (1989). Tissue-specific and hormonal regulation of the gene for rat prostatic steroid-binding protein in transgenic mice. *Mol. Cell. Biol.* **9**, 2254–2257.

Amato, A. A., Prior, T. W., Barohn, R. J., Snyder, P., Papp, A., and Mendell, J. R. (1993). Kennedy's disease: A clinicopathologic correlation with mutations in the androgen receptor gene. *Neurology* **43**, 791–794.

Andersson, S., Berman, D. M., Jenkins, E. P., and Russell, D. W. (1991). Deletion of steroid 5α-reductase 2 gene in male pseudohermaphroditism. *Nature (London)* **354**, 159–161.

Arnold, A. P., and Gorski, R. A. (1984). Gonadal steroid induction of structural sex differences in the central nervous system. *Annu. Rev. Neurosci.* **7**, 413–442.

Attardi, B., and Ohno, S. (1974). Cytosol androgen receptor from kidney of normal and testicular feminized (Tfm) mice. *Cell (Cambridge, Mass.)* **2**, 205–212.

Attardi, B., and Ohno, S. (1978). Physical properties of androgen receptors in brain cytosol from normal and testicular feminized (Tfm/y) mice. *Endocrinology (Baltimore)* **103**, 760–770.

Baarends, W. M., Themmen, A. P. N., Blok, L. J., Mackenbach, P., Brinkmann, A. O., Meijer, D., Faber, P. W., Trapman, J., and Grootegoed, J. A. (1990). The rat androgen receptor gene promoter. *Mol. Cell. Endocrinol.* **74**, 75–84.

Barrack, E. R., Bujnovszky, P., and Walsh, P. C. (1983). Subcellular distribution of androgen receptors in human normal, benign hyperplastic, and malignant prostatic tissues: Characterization of nuclear salt-resistant receptors. *Cancer Res.* **43**, 1107–1116.

Batch, J. A., Evans, B. A. J., Hughes, I. A., and Patterson, M. N. (1993). Mutations of the androgen receptor gene identified in perineal hypospadias. *J. Med. Genet.* **30**, 198–201.

Baum, M. J . (1992). Neuroendocrinology of sexual behavior in the male. *In* "Behavioral Endocrinology" (J. Beeker, S. M. Breedlove, and D. Crews, eds.), pp. 97–130. MIT Press, Cambridge, MA.

Beato, M. (1989). Gene regulation by steroid hormones. *Cell (Cambridge, Mass.)* **56**, 335–344.

Berg, J. M. (1990). Zinc fingers and other metal-binding domains. *J. Biol. Chem.* **265**, 6513–6516.

Berry, S. J., Coffey, D. S., Walsh, P. C., and Ewing, L. L. (1984). The development of human benign prostatic hyperplasia with age. *J. Urol.* **132**, 474–479.

Biancalana, V., Serville, F., Pommier, J., Julien, J., Hanauer, A., and Mandel, J. L. (1992). Moderate instability of the trinucleotide repeat in spino bulbar muscular atrophy. *Hum. Mol. Genet.* **1**, 255–258.

Blok, L. J., Mackenbach, P., Trapman, J., Themmen, A. P. N., Brinkmann, A. O., and Grootegoed, J. A. (1989). Follicle-stimulating hormone regulates androgen receptor mRNA in Sertoli cells. *Mol. Cell. Endocrinol.* **63**, 267–271.

Blok, L. J., Hoogbrugge, J. W., Themman, A. P. N., Baarends, W. M., Post, M., and Grootegoed, J. A. (1992a). Transient down-regulation of androgen receptor messenger ribonucleic acid (mRNA) expression in Sertoli cells by follicle-stimulating hormone is followed by up-regulation of androgen receptor mRNA and protein. *Endocrinology (Baltimore)* **131**, 1343–1349.

Blok, L. J., Themmen, A. P., Peters, A. H., Trapman, J., Baarends, W. M., Hoogbrugge, J. W., and Grootegoed, J. A. (1992b). Transcriptional regulation of androgen receptor gene expression in Sertoli cells and other cell types. *Mol. Cell. Endocrinol.* **88**, 153–164.

Bonne, C., and Raynaud, J.-P. (1977). Characterization and hormonal control of the androgen receptor in the hamster sebaceous glands. *J. Invest. Dermatol.* **68**, 215–220.

Breedlove, S. M. (1992). Sexual differentiation of the brain and behavior. *In* "Behavioral Endocrinology" (J. B. Becker, S. M. Breedlove, and D. Crews, eds.), pp. 39–68.

Brinkmann, A. O., Klaasen, P., Kuiper, G. G. J. M., van der Korput, G. M., Bolt, J., de Boer, W., Smit, A., Faber, P. W., van Rooij, H. C. J., Geurts van Kessel, A., Voorhorst, M. M., Mulder, E., and Trapman, J. (1989). Structure and function of the androgen receptor. *Urol. Res.* **17**, 87–93.

Brinkmann, A. O., Kuiper, G. G. J. M., Ris-Stalpers, C., van Rooij, H. C. J., Romalo, G., Trifiro, M., Mulder, E., Pinsky, L., Schweikert, H. U., and Trapman, J. (1991). Androgen receptor abnormalities. *J. Steroid Biochem. Mol. Biol.* **40**, 349–352.

Brown, T. R., Lubahn, D. B., Wilson, E. M., French, F. S., Migeon, C. J., and Corden, J. L. (1990). Functional characterization of naturally occurring mutant androgen receptors from subjects with complete androgen insensitivity. *Mol. Endocrinol.* **4**, 1759–1772.

Brown, T. R., Scherer, P. A., Chang, Y. T., Migeon, C. J., Ghirri, P., Murono, K., and Zhou, Z. (1993). Molecular genetics of human androgen insensitivity. *Eur. J. Pediatr.* **152**, S62–S69.

Burkick, K. H., and Hill, R. (1970). The topical effect of the antiandrogen chlormadinone acetate and some of its chemical modifications on the hamster costovetebral organ. *Br. J. Dermatol.* **82**, Suppl. 6, 19–25.

Carter, C. S. (1992). Hormonal influences on human sexual behavior. *In* "Behavioral Endocrinology" (J. B. Becker, S. M. Breedlove, and D. Crews, eds.), pp. 131–142. MIT Press, Cambridge, MA.

Celotti, F., and Negri-Cesi, P. (1992). Anabolic steroids: A review of their effects on the muscles, of their possible mechanisms of action and of their use in athletics. *J. Steroid Biochem. Mol. Biol.* **43**, 469–477.

Chan, L., Johnson, M. P., and Tindall, D. J. (1989). Steroid hormone action. *In* "Pediatric Endocrinology" (R. Collu, J. R. Ducharme, and H. J. Guyda, eds.), pp. 81–124. Raven Press, New York.

Chang, C., Kokontis, J., and Liao, S. (1988a). Molecular cloning of human and rat complementary DNA encoding androgen receptors. *Science* **240**, 324–326.

Chang, C., Kokontis, J., and Liao, S. (1988b). Structural analysis of complementary DNA and amino acid sequences of human and rat androgen receptors. *Proc. Natl. Acad. Sci. U.S.A.* **85**, 7211–7215.

Charest, N. J., Zhou, Z. X., Lubahn, D. B., Olsen, K. L., Wilson, E. M., and French, F. S. (1991). A frameshift mutation destabilizes androgen receptor messenger RNA in the Tfm mouse. *Mol. Endocrinol.* **5**, 573–581.

Chodak, G. W., Kranc, D. M., Puy, L. A., Takeda, H., Johnson, K., and Chang, C. (1993). Nuclear localization of androgen receptor in heterogeneous samples of normal, hyperplastic and neoplastic human prostate. *J. Urol.* **147**, 798–803.

Choi, W.-T., MacLean, H. E., Chu, S., Warne, G. L., and Zajac, J. D. (1993). Kennedy's disease: Genetic diagnosis of an inherited form of motor neuron disease. *Aust. N.Z. J. Med.* **23**, 187–192.

Choo, J. J., Emery, P. W., and Rothwell, N. J. (1991). Dose-dependent effects of an anabolic steroid, nondrolone phenylpropionate (Durabolin) on body composition and muscle protein metabolism in female rats. *Ann. Nutr. Metab.* **35**, 141–147.

Choudhry, R., Hodgins, M. B., Van der Kwast, T. H., Brinkmann, A. O., and Boersma, W. J. A. (1992). Localization of androgen receptors in human skin by immunohistochemistry: Implications for the hormonal regulation of hair growth, sebaceous glands and sweat glands. *J. Endocrinol.* **133**, 467–475.

Claessens, F., Celis, L., Peeters, B., Heyns, W., Verhoeven, G., and Rombauts, W. (1989). Functional characterization of an androgen response element in the first intron of the C3(1) gene of prostatic binding protein. *Biochem. Biophys. Res. Commun.* **164**, 833–840.

Claessens, F., Rushmere, N., Celis, L., Peeters, B., Davies, P., and Rombauts, W. (1990a). Functional characterization of an androgen response element. *Biochem. Soc. Trans.* **18**, 561–562.

Claessens, F., Rushmere, N. K., Davies, P., Celis, L., Peeters, B., and Rombauts, W. A. (1990b). Sequence-specific binding of androgen–receptor complexes to prostatic binding protein genes. *Mol. Cell. Endocrinol.* **74**, 203–212.

Claessens, F., Celis, L., De Vos, P., Peeters, B., Heyns, W., Verhoeven, G., and Rombauts, W. (1993a). Intronic androgen response elements of prostatic binding protein genes. *Biochem. Biophys. Res. Commun.* **191**, 688–694.

Claessens, F., Celis, L., De Vos, P., Heyns, W., Verhoeven, G., Peeters, B., and Rombauts, W. (1993b). Functional androgen response elements in the genes encoding for prostatic binding protein. *Ann. N. Y. Acad. Sci.* **684**, 199–201.

Clements, J. A. (1989). The glandular kallikrein family of enzymes: Tissue specific expression and hormonal regulation. *Endocr. Rev.* **10**, 393–419.

Coffey, D. S. (1993). Prostate cancer. An overview of an increasing dilemma. *Cancer (Philadelphia)* **71**, 880–886.

Coffey, D. S., and Walsh, P. C. (1990). Clinical and experimental studies of benign prostatic hyperplasia. *Urol. Clin. North Am.* **17**, 461–475.

Cox, B. J., and Robins, D. M. (1988). Tissue-specific variation in C4 and Slp gene regulation. *Nucleic Acids Res.* **16**, 6857–6870.

Crews, D. (1993). The organization concept and vertebrates without sex chromosomes. *Brain, Behav. Evol.* **42**, 202–214.

Cunha, G. R., Chung, L. W. K., Shannon, J. M., and Reese, B. A. (1980). Stromal–epithelial interactions in sex differentiation. *Biol. Reprod.* **22**, 19–42.

Cunha, G. R., Donjacour, A. A., Cooke, P. S., Mee, S., Bigsby, R. M., Higgins, S. J., and Sugimura, Y. (1987). The endocrinology and developmental biology of the prostate. *Endocr. Rev.* **8**, 338–362.

Cunliffe, W. J., Shuster, S., and Smith, A. J. (1969). The effect of topical cyproterone acetate on sebum secretion in patients with acne. *Br. J. Dermatol.* **81**, 200–201.

Daneshgari, F., and Crawford, E. D. (1993). Endocrine therapy of advanced carcinoma of the prostate. *Cancer (Philadelphia)* **71**, Suppl., 1089–1097.

Danielson, M., Hinck, L., and Ringold, G. M. (1989). Two amino acids within the knuckle of the first zinc finger specify DNA response element activation by the glucocorticoid receptor. *Cell (Cambridge, Mass.)* **57**, 1131–1138.

DeBellis, A., Quigley, C. A., Cariello, N. F., el-Awady, M. K., Sar, M., Lane, M. V., Wilson, E. M., and French, F. S. (1992). Single base mutations in the human androgen receptor gene causing complete androgen insensitivity: Rapid detection by a modified denaturing gradient gel electrophoresis technique. *Mol. Endocrinol.* **6,** 1909–1920.

Delaey, B., Dirckx, L., Decourt, J.-L., Claessens, F., Peeters, B. and Rombauts, W. (1987). Rat prostatic binding protein: The complete sequence of the C2 gene and its flanking regions. *Nucleic Acids Res.* **15,** 1627–1641.

Denis, L. (1993). Prostate cancer. Primary hormonal treatment. *Cancer (Philadelphia)* **71,** Suppl., 1050–1058.

Denner, L. A., Weigel, N. L., Maxwell, B. L., Schrader, W. T., and O'Malley, B. W. (1990). Regulation of progesterone receptor-mediated transcription by phosphorylation. *Science* **250,** 1740–1743.

De Vos, P., Claessens, F., Winderickx, J., Van Dijck, P., Celis, L., Peeters, B., Rombauts, W., Heyns, W., and Verhoeven, G. (1991). Interaction of androgen response elements with the DNA-binding domain of the rat androgen receptor expressed in *Escherichia coli. J. Biol. Chem.* **266,** 3439–3443.

De Vos, P., Claessen, F., Peeters, B., Rombauts, W., Heyns, W., and Verhoeven, G. (1993a). Interaction of androgen and glucocorticoid receptor DNA-binding domains with their response elements. *Mol. Cell. Endocrinol.* **90,** R11–R16.

De Vos, P., Claessen, F., Celis, L., Heyns, W., Rombauts, W., and Verhoeven, G. (1993b). Nuclear extracts enhance gel retardation with androgen and glucocorticoid receptor DNA-binding domains. *Ann. N. Y. Acad. Sci.* **684,** 202–204.

Dryden, G. L., and Conaway, C. H. (1967). The origin and hormonal control of scent production in *Suncus murinus. J. Mammal.* **48,** 420–428.

Dube, J. Y., Lesage, R., and Tremblay, R. R. (1976). Androgen and estrogen binding in rat skeletal and perineal muscles. *Can. J. Biochem.* **54,** 50–55.

Ebling, F. J., and Skinner, J. (1983). The local effects of topically applied estradiol, cyproterone acetate and ethanol on sebaceous secretion in intact male rats. *J. Invest. Dermatol.* **81,** 448–451.

Ekman, P., Snochowski, M., Zetterberg, A., Hogberg, B., and Gustafsson, J. A. (1979). Steroid receptor content in human prostatic carcinoma and response to endocrine therapy. *Cancer (Philadelphia)* **44,** 1173–1181.

Faber, P. W., Kuiper, G. G. J. M., van Rooij, H. C. J., van der Korput, J. A. G. M., Brinkmann, A. O., and Trapman, J. (1989). The N-terminal domain of the human androgen receptor is encoded by one, large exon. *Mol. Cell. Endocrinol.* **61,** 257–262.

Faber, P. W., van Rooij, H. C. J., van der Korput, A. G. M., Baarends, W. M., Brinkmann, A. O., Grootegoed, J. A., and Trapman, J. (1991a). Characterization of the human androgen receptor transcription unit. *J. Biol. Chem.* **266,** 10748–10749.

Faber, P. W., King, A., van Rooij, H. C. J., Brinkmann, A. O., de Both, N. J., and Trapman, J. (1991b). The mouse androgen receptor. Functional analysis of the protein and characterization of the gene. *Biochem. J.* **278,** 269–178.

Faber, P. W., van Rooij, H. C. J., Schipper, H. J., Brinkmann, A. O., and Trapman, J. (1993). Two different, overlapping pathways of transcription initiation are active on the TATA-less human androgen receptor promoter. *J. Biol. Chem.* **268,** 9296–9301.

Fentie, D. D., Lakey, W. H., and McBlain, W. A. (1986). Applicability of nuclear androgen receptor quantification to human prostatic adenocarcinoma. *J. Urol.* **135,** 167–173.

Fernandez-Gonzalez, A. L., Marzo, F., Tosar, A., Fruhbeck, G., and Santidrian, S. (1989). Effect of anabolic agents on muscle protein metabolism in growing rats treated with glucocorticoids: A short review. *Rev. Esp. Fisiol.* **45,** 61–64.

Fishman, R. B., and Breedlove, S. M. (1988). Sexual dimorphism in the developing nervous system. *In* "Handbook of Human Growth and Developmental Biology" (E. Meisami and P. Timiras, eds.), Vol. 17, pp. 45–57. CRC Press, Boca Raton, FL.

Frost, P., and Gomez, E. C. (1972). Inhibitors of sex hormones: Development of experimental models. *Adv. Biol. Skin* **12**, 403–413.

Gaspar, M. L., Meo, T., and Tosi, M. (1990). Structure and size distribution of the androgen receptor mRNA in wild-type and Tfm/y mutant mice. *Mol. Endocrinol.* **6**, 1600–1610.

Gaspar, M. L., Meo, T., Bourgarel, P., Guenet, J. L., and Tosi, M. (1991). A single base deletion in the Tfm androgen receptor gene creates a short-lived messenger RNA that directs internal translation initiation. *Proc. Natl. Acad. Sci. U.S.A.* **88**, 8606–8610.

Geller, J. (1993). Basis for hormonal management of advanced prostate cancer. *Cancer (Philadelphia)* **71**, 1039–1045.

Georgatsou, E., Bourgarel, P., and Mio, T. (1993). Male-specific expression of mouse sex-limited protein requires growth hormone, not testosterone. *Proc. Natl. Acad. Sci. U.S.A.* **90**, 3626–3630.

Golsteyn, E. J., Goren, H. J., Lehoux, J. G., and Lefebvre, Y. A. (1990). Phosphorylation and nuclear processing of the androgen receptor. *Biochem. Biophys. Res. Commun.* **171**, 336–341.

Gonzalez-Cadavid, N. F., Swerdloff, R. S., Lemmi, C. A. E., and Rajfer, J. (1991). Expression of the androgen receptor gene in rat penile tissue and cells during sexual maturation. *Endocrinology (Baltimore)* **129**, 1671–1678.

Gonzalez-Cadavid, N. F., Vernet, D., Navarro, A. F., Rodriguez, J. A., Swerdloff, R. S., and Rajfer, J. (1993). Up-regulation of the levels of androgen receptor and its mRNA by androgens in smooth-muscle cells from rat penis. *Mol. Cell. Endocrinol.* **90**, 219–229.

Gorski, R. A. (1984). Critical role for the medial preoptic area in the sexual differentiation of the brain. *In* "Progress in Brain Research: Sex Differences in the Brain" (G. J. DeVries, G. J. DeBruin, H. M. B. Uylings, and M. A. Corner, eds.), pp. 129–146. Elsevier, Amsterdam.

Goueli, S. A., Holtzman, J. L., and Ahmed, K. (1984). Phosphorylation of the androgen receptor by a nuclear cAMP-independent protein kinase. *Biochem. Biophys. Res. Commun.* **123**, 778–784.

Green, H. (1993). Human genetic diseases due to codon reiteration: Relationship to an evolutionary mechanism. *Cell (Cambridge, Mass.)* **74**, 955–956.

Griffin, J. E., and Wilson, J. D. (1989). The androgen resistance syndromes: 5α-Reductase deficiency, testicular feminization, and related disorders. *In* "The Metabolic Basis of Inherited Disease" (C. R. Scriver, A. L. Beaudet, W. S. Sly, and D. Valle, eds.), 6th ed., pp. 1919–1944. McGraw-Hill, New York.

Grino, P. B., Griffin, J. E., and Wilson, J. D. (1990). Testosterone at high concentrations interacts with the human androgen receptor similarly to dihydrotestosterone. *Endocrinology (Baltimore)* **126**, 1165–1172.

Grumbach, M. M., and Conte, F. A. (1992). Disorders of sex differentiation. *In* "Textbook of Endocrinology" (J. D. Wilson and D. W. Foster, eds.), 8th ed., pp. 853–951. Harcourt Brace Jovanovich, Philadelphia.

Hagström, J. E., Harvey, S., Madden, B., McCormick, D., and Wieben, E. D. (1989). Androgens affect the processing of secretory protein precursors in the guinea pig seminal vesicle. II. Identification of conserved sites for protein processing. *Mol. Endocrinol.* **3**, 1797–1806.

Hagström, J., Harvey, S., and Wieben, E. (1992). Androgens are necessary for the establishment of secretory protein expression in the guinea pig seminal vesicle epithelium. *Biol. Reprod.* **47,** 768–775.

Hamilton, J. B., and Montagna, W. (1950). The sebaceous glands of the hamster. I. Morphological effects of androgens on integumentary structures. *Am. J. Anat.* **86,** 191–233.

Harding, A. E., Thomas, P. K., Baraister, M., Bradbury, P. G., Morgan-Hughes, J. A., and Ponsford, J. R. (1982). X-linked recessive bulospinal neuronopathy: A report of ten cases. *J. Neurol., Neurosurg. Psychiatry* **45,** 1012–1019.

Harris, S. E., Smith, R. G., Zhou, H., Mansson, P. E., and Malark, M. (1989). Androgens and glucocorticoids modulate heparin-binding growth factor I mRNA accumulation in DDT1 cells as analyzed by *in situ* hybridization. *Mol. Endocrinol.* **3,** 1839–1844.

Hayden, J. M., Bergen, W. G., and Merkel, R. A. (1992). Skeletal muscle protein metabolism and serum growth hormone, insulin, and cortisol concentrations in growing steers implanted with estradiol-17 beta, trenbolone acetate or estradiol-17 beta plus trenbolone acetate. *J. Anim. Sci.* **70,** 2109–2119.

He, W. W., Fischer, L. M., Sun, S., Bilhartz, D. L., Zhu, X., Young, C. Y. F., Kelley, D. B., and Tindall, D. J. (1990). Molecular cloning of androgen receptors from divergent species with a polymerase chain reaction technique: Complete cDNA sequence of the mouse androgen receptor and isolation of androgen receptor cDNA probes from dog, guinea pig, and clawed frog. *Biochem. Biophys. Res. Commun.* **171,** 697–704.

He, W. W., Kumar, M. V., and Tindall, D. J. (1991). A frame-shift mutation in the androgen receptor gene causes complete androgen insensitivity in the testicular-feminized mouse. *Nucleic Acids Res.* **19,** 2372–2378.

He, W. W., Lindzey, J., Kumar, M. V., and Tindall, D. J. (1994). The androgen receptor in the testicular feminized (Tfm) mouse is a product of internal translation initiation. *Receptor* (in press).

Henttu, P., and Vihko, P. (1993). Prostatic gene expression is modulated by protein kinase C in LNCaP cancer cells. *Endocr. Soc. Program,* Abstract, June, p. 364.

Husmann, D. A., Wilson, C. M., McPhaul, M. J., Tilley, W. D., and Wilson, J. D. (1990). Antipeptide antibodies to two distinct regions of the androgen receptor localize the receptor protein to the nuclei of target cells in the rat and human prostate. *Endocrinology (Baltimore)* **126,** 2359–2368.

Ibaraki, K., Robey, P. G., and Young, M. F. (1993). Partial characterization of a novel "GGA" factor which binds to the osteonectin promoter in bovine bone cells. *Gene* **130,** 225–232.

Igarashi, S., Tanno, Y., Onodera, O., Yamazaki, M., Sato, S., Ishikawa, A., Miyatani, N., Nagashima, M., Ishikawa, Y., Sahashi, K., Ibi, T., Myiatake, T., and Tsuji, S. (1992). Strong correlation between number of CAG repeats in androgen receptor genes and the clinical onset of features of spinal and bulbar muscular atrophy. *Neurology* **42,** 2300–2302.

Imperato-McGinley, J., Miller, M., Wilson, J. D., Peterson, R. E., Shackleton, C., and Gajdusek, D. C. (1991). A cluster of male pseudohermaphrodites with 5α-reductase deficiency in Papua New Guinea. *Clin. Endocrinol. (Oxford)* **34,** 293–298.

Imperato-McGinley, J., Gautier, T., Cai, L. Q., Yee, B., Epstein, J., and Pochi, P. (1993). The androgen control of sebum production. Studies of subjects with dihydrotestosterone deficiency and complete androgen insensitivity. *J. Clin. Endocrinol. Metab.* **76,** 524–528.

Isaacs, J. T., and Kyprianou, N. (1987). Development of androgen-independent tumor cells and their implication for the treatment of prostate cancer. *Urol. Res.* **15**, 4716–4720.

Jakubiczka, S., Werder, E. A., and Wieacker, P. (1992). Point mutation in the steroid-binding domain of the androgen receptor gene in a family with complete androgen insensitivity syndrome (CAIS). *Hum. Genet.* **90**, 311–312.

Jenkins, E. P., Hsieh, C. L., Milatovich, A., Normington, K., Berman, D. M., Francke, U., and Russell, D. W. (1991). Characterization and chromosomal mapping of human steroid 5α-reductase gene and pseudogene and mapping of the mouse homologue. *Genomics* **11**, 1102–1112.

Jenkins, E. P., Andersson, S., Imperato-McGinley, J., Wilson, J. D., and Russell, D. W. (1992). Genetic and pharmacological evidence for more than one human steroid 5α-reductase. *J. Clin. Invest.* **89**, 293–300.

Jenster, G., van der Korput, H. A. G. M., van Vroonhoven, C., van der Kwast, T. H., Trapman, J., and Brinkmann, A. O. (1991). Domains of the human androgen receptor involved in steroid binding, transcriptional activation, and subcellular localization. *Mol. Endocrinol.* **5**, 1396–1404.

Jenster, G., Trapman, J., and Brinkmann, A. O. (1993). Nuclear import of the human androgen receptor. *Biochem. J.* **293**, 761–768.

Johnson, M. P., Rowley, D. R., Young, C. Y. F., and Tindall, D. J. (1988). Characterization of the androgen receptor. *In* "Steroid Receptors and Disease" (P. J. Sheridan, ed.), pp. 207–228. Dekker, New York.

Josso, N., Picard, J. Y., and Tran, D. (1977). The antimullerian hormone. *Recent Prog. Horm. Res.* **33**, 117–167.

Jost, A. (1953). Problems of fetal endocrinology: The gonadal and hypophyseal hormones. *Recent Prog. Horm. Res.* **8**, 379–418.

Jost, A. (1971). Embryonic sexual differentiation (morphology, physiology, abnormalities). *In* "Hermaphroditism, Genital Anomalies and Related Endocrine Disorders" (H. W. Jones, Jr. and W. W. Scott, eds.), pp. 16–64. Williams & Wilkins, Baltimore, MD.

Kasumi, H., Komori, S., Yamasaki, N., Shima, H., and Isojima, S. (1993). Single nucleotide substitution of the androgen receptor gene in a case with receptor-positive androgen insensitivity syndrome (complete form). *Acta Endocrinol. (Copenhagen)* **128**, 355–360.

Kazemi-Esfarjani, P., Beitel, L. B., Trifiro, M., Kaufman, M., Rennie, P., Sheppard, P., Matusik, R., and Pinsky, L. (1993). Substitution of valine-865 by methionine or leucine in the human androgen receptor causes complete or partial androgen insensitivity, respectively, with distinct androgen receptor phenotypes. *Mol. Endocrinol.* **7**, 37–46.

Kelley, D. B. (1986). Neuroeffectors for vocalization in *Xenopus laevis:* Hormonal regulation of sexual dimorphism. *J. Neurobiol.* **17**, 231–248.

Kelley, D. B., Sassoon, D., Segil, N., and Scudder, M. (1989). Development and hormone regulation of androgen receptor levels in the sexually dimorphic larynx of *Xenopus laevis. Dev. Biol.* **131**, 111–118.

Kemppainen, J. A., Lane, M. V., Sar, M., and Wilson, E. M. (1992). Androgen receptor phosphorylation, turnover, nuclear transport, and transcriptional activation. *J. Biol. Chem.* **267**, 968–974.

Klein, H., Bressel, M., Kastendieck, H., and Voigt, K.-D. (1991). Biochemical endocrinology of prostate cancer. *In* "Endocrine Dependent Tumors" (K.-D. Voigt, ed.), pp. 131–163. Raven Press, New York.

Klocker, H., Kasper, F., Eberle, J., Uberreiter, S., Radmayr, C., and Bartsch, G. (1992).

Point mutation in the DNA binding domain of the androgen receptor in two families with Reifenstein syndrome. *Am. J. Hum. Genet.* **52**, 1318–1327.

Koenig, H., Goldstone, A., and Lu, C. Y. (1980). Androgens regulate mitochondrial cytochrome c oxidase and lysosomal hydrolases in mouse skeletal muscle. *Biochem. J.* **192**, 349–353.

Komada, S., Itami, S., Kurata, S., and Takayasu, S. (1989). Side gland of *Suncus murinus* as a new model of sebaceous gland: 5α-Reductase, androgen receptor and nuclear androgen content in male and female animals. *Arch. Dermatol. Res.* **280**, 487–493.

Komi, P. V., and Karlsson, J. (1978). Skeletal muscle fibre types, enzyme activities and physical performance in young males and females. *Acta Physiol. Scand.* **103**, 210–218.

Krieg, M., Bartsch, W., Janssen, W., and Voigt, K. D. (1979). A comparative study of binding, metabolism and endogenous levels of androgens in normal, hyperplastic and carcinomatous human prostate. *J. Steroid Biochem.* **11**, 615–624.

Krongrad, A., Wilson, C. M., Wilson, J. D., Allman, D. R., and McPhaul, M. J. (1991). Androgen increases androgen receptor protein while decreasing receptor mRNA in LNCaP cells. *Mol. Cell. Endocrinol.* **76**, 79–88.

Kuiper, G. G. J. M., de Ruiter, P. E., Trapman, J., Boersma, W. J. A., Grootegoed, J. A., and Brinkmann, A. O. (1993). Localization and hormonal stimulation of phosphorylation sites in the LNCaP-cell androgen receptor. *Biochem. J.* **291**, 95–101.

Kumar, M. V., Grossmann, M. E., Leo, M. E., Jones, E. A., and Tindall, D. J. (1992). Isolation and characterization of the 5′-flanking region of the mouse androgen receptor gene. *J. Cell. Biochem.* **Suppl. 16C**, L123.

Kupfer, S. R., Marschke, K. B., Wilson, E. M., and French, F. S. (1993). Receptor accessory factor enhances specific DNA binding of androgen and glucocorticoid receptors. *J. Biol. Chem.* **268**, 17519–17527.

Labrie, F., Bélanger, A., Simard, J., Labrie, C., and Dupont, A. (1993). Combination therapy for prostate cancer. Endocrine and biologic basis of its choice as new standard first-line therapy. *Cancer (Philadelphia)* **71**, Suppl., 1059–1067.

La Spada, A. R., Wilson, E. M., Lubahn, D. B., Harding, A. E., and Fischbeck, K. H. (1991). Androgen receptor gene mutations in X-linked spinal and bulbar muscular atrophy. *Nature (London)* **352**, 77–79.

Lilja, H. (1985). A kallikrein-like serine protease in prostatic fluid cleaves the predominant seminal vesicle protein. *J. Clin. Invest.* **76**, 1899–1903.

Lin, M. C., Rajfer, J., Swerdloff, R. S., and Gonzalez-Cadavid, N. F. (1993). Testosterone down-regulates the levels of androgen receptor mRNA in smooth muscle cells from the rat corpora cavernosa via aromatization to estrogens. *J. Steroid Biochem. Mol. Biol.* **45**, 333–343.

Lindzey, J., Grossman, M., Kumar, M. V., and Tindall, D. J. (1993). Regulation of the 5′-flanking region of the mouse androgen receptor gene by cAMP and androgen. *Mol. Endocrinol.* **7**, 1530–1540.

Lobaccaro, J.-M., Lumbroso, S., Ktari, R., Dumas, R., and Sultan, C. (1993). An exonic point mutation creates a MaeIII site in the androgen receptor gene of a family with complete androgen insensitivity syndrome. *Hum. Mol. Genet.* **2**, 1041–1043.

Lookingbill, D. P., Horton, R., Demers, L. M., Eagan, N., Marks, J. G., and Santen, R. J. (1985). Tissue production of androgens in women. *J. Am. Acad. Dermatol.* **12**, 481–487.

Lubahn, D. B., Joseph, D. R., Sullivan, P. M., Willard, H. F., French, F. S., and Wilson, E. M. (1988). Cloning of human androgen receptor complementary DNA and localization to the X chromosome. *Science* **240**, 327–330.

Lubahn, D. B., Tan, J. A., Quarmby, V. E., Sar, M., Joseph, D. R., French, F. S., and Wilson, E. M. (1989a). Structural analysis of the human and rat androgen receptors and expression in male reproductive tract tissues. *Ann. N. Y. Acad. Sci.* **564**, 48–56.

Lubahn, D. B., Brown, T. R., Simental, J. A., Higgs, H. N., Migeon, C. J., Wilson, E. M., and French, F. S. (1989b). Sequence of the intron/exon junctions of the coding region of the human androgen receptor gene and identification of a point mutation in a family with complete androgen insensitivity. *Proc. Natl. Acad. Sci. U.S.A.* **86**, 9534–9538.

Lukas, S. E. (1993). Current perspective on anabolic-androgen steroid abuse. *Trends Pharmacol. Sci.* **14**, 61–68.

Lutsky, B. N., Budak, M., Koziol, P., Monahan, M., and Neri, R. O. (1975). The effects of a nonsteroid antiandrogen, flutamide, on sebaceous gland activity. *Invest. Dermatol.* **64**, 412–417.

Lyons, F., and Shuster, S. (1982). Sex difference in response of the human sebaceous gland to topical flutamide. *Br. J. Dermatol.* **107**, 697–699.

MacLean, H. E., Chu, S., Warne, G. L., and Zajac, J. D. (1993). Related individuals with different androgen receptor gene deletions. *J. Clin. Invest.* **91**, 1123–1128.

Marcelli, M., Tilley, W. D., Wilson, C. M., Griffin, J. E., Wilson, J. D., and McPhaul, M. J. (1990a). Definition of the human androgen receptor gene structure permits the identification of mutations that cause androgen resistance: Premature termination of the receptor protein at amino acid residue 588 causes complete androgen resistance. *Mol. Endocrinol.* **4**, 1105–1116.

Marcelli, M., Tilley, W. D., Wilson, C. M., Wilson, J. D., Griffin, J. E., and McPhaul, M. J. (1990b). A single nucleotide substitution introduces a premature termination codon into the androgen receptor gene of a patient with receptor-negative androgen resistance. *J. Clin. Invest.* **85**, 1522–1528.

Marcelli, M., Tilley, W. D., Zoppi, S., Griffin, J. E., Wilson, J. D., and McPhaul, M. J. (1991a). Androgen resistance associated with a mutation of the androgen receptor at amino acid 772 (Arg-Cys) results from a combination of decreased messenger ribonucleic acid levels and impairment of receptor function. *J. Clin. Endocrinol. Metab.* **73**, 318–325.

Marcelli, M., Zoppi, S., Grino, J. E., Wilson, J. D., and McPhaul, M. J. (1991b). A mutation in the DNA-binding domain of the androgen receptor gene causes complete testicular feminization in a patient with receptor-positive androgen resistance. *J. Clin. Invest.* **87**, 1123–1126.

Martin, J. (1962). Effect of castration and androgen replacement on the supracaudal gland of the male guinea pig. *J. Morphol.* **110**, 285–293.

Matsuura, T., Demura, T., Aimoto, Y., Mizuno, T., Moriwaka, F., and Tashiro, K. (1992). Androgen receptor abnormality in X-linked spinal and bulbar muscular atrophy. *Neurology* **42**, 1724–1726.

Matusik, R. J., Kreis, C., McNicol, R., Sweetland, R., Mullin, C., Fleming, W. H., and Dodd, J. G. (1986). Regulation of prostatic genes: Roles of androgens and zinc in expression. *J. Biochem. Cell Biol.* **64**, 601–607.

Max, S. R., Mufti, S., and Carlson, B. M. (1981). Cytosolic androgen receptor in regenerating rat levator ani muscle. *Biochem. J.* **200**, 77–82.

McLeod, D. G. (1993). Antiandrogenic drugs. *Cancer (Philadelphia)* **71**, Suppl., 1046–1049.

McPhaul, M. J., Marcelli, M., Tilley, W. D., Griffin, J. E., Isidro-Gutierrez, R. F., and Wilson, J. D. (1991). Molecular basis of androgen resistance in a family with a qualitative abnormality of the androgen receptor and responsive to high-dose androgen therapy. *J. Clin. Invest.* **87**, 1413–1421.

McPhaul, M. J., Marcelli, M., Zoppi, S., Wilson, C. M., Griffin, J. E., and Wilson, J. D. (1992). Mutations in the ligand-binding domain of the androgen receptor gene cluster in two regions of the gene. *J. Clin. Invest.* **90**, 2097–2101.

McPhaul, M. J., Marcelli, M., Zoppi, S., Griffin, J. E., and Wilson, J. D. (1993). Genetic basis of mutations in the androgen receptor gene that causes androgen resistance. *J. Clin. Endocrinol. Metab.* **76**, 17–23.

Michel, G., and Baulieu, E. E. (1976). An approach to the anabolic action of androgens by an experimental system. *Environ. Qual. Saf., Suppl.* **5**, 54–59.

Miyagoer, Y., Georgatsou, E., Varin-Blank, N., and Meo, T. (1993). The androgen-dependent C4-Slp gene is driven by a constitutively competent promoter. *Proc. Natl. Acad. Sci. U.S.A.* **90**, 5786–5790.

Miyamoto, K. K., McSherry, S. A., Dent, G. A., Madhabananda, S., Wilson, E. M., French, F. S., Sharief, Y., and Mohler, J. L. (1993). Immunohistochemistry of the androgen receptor in human benign and malignant prostate tissue. *J. Urol.* **149**, 1015–1019.

Mooradian, A. D., Morley, J. E., and Korenman, S. G. (1987). Biological actions of androgens. *Endocr. Rev.* **8**, 1–28.

Moore, M. C., and Lindzey, J. (1992). The physiological basis of sexual behavior in male reptiles. *In* "Biology of the Reptilia" D. Crews, ed.), pp. 70–113. Univ. of Chicago Press, Chicago.

Morris, B. J. (1989). hGK-1: A kallikrein gene expressed in human prostate. *Clin. Exp. Pharmacol. Physiol.* **16**, 345–351.

Moudgil, V. K. (1990). Phosphorylation of steroid hormone receptors. *Biochim. Biophys. Acta* **1055**, 243–258.

Mowszowicz, I., Lee, H.-J., Chen, H.-T., Mestayer, C., Portois, M.-C., Cabrol, S., Mauvais-Jarvis, P., and Chang, C. (1993). A point mutation in the second zinc finger of the DNA-binding domain of the androgen receptor gene causes complete androgen insensitivity in two siblings with receptor-positive androgen resistance. *Mol. Endocrinol.* **7**, 861–869.

Murtha, P., Tindall, D. J., and Young, C. Y. F. (1993). Androgen induction of a human prostate-specific kallikrein, hKLK2: Characterization of an androgen response element in the 5′ promoter region of the gene. *Biochemistry* **32**, 6459–6464.

Nakao, R., Haji, M., Yanase, T., Ogo, A., Takayanagi, R., Katsube, T., Fukumaki, Y., and Nawata, H. (1992). A single amino acid substitution (Met[786]-Val) in the steroid-binding domain of human androgen receptor leads to complete androgen insensitivity syndrome. *J. Clin. Endocrinol. Metab.* **74**, 1152–1157.

Nakao, R., Yanase, T., Sakai, Y., Haji, M., and Nawata, H. (1993). A single amino acid substitution (Gly743-Val) in the steroid-binding domain of the human androgen receptor leads to Reifenstein syndrome. *J. Clin. Endocrinol. Metab.* **77**, 103–107.

Neumann, F., and Kalmus, J. (1991). Cyproterone acetate in the treatment of sexual disorders: Pharmacological base and clinical experience. *Exp. Clin. Endocrinol.* **98**, 71–80.

Neumann, F., von Berswordt-Wallrabe, R., Elger, W., Steinbeck, H., Hahu, J. D., and Krause, M. (1970). Aspects of androgen-dependent events as studied by anti-androgens. *Recent Prog. Horm. Res.* **26**, 337–410.

Norvitch, M. E., Harvey, S., Smith, S., Hagström, J. E., and Wieben, E. D. (1989). Androgens affect the processing of secretory protein precursors in the guinea pig seminal vesicle. I. Evidence for androgen-regulated proteolytic processing. *Mol. Endocrinol.* **3**, 1788–1796.

Norvitch, M. E., Harvey, S., Hagström, J. E., Toft, J., and Wieben, E. D. (1991). Post-transcriptional regulation of secretory protein production during the development of the guinea pig seminal vesicle. *Biol. Reprod.* **45**, 797–803.

Page, M. J., and Parker, M. G. (1982). Effect of androgen on the transcription of rat prostatic binding protein genes. *Mol. Cell. Endocrinol.* **27**, 343–355.

Parker, M. G., Webb, P., Needham, M., White, R., and Ham, J. (1987). Identification of androgen response elements in mouse mammary tumor virus and the rat prostate C3 gene. *J. Cell Biochem.* **35**, 285–292.

Pochi, P. E., and Strauss, J. S. (1974). Endocrinologic control of the development and activity of the human sebaceous gland. *J. Invest. Dermatol.* **62**, 191–201.

Pochi, P. E., Strauss, J. S., and Downing, D. T. (1979). Age-related changes in sebaceous gland activity. *J. Invest. Dermatol.* **73**, 108–111.

Prior, L., Bordet, S., Trifiro, M. A., Mhatre, A., Kaufman, M., Pinsky, L., Wrogeman, K., Belsham, D. D., Pereira, F., Greenberg, C., Trapman, J., Brinkmann, A. O., Chang, C., and Liao, S. (1992). Replacement of arginine 773 by cysteine or histidine in the human androgen receptor causes complete androgen insensitivity with different receptor phenotypes. *Am. J. Hum. Genet.* **51**, 143–155.

Pugh, B. F., and Tjian, R. (1990). Mechanism of transcriptional activation by Sp1: Evidence for coactivators. *Cell (Cambridge, Mass.)* **61**, 1187–1197.

Pugh, B. F., and Tjian, R. (1991). Transcription from a TATA-less promoter requires a multisubunit TFIID complex. *Genes Dev.* **5**, 1935–1945.

Pye, R. J., Burton, J. L., and Harris, J. I. (1976). Effect of 1% cyproterone acetate in Cetomacrogol cream BPC (formula A) on sebum excretion rate in patients with acne. *Br. J. Dermatol.* **95**, 427–428.

Quarmby, V. E., Yarbrough, W. G., Lubahn, D. B., French, F. S., and Wilson, E. M. (1990). Autologous down-regulation of androgen receptor messenger ribonucleic acid. *Mol. Endocrinol.* **4**, 22–28.

Quigley, C. A., Friedman, K. J., Johnson, A., Lafreniere, R. G., Silverman, L. M., Lubahn, D. B., Brown, T. R., Wilson, E. M., Willard, H. F., and French, F. S. (1992a). Complete deletion of the androgen receptor gene: Definition of the null phenotype of the androgen insensitivity syndrome and determination of carrier status. *J. Clin. Endocrinol. Metab.* **74**, 927–933.

Quigley, C. A., Evans, B. A. J., Simental, J. A., Marschke, K. B., Sar, M., Lubahn, D. B., Davies, P., Hughes, L. A., Wilson, E. M., and French, F. S. (1992b). Complete androgen insensitivity due to deletion of exon c of the androgen receptor gene highlights the functional importance of the second zinc finger of the androgen receptor *in vivo*. *Mol. Endocrinol.* **6**, 1103–1112.

Rennie, P. S., Bruchovsky, N., and Goldenberg, S. L. (1988). Relationship of androgen receptors to the growth and regression of the prostate. *Am. J. Clin. Oncol.* **11**, S13–S17.

Rennie, P. S., Bruchovsky, N., Leco, K. J., Sheppard, P. C., McQuenn, S. A., Cheng, H., Snoek, R., Hamel, A., Bock, M. E., MacDonald, B. S., Nickel, B. E., Chang, C., Liao, S., Cattini, P. A., and Matusik, R. J. (1993). Characterization of two cis-acting DNA elements involved in the androgen regulation of the probasin gene. *Mol. Endocrinol.* **7**, 23–36.

Ricciardelli, C., Horsfall, D. J., Skinner, J. M., Henderson, D. W., Marshall, V. R., and Tilley, W. D. (1989). Development and characterization of primary cultures of smooth muscle cells from the fibromuscular stroma of the guinea pig prostate. *In Vitro Cell Dev. Biol.* **25**, 1016–1024.

Riegman, P. H. J., Vlietstra, R. J., Klaasen, P., van der Korput, J. A. G. M., Geurtsvan Kessel, A., Romijn, J. C., and Trapman, J. (1989). The prostate-specific antigen gene and the human glandular kallikrein gene are tandemly located on chromosome 19. *FEBS Lett.* **247,** 123–126.

Riegman, P. H. J., Vlietstra, R. J., van der Korput, J. A. G. M., Brinkmann, A. O., and Trapman, J. (1991). The promoter of the prostate-specific antigen contains a functional androgen responsive element. *Mol. Endocrinol.* **5,** 1921–1930.

Ris-Stalpers, C., Kuiper, G. G. J. M., Faber, P. W., Schweikert, H. U., van Rooij, H. C. J., Zegers, N. D., Hodgins, M. B., Degenhart, H. J., Trapman, J., and Brinkmann, A. O. (1990). Aberrant splicing of androgen receptor mRNA results in synthesis of a nonfunctional receptor protein in a patient with androgen insensitivity. *Proc. Natl. Acad. Sci. U.S.A.* **87,** 7866–7870.

Ris-Stalpers, C., Trifiro, M. A., Kuiper, G. G. J. M., Jenster, G., Romalo, G., Sai, T., van Rooij, H. C. J., Kaufman, M., Rosenfield, R. L., Liao, S., Schweikert, H.-U., Trapman, J., Pinsky, L., and Brinkmann, A. O. (1991). Substitution of aspartic acid-686 by histidine or asparagine in the human androgen receptor leads to a functionally inactive protein with altered hormone-binding characteristics. *Mol. Endocrinol.* **5,** 1562–1569.

Ris-Stalpers, C., Turberg, A., Verleun-Mooyman, M. C. T., Romalo, G., Schweikert, H. U., Trapman, J., and Brinkmann, A. O. (1992). Expression of an aberrantly spliced androgen receptor mRNA in a family with complete androgen insensitivity. *Ann. N. Y. Acad. Sci.* **684,** 239–242.

Ris-Stalpers, C., Verleun-Mooijman, M. C. T., Trapman, J., and Brinkmann, A. O. (1993). Threonine on amino acid position 868 in the human androgen receptor is essential for androgen binding specificity and functional activity. *Biochem. Biophys. Res. Commun.* **196,** 173–180.

Roche, P. J., Hoare, S. A., and Parker, M. G. (1992). A consensus DNA-binding site for the androgen receptor. *Mol. Endocrinol.* **6,** 2229–2235.

Rushmere, N. K., Parker, M. G., and Davies, P. (1987). Androgen-binding regions of an androgen-responsive gene. *Mol. Cell. Endocrinol.* **51,** 259–265.

Rushmere, N. K., Claessen, F., Peeters, B., Rombauts, W., and Davies, P. (1990). Intronic steroid response elements in prostate binding protein genes. *Biochem. Soc. Trans.* **18,** 560–561.

Sai, T., Seino, S., Chang, C., Trifiro, M., Pinsky, L., Mhatre, A., Kaufman, M., Lambert, B., Trapman, J., Brinkmann, A. O., Rosenfield, R. L., and Liao, S. (1990). An exonic point mutation of the androgen receptor gene in a family with complete androgen insensitivity. *Am. J. Hum. Genet.* **46,** 1095–1100.

Sassoon, D., and Kelley, D. B. (1986). The sexually dimorphic larynx of *Xenopus laevis:* Development and androgen regulation. *Am. J. Anat.* **177,** 457–472.

Saunders, P. T. K., Padayachi, T., Tincello, D. G., Shalet, S. M., and Wu, F. C. W. (1992). Point mutations detected in the androgen receptor gene of three men with partial androgen insensitivity syndrome. *Clin. Endocrinol. (Oxford)* **37,** 214–220.

Schroder, H. G., Ziegler, M., Nickisch, K., Kaufmann, J., and El Etreby, M. F. (1989). Compounds of sebaceous glands of hamster ear and flank organs. *J. Invest. Dermatol.* **92,** 769–773.

Shan, L.-X., Rodriguez, M. C., and Janne, O. A. (1990). Regulation of androgen receptor protein and mRNA concentrations by androgens in rat ventral prostate and seminal vesicles and in human hepatoma cells. *Mol. Endocrinol.* **4,** 1636–1646.

Siiteri, P. K., and Wilson, J. D. (1970). Dihydrotestosterone in prostatic hypertrophy. *J. Clin. Invest.* **49,** 1737–1745.

Siiteri, P. K., and Wilson, J. D. (1974). Testosterone formation and metabolism during male sexual differentiation in the human embryo. *J. Clin. Endocrinol. Metab.* **38,** 113–125.

Simental, J. A., Sar, M., Lane, M. V., French, F. S., and Wilson, E. M. (1991). Transcriptional activation and nuclear targeting signals of the human androgen receptor. *J. Biol. Chem.* **266**, 510–518.

Smale, S. T., and Baltimore, D. (1989). The "initiator" as a transcription control element. *Cell (Cambridge, Mass.)* **57**, 103–113.

Soloway, M. S., and Matzkin, H. (1993). Antiandrogenic agents as monotherapy in advanced prostatic carcinoma. *Cancer (Philadelphia), Suppl.* **71**, 1083–1088.

Sonada, T., Itami, S., Kurata, S., and Takayasu, S. (1991). Influences of gonadal and adrenal androgens on the side glands of *Suncus murinus*. *Endocrinol. Jpn.* **38**, 253–258.

Spence, A. M., Sheppard, P. C., Davie, J. R., Matuo, Y., Nishi, N., McKeehan, W. L., Dodd, J. G., and Matusik, R. J. (1989). Regulation of a bifunctional mRNA results in synthesis of secretal and nuclear probasin. *Proc. Natl. Acad. Sci. U.S.A.* **86**, 7843–7847.

Stavenhagen, J. B., and Robins, D. M. (1986). An ancient provirus has imposed androgen regulation on the adjacent mouse sex-limited protein gene. *Cell (Cambridge, Mass.)* **55**, 247–255.

Steiner, M. S. (1993). Role of peptide growth factors in the prostate: A review. *Urology* **42**, 99–110.

Sweet, C. R., Behzadian, M. A., and McDonough, P. G. (1992). A unique point mutation in the androgen receptor gene in a family with complete androgen insensitivity syndrome. *Fertil. Steril.* **58**, 703–707.

Sweetland, R., Sheppard, P. C., Dodd, J. G., and Matusik, R. J. (1988). Post-castration rebound of an androgen regulated prostate gene. *Mol. Cell. Biochem.* **84**, 3–15.

Syms, A. J., Nag, A., Norris, J. S., and Smith, R. G. (1987). Glucocorticoid effects on growth, and androgen receptor concentrations in DDT1MF-2 cell lines. *J. Steroid Biochem.* **28**, 109–116.

Takayasu, S., and Adachi, K. (1970). Hormonal control of metabolism in hamster costovertebral glands. *J. Invest. Dermatol.* **55**, 13–19.

Takayasu, S., and Adachi, K. (1972). The *in vivo* and *in vitro* β conversion of testosterone to 17-βhydroxy-5α-androstane-3-one (dihydrotestosterone) by the sebaceous gland of hamsters. *Endocrinology (Baltimore)* **90**, 73–80.

Tan, J., Joseph, D. R., Quarmby, V. E., Lubahn, D. B., Sar, M., French, F. S., and Wilson, E. M. (1988). The rat androgen receptor: Primary structure, autoregulation of its messenger ribonucleic acid, and immunocytchemical localization of the receptor protein. *Mol. Endocrinol.* **12**, 1276–1285.

Tan, J., Marschke, K. B., Ho, K.-C., Perry, S. T., Wilson, E. M., and French, F. S. (1992). Response elements of the androgen-regulated C3 gene. *J. Biol. Chem.* **267**, 4456–4466.

Thiessen, D. D., Blum, S. L., and Lindzey, G. (1969). A scent marking response associated with the ventral sebaceous gland of the Mongolian gerbil (*Meriones unguiculatus*). *Anim. Behav.* **18**, 26–30.

Thigpen, A. E., Davis, D. L., Milatovich, A., Mendonca, B. B., Imperato-McGinley, J., Griffin, J. E., Francke, U., Wilson, J. D., and Russell, D. W. (1992). Molecular genetics of steroid 5α-reductase 2 deficiency. *J. Clin. Invest.* **90**, 799–809.

Thigpen, A. E., Silver, R. I., Guileyardo, J. M., Casey, M. L., McConnell, J. D., and Russell, D. W. (1993). Tissue distribution and ontogeny of steroid 5α-reductase isozyme expression. *J. Clin. Invest.* **92**, 903–910.

Thody, A. J., and Shuster, S. (1989). Control and function of sebaceous glands. *Physiol. Rev.* **69**, 383–416.

Tilley, W. D., Marcelli, M., Wilson, J. D., and McPhaul, M. J. (1989). Characterization and expression of a cDNA encoding the human androgen receptor. *Proc. Natl. Acad. Sci. U.S.A.* **86**, 327–331.

Tilley, W. D., Marcelli, M., and McPhaul, M. J. (1990). Expression of the human androgen receptor gene utilized a common promoter in diverse human tissues and cell lines. *J. Biol. Chem.* **265**, 13776–13781.

Trachtenberg, J., Hicks, L. L., and Walsh, P. C. (1980). Androgen- and estrogen-receptor content in spontaneous and experimentally induced canine prostatic hyperplasia. *J. Clin. Invest.* **65**, 1051–1059.

Trapman, J., Klaasen, P., Kuiper, G. G. J. M., van der Korput, J. A. G. M., Faber, P. W., van Rooij, H. C. J., van Kessel, A. G., Voorhorst, M. M., Mulder, E., and Brinkmann, A. O. (1988). Cloning, structure and expression of a cDNA encoding the human androgen receptor. *Biochem. Biophys. Res. Commun.* **153**, 241–248.

van Laar, J. H., Berrevoets, C. A., Trapman, H., and Zegers, N. (1991). Hormone-dependent androgen receptor phosphorylation is accompanied by receptor transformation in human lymph node carcinoma of the prostate cells. *J. Biol. Chem.* **266**, 3734–3738.

Veldscholte, J., Ris-Stalpers, C., Kuiper, G. G. J. M., Jenster, G., Berrevoets, C., Claassen, E., van Rooij, H. C. J., Trapman, J., Brinkmann, A. O., and Mulder, E. (1990a). A mutation in the ligand binding domain of the androgen receptor of human LNCaP cells affects steroid binding characteristics and response to antiandrogens. *Biochem. Biophys. Res. Commun.* **173**, 534–540.

Veldscholte, J., Voorhorst-Ogink, M. M., Bolt-de Vries, J., van Rooij, H. C. J., Trapman, J., and Mulder, E. (1990b). Unusual specificity of the androgen receptor in the human prostate tumor cell line LNCaP: High affinity for progestagenic and estrogenic steroids. *Biochim. Biophys. Acta* **1052**, 187–194.

Venable, J. H. (1966). Morphology of the cells of normal, testosterone-deprived and testosterone-stimulated levator ani muscles. *Am. J. Anat.* **119**, 271–302.

Verhoeven, G., and Cailleau, J. (1988). Follicle-stimulating hormone and androgens increase the concentration of the androgen receptor in Sertoli cells. *Endocrinology (Baltimore)* **122**, 1541–1550.

Vermorken, A. J. M., Goos, C. M. A., Sultan, C., Vermeesch-Markslag, A. M. G., and Dijstra, A. C. (1986). Studies on the local activity of antiandrogens at the molecular and histologic level. *Mol. Biol. Rep.* **11**, 99–105.

Walsh, P. C., Hutchins, G. M., and Ewing, L. L. (1983). Tissue content of dihydrotestosterone in human prostatic hyperplasia is not supranormal. *J. Clin. Invest.* **72**, 1772–1777.

Walton, S., Cunliffe, W. J., Lookingbill, P., and Keczkes, K. (1986). Lack of effect of topical spironolactone on sebum excretion. *Br. J. Dermatol.* **114**, 261–269.

Weismann, A., Bowden, J., Frank, B. L., Horwitz, S. N., and Frost, P. (1985). Antiandrogenic effects of topically applied spironolactone on the hamster flank organ. *Arch. Dermatol.* **121**, 57–62.

West, N. B., Chang, C. S., Liao, S. S., and Brenner, R. M. (1990). Localization and regulation of estrogen, progestin and androgen receptors in the seminal vesicle of the rhesus monkey. *J. Steroid Biochem. Mol. Biol.* **37**, 11–21.

Wolf, D. A., Herzinger, T., Hermeking, H., Blaschke, D., and Horz, W. (1993). Transcriptional and posttranscriptional regulation of human androgen receptor expression by androgen. *Mol. Endocrinol.* **7**, 924–936.

Wong, C., Zhou, Z., Sar, M., and Wilson, E. M. (1993). Steroid requirement for androgen receptor dimerization and DNA binding. *J. Biol. Chem.* **268**, 19004–19012.

Wooster, R., Mangion, J., Eeles, R., Smith, S., Dowsett, M., Averill, D., Barrett-Lee, P., Easton, D. F., Ponder, B. A. J., and Stratton, M. R. (1992). A germline mutation in the androgen receptor gene in two brothers with breast cancer and Reifenstein syndrome. *Nat. Genet.* **2**, 132–134.

Yarbrough, W. G., Quarmby, V. E., Simental, J. A., Joseph, D. R., Sar, M., Lubahn, D. B., Olsen, K. L., French, F. S., and Wilson, E. M. (1990). A single base mutation in the androgen receptor gene causes androgen insensitivity in the testicular feminized rat. *J. Biol. Chem.* **265**, 8893–8900.

Young, C. Y. F., Johnson, M, P., Prescott, J. L., and Tindall, D. J. (1989). The androgen receptor of the testicular-feminized (Tfm) mutant mouse is smaller than the wild-type receptor. *Endocrinology (Baltimore)* **124**, 771–775.

Young, C. Y. F., Qui, S., Prescott, J. L., and Tindall, D. J. (1990). Overexpression of a partial human androgen receptor in *E. coli:* Characterization of steroid binding, DNA binding, and immunological properties. *Mol. Endocrinol.* **4**, 1841–1849.

Young, C. Y. F., Montgomery, B. T., Andrews, P. E., Qui, S., Bilhartz, D. L., and Tindall, D. J. (1991). Hormonal regulation of prostate-specific antigen messenger RNA in human prostatic adenocarcinoma cell line LNCaP. *Cancer Res.* **51**, 3748–3752.

Young, C. Y. F., Andrews, P. E., Montgomery, B. T., and Tindall, D. J. (1992). Tissue-specific and hormonal regulation of human prostate-specific glandular kallikrein. *Biochemistry* **31**, 818–824.

Zhang, J., Dirchs, L., Marynen, P., Rombauts, W., Delaey, B., Van den Berghe, H., and Cassiman, J. J. (1988). Mapping of rat prostatic binding protein genes C1, C2, and C3 to rat chromosome 5 by *in situ* hybridization. *Cytogenet. Cell Genet.* **48**, 121–123.

Zoppi, S., Marcelli, M., Deslypere, J.-P., Griffin, J. E., Wilson, J. D., and McPhaul, M. J. (1992). Amino acid substitutions in the DNA-binding domain of the human androgen receptor are a frequent cause of receptor-binding positive androgen resistance. *Mol. Endocrinol.* **6**, 409–415.

Zoppi, S., Wilson, C. M., Harbison, M. D., Griffin, J. E., Wilson, J. D., McPhaul, M. J., and Marcelli, M. (1993). Complete testicular feminization caused by an amino-terminal truncation of the androgen receptor with downstream initiation. *J. Clin. Invest.* **91**, 1105–1112.

VITAMINS AND HORMONES, VOL. 49

Role of Androgens in Prostatic Cancer

JOHN T. ISAACS

The Johns Hopkins Oncology Center
and Department of Urology
The Johns Hopkins School of Medicine
Baltimore, Maryland 21231

I. Introduction: Clinical and Epidemiological Observations

During the last decade, the incidence of prostatic cancer has risen so dramatically that in 1990, prostatic cancer became the most commonly diagnosed malignancy in the American male (Boring *et al.*, 1993). Presently, one of every four cancers diagnosed in American males is of prostatic origin. The rise in annual incidence rates for prostatic cancer is continuing within the United States with an even more enhanced rate of increase among the African-American male population (Sondik, 1988). It has been calculated that 1 out of every 12 American white males will have prostatic cancer detected during their lifetime versus 1 out of every 8 African-American males (Seidman et al., 1985). Besides a rise in incidence rates, there has been a twofold rise in the annual number of prostatic cancer deaths since 1930 (Boring et al., 1993). In addition, there is a nearly twofold higher frequency of prostatic cancer deaths per year among African-American males as compared to white males in the United States (Sondik, 1988).

Prostatic cancer incidence increases with age more rapidly than any other type of cancer; fewer than 1% of prostatic cancers are diagnosed in men under 50 years of age (Young et al., 1981). Because of its age-related nature, prostatic cancer is often considered a disease of the very elderly. Of the 165,000 new cases of prostatic cancer diagnosed in the United States in 1993, 20% (33,000 cases) will occur in men under the age of 65 (Boring et al., 1993). These cases under 65 years of age represent more than all the renal cancers and leukemias in men of all ages, more than all brain and central nervous system tumors of men and women of all ages, and will almost equal the number of buccal cavity/pharynx and rectal cancers in men of all ages (Boring et al., 1993). Although 80% of the cases detected each year are in men over the age of 65, the impact of this disease in this later group is still significant. This is emphasized by the fact that the average life span of a man who dies of metastatic prostate cancer is reduced by 9.2 years (Horm and Sondik, 1989).

If completely localized (i.e., within the prostatic capsule), prostatic cancer can be cured by surgery (i.e., radical prostatectomy) (Walsh and Jewett, 1980). Unfortunately, if diagnosed when not confined to the prostate, prostatic cancer is a fatal disease for which there is no curative treatment (Raghaven, 1988; Crawford et al., 1989). Because of the poor prognosis for men with metastatic prostatic cancer, aggressive screening programs have been suggested for men starting at age 50 to permit early detection of prostate cancer while localized and potentially curable. Screening for early-stage prostatic cancer is logical, however, there are complications with the presently proposed screening methods. Their basic limitation is that they are unable to predict the malignant behavior of prostatic cancer based solely on histological grading of a biopsy of the primary tumor. The histological grade of the primary prostatic cancer biopsy is evaluated using the Gleason grading system. In this system, grading is based on the degree of glandular differentiation and growth pattern of the tumor as it relates to the prostatic stroma (Gleason, 1966). The pattern may vary from well differentiated (Grade 1) to poorly differentiated (Grade 5). This system takes into account tumor heterogeneity by scoring both the primary and secondary tumor growth patterns. For example, if the majority of the tumor is well differentiated (Grade 1) and the secondary growth pattern is poorly differentiated (Grade 5), the combined score would be a Gleason Grade of 6. This system is predictive for tumors with either very low (i.e., <5), or very high (i.e., 8–10) scores; however, it is of limited use for tumors with intermediate (5–7) scores. This limitation is of particular importance as the majority of tumors (76%) fall into this intermediate Gleason grade (Gleason et al., 1974).

In addition, predicting the biological potential of prostatic cancer in asymptomatic patients based on histology alone is problematic. For example, autopsy studies indicate that 30% of men 50–60 years old and 50% of those 70–80 years old have histological prostatic cancers that produced no clinical symptoms during their lifetime (Ashley, 1965; Harbitz and Haugen, 1972; Breslow et al., 1977). These asymptomatic cancers found at autopsy have been referred to as latent, microscopic, incidental, dormant, etc. There are problems with all of these various labels. For example, latent implies that the biological potential of these asymptomatic cancers is known. However, presently it is not possible to predict with accuracy in an individual patient which of these cancers will produce clinical disease and which will not. The term microscopic is misleading because the lesions found at autopsy are by no means always microscopic. For example, data from the German Prostate Cancer Registry reveal that a third of these autopsy cancers are greater than 1 cm in diameter (Dhom, 1978). In addition, these tumors are not always well differentiated histologically and in one study only 58% of the prostate cancers found at autopsy were well differentiated (Hohbach and Dhom, 1980). Therefore, the term histological cancer, which implies nothing about the biological potential of the tumor, will be used to describe the asymptomatic prostate cancers that exist in most older men.

It has been estimated that approximately 10 million American men over the age of 50 have such asymptomatic histological prostatic cancer (H. B. Carter and Coffey, 1988). It is estimated that in only 20% of these men will these lesions eventually become clinically important prostatic cancers (i.e., life-altering or life-threatening during the lifetime of the host) (Scardino et al., 1992). Of those that do, approximately one-half (i.e., 1 million) are predicted to be detectable at an organ-confined stage curable by surgery alone (Scardino et al., 1992). The remaining one-half (i.e., 1 million) would be diagnosed at a non-organ-confined state not curable by local therapy alone (Fig. 1). The

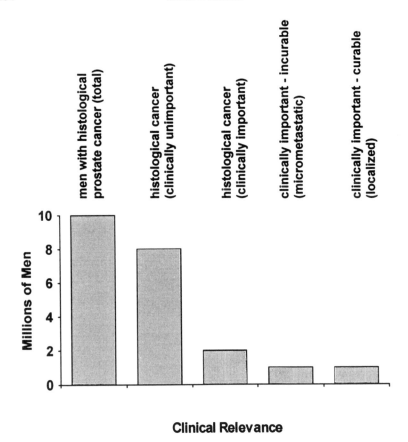

Clinical Relevance

FIG. 1. Summary of distribution of subtypes within the 10 million American males older than 50 within the United States with histological prostatic cancer.

remaining 80% of men with asymptomatic histological prostatic cancer (i.e., 8 million) will never develop clinical symptoms from these prostatic lesion (i.e., they are clinically unimportant; Fig. 1). Given that lifetime possibility of dying from prostate cancer in the United States is 2.9% (Seidman *et al.,* 1985), this means that only ≈7% of American men with initially asymptomatic histological prostatic cancer eventually die from the progression of their disease.

This low estimate is also supported by a series of international epidemiological observations that have demonstrated that the high age-related prevalence of asymptomatic histological prostatic cancer is observed consistently throughout the world's male populations (Fig. 2). In contrast, the incidence of clinically manifested (i.e., symptomatic)

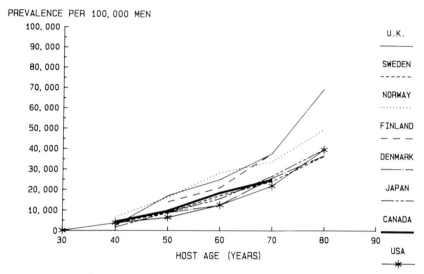

PREVALENCE PER 100, 000 MEN

FIG. 2. Age-related prevalence of histological prostatic cancer in various international male populations (for references for the original data, see H. B. Carter *et al.*, 1990).

prostatic cancer (Fig. 3) varies by more than 15-fold between the high incidence rate in Americans versus the low incidence rate in Japanese, even though their average life spans are equal [Ohno *et al.*, 1984; International Agency for Research on Cancer (IARC), 1987]. These results demonstrate that <10% of histologically detectable, asymptomatic prostatic cancers in Americans and <1% in Japanese men become life threatening during the lifetime of the host (H. B. Carter *et al.*, 1990; Scardino *et al.*, 1992).

A. MULTISTEP NATURE OF PROSTATIC CARCINOGENESIS

An explanation for the clinical and epidemiological observations discussed here is provided by the multistep nature of carcinogenesis. It is well established that the conversion of a normal cell to a fully malignant metastatic cancer cell involves a series of phenotypic changes. These phenotypic changes are themselves a result of a series of genetic changes in two basic classes of genes, oncogenes and tumor suppressor genes (Weinberg, 1989; Fearon and Vogelstein, 1990; Stanbridge, 1990). Carcinogenesis is a competing process whose outcome is determined by the balance between the expression of oncogenes driving malignant conversion and the expression of tumor suppressor genes

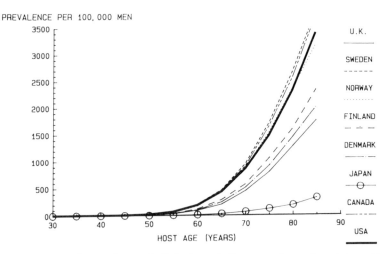

F IG. 3. Age-related prevalence of clinical prostatic cancer in various international male populations (calculations based on incidence data of IARC, 1987, and Ohno *et al.*, 1984, as described by H. B. Carter *et al.*, 1990).

inhibiting this process. Oncogenes are genes that function to enhance the net accumulation of the cell. Normally, the expression of oncogenes is tightly regulated within a particular cell type so that the net accumulation of cells neither continuously increases nor decreases with time. Adult tissues are normally self-renewing but do not grow because the rate of cell proliferation is balanced by an equal rate of cell death. Any gene that can either increase the rate of cell proliferation or decrease the rate of cell death can be an oncogene.

Any alteration in the tightly regulated control of the expression of an oncogene can enhance the malignant phenotype of the cell. Qualitative changes involve genetic mutations in the oncogenes resulting in an altered oncogenic protein with a deregulated function. The best example of this type of genetic change is the mutation in the *ras* family of oncogenes (i.e., H-, K-, and N-*ras*). For prostatic cancer, such *ras* mutations have been identified at a relatively low frequency, at least in localized prostatic cancer (B. S. Carter *et al.*, 1990a; Gumerlock *et al.*, 1991). Experimentally, expression of mutated H-*ras* oncogene in prostatic cancer cells results in an enhanced genetic instability and acquisition of high metastatic ability (Ichikawa *et al.*, 1991). Besides qualitative changes, quantitative derangements resulting in the continuous inappropriate overexpression of the oncogene can also occur. Quantitative overexpression in the H-*ras* oncogenes appears to be more common in metastatic prostatic cancer (Sumiya *et al.,* 1990). Experimentally, this overexpression can greatly enhance the motility of prostatic cancer, an effect that can directly affect the metastatic potential of these cells (Partin *et al.,* 1988).

In contrast to oncogenes whose derangement leads to a gain in oncogenic function, tumor suppressor genes are constitutively expressed and function to prevent the net continuous growth of cells (Fearon and Vogelstein, 1990; Stanbridge, 1990). A partial list of previously identified tumor suppressor genes includes the retinoblastoma (Rb), p53, Wilm's tumor (WT-1), deleted in colon cancer (DCC), adenomatous polyposis coli (APC), and van Hippel–Lindau disease tumor suppressor genes. Presently, the list of tumor suppressor genes is rapidly increasing. Within a particular cell type (e.g., prostatic glandular cells) not all of these tumor suppressor genes are constitutively expressed. The particular tumor suppressor genes involved in regulating a particular cell type is cell differentiation specific. It appears that for each differentiated cell type there is more than a single tumor suppressor gene preventing its malignancy.

For the malignant phenotype to be expressed, the function of the particular cell type-specific tumor suppressor genes must be abrogated (Fearon and Vogelstein, 1990; Stanbridge, 1990). This occurs frequently by the physical loss of chromosomal regions containing the tumor suppressor gene. It can be detected either cytogenetically or by using a molecular biological approach called allelo-typing. Using this technique, DNA is isolated from both normal and tumor tissue from the same patient. Matched DNAs are analyzed for the presence of the maternal and paternal alleles for particular reference genes chosen on the basis of their high degree of polymorphism and the fact that their chromosomal locations are known. If the analysis of normal tissue DNA from the patient is informative for a particular reference gene (i.e., polymorphic), both the paternally and maternally derived alleles are detectable. If the analysis of the matched DNA from the patient's cancer retains both polymorphic alleles for the informative gene, then no loss of heterozygosity at the particular chromosome region of this reference gene has occurred. In contrast, if one or both of the alleles are missing, then heterozygous or homozygous deletion, respectively, has occurred in the DNA of the cancer cell at the particular chromosomal region of the informative reference gene. Using this loss of heterozygosity (LOH) method, a variety of candidate chromosomal regions harboring tumor suppressor genes for prostatic cancer have been identified. In particular, the Rb tumor suppressor gene on chromosome 13q, the p53 tumor suppressor gene on chromosome 17p, and unidentified genes on chromosomes 17p, 8p, 10q, and 16q appear to be involved in prostatic carcinogenesis (B. S. Carter *et al.*, 1990b; Bergerheim *et al.*, 1991; Gao *et al.*, 1993; Bova *et al.*, 1993).

Although homozygous loss of tumor suppressor genes would completely abrogate the inhibitory effects of the gene, such a homozygous loss of chromosomal regions is rare in prostatic cancer. Heterozygous loss of a particular tumor suppressor gene by itself reduces (i.e., gene dosage effect) but does not entirely eliminate the expression of the particular tumor-suppressive gene. Such reduced expression may be sufficient to allow the malignant phenotype to be expressed. At least for the Rb and tumor p53 suppressor genes, LOH is often combined with a mutation in the remaining allele that results in either no production of the suppressor protein or production of a defective suppressor protein. For Rb, LOH coupled with a mutation producing complete loss of the Rb tumor suppressor function has been detected *in vitro* in a prostatic cancer cell line established from metastases from prostatic cancer patients (Bookstein *et al.*, 1990). Experimentally, replacing this lost Rb function by reintroduction of an exogenous normal Rb gene results in the loss of tumorigenicity of these human prostatic cancer cells (Bookstein *et al.*, 1990).

In prostatic cancer cells, LOH for p53 can be coupled with a mutation that prevents any p53 protein from being produced (W. B. Isaacs *et al.*, 1991). Such a null mutation leads to complete loss of the p53 tumor suppressor function. Alternatively, LOH for p53 can be coupled with a mutation in the p53 gene that results in the expression of an alternative protein with positive oncogenic abilities (i.e., dominant negative mutation producing an oncogene from what was originally a tumor suppressor gene).

How any of these tumor suppressor genes function to "suppress" the malignant phenotype has not been established definitively. For a cell to proliferate, it must be recruited from a resting state, called G_0, into the proliferative cell cycle. Once in the cycle, the cell must process through an initial stage, called G_1, in which specific events must occur to initiate DNA replication. Once DNA synthesis is initiated and completed during the S phase of the cycle, additional events must occur, called G_2, for mitosis to begin and be completed. Entrance, transit, and completion of the proliferative cell cycle (i.e., $G_0 \rightarrow G_1 \rightarrow S \rightarrow G_2 \rightarrow$ mitosis) is regulated by a series of "check points" that requires the functional activity of specific proteins. Increased synthesis and/or modification of these proteins can affect function. One theory of how tumor suppressor genes function is via their ability to bind to and inactivate needed check point proteins, thereby preventing progression through the check point and thus inhibiting cell proliferation (Nevins, 1992). A second mechanism for how tumor suppressor genes function is based on the demonstration that the protein encoded by the prototype p53 tumor suppressor gene is a positive transcription factor (El-Deiry *et al.*, 1993). Binding of this p53 transcription factor protein to specific DNA sequences in WAF-1/Cip1/Scd-1 gene induces the synthesis of the 21kd WAF-1 (i.e., also called Cip1 or Scd-1) protein, which binds to and inhibits cyclin-dependent kinase-2 (cdc-2) (El-Deiry *et al.*, 1993; Harper *et al.*, 1993). This cdc-2 inhibition functions to actively suppress the entrance and progression through the cell cycle. Regardless of the mechanism for tumor suppression, the physical loss and/or mutation of tumor suppressor genes results in a loss of ability to restrict the entrance and progression through the proliferative cell cycle.

Cytogenetic studies have demonstrated that the distal end of the long arm of chromosome 10 (i.e., 10q23 → terminus) is often deleted in metastatic but not nonmetastatic prostatic cancers (Atkin and Baker, 1985). This suggest that this chromosomal region contains genes that can suppress metastatic ability without suppressing growth rate. Experimentally, the presence of such "metastatic suppressor genes" for prostatic cancer has been demonstrated (Ichikawa *et al.*, 1992). Such genes do not suppress tumorigenicity or growth rates, but specifically inhibit the expression of the metastatic phenotype. Besides chromosomal 10q, these studies have identified the midsection of the short arm of chromosome 11 (i.e., 11p13-p12, not including the Wt 1 allele) as at least one site for such metastasis suppressor genes for prostatic cancer (Ichikawa *et al.*, 1992). Additional studies have also identified the long arm of chromosome 16 (i.e., 16q22.1) as a possible site of a metastasis suppressor gene as LOH analysis demonstrated that this region is deleted in 30% of primary and over 70% of metastatic prostate cancer cells (Umbas *et al.*, 1992). The E-cadherin gene is located in this specifically deleted region of chromosome 16q. E-cadherin is a Ca^{2+}-dependent cell-surface glycoprotein found on epithelial cells that functions as a Ca^{2+}-dependent epithelial cell–cell adhesion molecule. It is involved in the tight association between normal prostatic glandular cells. Experimentally, expression of the protein suppresses while loss of expression enhances the invasiveness and motility of epithelial cells. Using immunohistochemistry, a correlation between decreased E-cadherin expression and metastatic progression has been demonstrated, making it a strong candidate for a "metastatic suppressor" gene (Umbas *et al.*, 1992).

B. Genetic Instability and Prostatic Cancer Progression

The specific mechanisms involved in the genetic changes in oncogenes and tumor suppressor genes during prostatic carcinogenesis
are not fully understood. It is clear, however, that during this process
the progeny of the originally initiated cell become genetically unstable. Studies have demonstrated that mutation in certain genes encoding DNA repair enzymes results in their conversion to "mutator" proteins that increase, in a dominant negative fashion, the genetic
instability of the effected cell (Fishel et al., 1993). Thus, mutations in
such potential "mutators" genes are at least one mechanism for the
enhanced genetic instability of cancer cells. Regardless of its mechanism, this genetic instability results in the continuing acquisition of
further genetic changes in the progeny of the originally initiated cells.
Because of this genetic instability, genetically novel mutant clones are
continuously being produced within the developing tumor. If these
additional genetic changes induced in the new clones provide a growth
advantage to the cells, these new clones eventually become predomi-
nate cell types present within the primary tumor and thus the phenotype of the cancer can change over time. Such a change in phenotype
during continuous tumor growth is called progression.

A variety of experimental studies have demonstrated that this genetic instability producing novel transformed clones coupled with the
clonal selection of clones with a growth advantage is the driving force
behind the progression of prostatic cancer to a high metastatic state
(Isaacs et al., 1982). Based on a large series of studies by many independent investigators, an overview of the phenotypic and genotypic
changes associated with the progression of prostatic cancer is becom-
ing established (Fig. 4). Clinical and epidemiological data have demonstrated that progression of human prostatic cancer to a highly metastatic state occurs over an extremely long period of $\geq 30-40$ years and
that the frequency of the progression to a highly metastatic state
varies in different male populations throughout the world (H. B. Carter
et al., 1990; Sakr et al., 1993). In contrast to large international variation in the development of clinically relevant prostatic cancer (IARC,
1987), the genetic changes leading to the acquisition of morphological
changes producing prostate lesion detectable as either prostatic intraepithelial neoplasia (PIN) or histological prostatic cancer occur early
in this process and with a remarkably high prevalence that is rather
invariant throughout the world (Breslow et al., 1977; H. B. Carter et
al., 1990).

These observations demonstrate that (1) the factor(s) involved in the

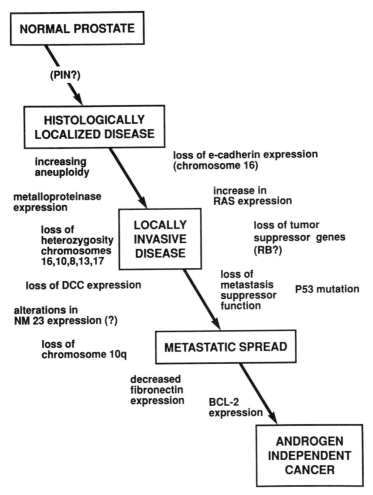

FIG. 4. Summary of the phenotypic and genotypic changes associated with the progression of prostatic cancer.

development of histological prostatic cancer are present universally with an equally high frequency in all geographic male populations and (2) the progression of histological cancer to a clinical disease involves additional malignant steps brought about by factor(s) that are not present with an equal frequency in all male populations (Ashley, 1965; Isaacs, 1988). Because there is a dramatic rise in the incidence of clinical prostatic cancer when Japanese men migrate to the United States (Haenzel and Kurihara, 1965), it is clear that environmental, not simply genetic factors, are critically involved in the progression of histological to clinically manifest prostatic cancer (Isaacs et al., 1983). The fact that even within the United States very few (i.e., >20%) of these histological prostatic cancers progress during the lifetime of their hosts to produce symptomatic disease raises the possibility that chemoprevention therapy could be developed to block the progression of these lesions to a life-altering state. If such therapy could be developed, it should be possible to minimize this progression, thereby reducing by over 10-fold the high incidence rate of clinical prostatic cancer among American males to at least as low a rate as that found for Japanese men in Japan.

Because androgens are required for the development of the normal prostate and because many prostatic cancer cells retain a responsiveness to androgen for their growth even when they become metastatic, the possibility that modulation of androgen within aging men might be chemopreventive for progression of prostatic cancer is currently being studied (Bostwick et al., 1992). In addition, because there is no curative therapy for metastatic prostatic cancer once it has developed clinically (Raghagen, 1988), major attention has been directed at how to improve on androgen ablation therapy. Therefore, the focus of the present review will be the role of androgens in both the etiology and progression of prostatic carcinogenesis and as targets for improved therapy for prostatic cancer. As background, the normal development and physiology of the prostate and the role of androgens in the processes will be discussed. This information will be followed by what is known about the role of androgen in prostatic carcinogenesis and therapy.

II. Normal Physiology of the Prostate

Despite the major progress that has occurred in the biological sciences during the last 50 years, it is rather remarkable that we are about to enter the twenty-first century and still the specific function of the prostate gland is unknown. Indeed, the prostate is the largest organ of unknown specific function in the human body. Although it is believed that the prostate is important in protecting the lower urinary tract from infection and in fertility, it is frequently the site of infection and inflammation and sperm harvested from the epididymis without exposure to seminal or prostatic fluid can produce fertilization and successful birth (Silber et al., 1988). The fact that the specific *in vivo* function of prostate is not fully understood might not be so problematic if it were not for the fact that the prostate is the most common site of neoplastic transformation in men. It is the site of the most commonly diagnosed cancer in American males (Boring *et al.*, 1993). Furthermore, the prostate is the most common site of benign neoplastic disease in males (Berry *et al.*, 1984). More than 50% of all men above the age of 50 have benign prostatic hyperplasia (BPH), with ≈25% of American men eventually requiring surgery for this condition (Berry *et al.*, 1984). Thus, it is remarkable that despite how common prostatic diseases are, the etiologies of neither prostatic cancer nor BPH are known.

A major reason why both the specific function of the prostate and etiology of the prostatic neoplasms have been difficult to elucidate is that the gross structure and histological appearance of this gland vary widely in the animal kingdom and thus comparative animal studies have been problematic. All placental (i.e., eutherial) mammals have male sex accessory tissues that minimally include the prostate gland (Price and Williams-Ashman, 1961). The term prostate is derived from the Latin word "to stand before." Thus, the gland that in males of placental mammals "stands before" the base of the bladder and produces and release secretion into the male ejaculate is defined as the prostate. In males of most placental mammals, there are additional glands that likewise release secretion into the ejaculate and these glands are given a variety of names depending on the species (e.g., seminal vesicles, bulbourethral glands, periurethral glands, preputial glands, etc.). These glands along with the prostate are thus called male accessory sex tissues. No organ system varies so widely among the animal species as the male sex accessory tissues (Price and Williams-Ashman, 1961). In humans, these include the prostate, seminal vesicles, bulbourethral gland, Cowper's glands, and glands of Littre. The dog is the only other species other than man which spontaneously develops both BPH and prostatic cancer with aging (Isaacs, 1984a). The dog has a well-developed prostate but completely lacks seminal vesicles. In contrast, the rat has a prostate that is composed of four anatomically and biochemically distinct prostatic lobes (i.e., the ventral, dorsal, lateral, and anterior lobes, the latter lobe also called the coagulating gland). In addition, the rat has seminal vesicles and preputial glands. Besides this anatomical variation, there is a large variation among the different species in the secretory products produced and released by the prostate into the ejaculate (Mann and Mann, 1981).

For example, the human prostatic epithelial cells synthesize and secrete a series of unique proteins into the male ejaculate (Coffey, 1992). These include serine protease, prostate-specific antigen (PSA), and prostatic-specific acid phosphatase (PAP). The essentially exclusive production of these two proteins by normal and malignant prostatic cells has allowed the abnormal detection of these proteins in the serum of men to be useful as a means of (1) initially detecting prostatic cancer in asymptomatic men, (2) monitoring residual presence of systemic micrometastatic disease in men who have undergone radical prostatectomy for presumed localized disease, and (3) monitoring the response of clinical detected metastatic disease to systemic therapy. Although other animal species do secrete prostate-specific proteins (e.g., prostatein secreted by rat ventral prostate and the arginine esterase secreted by the dog prostate), there are no genes directly homologous to PSA, based on DNA sequence, in the dog or rat genome. There is a homologous PAP gene in rat, however, the level of expression is nearly 1000-fold lower in rat versus human prostate epithelial cells (Coffey, 1992).

The human prostate is also unique in that it synthesizes and secretes large amounts of citrate (Coffey, 1992). Indeed the concentration of citrate in prostatic secretory fluid (i.e., 75 mM) is 615 times higher than that of blood serum. Likewise the human prostatic epithelial cells concentrate Zn^{2+} from the blood and transport it into the prostate secretion. As a result of this activity, the prostate has one of the highest tissue concentrations of Zn in the human body. It is believed that the role of such a high Zn^{2+} concentration in the prostate and its secretion is to function as a natural bactericidal compound (Coffey, 1992). Likewise the prostate is one of the richest sources of the highly charged, basic aliphatic polyamines (e.g., spermine). The biological role of polyamines has not been fully resolved, however, it is definitively known that polyamine metabolism is correlated with growth and that polyamines bind tightly to DNA and effect its conformation and template ability for DNA replication and transcription.

Based on such varied anatomy and biochemistry, it has been difficult to define the etiology of either BPH or prostatic cancer. It is known, however, that androgen, particularly the 5α-reductase metabolite of testosterone, 5α-dihydrotestosterone (DHT), has at least a permissive role, if not an inductive one, in both of these prostatic neoplasms. To appreciate the role of androgen, particularly DHT, in these neoplastic diseases, an understanding of the role of androgen in the normal development and physiology of the prostate is required.

A. ROLE OF ANDROGEN IN THE DEVELOPMENT OF THE PROSTATE

The urogenital sinus is the embryonic enlargen from which the prostate develops *in utero*. For the prostate to develop normally, a critical level of androgenic stimulation is required at specific times during its development *in utero* (Wilson, 1984). In the developing male, the fetal testis secretes testosterone into the fetal circulation at sufficient levels to stimulate the differentiation and growth of a portion of the urogenital sinus tissue, producing the definitive prostate gland. This usually begins during the first 3 months of fetal growth. If sufficient serum testosterone is not present at this critical state of intrauterine development, the prostate does not develop (Wilson, 1984).

After birth, the serum testosterone levels decrease to a low baseline value until puberty, when the serum testosterone levels rise to the adult range (Frasier *et al.*, 1969). Until puberty, the prostate remains small (approximately 1–2 g) (Isaacs, 1984a). During puberty, the prostate grows to its adult size of approximately 20 g (Isaacs, 1984a). Between the age of 10 and 20 years, the rate of prostatic growth is exponential with a prostatic weight doubling time of 2.78 years (Isaacs, 1984a). This period of exponential growth corresponds to the time period when the serum testosterone levels are rising from their initially low levels seen before the age of 10 to the high levels seen in an adult male (Frasier *et al.*, 1969). If an individual is castrated before the age of 10, the serum testosterone levels do not rise to their normal adult level and the proliferative growth of the human prostate between 10 and 20 years of life is completely blocked (Moore, 1944; Huggins and Johnson, 1947). These results demonstrate that a physiological level of androgen is chronically required for the normal growth of the human prostate. This chronic requirement for androgen is due to the fact that androgens regulate the total prostatic cell number by affecting both the rate of cell proliferation and cell death. Androgen does this by chronically stimulating the rate of cell proliferation (i.e., agonistic ability of androgen) while simultaneously inhibiting the rate of cell death (antagonistic ability of androgen) (Isaacs, 1984b). Because of this dual agonist/antagonist effect of androgen on the prostate, the rate of cell proliferation is greater than the rate of cell death during the normal prostatic growth period occurring between 10 and 20 years of age. Having reached its maximum adult size by 20 years of age, the prostate normally ceases its continuous net growth (Isaacs, 1984a). This does not mean, however, that the cells of the adult prostate in men over 20 years of age are not continuously turning over with time, only that the rate of prostatic cell proliferation is balanced by an equal ate of prostatic cell death such that neither involution nor overgrowth of the gland normally occurs with time. Thus, the adult prostate in men over 20 is an example of a steady-state self-renewing tissue. If an adult male whose prostate is in this steady-state maintenance condition is castrated, the serum testosterone levels rapidly decrease to low values comparable to those seen in the intact males younger than 10 years of age. As a result, the prostate rapidly involutes. This involution demonstrates that a physiological level of androgen is chronically required to maintain the normal size of the prostate.

B. ANDROGEN METABOLISM IN THE PROSTATE

Quantitatively, the major circulating androgen in the blood is testosterone. Within the prostate, however, testosterone is enzymatically coverted (Fig. 5) to 5α-dihydrotestosterone (Wilson, 1984). The enzyme responsible for the irreversible conversion of testosterone to DHT is the membrane-bound NADPH-dependent Δ^4-3-ketosteroid 5α-oxidoreductase (i.e., 5α-reductase) enzyme (Bruchovsky and Wilson, 1968). Biochemical studies have demonstrated that the irreversible conversion of testosterone to DHT by the 5α-reductase enzyme (Fig. 5) involves a sequential series of steps (Levy *et al.*, 1990). Initially, reduced nicotinamide-adenine dinucleotide phosphate (NADPH) cofactor binds to the 5α-reductase enzyme to form a 5α-reductase–NADPH complex. Once formed, testosterone binds to this 5α-reductase–NADPH complex. Electrons are stereospecifically transferred from NADPH to reduce the Δ^4 double bond of testosterone, producing a 5α-reductase-oxidized $NADP^+$–5α-DHT complex. After 5α-DHT is produced, it must leave this complex before the bound $NADP^+$ is able to leave, thus regenerating active 5α-reductase enzyme for another catalytic cycle (Levy *et al.*, 1990).

Based on this mechanism of enzymatic action, two types of reversible 5α-reductase inhibitors have been synthesized. The first type of 5α-reductase inhibitors are substrates (testosterone analogs) that compete and thus inhibit testosterone binding to the 5α-reductase–NADPH complex. The prototype of such reversible substrate 5α-reductase inhibitors is finasteride [17β-*N*-(2-methyl-2-propyl)-carbamoyl-4-aza-5α-androst-1-en-3-one] (Liang *et al.*, 1985) (Fig. 6). The second type of 5α-reductase inhibitors are compounds that are product (DHT) analogs. The product analog binds to the 5α-reductase–$NADP^+$ complex after DHT has been released and thus blocks the release of $NADP^+$ from the complex. This 5α-reductase–$NADP^+$-product analog complex

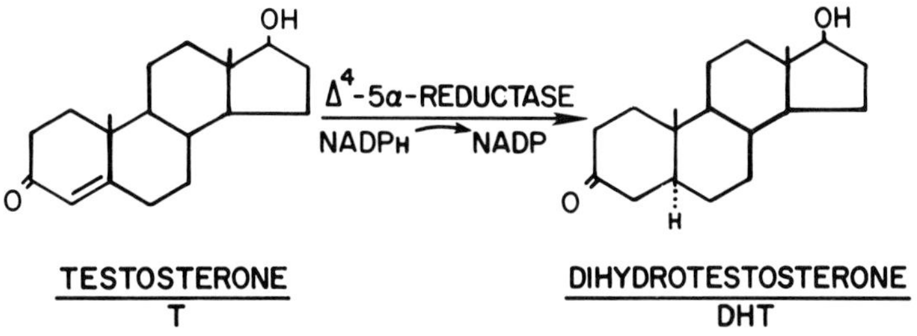

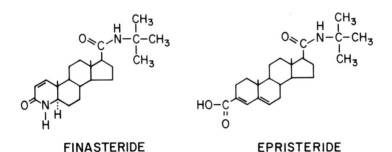

Fig. 5. Irreversible conversion of testosterone to DHT catalyzed by the NADPH-dependent 5α-reductase enzyme.

is thus inhibited from reentering another catalytic cycle (i.e., preventing further DHT production). The prototype of such reversible product 5α-reductase inhibitors is episteride [17β-N-(2-methyl-2-propyl)-carbamoyl-androst-3,5-dione-3-carboxylic acid] (Levy *et al.*, 1990; Lamb *et al.*, 1992) (Fig. 6).

There are two distinct 5α-reductase genes in man, each encoding a biochemically distinct isozyme. Both of these distinct isozymes have been cloned and the complete DNA-based sequence and amino acid composition are now known (Andersson and Russell, 1990; Jenkins *et al.*, 1991; Thigpen *et al.*, 1992; Labrie *et al.*, 1992). For both 5α-reductase type 1 and 2 isozymes, the genes encoding the proteins have a similar structure containing five exons separated by four introns. The two genes share ≈46% DNA sequence homology and encode for a protein of ≈29,000 molecular weight. The type 1 isozyme is encoded by a gene on human chromosome 5p15 (Jenkins *et al.*, 1991). This type 1 isozyme has a neutral pH optimum, a requirement for high concentration of testosterone to saturate the enzyme (high $K_m = 3\mu M$), and is

FINASTERIDE   EPRISTERIDE

Fig. 6. Chemical structures of the reversible substrate 5α-reductase inhibitor finasteride and the reversible product 5α-reductase inhibitor episteride.

rather insensitive to finasteride inhibition ($K_i \sim 300$ nM) (Andersson and Russell, 1990; Jenkins *et al.*, 1991). The type 1 isozyme is present at low levels in the prostate but is the predominant 5α-reductase isozyme in skin; it is also present in the liver (Jenkins *et al.*, 1992; Normington and Russell, 1992).

The type 2 isozyme is encoded by a gene on human chromosome 2p23 (Thigpen *et al.*, 1992). This type 2 isozyme has an acidic (pH 5.0) optimum, has a lower K_m (0.5 μM) for testosterone, and is sensitive to finasteride inhibitor ($K_i = 23$ nM). The type 2 isozyme is the predominant 5α-reductase in androgen target tissue, including the prostate. Analysis of individuals with male pseudohermaphroditism caused by 5α-reductase deficiency has revealed no mutation in the type 1 isozyme gene (Jenkins *et al.*, 1992). In contrast, molecular analysis demonstrated that mutations in the 5α-reductase type 2 gene account for this disorder (Thigpen *et al.*, 1992; Andersson *et al.*, 1991). Based on these results, it has been suggested that the type 1 isozyme functions in a catabolic manner in the metabolic removal of androgens by nontarget tissue whereas the type 2 isozyme functions in an anabolic role to amplify the androgenicity of testosterone by effectively converting it to DHT within androgen target tissue (Normington and Russell, 1992).

After DHT is formed from testosterone in the prostate, it is subjected to further reversible metabolic conversion to form 3α-diol (5α-androstane-3α,17β-diol), 3β-diol (5α-androstane-3β,17β-diol), or 5α-andostane-3,17-dione (Fig. 7). The enzymes that perform these transformations of DHT are the 17β-, 3α-, and 3β-hydroxysteroid oxidoreductases (3α-HSOR, 3β-HSOR, and 17β-HSOR) (Isaacs *et al.*, 1983). These reversible enzymes utilize NADP(H) as a cofactor, but in contrast to 5α-reductase they can also utilize NAD(H). The clinical equilibrium for the metabolism of DHT favors the formation of DHT, that is, the oxidation of the 3-hydroxy group of 3α- and 3β-diol to the 3-ketone that is present in DHT. It is known that administering 3α-diol to an animal results in the strong androgenic effect through its rapid conversion to the effective DHT. On the other hand, 3β-diol is not very effective as an androgen because it is rapidly and irreversibly converted to the triol form by hydroxylation in the 6α position or the 7α position (Isaacs *et al.*, 1979). The triols are "dead end" products of testosterone metabolism and are very water soluble and inactive as androgens and cannot re-form DHT (Isaacs *et al.*, 1979). Steroids also can form glucuronide or sulfate conjugates and be secreted in a more soluble form (Coffey, 1992). In summary, testosterone is irreversibly metabolized to DHT that is in a steady-state relationship with other reduced steroids primarily through

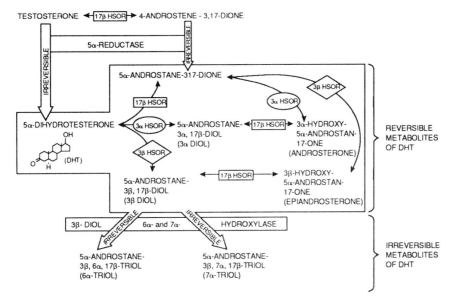

FIG. 7. Summary of the enzymatic pathway for androgen metabolism within the prostate (HSOR = hydroxysteroid oxidoreductase).

oxidation and reduction at the 3 position. The steroids are inactivated by being irreversibly hydroxylated to the inactive triols.

The extensive metabolic pathway for androgen within the prostate functions as a means of autoregulating so that the level of DHT remains constant during the diurnal periods of elevated versus depressed serum testosterone. Because growth versus regression (i.e., death) of the prostate glandular cell is determined by the specific level of prostatic DHT (Kyprianou and Isaacs, 1987), chronic maintenance of a constant prostatic DHT level is critical. The chronic requirement for a critical level of DHT is due to the dose-dependent ability of DHT to bind to the intracellular androgen receptor (Liao et al., 1972). Once formed, DHT–androgen receptor complex functions as a nuclear transcription factor to regulate the expression of a series of genes resulting in the complex differentiation and growth of the tissue into a prostate (Coffey, 1992). The critical importance of DHT in this developmental process has been demonstrated by the fact that the prostate does not develop normally in individuals who have inherited a mutation in the 5α-reductase gene, thus preventing DHT formation, even though serum testosterone levels are normal in these individuals (Inperato-McGinley et al., 1980).

C. Role of DHT in the Prostate

A series of observations suggest that within the prostate DHT, not testosterone, is the major active intracellular androgen regulating prostatic cell proliferation of death. These include the observation that (1) the androgen receptor has a higher affinity for DHT than testosterone (Liao et al., 1972); (2) both the total intracellular, and particularly the nuclear, concentration of DHT exceeds that of testosterone or any other androgen metabolite in the prostate (Krieg et al., 1979; Klein et al., 1988); (3) regardless of which metabolite is exogenously given to previously castrated rats, the major androgen retained in the nuclei of the prostatic cells is always DHT (Bruchovsky, 1971); (4) in adult males in which there is a genetic deficiency in the 5α-reductase enzyme, and thus low tissue levels of DHT, the prostate is vestigial even though these individuals have normal serum testosterone levels (Imperato-McGinley et al., 1980); and (5) there is a quantal dose–response relationship between prostatic DHT content and prostatic cell number (Kyprianou and Isaacs, 1987). Additional support for the unique role of DHT in the prostate is the demonstration that treatment with 5α-reductase inhibitors, which inhibits the production of DHT during the early postnatal period in the rat, leads to a retardation of the normal prostatic growth response (George et al., 1989). Treatment of adult dogs (Wenderoth et al., 1983; Brooks et al., 1986a) or rats (Blohm et al., 1986; Brooks et al., 1986b; Lamb et al., 1992) with 5α-reductase inhibitors (presented in Fig. 6) likewise decreases the size of the prostate. The results of these studies demonstrate the critical role of 5α-reductase-catalyzed irreversible production of DHT in the normal development and maintenance of the adult prostate.

Although testosterone itself without conversion to DHT has androgenic abilities, conversion of testosterone to DHT appears to function as a means of amplifying androgenic stimulation in the prostate (Grino et al., 1990). This is due to higher affinity of DHT than testosterone for the androgen receptor (Liao et al., 1972). Once DHT is bound to the androgen receptor within the nucleus of prostatic cells, this complex functions as a nuclear transcription factor for the regulation of the mRNA transcription of a series of androgen-regulated genes. This regulation is due to the interaction of the DHT–androgen receptor complex with specific genomic DNA sequences known as androgen response elements present in the noncoding regulatory regions (e.g., promoter or enhancer region) of the androgen-regulated gene (Coffey, 1992). This interaction can result in either a positive enhancement or negative repression in the transcription of the particular androgen-regulated gene. It is the expression of these androgen-regulated genes that determines the balance between prostatic cell proliferation and death and produces the unique differentiation phenotype of the prostate (e.g., synthesis of prostate-specific antigen and prostatic acid phosphatase) (Coffey, 1992).

III. Normal Response of the Prostate to Androgen

To understand the response of the prostate to androgen, an appreciation of the cellular organization of the gland is required. The prostate is a tubuloalveolar gland composed of multiple secretory acini that are lined by epithelial cells. These acini drain into a system of branching epithelial ducts and tubules that eventually end in the prostatic urethra. The functional unit is the glandular acinus, which is composed of both stromal and epithelial compartments (Fig. 8). The epithelial compartment consists of secretory epithelial (i.e., glandular) cells, basal epithelial cells, neuroendocrine cells, and nonepithelial fixed macrophages and intra-acinar lymphocytes. The secretory glandular cells are the major cell type in the epithelial compartment (Isaacs et al., 1992). These exocrine cells synthesize and secrete prostate-specific antigen and prostate-specific acid phosphatase into the glandular lumen, producing the prostatic fluid of the ejaculate (Coffey, 1992). They express androgen receptors and are chronically dependent on androgenic stimulation for their secretory ability and viability (Isaacs et al., 1992). If an adequate level of androgen is not maintained (e.g., following surgical castration or medical androgen ablation therapy), these glandular cells undergo a process of active cellular suicide called apoptosis or programmed cell death. The induced programmed death and subsequent elimination of these glandular cells result in the involution of the gland following androgen ablation (Isaacs et al., 1992).

In contrast to the androgen-dependent glandular cells, the basal epithelial cells rarely express the androgen receptor (Nakada et al., 1993) and do not undergo programmed death following withdrawal of androgen (Isaacs et al., 1992). A subset of these basal cells are stem cells for the prostatic epithelial compartment and are responsible for the self-renewing capacity of the gland (Isaacs and Coffey, 1989). Basal cells also differ from glandular epithelial cells by the type of intracellular structural proteins, cytokeratins, used to organize and maintain their characteristic shape. Cytokeratins are a complex class of epithelial cell-specific intracellular structural proteins that interact to form the intermediate filaments characteristic of epithelial cells.

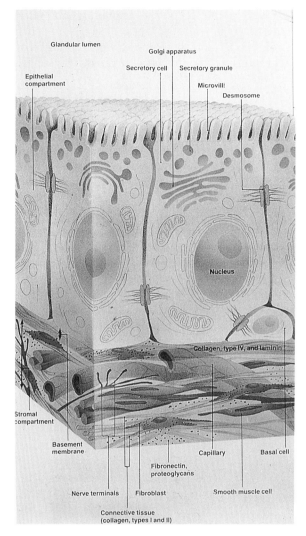

FIG. 8. Schematic overview of the composition of the prostatic functional unit.

There are at least 20 distinct cytokeratins produced by various epithelial cells throughout the body. Some are produced only by the prostatic glandular cells (e.g., cytokeratin 18) and others are produced only by the prostatic basal cells (e.g., cytokeratin 14). Thus, specific anti-keratin immunohistochemical staining can distinguish between these two types of epithelial cells (Verhagen *et al.*, 1992). Based on such keratin staining, it is clear that the glandular cells give rise to the majority of prostatic adenocarcinomas (Verhagen *et al.*, 1992). The epithelial compartment also consists of neuroendocrine cells. These cells contain and secrete serotonin and a variety of peptides, including chromogranin A, a secretory granule matrix glycoprotein. Though the role of these neuroendocrine cells is not understood, they do express the androgen receptor and appear to have a role in the regulation of prostatic secretion (Nakada *et al.*, 1993). In addition, these cells give rise to the minor subset of prostatic cancers called small-cell carcinoma (Logothetis and Hoosein, 1992).

A well-developed basement membrane separates the epithelial acini and ducts from the surrounding fibromuscular stromal tissue (Fig. 8). The prostatic stromal tissue is composed of fibroblasts, smooth muscle cells, endothelial cells, nerve cells, and assorted infiltrating cells (e.g., mast cells, lymphocytes). Thus, the prostatic basal and glandular cells receive nutrients, trophic nerve factors, and androgens that have traversed the endothelial cells, stromal cells, extracellular matrix, and the acinar basement membrane. Consequently, the stromal compartment has ample opportunity to modify the epithelial cell microenvironment.

A. PROSTATIC GROWTH CONTROL

The growth of any tissue, whether normal or malignant, depends on the quantitative relationship between the rate of cell proliferation and cell death. In the normal adult, the prostate is neither growing nor involuting because these rates are balanced such that a steady-state (i.e., self-renewing) relationship is maintained. If sufficient systemic levels of androgen are not chronically maintained (e.g., following castration), then the prostate rapidly involutes. This involution is due to the fact that glandular epithelial cells chronically require androgen to regulate their total number by affecting both cell proliferation and death (Isaacs *et al.*, 1992). Androgen does this by chronically stimulating the rate of cell proliferation (i.e., agonistic ability of androgen) while simultaneously inhibiting the rate of cell death (i.e., antagonistic ability of androgen) of prostatic glandular cells (Isaacs, 1984b). If a sufficient systemic level of androgen is not chronically maintained (e.g., following castration), then prostatic glandular cells rapidly die.

The death of a cell can occur through several biochemically and morphologically distinct pathways (Wyllie *et al.*, 1980). One of these pathways has been called necrotic cell death. Necrotic cell death is elicited by any of a large variety of factors that lead to the permeabilization of the plasma membrane, with the resultant osmotic lysis of the cell and its internal membranes. In necrotic cell death, the cell has a passive role in initiating death. In contrast, there is a second pathway for cell death, called programmed cell death, in which the cell actively participates in the initiation of its own death.

In programmed cell death, specific intracellular signals induce the cell to undergo an active, energy-dependent process of cell death that does not initially require a change in plasma membrane permeability (Wyllie *et al.*, 1980). Biochemical and morphological studies have demonstrated that the involution of the normal prostate after castration is not the result of necrotic cell death, but is an active process brought about by the initiation of a series of specific biochemical steps that lead to the programmed death (apoptosis) of androgen-dependent glandular epithelial cells within the prostate (Isaacs *et al.*, 1992).

B. Programmed Cell Death Induced by Androgen Ablation

The programmed death induced in the prostate by androgen ablation is cell type specific. Only the prostatic glandular epithelial cells and not the basal epithelial cells or stromal cells are androgen dependent and thus undergo programmed cell death following castration (Isaacs *et al.*, 1992). These glandular cells constitute approximately 80% of the total cells in the ventral prostate of an intact adult rat and approximately 70% of these glandular cells die by 7 days postcastration (Isaacs *et al.*, 1992). Using the ventral prostate of the rat as a model system, the temporal sequence of events involved in the programmed cell death pathway induced by androgen ablation has begun to be defined (Isaacs *et al.*, 1992) (Fig. 9).

Using the terminal transferase end-labeling technique of Gavrieli *et al.* to histologically detect prostatic glandular cells undergoing programmed death and adjusting for the half-life of the detection of these dying cells, it was determined that 1.2% of the glandular cells die per day via programmed death in the prostate of intact rats (Berges *et al.*, 1993). Using standard *in vivo* [^{3}H]thymidine pulse labeling to determine the percentage of glandular cells entering the S phase and the fact that S phase is of 9 hr duration in these prostatic cells, the daily rate of glandular cell proliferation is calculated to be ~1.3% per day in intact control rats (Berges *et al.*, 1993). This calculated daily rate of proliferation is essentially identical to the calculated daily rate of programmed death of prostatic glandular cells, which is consistent with the fact that the prostates of these adult, noncastrated male rats are neither continuously growing nor regressing.

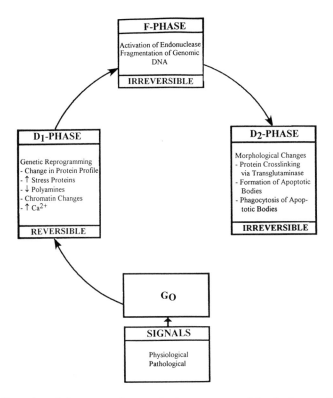

FIG. 9. Overview of the temporal events in the programmed death of prostatic glandular cells induced by androgen ablation.

Using the terminal transferase end-labeling technique, it was demonstrated that within the first day following castration, the percentage of prostatic cells dying increases and between days 2–5 postcastration, ~17–21% of these glandular cells die per day via programmed death (Berges *et al.*, 1993). These results again demonstrate that both the normal constitutive and androgen ablation-induced elimination of glandular cells in the prostate is due to programmed cell death and not to cellular necrosis. Using [³H]thymidine pulse labeling, it was demonstrated that within 1 day following castration, there is an 80% decrease ($P < 0.05$) in the percentage of glandular cells entering S phase. By 4 days following castration, there is more than a 90% reduction in this value. These results confirm the previous studies of Stiens and Helpap (1981) and Evans and Chandler (1987), which likewise demonstrated a decrease in the percentage of prostatic glandular cells in S phase following castration. Comparison of these data demonstrates that greater than 98% of prostatic glandular cells that die following castration never enter the proliferative cell cycle (Berges *et al.*, 1993).

If animals are castrated, the serum testosterone level drops to less than 10% of the intact control value within 2 hr. By 6 hr postcastration the serum testosterone level is only 1.2% of the intact control level. By 12 to 24 hr following castration, the prostatic dihydrotestosterone level is only 5% of intact control values. This lowering of prostatic DHT leads to changes in nuclear androgen receptor function (i.e., by 12 hr after castration, androgen receptors are no longer retained in biochemically isolated ventral prostatic nuclei). These nuclear receptor changes result in a major epigenetic reprogramming within the non-proliferating glandular cells resulting in the activation phase, called the D_1 phase of the programmed death process (Fig. 9).

During this D_1-activation phase, the expression of a series of genes is up-regulated during the period of programmed death by prostatic glandular cells induced by castration. These genes include c-*myc*, glutathione S-transferase subunit Yb_1, testosterone-repressed prostatic message-2 (TRPM-2) (also called sulfated glycoprotein-2), transforming growth factor-β_1, H-*ras*, calmodulin, α-prothymosin, and tissue transglutaminase (Furuya and Isaacs, 1993). TRPM-2, calmodulin, and tissue transglutaminase previously have been demonstrated to be induced in a variety of other cell types undergoing programmed cell death. Several of the genes (i.e., c-*myc*, H-*ras*,) previously have been demonstrated to be involved in cell proliferation. Thus as a comparison, the relative level of expression of these same genes during the androgen-induced proliferation regrowth of the involuted prostate in animals castrated 1 week before beginning androgen replacement was determined. Previous studies have demonstrated that between 2 and 3 days post-androgen replacement in 1 week castrated rats, the prostatic glandular cells are maximally undergoing DNA synthesis and cell proliferation (Coffey *et al.*, 1968). These comparative results demonstrate that the expression of c-*ras*, H-*ras*, and tissue transglutaminase is enhanced in both prostatic cell death and proliferation (Furuya and Isaacs, 1993). In contrast, the expression of calmodulin, TRPM-2, TGFβ_1, glutathione S-transferase subunit Yb_1 and α-prothymosin is enhanced only during prostatic cell death and not prostatic cell proliferation (Furuya and Isaacs, 1993).

Additional analysis demonstrated that the expression of a series of genes is decreased during the D_1-activation phase induced by castration. For example, the C_3 subunit of the prostatein gene (i.e., the major secretory protein of the glandular cells), orithine decarboxylase

(ODC), histone H_4, p53, and glucose-regulated protein 78 all decrease following castration. In contrast to the decrease in the mRNA expression of these latter genes during programmed cell death in the prostate following castration, the expression of each of these genes is enhanced during the androgen-induced prostatic cell proliferation (Furuya and Isaacs, 1993). These results demonstrate that activation of prostatic programmed cell death induces a distinct epigenetic reprogrammed cell proliferation involving differential gene expression. The result of this epigenetic reprogramming is that during the D_1-activation phase of the programmed death process, there is a change in the profile of proteins that are synthesized, which is coupled with a decrease in polyamine levels and increase in intracellular free Ca^{2+} levels (Isaacs et al., 1992; Berges et al., 1993; Furuya and Isaacs, 1993). The increase in intracellular free Ca^{2+} occurring following castration is derived from the extracellular Ca^{2+} pool. The mechanism for this induced elevation in intracellular free Ca^{2+} is not fully known. There are indications that enhanced expression of $TGF\beta_1$ mRNA and protein, as well as the receptor for $TGF\beta_1$ following castration, is somehow involved in the elevation in the intracellular free Ca^{2+} (Ca_i) level.

Once the Ca_i reaches a critical level, Ca^{2+}-Mg^{2+}-dependent endonucleases present within the nuclei of the prostatic glandular cells are enzymatically activated (Isaacs et al., 1992). Normally, histone H_1 binds to genomic DNA in the linker region between nucleosomes and this binding is involved in packing of the DNA nucleosome into solenoid structures. Likewise, DNA binding of polyamines, particularly spermine, owing to their negative charge, is involved in maintaining the spatial constraint of genomic DNA in a compacted form. When the normal content of histone H_1 and polyamines is bound to genomic DNA, the DNA is compacted and is not an efficient substrate for the activated Ca^{2+}-Mg^{2+} endonucleases. During this D_1-activation phase, there is a decrease in polyamine levels and the nuclear content of histone H_1. Also during this phase there is a rise in the expression of the highly acidic (pI 3.5) α-prothymosin (Furuya and Isaacs, 1993). The combined results of these changes are that the genomic DNA conformation opens up in the linker region between nucleosomes in the glandular cells, which enhances the accessibility of the linker DNA to the activated Ca^{2+}-Mg^{2+}-dependent endonuclease. Once this occurs, DNA fragmentation begins at sites located between nucleosomal units (i.e., F phase of the programmed death process) and cell death is no longer reversible (Fig. 9). Recent unpublished studies using inverted pulse-gel electrophoresis have demonstrated that the initial DNA fragmentation produces $\approx$300- to 50-kb DNA pieces. Once formed, these 300- to

50-kb pieces are further degraded into nucleosomal-sized pieces (i.e., >1 kb).

During this F phase, the nuclear morphology changes (i.e., chromatin condensation with nuclear margination) even though the plasma and lysosomal membranes are still intact and mitochondria are still functional (Isaacs et al., 1992). During the subsequent portion of the death process, called the D_2 phase, the Ca^{2+}-dependent tissue transglutaminase actively cross-links various membrane proteins (unpublished data) and cell-surface blebbing, nuclear disintegration, and eventually cellular fragmentation into clusters of membrane-bound apoptotic bodies occur (Fig. 10A). Once formed, these apoptotic bodies are rapidly phagocytized by macrophages and/or neighboring epithelial cells (Fig. 10B). Thus, within 7–10 days postcastration, $\approx 80\%$ of the glandular epithelial cells die and are eliminated from the rat prostate (Isaacs et al., 1992).

Because double-stranded fragmentation of genomic DNA is induced during programmed cell death, this raises the issue of whether DNA repair is activated during the process and whether such a futile process is required for G_0 cell killing. To detect DNA repair in prostatic glandular cells, a high-dose/long-exposure bromodeoxyuridine (BrdU) labeling method was used. Instead of pulse labeling animals with a short exposure to a small dose of [^{3}H]thymidine, animals were given 50 mg/kg of BrdU and 6 hr later prostates were removed for immunohistological detection of BrdU-labeled prostatic glandular cells. This is a total dose of ~42 μmol of BrdU per rat, which is 3360 times higher than the nucleotide precursor dose used in the [^{3}H]thymidine studies reported. When animals are injected with such a high dose of BrdU, incorporation of the nucleotide precursor is not limited to the first 30 min but continues for several hours following injection. The use of such a high dose of BrdU and such a long period of exposure before dissecting prostatic tissue, coupled with the use of the highly sensitive immunocytochemical detection of BrdU, maximizes the possibility of detecting both scheduled S-phase DNA synthesis and unscheduled DNA repair (Berges et al., 1993).

The percentage of glandular cells incorporating BrdU into DNA was determined on prostatic tissue removed 6 hr after IP injection of 50 mg/kg of BrdU at various times following castration. These data demonstrate that there is a 3- to 4-fold increase in BrdU labeling by Day 2 postcastration. By Day 3 postcastration, there is a >10-fold increase, which peaks at a >20-fold increase in BrdU labeling on Day 4 postcastration, before decreasing on Day 5 postcastration (Berges et al., 1993). The distinguishing feature between scheduled S-phase-specific

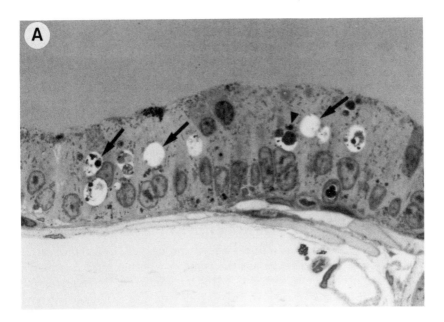

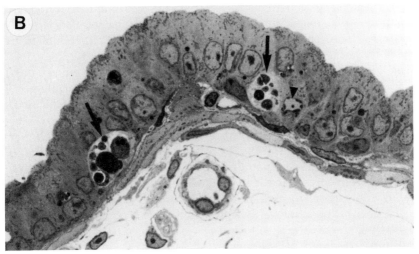

FIG. 10. Morphology of the rat ventral prostate following androgen ablation. (A) Two days postcastration, apoptotic bodies (arrows) are frequently observed. (B) Three days postcastration, apoptotic bodies phagocytosed by fixed macrophages (arrows). Also shown is a nucleus that displays peripheral areas of condensed chromatin (arrowhead), which is an early event in cellular fragmentation (taken from English *et al.*, 1989).

DNA synthesis and unscheduled DNA repair is that during S-phase DNA synthesis, there is a net accumulation of nuclear DNA content (i.e., cells have greater diploid content of DNA). In contrast, during G_0 DNA repair, no net accumulation occurs and the cells have a diploid content of DNA. To confirm that the majority of BrdU incorporation occurred as part of a futile G_0 DNA repair process and not S-phase-specific DNA synthesis, a flow cytometer was used to sort the prostatic cells from 3 day castrated rats that have incorporated BrdU. These BrdU positively labeled prostatic cells were then stained with propidium iodide and analyzed by flow cytometry for their DNA content. These studies demonstrated that BrdU-positive prostatic cells had a diploid (G_0) complement of DNA (Berges *et al.*, 1993).

The previous data demonstrate that during the programmed death of the prostatic glandular cells activated by androgen ablation, DNA fragmentation occurs that induces a futile process of DNA repair while these cells are in G_0. This raises the issue of whether such a futile G_0 DNA repair process is associated with, but not causally required for, prostatic cell death. To resolve this issue, rats were injected IP with 500 mg of hydroxyurea (HU)/kg every 8 hr for 5 days. This dose of HU was chosen based on demonstration that this treatment inhibited both prostatic S-phase-specific DNA synthesis and unscheduled G_0 DNA repair by more than 90% for 8 hr following an IP injection. When intact male rats were treated with this tridaily HU regimen for 1 week, there was no indication of an increase in programmed cell death in the prostate based on the lack of an increase in morphologically detectable apoptotic bodies or terminal transferase end-labeled cells, or loss of DNA content.

Based on these combined results, it is clear that tridaily treatment with 500 mg of HU/kg inhibits by at least 90% both scheduled S-phase DNA synthesis and unscheduled G_0 DNA repair without itself inducing programmed cell death in the prostate. Therefore, rats were castrated and either injected IP with 500 mg of HU/kg every 8 hr or injected with the saline vehicle every 8 hr as a control. Five days after castration, the prostatic cell content was reduced $51 \pm 2\%$ in the animals untreated with HU versus $47 \pm 3\%$ in castrated rats receiving the tridaily HU treatment (Berges *et al.*, 1993). Histological analysis also demonstrated an identical atrophic morphology for the prostates from both groups and an identical frequency of glandular cells was detected as apoptotic bodies in both groups of prostates. These data demonstrate that the programmed death of prostatic glandular cells induced by androgen ablation does not require either entrance into S phase or G_0 DNA repair.

Because enhanced expression of the p53 tumor suppressor gene has been suggested to be involved in programmed death in several other cell systems, the extent of programmed death of androgen-dependent cells in the prostate and seminal vesicles following castration was compared between wild-type and p53-deficient mice. The mutant mice were established using homologous recombination to produce null mutation in both of the p53 alleles. These homozygous null mutations prevent any production of p53 protein in these mice. Wild-type (i.e., p53 expressing) mice and p53-deficient mice were castrated and after 10 days the animals were killed, their seminal vesicles and prostates removed and weighed, and DNA content determined. Histological sections were also prepared from each of these tissues. These analyses demonstrated that there is an identical decrease in the wet weight and DNA content in both the seminal vesicles and prostate from wild-type and p53-deficient mice. Histological analysis likewise demonstrated an identical degree of cellular regression in these tissues in the two types of mice (i.e., similar percentage of terminal transferase end-labeled prostatic glandular cells in the two groups of animals). These studies demonstrate that androgen ablation-induced programmed death of androgen-dependent cells does not require any involvement of p53 protein expression (Berges *et al.*, 1993).

IV. Redefining the Prostatic "Cell Cycle"

With the realization of the importance of programmed cell death, the older idea that prostatic cell number is determined by the proliferative cell cycle alone has been modified. Based on this modification, a redefined "cell cycle" has been proposed (Berges *et al.*, 1993). The overall "cell cycle" controlling cell number is thus composed of a multicompartment system in which the prostatic glandular cell has at least three possible options (Fig. 11). The cell can be (1) metabolically active but not undergoing either proliferation or death (i.e., G_0 cell), (2) undergoing cell proliferation (i.e., $G_0 \rightarrow G_1 \rightarrow S \rightarrow G_2 \rightarrow$ mitosis), or (3) undergoing cell death by either the programmed pathway (i.e., $G_0 \rightarrow D_1 \rightarrow F \rightarrow D_2 \rightarrow$ apoptotic cellular fragmentation) or the nonprogrammed pathway (i.e., necrotic).

A. Stem Cell Model for the Organization of the Prostate

In many other steady-state self-renewing tissue systems (i.e., skin, bone marrow, testes, gut, etc.), control over the total cell content of a

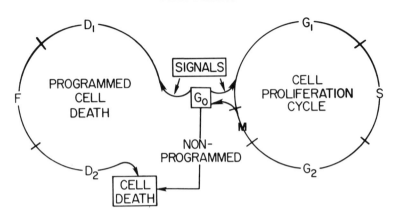

FIG. 11. Revised cell cycle denoting the options of a G_0 prostatic glandular cell.

tissue is determined by the number of self-renewing stem cells contained within the tissue itself. A stem cell is defined as a cell type capable of extensive self-renewal (i.e., proliferative) in spite of physiological or accidental removal or loss of cells from the population. The fraction of the proliferative pool of epithelial cells in the renewing prostate that are stem cells is unknown. It had been assumed that the epithelial stem cell fraction is close to 1 on the basis of thymidine labeling of the epithelial cells of the involuted prostate of long-term castrated animals in response to exogenous androgenic stimulation (Lesser and Bruckovsky, 1973). Androgen-cycling experiments, however, suggest that the fraction of prostatic epithelial cells that are stem cells is low, with the majority of the epithelial cell production within the gland being due instead to the proliferation of subclasses of epithelial cells that have only a limited self-renewal capacity (Isaacs and Coffey, 1989).

This conclusion is based on the following results obtained using a large cohort of age-matched (6 months old at start of experiment) male rats. Initially, the starting number of ventral prostatic cells was determined on a group of these intact (i.e., control) rats, and the remaining animals were castrated. One week of involution was allowed for the total number of prostatic cells in these castrated animals to decrease to a value only 25% of that of the intact control animals. After this week of involution, the castrated animals were then implanted subcutaneously with a testosterone-filled silastic capsule, which restored the serum testosterone to intact control values (i.e., approx. 3 ng/ml). One week after this testosterone treatment, the total number of prostatic cells increased by more than fourfold (i.e., two population doublings per cycle) back to a value identical with that of the intact control animals. After 1 week of exogenous androgen, the testosterone implants were surgically removed from each castrated animal (i.e., the 2-week involution/restoration cycle was completed). The castrated animals were than begun on a second cycle consisting of 1 week without exogenous androgen to induce prostatic involution followed by 1 week with exogenous androgen to induce two population doublings. This cycling was continued and after 10, 20, and 30 of such involution/restoration cycles, the total number of ventral prostatic cells was determined. These results documented that even after 60 population doublings (i.e., 30 cycles), the ventral prostate is completely able to repopulate itself normally (Isaacs, 1987).

These *in vivo* findings are in direct contrast to the findings obtained with *in vitro* cultured cells. It has been well documented that normal (i.e., nonneoplastic) cells have only a limited number of population doublings when they are cultured *in vitro* before they lose their proliferative potential (Hayflick, 1980). In addition, there is a good correlation between maximum life span of various vertebrate species and the maximum number of population doublings that their fibroblasts can undergo *in vitro* before entering a state of senescence in which the cell can still remain metabolically viable but be unable to proliferate.

For rodent fibroblasts, this senescence occurs after fewer than 10 population doublings; for human fibroblasts, it occurs after $50\pm$ population doublings (Hayflick, 1980). How then can the rat epithelial prostate cell undergo more than 60 *in vivo* population doublings and still retain its normal vigorous ability to restore its total cell number following androgen withdrawal? These results are highly paradoxical if restoration of the involuted prostate is due solely to the continuous proliferation of only a single class of epithelial cells (i.e., stem cells) in the gland. Alternatively, these results suggest that additional subclasses of amplifying epithelial cells capable of limited proliferation must also be present along with epithelial stem cells in the involuted prostate. Although these amplifying cells originate from epithelial stem cells and can proliferate for only limited numbers of cell division, these proliferations result in a major amplification in the total number of epithelial cells present. This amplification can be extensive; for example, if the amplifying cells can divide 5 times, this produces a 32-fold amplification in total cell number, and if they can divide 10 times, this produces a 1000-fold amplification. Such amplification results in the epithelial stem cells being a minority population in the tissue.

Thus, restoration of the involuted prostate after castration by exogenous androgen probably involves only a small, if any, increase in the rate of epithelial stem cell renewal, the major restoration in cell number being due to the increased proliferation of the pool of preexisting epithelial amplifying cells. Such a model for prostatic epithelial cell renewal is presented in Fig. 12. In this model, attention is drawn to the fact that the amplifying cells are androgen independent because they are able to exist even in long-term castrated animals. This point is emphasized by the fact that it is possible to castrate adult male rats and allow an extended period (i.e., >3 years) before replacing androgen and still fully restore the gland (Isaacs and Coffey, 1989). These results demonstrated that both stem cells and amplifying cells are able to maintain themselves (i.e., renew themselves) during long-term androgen withdrawal (i.e., they are androgen independent). This androgen independence does not mean, however, that these cells are completely insensitive to androgenic stimulation, only that they do not absolutely require such androgenic stimulation for their continuous maintenance.

There are cells within the prostate of an intact adult host (i.e., prostatic glandular cells) that absolutely do require a critical level of androgen stimulation for their continuous presence, as these cells are rapidly eliminated from the gland following castration (i.e., they are androgen dependent) (Isaacs *et al.*, 1992). These glandular cells, called transit cells, are derived from the pool of amplifying cells (Fig. 12).

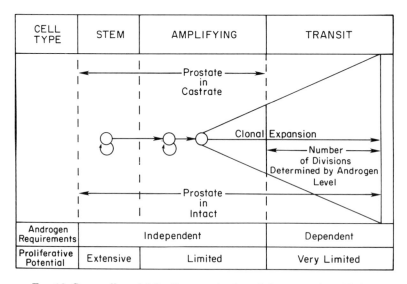

FIG. 12. Stem cell model for the organization of the prostatic epithelium.

Once such a transit cell is produced, it can proliferate only a limited number of times before it dies. A series of studies have suggested that the epithelial stems are a subset of prostatic basal epithelial cells (English *et al.*, 1989; Verhagen *et al.*, 1988; Walensky *et al.*, 1993). In addition, the three classes of prostatic epithelial cells (i.e., stem, amplifying, and transit) can be identified on the basis of the expression of a 32-kDa nuclear phosphoprotein called pp32 (Walensky *et al.*, 1993). These studies have demonstrated that prostatic epithelial stem cells express both the pp32 mRNA and protein, the amplifying cells express only the pp32 mRNA, and the transit (i.e., glandular) cells express neither pp32 mRNA nor protein.

B. ANDROGEN DEPENDENCE OF PROSTATIC TRANSIT (GLANDULAR) CELLS

The total number of proliferations (i.e., population doublings) that prostatic transit cells can undergo is determined by the level of androgenic stimulation within certain physiological limits (Kyprianou and Isaacs, 1987). This is demonstrated by the following observations. Intact 1-year-old adult male rats (i.e., an age by which the ventral prostate has reached its maximal normal size) (Berry and Isaacs, 1984) were castrated and their ventral prostates were allowed to involute for 2 weeks. The animals were then divided into groups. One group was adrenalectomized and not treated further (castrated andrenalectomized −no implant group), one group was not treated further (castrated−no implant group), and each of the remaining groups was implanted subcutaneously in the flank with a testosterone-filled silastic capsule of varying length.

These treatments were chosen because they chronically maintained the serum testosterone at values ranging from below detection obtained in the castrated/adrenalectomized animals to pharmacologically high levels of greater than 6 ng/ml obtained in the castrated rats given large implants. At 2 and 4 weeks of treatment, animals from each group were bled for serum testosterone determination; then they were killed and their ventral prostates removed for the determination of their total prostatic cell number and DHT content. Figure 13 compares the serum testosterone concentration in the various treatment groups with the resulting prostatic DHT concentration after 4 weeks of treatment. Between serum testosterone levels of 0 and 1.2 ng/ml, there is a linear increase in the prostatic DHT content from undetectable levels to approximately 3.2 ng/10^8 cells. Regression analysis revealed that there is a highly significant (i.e., $P < 0.05$) linear correla-

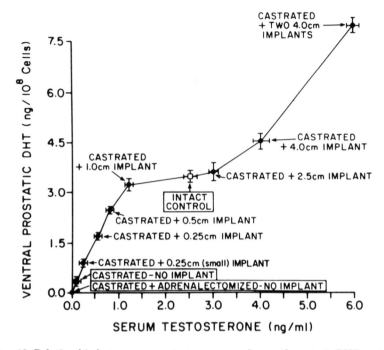

FIG. 13. Relationship between serum testosterone and ventral prostatic DHT content.

tion (i.e., correlation coefficient = 0.99) between serum testosterone and prostatic DHT in this portion of the curve. In contrast, between 1.2 and 3.0 ng/ml of serum testosterone, there is little further increase in the prostatic DHT content. At these serum testosterone levels, the prostatic DHT is approximately 3.2–3.7 ng/10^8 cells. To further elevate the prostatic DHT level above this range, the serum testosterone levels must be increased to a value greater than 4 ng/ml. These results demonstrate that when the serum testosterone concentration is in the range of 1–3 ng/ml, the prostate has the ability to maintain its DHT content at a rather fixed value of approximately 3.5 ng/10^8 cell. This suggests that the prostate has an autoregulatory ability via its extensive androgen metabolism (Fig. 7) to buffer the effects of varying androgen levels supplied to it with regard to its DHT content. Thus the prostatic DHT content is rather insensitive to changes within the serum testosterone range of 1 to 3 ng/ml. This may be significant because in the rat there is a diurnal variation in the serum testosterone concentration ranging from a nadir value of 0.8–1 ng/ml to a maximal value of approximately 3–3.5 ng/ml, even though there is no diurnal change within prostatic cell content. In contrast, at serum testosterone levels above 3.5 ng/ml or below 1 ng/ml, the prostatic DHT content is extremely sensitive to changes in serum testosterone (Kyprianou and Isaacs, 1987).

The ability of the various treatments to affect the growth of the involuted prostate is presented in Fig. 14. There is no difference in the number of prostatic cells in the castrated–andrenalectomized animals or the castrated-only animals. In contrast, when castrated rats are given testosterone-filled silastic capsules that elevate the serum testosterone to 0.25 ng/ml or above, an increase in prostatic cell number is always induced, with the response being completed by 2 weeks of treatment. The absolute magnitude of the response, however, is highly related to the level of serum testosterone that each implant produces, up to the 2.5 cm size (Fig. 14). Implants 2.5 cm or longer produced the same maximal level of cellular increase. In these latter three groups, the implants fully restored the involuted prostate back to the maximal normal size obtained in age-matched intact controls.

The prostatic DHT content was compared with the corresponding number of ventral prostatic cells in animals from the castrated–

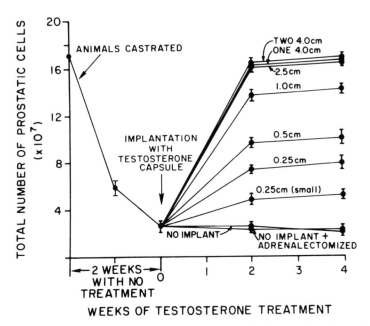

FIG. 14. Effect of various androgen treatments on ventral prostatic cell number.

adrenalectomized group, the castrated-only group, and the castrated groups treated with various-sized testosterone capsules at 4 weeks of treatment (Fig. 15). These results demonstrate that only in the castrated–adrenalectomized group was there no measurable prostatic DHT. Whereas in the castrated group there is measurable prostatic DHT (i.e., 0.4 ng/10^8 cells, which is 11% of intact control value), there is no difference in the number of prostatic transit (i.e., glandular) cells between this group and the group castrated and adrenalectomized. This demonstrates that until the prostatic DHT level is raised to a value higher than 0.4 ng/10^8 cells, no effect is induced in the prostate. There is a statistically significant ($P < 0.05$) linear correlation (i.e., correlation coefficient = 0.99) between the prostatic DHT content and total prostatic cell number between the range of prostatic DHT of 0.4 to 3.7 ng/10^8 cells. In this portion of the curve, the number of ventral prostatic cells is directly proportional to prostatic DHT content. Above 3.7 ng of DHT/10^8 cells (i.e., castrated group given one or two 4-cm implants) there is no further increase in prostatic cell number even though the DHT content is further increased. These results demonstrate that below 0.4 ng and above 3.7 ng of DHT per 10^8 cells, the prostatic cell number is not continuously proportional to the prostatic DHT concentration (Kyprianou and Isaacs, 1988).

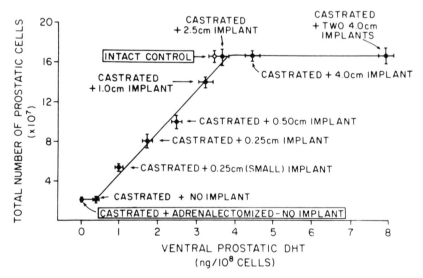

FIG. 15. Relationship between ventral prostatic DHT content and ventral prostatic cell number.

These combined results demonstrate that the prostatic transit (i.e., glandular) cell content in general is proportional to the prostatic content of DHT, at least between the range of DHT levels observed for intact control males. If larger (i.e., 6 cm) testosterone-filled implants are given to castrated male rats to maintain the serum testosterone level at greater than twice the normal level of untreated intact animals (i.e., >6 ng/ml) then the level of prostatic DHT can be increased to a value of 16 ± 2 ng/10^8 cells, a value more than twice that of intact control animals (i.e., 7.5 ± 0.6 ng/10^8 cells). Such treatment, though increasing the prostatic DHT content twofold above normal, did not induce any further increase in cell number in the normal prostate above that seen in the untreated intact control rats, even if such treatments were continued for several months (Kyprianou and Isaacs, 1987). As will be discussed later, in contrast to this normal situation, once prostatic cells have been initiated with a chemical carcinogen, such increased prostatic DHT concentration can result in an enhanced number of cancer cells (see Section VI,A).

These results demonstrate that the androgen-stimulated clonal expansion of normal prostatic epithelial transit (i.e., glandular) cells eventually results in the prostate growing to its maximal normal size. In reaching this maximal size, transit cells quantitatively become the vast majority of cells present in the prostate. For this reason, the total prostatic cell number is highly androgen sensitive because following castration these transit cells undergo programmed death (as described in Section III,B) and are thus rapidly eliminated from the prostate (Isaacs et al., 1992). Because these transit cells are the vast majority of cells present in the normal prostate, their elimination leads to the rapid prostatic involution even though the epithelial stem cells, epithelial amplifying cells, and the prostatic stromal cells are all maintained following castration (Isaacs et al., 1992).

C. Effect of Androgen on Prostatic Epithelial Stem Cell Number

For both the rat and dog, the total number of epithelial stem cells within the prostate is affected by material released from the testes. This was first demonstrated by the observation of Rajfer and Coffey (1978) that if male rats are castrated between birth and 50 days of age and allowed to go untreated until reaching adulthood (i.e., 70 days of age) before being subsequently treated with exogenous androgen, the prostate grows only to 40–50% of its normal adult size no matter how high a replacement dose of androgen is used (Chung and MacFadden, 1980).

These earlier observations have been repeated to test the effect of castration at different ages on the subsequent response of the rat ventral prostate to exogenous androgen (Isaacs and Coffey, 1989). To do this, animals were castrated at 20, 50, or 90 days of life and left either untreated or implanted at the time of castration or 40 days following castration with a testosterone-filled silastic capsule that can maintain the serum testosterone level in the physiological normal range. After 1 year of treatment, the total number of ventral prostatic cells was determined for each rat. These results demonstrated that if sufficient androgen is not chronically maintained up to 50 days of life in the rat, then it is not possible to stimulate the full development of the involuted prostate to reach its maximum normal size at 1 year of life. In additional studies, rats were castrated at 20 days of life and after 40 additional days (i.e., 60 days of age) were implanted with testosterone-filled implants that were able to elevate the serum testosterone more than twofold above normal (i.e., >6 ng/ml), and these rats were followed until 1 year of treatment. However, again the ventral prostate cell number was still only 41% of that of 1-year-old intact control rats.

In direct contrast to these findings, if adequate androgen is present up to Day 50 of life, it is possible to remove the testes and allow the prostate to involute and then treat with androgen 40 days later and still fully restore the normal size of the prostate at 1 year (Isaacs and Coffey, 1989). These results suggest that ~60% of the total ventral prostatic epithelial stem cells are produced between 20 and 50 days of life and, in the rat, prostatic epithelial stem cell number is dependent on androgen for its initial development. A similar observation is also true for the dog. If beagles are castrated at 14 months of age and allowed to go untreated for an additional 46 months (i.e., until 5 years of age) and then treated for 6 months with exogenous testosterone- and estradiol-filled silastic implants that restore the normal physiological levels of the steroids in the blood, the involuted prostate grows to only ~50% of size of the prostate of age-matched (i.e., 66 months old) beagles never castrated (Juniewicz et al., 1993). In contrast, if 2-year-old beagles are castrated and allowed to go untreated for 5 months and then given the same testosterone and estradiol treatment for 7 months, the involuted prostates grow to 100% of the size of the prostates of age-matched beagles never castrated.

These combined results demonstrate that normally in the rat the total epithelial stem cell number is reached by 50 days of life (i.e., 5% of total life span), whereas in the dog the total number of epithelial stem cells is not reached in a normal gland until 2 years of age (i.e., 20% of total life). In both species, however, the initial development of the total normal number of stem cells is androgen responsive. In the rat, once the number of prostatic epithelial stem cells is reached, this number normally becomes androgen dependent. This is demonstrated by the fact that if rats are castrated at 50 days of age and allowed to go untreated for 1 year before being treated with physiological levels of androgen, it is possible to restore fully the total cell number of these prostates within 1 month of treatment to the identical value obtained from 1-year-old intact male rats never castrated. If 20- or 50-day-old rats are not castrated but instead given pharmacologically high doses of androgen, the prostate increases its normal growth rate twofold (Berry and Isaacs, 1984). By 20 additional days of treatment, the total number of ventral prostate cells reaches the maximal value normally seen only after 1 year in an untreated control rat. Although the rate at which the maximal cell number is reached is greatly increased by pharmacologically high levels of androgen, continuous exposure to high levels of androgen, even for more than 600 days, is not able to increase the total prostatic cell number above that observed for 1-year-old untreated male rats (Berry and Isaacs, 1984). These results demonstrate that during the initial period of up to 50 days of life in the rat, the total number of epithelial stem cells is reached and that during this period the initial development of this number depends on a certain level of androgenic stimulation. Once this critical level of androgenic stimulation (i.e., physiological levels of androgen) is reached, further increases in androgenic stimulation (i.e., pharmacological level of androgen) cannot induce an overgrowth of epithelial stem cells to produce an abnormally enlarged gland unless additional malignant events occur (see Section VI,A).

V. Mechanisms of Action of Androgen in the Prostate

Presently, there is a series of major, unresolved questions concerning the mechanism of action of androgen within the prostate. The major question is whether androgen acts directly on the prostatic glandular cells themselves without the requirement for the cooperation of the prostatic stromal cells. If prostatic stromal cells are required, this would strongly suggest that paracrine growth factors (e.g., EGF, FGF, IGF, etc.) are produced by the prostatic stromal cells in response to androgen and that these paracine growth factors (called andromedins), not androgen, regulate the prostatic glandular cell number (i.e., glandular cells under paracine regulation). In contrast, if prostatic stromal cells are not required for prostatic glandular cell response to androgen, then the question arises as to whether androgen induces the glandular cells themselves to produce local prostatic growth factors (i.e., autocrine regulation) or whether androgen itself mediates all of the response directly (i.e., intracrine regulation).

A. ROLE OF STROMAL CELLS

Cunha and his associates (1987) have demonstrated the critical importance of prostatic stromal cells in the embryonic and neonatal growth of the prostatic epithelial cells. By recombining murine epithelium and mesenchyme from the embryonic urogenital sinus (i.e., prostatic anlage) of normal and testicular feminized mice, Cunha has demonstrated (1) that the morphogenesis and growth of the prostatic epithelium during the embryonic and neonatal period are dependent on androgen; (2) that at these early stages of growth, androgen receptors are present only in the prostatic stromal, not the epithelial, cells of the developing prostate; and (3) the normal androgen-dependent growth of the prostatic epithelium can be induced when androgen receptor-positive wild-type urogenital sinus stromal cells are recombined with prostatic epithelium derived from testicular feminized mice that are genetically deficient in androgen receptors. These results strongly suggest that the growth of the prostatic epithelium during embryonic and neonatal periods is regulated by local paracrine growth factors produced by the prostatic stromal cells in an androgen-dependent manner.

B. ROLE OF ANDROMEDINS

Whereas the androgen-dependent regulation of prostatic epithelial growth in early development via the prostatic stromal cells is well established, the effects of prostatic stromal cells on the prostatic epithelial cells in the adult prostate have not been resolved. McKeehan and associates (1984) have demonstrated that adult rat ventral prostatic epithelial cells can be grown in a serum-free media not containing any androgen if epidermal growth factor (EGF), fibroblast growth factor (FGF), and the systemic growth factors insulin and glucocorticoid are included in the serum-free media, which also contains cholera toxin to elevate the intracellular levels of cyclic-AMP. Addition of androgen to the serum-free media had no effect on prostatic epithelial cells number. The inability of androgen to stimulate prostatic epithelial cells cultured in serum-free media containing EGF, FGF, insulin, glucocorticoid, and cholera toxin has also been reported by Nishi *et al.* (1988) for normal rat dorsolateral prostatic epithelial cells and by Peehl and Stamey (1986) and Chaproniere and McKeehan (1986) for normal human, BPH, and malignant prostatic epithelial cells.

One possible explanation of these interesting findings is that, as discussed earlier, only the prostatic glandular epithelial (i.e., transit) cells are androgen dependent. In contrast, the prostatic basal epithelial cells are not dependent on continuous androgen for their maintenance *in vivo*. A series of studies has demonstrated that when prostatic tissue is established in primary culture, it is the prostatic basal cells, not the prostatic glandular cells, that grow *in vitro* (Merchant *et al.*, 1983). Thus, it is possible that the reason why the previous *in vitro* studies have failed to demonstrate an effect of androgen on the prostatic epithelial cells in *in vitro* culture is that these cultures contained only prostatic basal epithelial cells. If this is correct, it would not be unexpected that androgen has little, if any, effect on these basal cell cultures. There are presently available a wide range of monoclonal and polyclonal antibodies to the epithelial-specific keratin family of intermediate filament proteins. It has been demonstrated that the prostatic glandular and basal cells express both common and unique keratin proteins (Verhagen *et al.*, 1992). There are keratin antibodies that specifically react with only the prostatic basal and not the glandular, epithelial cells. Thus, using these basal cell-specific keratin antibodies, the issue of whether these previously studied cells are of prostatic basal cell origin should be directly answerable.

If these cultures are shown to be not of prostatic basal cell origin, but instead are derived from the prostatic glandular cell, this would suggest that either (1) androgen induces the autocrine production of EGF, FGF, and other local growth factors by normal and neoplastic prostatic epithelial cells directly, or (2) androgen induces the paracrine production by the supporting stromal cells of andromedins (e.g., EGF, FGF, and/or other local growth factors). In either case, however, the presence of exogenous EGF and FGF added to the *in vitro* culture media would abrogate the requirement for androgen to stimulate the growth of either normal or neoplastic prostatic epithelial cells in culture.

Kabalin *et al.* (1989) have demonstrated that the clonal growth of human prostatic epithelial cells in *in vitro* culture is stimulated by soluble growth factors released by fibroblasts. This stimulation was demonstrated by coculturing human prostatic epithelial cells with fibroblasts. Epithelial growth in coculture with fibroblasts was greater than could be obtained in isolated culture in serum-free media containing exogenous EGF, FGF, insulin, glucocorticoid, and cholera toxin. Not only human prostatic cells, but also adult human skin, human fetal lung, and mouse 3T3 fibroblast were capable of this stimulation. The fibroblasts were able to compensate for the decrease in stimulatory effect in culture when the exogenously added EGF and insulin, but not FGF, were deleted from the serum-free media (Kabalin *et al.*, 1989).

This last finding is rather unexpected as Story *et al.* (1989) demonstrated that when human prostatic-derived fibroblasts were cultured *in vitro*, they produced and released FGF. A possible explanation for the inability of fibroblasts to compensate for the deletion of FGF in the culture media in the studies of Kabalin *et al.* (1989) is that in these latter studies the fibroblasts remain viable for only 20–40 hr in *in vitro* coculture within the human prostatic epithelial cells. This possibility is also supported by the fact that in the studies of Kabalin *et al.* (1989), formalin-fixed fibroblast monolayers and extracellular matrix prepared from the fibroblast cultures failed to stimulate human prostatic epithelial growth.

An alternative hypothesis for the role of stromal paracrine factor has been developed by McKeehan *et al.* (Yan *et al.*, 1992). The group has demonstrated that the prostatic stromal cells synthesize and secrete kaeratinocyte growth factor (KGF). KGF (also called FGF-7) is a member of the heparin-binding growth factor family, but unlike acidic and basic FGF, KGF has a signal peptide directing it to be secreted extracellularly. McKeehan *et al.* demonstrated that the expression of KGF by prostatic stromal cells is androgen sensitive, however, the produced KGF is not mitogenic for these stromal cells as these cells lack the bek type of FGF receptor that specifically binds KGF. In contrast, KGF is not expressed by prostatic glandular cells but these cells do express the bek receptor, which has high affinity for KGF and whose binding induces mitogenic stimulation in the glandular cells. Thus androgen may function by stimulating the prostatic stromal cells to produce and secrete the andromedin KGF, which diffuses across the basement membrane to bind to the bek receptors of the prostatic glandular cells, thus regulating their rates of cell proliferation and death.

VI. Role of Androgen in Prostatic Carcinogenesis

Although it is unclear exactly how androgens, and in particular DHT, are involved in the multistep process of prostatic carcinogenesis, it is known that androgens have at least a permissive role in this malignant process because (1) prepubertal orchiectomy prevents the clinical development of prostatic cancer (Moore, 1944; Zuckerman, 1936) and (2) clinically established prostatic cancers often respond to androgen ablation (Huggins et al., 1947; Crawford et al., 1989). Because a critical supply of androgen is critically required to regulate the balance between prostatic epithelial cell death and proliferation, so that neither overgrowth nor regression of the prostatic epithelial cells normally occurs, the observation that androgen is chronically required for prostatic carcinogenesis does not establish whether or not, besides maintaining these cells, androgens have additional carcinogenic abilities for prostatic epithelial cells.

To address this latter possibility, epidemiological studies have focused on three approaches: (1) comparing the serum androgen levels in high- versus low-risk populations; (2) comparing serum androgen levels in patients at the time of diagnosis and in controls; and (3) comparing serum androgen levels in cases and controls from cohorts of men from when blood was previously drawn. Using the first approach, Hill et al. (1980) have compared hormone level in high- and low-risk populations for prostate cancer. Estradiol and estrone levels were higher in South African compared to North African blacks, whereas the inverse was true for testosterone. They have also examined the role of diet and hormones in populations at high and low risk for prostate cancer. They report that switching to a vegetarian diet by high-risk North American blacks is associated with decreased excretion of androgens and estrogens. When placed on a "Western" diet, low-risk South African blacks increased their excretion of androgens and estrogens (Hill et al., 1979).

Using the second approach, a series of studies have compared hormone levels in cases at the time of diagnosis to control groups. Studies of prolactin seem to have given the most consistent results. Three studies have found higher levels of prolactin in prostate cancer cases compared to normal controls and two have found higher levels in cancer cases compared to benign prostatic hyperplasia controls (B. S. Carter et al., 1990b). Two other studies found no difference in prolactin levels between cases and controls (B. S. Carter et al., 1990b). It has been proposed that prolactin exhibits its effects by acting in synergistic fashion with testosterone to have a cancer-promoting effect.

Studies of testosterone and DHT have given far more inconsistent results. Two studies have shown higher levels of testosterone in prostate cancer patients versus controls, four have shown lower levels, and three have shown no difference (B. S. Carter et al., 1990b). Meikle et al. (1985) highlighted familial factors in a study of hormones and prostate cancer and demonstrated that prostate cancer cases and case relatives under the age of 62 had lower testosterone levels than a group of unrelated controls of similar ages. DHT levels have been reported to be similar between prostate cancer cases and controls in four studies and lower in two studies (H. B. Carter *et al.*, 1990).

Using the third approach, a population-based case control study using serum collected in 1974 compared prediagnostic serum levels of testosterone, DHT, estrone, estradiol, prolactin, FSH, and LH among 103 cases of prostate cancer that developed between 1974 and 1986 and 103 age- and race-matched controls (Hsing and Comstock, 1989). Elevated levels of testosterone and LH were associated with an increased risk of prostate cancer. Persons whose serum testosterone was in the highest quartile were twofold more likely to develop prostate cancer than persons with a serum testosterone in the lowest quartile. In another study of 6860 men from whom serum samples had been previously obtained, there was a suggestion of lower DHT, higher testosterone/DHT ratio, and no difference in testosterone in a comparison of 98 subsequent cases of prostate cancer and 98 matched controls (Nomura *et al.*, 1988).

Even if future studies do demonstrate some type of positive correlation between elevated systemic blood levels of androgens and prostatic cancer development, any attempt to prove that such elevations have carcinogenic abilities for prostatic epithelial cells must explain why they do not have such abilities for the epithelial cells of the seminal vesicles. During the next year 165,000 new cases of prostate cancer will be diagnosed in the United States alone, yet it is unlikely that during this same time period even a single case of primary cancer of the seminal vesicles will be reported. Indeed, there are fewer than 50 documented cases of primary seminal vesicle cancer in the entire world literature (Isaacs, 1983). Why should the seminal vesicles— which (1) are anatomically in direct proximity to the prostate and therefore have much of the same local environment, (2) express the same metabolic pathways for DHT metabolism, and (3) express the same androgen receptor as the prostate—be so distinct in having low cancer incidence?

A. Role of Androgen in the Experimental Induction of Prostatic Cancers

Although the epidemiological data are not entirely consistent, there are consistent experimental animal data demonstrating that androgens, besides having a permissive role, also have a role in the promotion phase of prostatic carcinogenesis.

In general, both outbred stocks and the majority of inbred strains of male rats do not spontaneously develop prostatic cancer with any useful frequency, even if allowed to age several years (Bosland, 1987). In addition, it is also uncommon to be able to induce the development of prostate cancer in male rats even by treatment with potent carcinogens (Bosland et al., 1992). In contrast to these findings, Shain et al. (1975) initially reported that in a group of 41 aged (34–37 months old) male ACI inbred strain rats, 7 animals (17% incidence) had spontaneous carcinomas of the ventral lobe of the prostate as determined by histological appearance. The principal feature of these histological lesions was an intraglandular hyperplasia of the acinar epithelial cells. In some of these atypical prostate acini, the hyperplastic cells were markedly anaplastic and possessed nuclei that were pleomorphic, vesiculated, enlarged, and hyperchromatic. Mitotic figures were frequent in these acini. On this basis, Shain et al. classified these histological lesions as prostatic cancer. Unfortunately, however, none of the 7 rats that had histological lesions of the ventral prostate had any indications of gross tumor (i.e., enlarged prostatic mass). The only consistent gross evidence of possible neoplastic involvement in these 7 animals was the presence of intraprostatic hemorrhage. Shain et al. (1979) followed up this initial study with a more complex study in which they examined 33 aged ACI male rats (30–34 months of age). In this second study, 70% of the rats had histological prostatic cancer of the ventral lobe. Some of these histological adenocarcinomas involved multiple acini and in 25% of the affected animals the lesions were in both right and left ventral prostatic lobes. Shain et al. treated a small subgroup of these aged ACI rats with androgen for 42 days and noted that such treatment increased the malignant potential of these prostatic lesions as judged by their increased ability to be transplanted into syngeneic inbred ACI male rats (i.e., none of the prostatic lesions from host rats that were untreated with exogenous androgens was able to be transplanted into a syngeneic host, whereas 50% of the lesions from androgen-treated hosts took in appropriate male hosts). Indeed, by transplantation of tumors from animals treated with exogenous androgen, Shain et al. (1979) have been able to establish serially transplantable ACI tumor lines.

Ward et al. (1980) also studied the incidence of prostatic lesions in aged ACI rats. Their study examined 201 male virgin or breeder ACI rats, 24 to 40 months of age, for the gross and histological appearance of cancer of the ventral lobe of the prostate. The results indicated that at 24 months of age, 35 to 45% of all male rats had the earliest lesions, intra-alveolar atypical hyperplasias. These lesions progressed to intra-alveolar cribriform carcinomas that spread along alveoli and ducts. As the tumors enlarged, they became nodular and invaded the capsule or adjacent tissue. By 33 months, 95 to 100% of the rats had intra-alveolar prostatic atypical hyperplasias and 30 to 40% had what the authors called "invasive carcinomas." However, none of the 201 rats actually died of prostatic tumors, even though these rats were all over 2 years of age. Isaacs (1984c) also studied a large group of untreated aging ACI rats (267 animals) and confirmed that these animals are an appropriate animal model for the study of the spontaneous development of histological prostatic cancer. This is based on the observation that although male ACI rats develop high incidence rates of histological tumors of the ventral prostate with age, and often live longer than 3 years, these rats rarely die of, or with, grossly obvious prostatic cancer. These results again demonstrate that multiple malignant changes are required for the production of histological prostatic cancer and that additional malignant changes are required before these histological lesions can produce clinically important cancers (Isaacs, 1984c). Thus, these ACI rats are genetically unique and are useful as a model system to identify factors that can induce the progression of histological prostatic cancer into the fully malignant clinical disease.

The Noble rat (Nb) is an inbred strain developed by R. L. Noble of the Cancer Research Center in Vancouver. This particular strain of rat, like the ACI rat, is genetically unique in that it is markedly susceptible to induction of adenocarcinoma of the dorsal lobe of prostate by subcutaneous implantation of hormones in young rats. The original method of inducing the Noble rat prostatic carcinomas involves subcutaneously implanting a steroid pellet consisting of approximately 10 mg of a 90% steroid–10% cholesterol mixture (Noble, 1977). Grossly recognizable dorsal prostatic adenocarcinoma in Nb rats after treatment with steroids in the form of pellets was initially observed by Dr. Noble in several groups. Animals treated with only one testosterone pellet did not develop prostate carcinoma, whereas in animals treated with two pellets, 16.6% did develop prostatic carcinoma, with a mean duration of 57 weeks. The incidence increased to 20% when three testosterone propionate pellets were implanted; however, the duration necessary was 64 weeks as an average, with a range of 29 to 90 weeks. The most successful combination of estrogen pelleting and testosterone propionate included one estrone and three testosterone pellets. The mean duration necessary for tumor induction was 46 weeks, with an average in the range of 27 to 62 weeks. In such series, 18% of the animals developed tumors of the prostate.

Drago (1984) has also utilized hormonal treatment to induce the development of tumor of the dorsal prostate lobe in the Noble rat.

Drago did not use the pelleting technique of Noble, however, but instead used silastic implants as a means of increasing hormonal levels; the hormones used were testosterone and estradiol. Indeed, using these silastic steroid-containing implants, this group has been able to induce prostate adenocarcinomas reproducible in the Noble rat. The time of induction has ranged from 18 to 36 weeks; this is a considerably shorter period than that reported by Noble to induce similar tumors with hormonal pellets. Silastic implants thus seem to be a preferable method for tumor induction in the Noble rat model. The majority of such tumors are, however, only microscopically evident. Only a few animals (>15%) actually developed macroscopically obvious (i.e., palpable) prostatic cancer (Drago, 1984).

On the basis of the observation of Noble, Pollard et al. (1982) have induced the development of prostatic adenocarcinomas in male Lobund-Wistar (LW) outbred stock rats by means of long-term treatment with high levels of exogenous testosterone. Pollard et al. (1982) subcutaneously placed testosterone propionate-filled silastic capsules into 4-month-old Lobund-Wistar male rats and at 8 to 14 months of treatment autopsied each animal for the presence of grossly manifest and histological prostatic cancer. The results demonstrated a 40% incidence of histological and 16% incidence of grossly manifest cancer of the dorsal lobe of the prostate. In the same study, inbred ACI male rats were likewise treated with exogenous testosterone capsules for 12 to 14 months. Unlike in the LW animals, no grossly manifest prostatic cancers developed in ACI rats; however, 30% of animals did develop histological prostatic cancer. Pollard and Luckert (1985) additionally demonstrated that high-vegetable-fat (i.e., corn oil) diets had a promoting effect on the development of prostatic cancer induced by chronic androgen treatment of Lobund-Wistar male rats.

Pollard and Luckert (1986) also demonstrated that when 90-day-old male LW rats are inoculated intravenously with the potent clinical carcinogen nitroso-N-methylurea (MNU) at 30 mg/kg body weight and then 7 days later implanted subcutaneously with a testosterone propionate-filled silastic capsule, more than 70% of the rats developed grossly obvious prostatic and seminal vesicle adenocarcinomas if the testosterone propionate implants are replaced every 2 months. In addition, using this MNU initiation/testosterone propionate promotion protocol, ≈60% of the rats that develop primary cancers also developed distant lung metastases. The average time following exposure to MNU before the tumors became detectable was 10.6 months. Within the same time period, no rats given only testosterone capsules developed palpable prostatic cancers. These results have been confirmed by Hoover *et al.* (1990), except that in the latter study it was demonstrated that the cancer arose in the dorsolateral and anterior prostate (coagulating gland) and seminal vesicles. These studies of Hoover *et al.* (1990), as well as unpublished studies from our laboratory at Johns Hopkins, demonstrated that the majority (i.e., 60%) of cancer produced in the LW male rats using the described MNU initiation/testosterone propionate promotion protocol are derived from the seminal vesicles, with 20% being derived from the anterior prostate and 20% from the dorsolateral prostate. Thus, these combined results demonstrate that this MNU/testosterone promotion-induced LW rat model is useful for studying the progression from the earliest malignant steps through acquisition of metastatic ability by the epithelial cells of the male accessory sex tissues.

Additional studies from Dr. Pollard's lab demonstrated that if the silastic capsules are filled with DHT instead of testosterone propionate, less than 20% of the MNU exposed rats developed cancers within 1 year (Pollard *et al.*, 1989). The basis for this difference in the promoting ability of DHT versus testosterone propionate in this model, however, is not fundamental but chemical. Recently, unpublished studies from our laboratory at Hopkins has demonstrated that if 90-day-old LW male rats are given the MNU initiation and then repeatedly implanted every 2 months with silastic implants containing either DHT propionate, DHT, testosterone propionate, or testosterone, a high incidence of prostatic and seminal vesicle cancer occurs only when the propionates of the two steroids are used (i.e., ≤20% of the animals develop prostatic cancer when treated with either DHT- or testosterone-containing implants versus ≥80% indication with either DHT propionate or testosterone propionate). The reason for this is that by esterifying the compounds, the release rates of DHT propionate and testosterone propionate from the silastic capsules are at least 10- to 100-fold greater than those for the nonesterified DHT or testosterone. By esterifying DHT and testosterone, the blood levels of androgen are chronically maintained at 10- to 30-fold-higher values than those observed in nonimplanted rates and 5- to 15-fold higher than in animals treated with the nonesterified form of these steroids. These results demonstrate that extremely high levels of DHT can promote the prostatic and seminal vesicle cells initiated by MNU.

In summary, these experimental rodent models demonstrate two important requirements for the efficient development of prostate cancers. First, the host must have a genetic predisposition for the development of prostatic cancer, and second, when coupled with this predisposition, androgen can promote the progression of prostatic cells that are malignantly initiated by either the inheritance of a genetic deficient alone or the induction of further genetic changes induced by carcinogens. Interestingly, there is a comparable definitive genetic basis for the familial predisposition of humans to develop prostatic cancer. B. S. Carter *et al.* (1992) have demonstrated that in certain families the development of early age onset prostatic cancer is due to the Mendelian inheritance of a single autosomal dominant allele that predisposes the host to prostatic cancer. Whether androgen is a critical promoting agent in individuals that inherent this genetic predisposition is currently unknown. Because of the ability of 5α-reductase inhibitors (discussed in Section II,B) to inhibit the production of prostatic DHT without lowering serum testosterone, there are plans to test whether such inhibitors could be used to chemoprevent the development of prostate cancer (Boswick *et al.,* 1992). Testosterone itself, without conversion to DHT, is capable of maintaining the anabolic effects of androgen on muscle mass, male libido, and penile erection (George *et al.,* 1989). Theoretically, 5α-reductase inhibitors could inhibit DHT-induced stimulation of prostatic cancer growth without negating the effects of testosterone on muscle mass, libido, and erectile potence. Because of these quality of life issues, there is a great deal of clinical interest in testing these effective 5α-reductase inhibitors as chemopreventives since such chemoprevention will have to be given for many years. Initial testing of such long-term 5α-reductase inhibitor treatment should focus on the group of men whose family has been identified as genetically susceptible to early age onset prostatic cancer.

B. Mechanism for the Promoting Ability of Androgens

For cancer to develop from an initiated prostatic cell, the rate of proliferation of the initiated cell must exceed the rate of its death. Normally these rates are balanced so that continuous growth does not occur. Continuous net growth is the hallmark of cancer and thus during prostatic carcinogenesis, there must be a fundamental change in regulation of cell proliferation and/or death. Because androgen normally regulates both of these processes, this suggests that there is some type of dysfunction in these androgen-regulated processes during promotion. Indeed, there is a growing body of evidence demonstrating that "promoting" agents function to inhibit the rate of cell death, thus imbalancing the relationship between proliferation and death to allow net continuous growth (Isaacs, 1993). Though it remains unclear exactly how normal or even excess androgenic stimulation would function as a "promotion" for prostatic carcinogenesis, there are hints of possible mechanism(s) based on the ability of androgen to regulate the structure and function of the prostate. Studies by many investigators have been reviewed with regard to the basic structure and function of the prostate in an attempt to examine their relationship to prostatic cancer etiology (Isaacs *et al.,* 1983).

It is well recognized that prostatic glandular epithelial cells have an extensive exocrine secretory activity, that is, these cells synthesize, store, and secrete specific intracellular components into the glandular lumen, with PSA and prostatic acid phosphatase being good examples. It is often not realized, however, that these epithelial cells also additionally have the ability to transport a variety of substances initially derived from the blood transmurally into the glandular lumen (i.e., substances are passed across the epithelial basal cell membrane, through the cell, and then out the apical end into the glandular lumen, Zn^{2+} being an example). In addition, it is also often not realized that the prostatic epithelial cells are constantly secreting and transporting material into the glandular lumen, even when active ejaculation is not occurring (i.e., basal condition) (Isaacs *et al.,* 1983). Because of these functional activities, the prostatic fluid in the glandular lumen is a complex mixture of a variety of components derived not only from the synthetic activity of the glandular epithelial cells of the gland itself, but also from the blood serum. Besides natural compounds derived from the blood, foreign (i.e., xenobiotic) compounds also enter the prostatic fluid. The levels of these components are continuously modulated, not only by the frequency of active ejaculation but also under basal conditions by the continuous interaction with the glandular prostatic cells lining the acinar lumen and duct (Isaacs *et al.,* 1983). The initiation and/or promotion of prostatic carcinogenesis may well involve the chronic modulation/interaction of prostatic glandular cells with their lumenal fluid.

Experimentally, the prostatic fluid concentration of exogenously given carcinogens has been demonstrated to be directly related to the flow rate of prostatic fluid output (Smith and Hagopian, 1981). Along these lines, it is interesting that the basal prostatic output per prostatic weight varies considerably between species, such as the rat, dog, and man. Prostatic cancer is an extremely rare event in the rat (Isaacs *et al.,* 1983), and this species has the highest ratio of basal prostatic output per weight of the prostate. This suggests that the material in the acinar lumen and in the ducts of the rat prostate has the least amount of time relative to the dog and human both to be modified by and subsequently to have an effect on the glandular epithelium of the prostate by means of reabsorption. In addition, the rat rarely develops prostatic concretions (i.e., identified bodies), again in the glandular lumen, suggesting that the formation of these intraprostatic bodies requires modulation of the basal prostatic fluid, which, owing to the rapid flow rate of the rate prostate, is normally very limited (Isaacs *et al.*, 1983). Interestingly, in the inbred ACI strain of rat, modulation of prostatic fluid must occur because these rats develop prostatic concretions at an extremely high age-related incidence (i.e., 100% of ACI rat prostates have large numbers of concretions by 18 months of age) (Isaacs, 1984c). The ACI rat is also the only known rat strain in which the spontaneous incidence of histological prostatic cancer is high (i.e., >80% after 18 months of age) (Isaacs, 1984c). The dog has a higher incidence of prostatic cancer and concretions than the rat but a lower incidence than the human, and the ratio of basal prostatic output to canine prostatic weight is intermediate between that of rat and human. The human prostate has the lowest ratio of basal prostatic output per prostatic weight, thus maximizing, relative to rat and dog, the time for modulation/interaction between the components of prostatic fluid and glandular cells. This may be significant, because the human prostate also has the highest incidence of both prostatic cancer and concretions of the three species. These data suggest that the basal rate of flow through the prostatic acinar lumen and ducts, which determines how much modulation/interaction can occur between the components of prostatic fluid and glandular cells, may be involved in the differential susceptibility of the rat, dog, and human to prostatic carcinogenesis (Isaacs *et al.*, 1983).

The components of prostatic fluid that have initiating and/or promoting abilities for prostatic carcinogenesis are not necessarily limited to xenobiotic components transported from the blood. The prostate itself synthesizes and secretes a variety of endogenous components into the glandular fluid, which might also be involved in this malignant transformation. For example, the prostate, besides secreting acid phosphatase, secretes a series of proteases including PSA and plasminogen activator; in contrast, the seminal vesicle secretes a variety of protease inhibitors. This difference in secretory products may be significant, because there has been a series of observations suggesting that proteases are actively involved in the process of carcinogenesis. For example, it has been demonstrated that in a number of virally and chemically induced cancers, this transformation is associated with the production of plasminogen activator (Unbeless *et al.*, 1974). Likewise, it has been demonstrated that treatment of tissue culture cells with the potent tumor promoter phorbol ester induces the production of plasminogen activator (Wigler and Weinstein, 1976). In addition, protease inhibitors have the ability to block skin carcinogenesis induced by chemical agents (Troll *et al.*, 1970). These results suggest that the presence of endogenous proteases like plasminogen activator in the prostatic glandular lumen for extended periods of time might be involved in the high susceptibility of this organ to cancer, whereas the presence of endogenous protease inhibitors found in the seminal vesicle lumen might actually protect the seminal vesicle from cancer development.

VII. Response of Metastatic Prostatic Cancers to Androgen

Since the work of Charles Huggins in the 1940s, it has been known that prostatic cancers retain androgen sensitivity. Nearly all men with metastatic prostatic cancer given androgen ablation therapy have an initial, often dramatic, beneficial response to such androgen withdrawal. Presently such androgen ablation therapy can be either irreversibly induced by surgical means (i.e., castration or hypophysectomy to remove testicular androgens or adrenalectomy to remove adrenal androgens) or reversibly induced by chemical means. Reversible chemical means induce (1) suppression of luteinizing hormone release via either pharmacological levels of estrogens or LH releasing hormone (LHRH) antagonists or agonists, (2) inhibition of androgen synthesis in testes and adrenal (e.g., via 5α-reductase inhibitors), or (3) inhibition of androgen action itself at the level of the androgen receptor via the use of direct acting antiandrogens (e.g., flutamide) (Steinberg and Isaacs, 1993). Though the initial response to any of these effective forms of androgen ablation is of substantial palliative value, virtually all treated patients eventually relapse to an androgen-insensitive state and succumb to the progression of their cancer unless they die of intercurrent disease first. Cures, if any, are rare no matter how complete the androgen ablation (Lepor *et al.*, 1982; Crawford *et al.*, 1989).

The major cause of relapse to androgen ablation is that prostatic cancers in an individual patient are heterogeneously composed of a variety of phenotypically distinct cell clones prior to treatment. For example, Kastendieck (1980) demonstrated that of 180 clinically manifest prostatic cancers removed surgically from hormonally untreated patients, 60% were already histologically heterogeneous, including a variety of cell types of varying degrees of differentiation (admixture of glandular, cribriform, and anplastic morphology within the same cancer). Thus, prostatic cancer cell heterogeneity occurs early in the clinical course and can occur in the setting of physiological levels of testosterone. Along these lines, Viola *et al.* (1986) examined the cellular distribution of prostate-specific antigen, carcinoemybryonic antigen, and p21 Harvey-*ras* oncogene protein in metastatic foci from hormone-naive patients by immunohistochemistry. Marked heterogeneity of expression was observed. Similar findings were observed using antibodies to the androgen receptor.

With regard to the response of prostate cancer cells to androgen, this too is heterogeneous. Three distinct cellular phenotypes are possible: androgen dependent, androgen sensitive, or androgen independent (Isaacs, 1982). Androgen-dependent cancer cells chronically require a critical level of androgenic stimulation for their continued maintenance and growth. Without adequate androgenic stimulation these cells undergo the same programmed cell death pathway described for normal prostatic glandular cells (Kyprianou *et al.*, 1990). Again, the programmed death of these androgen-dependent prostatic cancer cells does not require them to enter the proliferative cell cycle (i.e., cancer cells die in G_0) (Isaacs *et al.*, 1992). In contrast, androgen-dependent cells can progress to become androgen-sensitive cancer cells that do not die, even if no androgen is present. Growth rates of these androgen-sensitive cancer cells, however, are decreased following androgen ablation. Because of the progression of either androgen-dependent or -sensitive prostatic cancer cells, androgen-independent cells arise that neither die nor slow their continuous growth following androgen ablation, no matter how complete; these cells are completely autonomous to androgenic effects on growth. Thus, of the three phenotypes of prostatic cancer, the only one that is eliminated by androgen ablation, even if complete, is the androgen-dependent cancer cell. By the time metastatic prostatic cancer is diagnosed, late in the natural course of the disease, the cancers are heterogeneously composed of varying proportions of malignant clones possessing each of these phenotypes.

The progression of initial androgen-dependent prostatic cancer cells into sensitive and independent cells can occur by a variety of mechanisms (e.g., multifocal origin, adaptation, and/or genetic instability) (Isaacs, 1981). With regard to a genetic mechanism, it has been shown that progression to androgen independence coincides with the enhanced expression of the bcl-2 oncogene (McDonnell *et al.*, 1992) This oncogene is unique in that its expression inhibits the rate of death of cancer cells without affecting their rate of cell proliferation. Inhibiting the programmed cell death normally induced by androgen ablation could allow these cells to become androgen independent. Regardless of the mechanisms involved, the loss of androgen dependence and the resultant development of tumor cell heterogeneity render the patient incurable by androgen withdrawal therapy no matter how complete (Crawford *et al.*, 1989).

To affect all the heterogeneous prostatic cancer cell populations within an individual patient, effective chemotherapy, specifically targeted against the androgen-independent and -sensitive cancer cells, must be simultaneously combined with androgen ablation to affect the androgen-dependent cells. The validity of each of these points has been demonstrated by a series of animal studies (Isaacs *et al.*, 1982; Isaacs, 1982, 1989; Isaacs and Hukku, 1988), which demonstrated that only by giving such a combined chemohormonal treatment is it possible to produce any reproducible level of cures in animals bearing prostatic cancers (Isaacs, 1989). To produce cures, however, treatment must be started early in the course of the disease, the chemotherapy must have definitive efficacy against androgen-independent cells, it must be given for a crucial period, and it must be begun simultaneously with, not sequential to, androgen ablation. Even under these ideal conditions, the cure rate of animals is not high (Isaacs, 1989). Although the concept of early combinational chemohormonal therapy for prostatic cancer is valid, for such an approach to be therapeutically effective in humans a chemotherapeutic agent that can effectively control the growth of the androgen-independent prostatic cancer cells must be available. There are at present no effective chemotherapeutic agents that can control the growth of androgen-independent prostatic cancer cells (Raghagen, 1988).

A. PROGRAMMED DEATH OF PROLIFERATING ANDROGEN-INDEPENDENT PROSTATIC CANCER CELLS

The inability to control androgen-independent prostate cancer cells in human and rodent tumors by standard chemotherapeutic methods has led to a search for new approaches. Growth of a cancer is determined by the relationship between the rate of cell proliferation and the rate of cell death. Only when the rate of cell proliferation is greater than cell death does tumor growth continue. If the rate of cell proliferation is lower than the rate of cell death, then involution of the cancer occurs. Therefore, a successful treatment of cancer can be obtained by either lowering the rate of proliferation or by raising the rate of cell death so that the rate of cell proliferation is lower than the rate of cell death. Although androgen-independent prostatic cancer cells do not activate the program of cell death after androgen ablation, these cells still retain the major portion of the programmed cell death pathway. This has been demonstrated using a series of Dunning

R-3327 androgen-independent prostatic cancers established as continuously growing *in vitro* cell lines (Kyprianou and Isaacs, 1989). For example, Dunning AT-3 androgen-independent, high-metastatic, anaplastic prostatic cancer cells have been treated *in vitro* with a variety of non-androgen-ablative agents that induce "thymineless death" of cells [e.g., cells treated with 5-fluorodeoxyuridine (5-FrdU) or trifluorothymidine (TFT)]. Analysis has revealed that thymineless death results in an increase in the expression of the TRPM-2 gene and an activation of the nuclear Ca^{2+}/Mg^{2+}-dependent endonuclease, with the resultant fragmentation of the genomic DNA of the AT-3 cells into a nucleosomal ladder similar to that seen in the death of androgen-dependent prostatic cells after castration (Kyprianou and Isaacs, 1989). This cascade of events required 6 to 18 hr before fragmentation of the DNA is complete. The AT-3 cells are not dead, as defined by their ability to metabolize a mitochondrial vital dye, until 24 hr of treatment. These results demonstrate that even androgen-independent prostatic cancer cells can be induced to undergo programmed death.

B. Programmed Death of Nonproliferating Androgen-Independent Prostatic Cancer Cells

Although the exact magnitude of either the cell proliferation rate or death rate has not been determined precisely for many human prostatic cancers, available data on the thymidine labeling index suggest that they have both a low cell proliferation rate and a low cell death rate (Meyer *et al.*, 1982). Successful treatment of slow-growing prostatic cancers probably will require simultaneous antiproliferation chemotherapy targeted at the small number of dividing cancer cells and some type of additional therapy targeted at increasing the low cell death rate of the majority of androgen-independent cancer cells not proliferating within the prostatic cancer.

The problem with agents like 5-FrdU and TFT is that cell proliferation is required for the thymineless state to activate programmed death in these AT-3 cells. In this regard, it is interesting that in thymineless programmed death in the AT-3 cells, TGF-β_1 mRNA is not enhanced. This may be significant in that, *in vitro,* untreated exponentially growing AT-3 cells produce and secrete substantial amounts of TGF-β_1. At a concentration of 5 ng/ml, TGF-β_1 has only a slight inhibitory effect on the growth of AT-3 cells in culture (i.e., <30% reduction in growth rate). However, TGF-β_1, even at a concentration of 20 ng/ml, does not induce death of AT-3 cells (Isaacs *et al.*, 1992). In contrast, both *in vivo* and *in vitro* studies have demonstrated that TGF-β_1 can induce the programmed death of androgen-dependent normal rat prostatic glandular cells, even when physiological levels of androgen are present (Martikainen *et al.*, 1990). This TGF-β_1-induced programmed death does not require cell proliferation. These results suggest that the programmed death of nonproliferating androgen-dependent normal and malignant prostate cells induced after androgen ablation may involve TGF-β_1 response, and that androgen-independent prostatic cancer cells have lost this coupled responsiveness.

One explanation for the inability of androgen ablation to induce programmed death in androgen-independent prostatic cancer cells is that such ablation is not coupled to TGF-β_1 changes, and thus a sustained elevation in the intracellular free Ca^{2+} levels is not induced in these cells. This raises the issue of whether interphase (i.e., nonproliferating) androgen-independent prostatic cancer cells can be induced to undergo programmed death if an elevation of the intracellular free Ca^{2+} is sufficiently sustained. To test this possibility, chronic exposure to a calcium-specific ionophore was used to examine the temporal response of androgen-independent, AT-3 rat prostatic cancer cells to a sustained elevation in intracellular free Ca^{2+} (Martikainen *et al.*, 1991).

If AT-3 cells are treated *in vitro* with 10 μM of the calcium-specific ionophore ionomycin, cell death can be induced. Using microfluorescence image analysis of AT-3 cells loaded with the fluorescent dye fura-2 to measure intracellular free Ca^{2+} levels (Ca_i), such ionomycin treatment has been demonstrated to elevate Ca_i from less than 50 nmol/liter to more than 300 nmol/liter within the first minute of treatment. After the first few minutes, the Ca_i returned to approximately 100 to 200 nmol. Such sustained elevations in Ca_i result in cell proliferation stopping within 6 hr of treatment and the cells arresting in the G_0 phase of the cell cycle. These nonproliferating cells begin to die after about 48 to 72 hr of elevation in Ca_i. Biochemical analysis during this time course demonstrated that DNA fragmentation into nucleosomal oligomers begins as early as 6 hr after Ca^{2+} ionophore treatment (Martikainen *et al.*, 1991).

The temporal pattern of DNA fragmentation is coincident with these AT-3 cells losing their clonogenic ability. This demonstrates that once DNA fragmentation has occurred, the cell has lost its proliferating ability and is irreversibly committed to death. DNA fragmentation is initiated at a time when plasma membrane permeability, mitochondrial function, and cellular ATP content all are still maintained. Thus, the nonproliferative death of AT-3 cells induced by 10 μM ionomycin treatment occurs through programmed, not necrotic, cell death (Martikainen *et al.*, 1991).

Thapsigargin (TG), a sesquiterpene γ-lactone isolated from the root of the umbelliferous plant *Thapsia garganica* (Fig. 16), is highly lipophilic and irreversibly inhibits the sarcoplasmic reticulum and endoplasmic reticulum Ca^{2+}-dependent ATPase (SERCA) enzymes with an IC_{50} value of ≈ 30 nM (Furuya *et al.*, 1994). TG is highly selective as plasma membrane Ca^{2+}-dependent ATPases (PMCA) are not inhibited by TG even at micromolar concentrations (Thastrup *et al.*, 1990). Cytoplasmic Ca^{2+} levels are maintained on the order of 20–40 nM by both classes of Ca^{2+}-pumping ATPases, which either transport Ca^{2+} out of the cell across the plasma membrane (PMCA) or sequester Ca^{2+} within internal pools such as those located in the ER (SERCA). Although the absolute concentration of Ca^{2+} within the ER is not known, it is likely to exceed 100 μM and it is apparent that the SERCA family of Ca^{2+}-ATPases constantly compensates for the release of Ca^{2+} across the ER membrane, either by passive leak or by second messenger-activated Ca^{2+} channels. Interestingly, depletion of the ER Ca^{2+} pool results in the generation of a signal, possibly an ER-derived diffusible messenger, which induces an increased permeability of the plasma membrane to divalent cations and a sustained elevation of Ca_i (Randriamampita and Tsian, 1993; Parekh *et al.*, 1993).

Treatment of androgen-independent rodent (i.e., Dunning rat AT-3) and human prostatic cancer cells (i.e., PC-3, DU-145, TSU) with 500 nM thapsigargin results in Ca_i elevation without hydrolysis of the inositol phospholipids. Within 1 day of thapsigargin exposure, the prostatic cancer cells arrest in the G_0 phase of the cell cycle. During

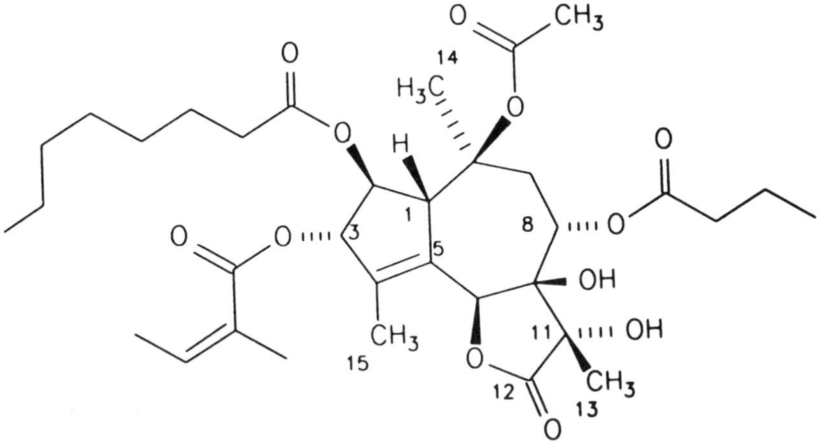

FIG. 16. Chemical structures of thapsigargin.

the period from 24 to 48 hr of exposure they lose their clonogeneic ability and the cells fragment their DNA into nucleosomal ladders. Between 72 and 96 hr, the cells lyse into apoptotic bodies. These results demonstrate that both rat and human androgen-independent prostatic cancer cells can be induced to undergo programmed death by means of specific inhibition of the ER Ca^{2+}-ATPase pump. This programmed death does not require these cancer cells to proliferate. These studies have identified the ER Ca^{2+}-ATPase pump as a new therapeutic target for cell proliferation-independent cytotoxic chemotherapy of androgen-independent prostatic cancer (Furuya et al., 1994).

Thapsigargin is cytotoxic to a range of both normal and malignant cells in in vitro cell culture. Therefore, thapsigargin itself probably will not be useful as an in vivo systemic agent for prostatic cancer owing to its nonspecific toxicity. We are derivating thapsigargin to produce a prodrug form that requires metabolic activation before it can inhibit the ER Ca^{2+}-ATPase pump and become cytotoxic. Initial attempts to target this prodrug to prostate cells, thus limiting systemic toxicity, have focused on making this prodrug a unique substrate for the serine protease activity of prostate-specific antigen PSA. Because PSA is produced only by normal prostatic glandular cells and prostatic cancer, this could allow targeting of the cytotoxic effects of thapsigargin.

VIII. CONCLUSIONS

To increase the survival of men with metastatic prostatic cancer, a modality that can effectively eliminate androgen-independent cancer cells is desperately needed. By combining such an effective modality with androgen ablation, all the heterogeneous populations of tumor cells within a prostatic cancer patient can be affected, thus optimizing the chances of cure. Unfortunately, such effective therapy for the androgen-independent prostatic cancer cell is not yet available. This therapy will probably require two types of agents, one having anti-proliferative activity affecting the small number of dividing androgen-independent cells, and the other able to increase the low rate of cell death among the majority of nonproliferating androgen-independent prostatic cancer cells present. Androgen-dependent prostatic cancer cells can be made to undergo programmed death by means of androgen ablation, even if the cells are not proliferating. Androgen-independent prostatic cancer cells retain the major portion of this programmed cell death pathway, but there is a defect in the pathway such that it is no longer activated by androgen ablation. If the intracellular free Ca^{2+} is sustained at an elevated level for a sufficient time, androgen-independent cells can be induced to undergo programmed death without proliferation. Therefore, a long-term goal is to develop some type of non-androgen-ablative method that can be used *in vivo* to induce a sustained elevation in Ca_i in androgen-independent prostatic cancer cells. To accomplish this task, a more complete understanding of the biochemical pathways involved in programmed cell death is needed. Studies are focusing on the mechanisms involved in the Ca^{2+} elevation in the normal and malignant androgen-dependent cell induced after androgen ablation.

REFERENCES

Akazaki, K., and Stemmerman, G. N. (1973). Comparative study of latent carcinoma of the prostate among Japanese in Japan and Hawaii. *J. Natl. Cancer Inst. (U.S.)* **50,** 1137–1144.

Andersson, S., and Russell, D. W. (1990). Structural and biochemical properties of cloned and expressed human and rat steroid 5α-reductases. *Proc. Natl. Acad. Sci. U.S.A.* **87,** 3640–3644.

Andersson, S., Berman, D. M., Jenkins, E. P., and Russell, D. W. (1991). Deletion of steroid 5α-reductase 2 gene in male pseudohermaphroditism. *Nature (London)* **345,** 159–161.

Ashley, D. J. (1965). On the incidence of carcinoma of the prostate. *J. Pathol. Bacteriol.* **90,** 217–224.

Atkin, N. B., and Baker, M. C. (1985). Chromosome study of five cancers of the prostate. *Hum. Genet.* **70,** 359–364.

Bergerheim, U. S. R., Kinimi, K., Collins, V. P., and Ekman, P. (1991). Deletion mapping of chromosomes 8, 10, and 16 in human prostatic carcinoma. *Genes, Chromosomes, Cancer* **3,** 215–220.

Berges, R. S., Furuya, Y., Jacks, T., English, H., and Isaacs, J. T. (1993). Cell proliferation, DNA repair, and p53 function are not required for programmed death of prostatic glandular cells induced by androgen ablation. *Proc. Natl. Acad. Sci. U.S.A.* **90,** 8910–8914.

Berry, S. J., and Isaacs, J. T. (1984). Comparative aspect of prostatic growth and androgen metabolism with aging in the dog versus the rat. *Endocrinology* **114,** 511–520.

Berry, S. J., Coffey, D. S., Walsh, P. C., and Ewing, L. L. (1984). The development of human benign prostatic hyperplasia with age. *J. Urol.* **132,** 474–479.

Blohm, T. R., Laughlin, M. E., Benson, D. H., Johnston, J. O., Wright, C. L., Schatzman, G. L., and Weintraub, P. M. (1986). Pharmacological induction of 5α-reductase deficiency in the rat: Separation of testosterone-mediated and 5α-dihydrotestosterone-mediated effects. *Endocrinology (Baltimore)* **119,** 959–966.

Bookstein, R., Rio, P., Madrepela, S. A., Hong, F., Allred, C., Grizzle, W. E., and Lee, W.-H. (1990). Promoter deletion and loss of retinoblastoma gene expression in human prostate carcinoma. *Proc. Natl. Acad. Sci. U.S.A.* **87,** 7762–7766.

Boring, C. C., Squires, T. S., and Tong, T. (1993). Cancer statistics 1993. *Ca—Cancer J. Clin.* **43,** 7–26.

Bosland, M. C. (1987). Adenocarcinoma, prostate, rat. *In* "Genital System" (T. C. Jones, U. Mohr, and R. D. Hunt, eds.), pp. 252–260. Berlin: Springer-Verlag.

Bosland, M. G., and Priusen, M. K. (1990). Induction of adenocarcinomas of the dorsolateral prostate induced in Wistar rate by *N*-methyl-*N*-nitrosourea, 7,12-dimethylbenz[a]anthracene, and 3,2′-dimethyl-4-aminobipheny following sequential treatment with cyproterone acetate and testosteone propionate. *Cancer Res.* **50,** 691–699.

Bostwick, D. G., Scardino, P. T., Kelloff, G. J., and Boone, C. W. (1992). Chemoprevention of premalignant and early malignant lesions of the prostate. *J. Cell. Biochem., Suppl.* **16H.**

Bova, G. S., Carter, B. S., Bussemakers, M. J. G., Emi, M., Fujiwara, Y., Kyprianou, N., Jacobs, S., Robinson, K. C., Epstein, J. I., Walsh, P. C., and Isaacs, W. B. (1993). Homozygous deletion and frequent loss of chromosome 8p22 loci in human prostate cancer. *Cancer Res.* **53,** 3869–3878.

Breslow, N., Chan, C., Dhom, G., Drury, R., Franks, L. M., Geller, B., Lee, Y., Lundberg, S., Sparke, B., Stemby, N., and Tilinius, H. (1977). Latent carcinoma of the prostate at autopsy in seven areas. *Int. J. Cancer* **10,** 680–688.

Brooks, J. R., Berman, C., Garnes, D., Giltman, D., Gordon, L. R., Malatesta, P. F., Primka, R. L., Reynolds, G. F., and Rasmusson, G. H. (1986a). Prostatic effects induced in dogs by chronic or acute oral administration of 5α-reductase inhibitors. *Prostate* **9,** 65–75.

Brooks, J. R., Berman, C., Primka, R. L., Reynolds, G. F., and Rasmusson, G. H. (1986b). 5α-reductase inhibitory and anti-androgenic activities of some 4-azasteroids in the rat. *Steroids* **47,** 1–19.

Bruchovsky, N. (1971). Comparison of the metabolites formed in rat prostate following the *in vivo* administration of seven natural androgens. *Endocrinology (Baltimore)* **89,** 1212–1218.

Bruchovsky, N., and Wilson, J. D. (1968). The conversion of testosterone to 5α-androstan-17-β-ol-3-one by the rat prostate *in vivo* and *in vitro. J. Biol. Chem.* **243,** 2012–2021.

Carter, B. S., Carter, H. B., and Isaacs, J. T. (1990). Epidemiologic evidence regarding predisposing factors to prostatic cancer. *Prostate* **16,** 187–198.

Carter, B. S., Epstein, J. I., and Isaacs, W. B. (1990a). ras gene mutations in human prostate cancer. *Cancer Res.* **50,** 6830–6832.

Carter, B. S., Ewing, C. M., Ward, W. S., Treiger, B. F., Aalders, T. W., Schalkin, J. A., Epstein, J. I., and Isaacs, W. B. (1990b). Allelic loss of chromosome 16q and 10q in human prostate cancer. *Proc. Natl. Acad. Sci. U.S.A.* **87,** 8751–8755.

Carter, B. S., Beaty, T. H., Steinberg, G. D., Childs, B., and Walsh, P. C. (1992). Mendelian inheritance of familial prostate cancer. *Proc. Natl. Acad. Sci. U.S.A.* **89,** 3367–3372.

Carter, H. B., and Coffey, D. S. (1988). Prostate cancer: The magnitude of the problem in the United States. *In* "A Multidisciplinary Analysis of Controversies in the Management of Prostate Cancer" (D. S. Coffey, M. I. Resnick, F. A. Dorr, and J. P. Karr, eds.), pp. 1–9. Plenum, New York.

Carter, H. B., Pianpadosi, S., and Isaacs, J. T. (1990). Clinical evidence for and implications of the multistep development of prostate cancer. *J. Urol.* **143,** 742–746.

Chaproniere, D. M., and McKeehan, W. L. (1986). Serial culture of single adult human prostatic epithelial cells in serum-free medium containing low calcium and a new growth factor from bovine brain. *Cancer Res.* **46,** 19–24.

Chung, L. W., and MacFadden, D. K. (1980). Sex steroids imprinting and prostatic growth. *Invest. Urol.* **17,** 337–342.

Coffey, D. S. (1992). The molecular biology, endocrinology and physiology of the prostate and seminal vesicles. *In* "Campbell's Textbook of Urology" (P. C. Walsh, A. B. Retik, T. A. Stamey, and E. D. Vaughan, eds.), pp. 221–266. Saunders, Philadelphia.

Coffey, D. S., Shimazaki, J., and Williams-Ashman, H. G. (1968). Polymerization of deoxyribonucleotides in relation to androgen-induced prostatic growth. *Arch. Biochem. Biophys.* **124,** 184–198.

Crawford, E. D., Eisenberger, M. A., McLeod, D. C., Spaulding, J., Benson, R., Dorr, F. A., Davis, M. A., and Goodman, P. J. (1989). A control randomized trial of leuprolide with and without flutamide in prostatic cancer. *N. Engl. J. Med.* **321,** 419–424.

Cunha, G. R., Donjacour, A. A., Cooke, P. S., Mee, S., Bigsby, R. M., Higgins, S. J., and Sugimura, Y. (1987). The endocrinology and developmental biology of the prostate. *Endocr. Rev.* **8,** 338–362.

Devesa, S. S., and Silverman, D. T. (1978). Cancer incidence and mortality trends in the United States: 1935–1974. *J. Natl. Cancer Inst. (U.S.)* **60,** 545–571.

Dhom, G. (1978). Das prostatacarzinom und die Bedeutung seiner Fruherkennung. *Med. Unserer Zeit* **5,** 134–140.

Drago, J. R. (1984). The induction of Nb rat prostatic carcinoma. *Anticancer Res.* **4,** 255–256.

El-Deiry, W. S., Tokino, T., Velculescu, V. E., Levy, D. B., Parsons, R., Trent, J. M., Lin, D., Mercer, W. E., Kenzler, K. W., and Vogelstein, B. (1993). WAF-1, a potential mediator of p53 tumor suppressor. *Cell (Cambridge, Mass.)* **75,** 817–825.

English, H., Santen, R., and Isaacs, J. T. (1987). Response of glandular versus basal rat prostatic epithelial cells to androgen withdrawal and replacement. *Prostate* **11,** 229–242.

English, H., Kyprianou, N., Isaacs, J. T. (1989). Relationship between DNA fragmentation and apoptosis in the programmed cell death in the rat prostate following castration. *Prostate* **15,** 233–251.

Evans, G. S., and Chandler, J. A. (1987). Cell proliferation studies in the rat prostate. II. The effects of castration and androgen replacement upon basal and secretory cell proliferation. *Prostate* **11,** 339–352.

Fishel, R., Lescoe, M. K., Rao, M. R. S., Copeland, N. G., Jenkins, N. A., Garber, J., Kane, M., and Kolodner, R. (1993). The human mutator gene homolog MSH2 and its association with hereditary nonpolyposis colon cancer. *Cell (Cambridge, Mass.)* **75,** 1027–1038.

Foaren, E. R., and Vogelstein, B. (1990). A genetic model for colorectal tumorigenesis. *Cell* **61,** 759–767.

Frasier, S. D., Gafford, F., and Horton, R. D. (1969). Plasma androgens in childhood and adolescence. *J. Clin. Endocrinol. Metab.* **29,** 1404–1408.

Furuya, Y., and Isaacs, J. T. (1993). Differential gene regulation during programmed death (apoptosis) versus proliferation of prostatic glandular cells. I. Induced by androgen manipulation. *Endocrinology (Baltimore)* **133,** 2660–2666.

Furuya, Y., Lundmo, P., Short, A. D., Gill, D. L., and Isaacs, J. T. (1994). Endoplasmic reticulum calcium-ATPase as a therapeutic target for activating programmed death of nonproliferating androgen-independent prostatic cancer cells. *Cancer Res.* (in press).

Gao, X., Honn, K. V. D., Grignon, W., Sakr, W. A., and Chen, Y. Q. (1993). Frequent loss of expression and loss of heterozygosity of the putative tumor suppressor gene DSS in prostatic carcinomas. *Cancer Res.* **53,** 2723–2727.

George, F. W., Johnson, L., and Wilson, J. D. (1989). Steroid 5α-reductase inhibitor on androgen physiology in the immature male rat. *Endocrinology (Baltimore)* **125,** 2434–2438.

Gleason, D. F. (1966). Classification of prostatic carcinomas. *Cancer Chemother. Rep.* **50,** 125–128.

Gleason, D. F., Mellinger, G. T., and The Veterans Administrative Cooperative Urological Research Group (1974). Prediction of prognosis for prostatic adenocarcinoma by combined histological grading and clinical staging. *J. Urol.* **111,** 58–64.

Grino, P. B., Griffin, J. E., and Wilson, J. D. (1990). Testosterone at high concentrations interacts with the human androgen receptor similarly to dihydrotestosterone. *Endocrinology (Baltimore)* **126,** 1165–1172.

Gumerlock, P. H., Poonamallee, U. R., Meyers, F. J., and de Vere White, R. W. (1991). Activated ras alleles in human carcinoma of the prostate are rare. *Cancer Res.* **51,** 1632–1637.

Haenszel, W., and Kurihara, M. (1965). Studies of Japanese migrants. I. Mortality from cancer and other diseases among Japanese in the United States. *J. Natl. Cancer Inst. (U.S.)* **40,** 43–68.

Harbitz, T. B., and Haugen, O. A. (1972). Histology of the prostate in elderly men. *Acta Pathol. Microbiol. Scand., Sect. A* **80A,** 756–768.

Harper, J. W., Adami, G. R., Wei, N., Keyomarsi, K., and Elledge, S. J. (1993). The p21 cdk-interacting protein Cip1 is a potent inhibitor of G1 cyclin-dependent kinases. *Cell (Cambridge, Mass.)* **75,** 805–816.

Hayflick, L. (1980). Recent advances in the cell biology of aging. *Mech. Ageing Dev.* **14,** 59–79.

Higgins, C., Stevens, R. E., and Hodges, G. V. (1943). Studies on prostatic cancer II, the effects of castration on advanced carcinoma of the prostate gland. *Arch. Surg.* **43,** 209–215.

Hill, P., Wynder, R., Garnes, H., and Walker, A. (1980). Environmental factors, hormone status, and prostatic cancer. *Prev. Med.* **9,** 657–666.

Hill, P., Wyndar, R., Garbaogowski, L., Garnes, H., and Walker, A. (1979). Diet and urinary steroids in black and white North American men and black South African men. *Cancer Res.* **39,** 4101–5105.

Hohbach, C., and Dhom, G. (1980). Pathology of prostatic cancer. *Scand. J. Urol. Nephrol. Suppl.* **55,** 37–42.

Hoover, D. M., Best, K. L., McKenney, B. K., Tamura, R. N., and Neubauer, B. L. (1990). Experimental induction of neoplasia in the accessory sex organs of male Lobund-Wistar rats. *Cancer Res.* **50,** 142–146.

Horm, J. W., and Sondik, E. J. (1989). Person-years of life lost due to cancer in the United States, 1970 and 1984. *Am. J. Public Health* **79,** 1490–1493.

Hsing, A., and Comstock, G. (1989). Serum hormone and risk of subsequent prostate cancer. *Am. J. Epidemiol.* **130,** 829–834.

Huggins, C., and Johnson, M. A. (1947). Carcinoma of the bladder and prostate. *JAMA, J. Am. Med. Assoc.* **135,** 1146–1152.

Ichikawa, T., Schalken, J. A., Ichikawa, Y., Steinberg, G. D., and Isaacs, J. T. (1991). H-*ras* expression, genetic instability and acquisition of metastatic ability by rat prostatic cancer cells following v-H-*ras* oncogene transfection. *Prostate* **18,** 163–172.

Ichikawa, T., Ichikawa, Y., Hawkins, A. L., Griffin, C. A., Isaacs, W. B., Oshimura, M., Barrett, J. C., and Isaacs, J. T. (1992). Localization of metastasis suppressor gene(s) for prostatic cancer to the short arm of human chromosome 11. *Cancer Res.* **52,** 3486–3490.

Imperato-McGinley, J., Peterson, R. E., Leshin, M., Griffin, J. E., Cooper, G., Draghi, S., Berenyi, M., and Wilson, J. D. (1980). Steroid 5α-reductase deficiency in a 65 year old male pseudohermaphrodite: The natural history ultrastructure of the testes and evidence for inherited enzyme heterogeneity. *J. Clin. Endocrinol. Metab.* **50**, 15–21.

International Agency for Research on Cancer (IARC) (1987). Cancer incidence in five continents. *IARC Sci. Publ.* **88**(5), 463–464.

Isaacs, J. T. (1981). Cellular factors in the development of resistance to hormonal therapy. *In* "Drug and Hormone Resistance in Neoplasia" (N. Burchovsky and J. Goldie, eds.), Vol. I, pp. 139–156. CRC Press, Boca Raton, FL.

Isaacs, J. T. (1982). Hormonally responsive *vs.* unresponsive progression of prostatic cancer to antiandrogen therapy as studied with the Dunning R-3327-AT and G rat prostatic adenocarcinoma. *Cancer Res.* **42**, 5010–5014.

Isaacs, J. T. (1983). Prostatic Structure and Function in Relation to the Etiology of Prostatic Cancer. *Prostate* **4**, 351–366.

Isaacs, J. T. (1984a). Common characteristics of human and canine benign prostatic hyperplasia. *In* "New Approaches to the Study of Benign Prostatic Hyperplasia" (F. A. Kimb, A. E. Buhl, and D. B. Carter, eds.), pp. 217–234. Liss, New York.

Isaacs, J. T. (1984b). Antagonistic effect of androgens on prostatic cell death. *Prostate* **5**, 545–557.

Isaacs, J. T. (1984c). The aging ACI/Seg *versus* Copenhagen male rat as a model system for the study of prostatic carcinogenesis. *Cancer Res.* **44**, 5785–5796.

Isaacs, J. T. (1987). Control of cell proliferation and cell death in the normal and neoplastic prostate: A stem cell model. *In* "Benign Prostatic Hyperplasis" (C. H. Rodgers, D. S. Coffey, G. Cunha, J. T. Grayhack, F. Hinman, Jr., and R. Horton, eds.), pp. 85–94. U.S. Department of Health and Human Services, NIH Publication #87-2881. Washington DC.

Isaacs, J. T. (1988). Prevention of prostatic carcinogenesis—Is it realistic? *In* "A Multi disciplinary Analysis of Controversies in the Management of Prostate Cancer" (D. S. Coffey, M. Resnick, A. A. Dorr, and J. Karr, eds.), pp. 25–34. Plenum, New York.

Isaacs, J. T. (1989). Relationship between tumor size and curability of prostatic cancer by combined chemohormonal therapy. *Cancer Res.* **49**, 6290–6294.

Isaacs, J. T. (1993). Role of programmed cell death in carcinogenesis. *Environ. Health Perspect.* **101**, Suppl. 5, 27–34.

Isaacs, J. T., and Coffey, D. S. (1989). Etiology and disease process of benign prostatic hyperplasia. *Prostate, Suppl.* **2**, 33–50.

Isaacs, J. T., and Hukku, B. (1988). Nonrandom involvement of chromosome 4 in the progression of rat prostatic cancer. *Prostate* **13**, 165–188.

Isaacs, J. T., McDermott, I. R., and Coffey, D. S. (1979). The identification and characterization of a new $C_{19}O_3$ steroid metabolite in the rat ventral prostate: 5α-Androstane-3β, 6α-17β-triol. *Steroids* **33**, 639–675.

Isaacs, J. T., Wake, N., Coffey, D. S., and Sandberg, A. A. (1982). Genetic instability coupled to clonal selection as a mechanism for tumor progression in the Dunning R-3327 rat prostatic adenocarcinoma system. *Cancer Res.* **42**, 2353–2361.

Isaacs, J. T., Brendler, C. B., and Walsh, P. C. (1983). Changes in the metabolism of dihydrotestosterone in the hyperplastic human prostate. *J. Clin. Endocrinol. Metab.* **56**, 139–146.

Isaacs, J. T., Lundmo, P. I., Berges, R., Martikainen, P., Kyprianou, N., and English, H. F. (1992). Androgen regulation of programmed death of normal and malignant prostatic cells. *J. Androl.* **13**, 457–464.

Isaacs, W. B., Carter, B. S., and Ewing, C. M. (1991). Wild-type p53 suppresses growth of human prostatic cancer cells containing mutant p53 alleles. *Cancer Res.* **51,** 4716–4720.

Jenkins, E. P., Hsieh, C.-L., Milatovich, A., Normington, K., Berman, D. M., Franke, U., and Russel, P. W. (1991). Characterization and chromosomal mapping of a human steroid 5α-reductase gene and pseudogene and mapping of the mouse homologue. *Genomics* **11,** 1102–1112.

Jenkins, E. P., Andersson, S., Imperato-McGinley, J., Wilson, J. D., and Russell, D. W. (1992). Genetic and pharmacologic evidence for more than one human steroid 5α-reductase. *J. Clin. Invest.,* 293–300.

Juniewicz, P. E., Berry, S. J., Coffey, D. S., Stranberg, J. D., and Ewing, L. L. (1993). The requirement of the testis in establishing the sensitivity of the canine prostate to develop benign prostatic hyperplasia. *J. Urol.,* in press.

Kabalin, J. N., Peehl, D. M., and Stamey, T. A. (1989). Clonal growth of human prostatic epithelial cells is stimulated by fibroblasts. *Prostate* **14,** 251–263.

Kastendieck, H. (1980). Correlation between atypical primary hyperplasia and carcinoma of the prostate. Histologic studies on 180 total prostatectomies due to manifest carcinoma. *Pathol. Res. Practice* **169,** 366–387.

Klein, H., Bressel, M., Kastendieck, K., and Voigt, K. D. (1988). Quantitative assessment of endogenous testicular and adrenal sex steroids and of steroid metabolizing enzymes in untreated human prostatic cancerous tissue. *J. Steroid Biochem.* **30,** 119–130.

Krieg, M., Bartsch, W., Janssen, W., and Voigt, K. D. (1979). A comparative study of binding, metabolism and endogenous levels of androgens in normal, hyperplastic and carcinomatous human prostate. *J. Steroid Biochem.* **11,** 615–624.

Kyprianou, N., and Isaacs, J. T. (1987). Quantal relationship between prostatic dihydrotestosterone and prostatic cell content: Critical threshold concept. *Prostate* **11,** 41–50.

Kyprianou, N., and Isaacs, J. T. (1988). Activation of programmed cell death in the rat ventral prostate following castration. *Endocrinology (Baltimore)* **122,** 552–562.

Kyprianou, N., and Isaacs, J. T. (1989). "Thymineless" death in androgen-independent prostatic cancer cells. *Biochem. Biophys. Res. Commun.* **65,** 73–81.

Kyprianou, N., English, H. P., and Isaacs, J. T. (1990). Programmed cell death during regression of PC-82 human prostate cancer following androgen ablation. *Cancer Res.* **50,** 3748–3753.

Labrie, F., Sugimoto, Y., Luu-The, Simand, J., Lachance, Y., Bachuarov, D., Leblanc, J., Durocher, F., and Paquet, N. (1992). Structure of human type II 5α-reductase gene. *Endocrinology (Baltimore)* **131,** 1571–1573.

Lamb, J. C., English, H., Levandoski, P. L., Rhodes, G. R., Johnson, R. K., and Isaacs, J. T. (1992). Prostatic involution in rats induced by a novel 5α-reductase inhibitor, SK&F 105657: Role for testosterone in the androgenic response. *Endocrinology (Baltimore)* **130,** 685–694.

Lepor, H., Ross, A., and Walsh, P. C. (1982). The influence of hormonal therapy on survival of men with advanced cancer. *J. Urol.* **12,** 335–340.

Lesser, B., and Bruchovsky, N. (1973). The effects of testosterone, 5α-dihydrothestosterone, and adenosine 3′,5′-menophosphate in cell proliferation and differentiatio on the rat prostate. *Biochim. Biophys. Acta.* **308,** 426–437.

Levy, M. A., Brandt, M., Heys, J. R., Holt, D. A., and Metcalf, B. W. (1990). Inhibition of rat liver steroid 5α-reductase by 3-androstene-3-carboxylic acids: Mechanism of enzyme–inhibitor interaction. *Biochemistry* **29,** 2815–2824.

Liang, T., Cascieri, A., Cheung, A. H., Reynolds, F. G., and Rasmusson, G. H. (1985).

Species difference in prostate steroid 5α-reductase of rat, dog and man. *Endocrinology (Baltimore)* **117**, 571–579.

Liao, S., Leong, T., and Tymocyko, J. L. (1972). Structural recognition in interactions of androgens and receptor proteins and in their association with nuclear components. *J. Steroid Biochem.* **3**, 401–407.

Logothetic, C., and Hoosein, N. (1992). The inhibition of the paracrine progression of prostate cancer as an approach of early therapy of prostatic carcinoma. *J. Cell. Biochem., Suppl.* **16H**, 128–134.

Mann, T., and Mann, C. L. (1981). "Male Reproductive Function and Semen." Springer-Verlag, New York.

Martikainen, P., Kyprianou, N., and Isaacs, J. T. (1990). Effect of transforming growth factor-β₁ on proliferation and death of rat prostatic cells. *Endocrinology (Baltimore)* **127**, 2963–2968.

Martikainen, P., Kyprianou, N., Tucker, R. W., and Isaacs, J. T. (1991). Programmed death of non-proliferating androgen independent prostatic cancer cells. *Cancer Res.* **51**, 4693–4700.

McDonnell, T. J., Troncoso, P., Brisbay, S. M., *et al.* (1992). Expression of the protoon-cogene bcl-2 in the prostate and its association with emergency of androgen-independent prostate cancer. *Cancer Res.* **52**, 6940–6944.

McKeehan, W. L., Adams, P. S., and Posser, M. P. (1984). Direct mitogenic effect of insulin, epidermal growth factor, glucocorticoid, cholera toxin, unknown pituitary factors and possibly prolactin, but not androgen, on normal rat prostate epithelial cells in serum-free primary cell culture. *Cancer Res.* **44**, 1998–2010.

Meikle, W., Smith, J., and West, D. (1985). Familial factors affecting prostatic cancer risk and plasma steroid levels. *Prostate* **6**, 121–128.

Merchant, D. J., Clarke, S. M., and Harris, S. (1983). Primary explant culture: An *in vitro* model of the human prostate. *Prostate* **4**, 523–542.

Meyer, J. S., Sufrin, G., and Martin, S. A. S. (1982). Proliferative activity of benign human prostate, prostatic adenocarcinoma and seminal vesicle evaluated by thymidine labeling. *J. Urol.* **128**, 1353–1356.

Moore, R. A. (1944). Benign hypertrophy and carcinoma of the prostate: Occurrence and experimental production in animals. *Surgery* **16**, 152–167.

Nakada, S. Y., di Sant'Agnese, P. A., Moynes, R. A., Hiipakka, R. A., Liao, S., Crockett, A. T. K., and Abrahamsson, P.-A. (1993). The androgen receptor status of neuroendocrine cells in human benign and malignant prostatic tissue. *Cancer Res.* **53**, 1967–1970.

Nevins, J. R. (1992). E2F: A link between the Rb tumor suppressor protein and viral oncoproteins. *Science* **158**, 424–429.

Nishi, N., Matuo, Y., Nakamoto, T., and Wada, F. (1988). Proliferation of epithelial cells derived from rat dorsolateral prostate in serum-free primary cell culture and their response to androgen. *In Vitro Cell Dev. Biol.* **24**, 778–786.

Noble, R. L. (1977). The development of prostatic adenocarcinoma in the Nb rat following prolonged sex hormone administration. *Cancer Res.* **37**, 1929–1933.

Nomura, A., Heilbrun, A., Stemmerman, G., and Judd, H. (1988). Prediagnostic serum hormones and the risk of prostate cancer. *Cancer Res.* **48**, 3515–3517.

Normington, K., and Russell, D. W. (1992). Tissue distribution and kinetic characteristics of rat steroid 5α-reductase isozymes. *J. Biol. Chem.* **267**, 19548–19554.

Ohno, Y., Aoki, K., Kuroishi, T., and Tominaga, S. (1984). Epidemiology of cancer of the urogenital organs in Japan. *Rinsho Hinyokika* **38**, 555–569.

Parekh, A. B., Terlau, H., and Staluner, W. (1993). Depletion of InsP₃ stores activates a

Ca^{2+} and K^+ current by means of a phosphatase and a diffusible messenger. *Nature* **364,** 814–818.

Partin, A. W., Isaacs, J. T., Treiger, B., and Coffey, D. S. (1988). Early cell motility changes associated with an increase in metastatic ability in rat prostatic cancer cells transfected with v-Harvey-ras oncogene. *Cancer Res.* **48,** 6050–6053.

Peehl, D. M., and Stamey, T. A. (1986). Growth response of normal benign hyperplastic, and malignant human prostatic epithelial cells *in vitro* to cholera toxin, pituitary extract, and hydrocortisone. *Prostate,* **8,** 51–62.

Pollard, M., and Luckert, P. H. (1985). Promotional effects of testosterone and dietary fat of prostatic carcinogenesis in genetically susceptible rats. *Prostate,* **6,** 1–6.

Pollard, M., and Luckert, P. H. (1986). Production of autochthonous prostate cancer in Lobund-Wistar rats by treatment with *N*-nitroso-*N*-methylurea and testosterone. *JNCI, J. Natl. Cancer Inst.* **77,** 538–589.

Pollard, M., Luckert, P. H., and Schmidt, M. A. (1982). Induction of prostatic adenocarcinomas in Lobund-Wistar rats by testosterone. *Prostate,* **3,** 563–568.

Pollard, M., Luckert, P. H., and Snyder, D. (1989). Prevention and treatment of experimental prostate cancer in Lobund-Wistar rat. I. Effects of estradiol, dihydrotestosterone and castration. *Prostate,* **15,** 95–103.

Price, D., and Williams-Ashman, G. H. (1961). The accessory reproductive glands of mammals. *In* "Sex and Internal Secretions" (W. C. Young, ed.), 3rd. ed., pp. 366–488. Williams & Wilkins, Baltimore, MD.

Raghaven, D. (1988). Non-hormone chemotherapy for prostate cancer: Principles of treatment and application to the testing of new drugs. *Semin. Oncol.* **15,** 371–389.

Rajfer, J., and Coffey, D. S. (1978). Sex steroid imprinting of the immature prostate. Long-term effect. *Invest. Urol.* **16,** 186–190.

Randriamampita, C., and Tsian, R. Y. (1993). Emptying of intracellular Ca^{2+} stores releases a novel small messenger that stimulates Ca^{2+} influx. *Nature* **364,** 809–814.

Sakr, W. A., Haas, G. P., Cassin, B. F., Pontes, J. E., and Crissman, J. D. (1993). The frequency of carcinoma and intraepithelial neoplasia of the prostate in young male patients. *J. Urol.* **150,** 379–385.

Scardino, P. T., Weaver, R., and Hudson, M. A. (1992). Early detection of prostate cancer. *Hum. Pathol.* **23,** 211–222.

Seidman, H., Mushinski, M. H., and Geib, S. K. (1985). Probabilities of eventually developing or dying of cancer—United States, 1985. *Ca—Cancer J. Clin.* **35,** 36–56.

Shain, S. A., McCullough, B., and Segaloff, A. (1975). Spontaneous adenocarcinoma of the ventral prostate of aged AXC rats. *J. Natl. Cancer Inst. (U.S.)* **55,** 177–180.

Shain, S. A., McCullough, B., and Nitchuk, W. N. (1979). Primary and transplantable adenocarcinomas of the AXC rat ventral prostate gland: Morphologic characterization and examination of C19-steroid metabolism of early-passage tumors. *J. Natl. Cancer Inst. (U.S.)* **62,** 313–322.

Silber, S. J., Balmacoda, J., and Borrero, C. (1988). Pregnancy with sperm aspiration from the proximal head of the epididymis: a new treatment for congenital absence of the vas deferens. *Fertil. Steril.* **50,** 525–530.

Smith, E. R., and Hagopian, M. (1981). Uptake and secretion of carcinogenic chemicals by the dog and rat prostate. *In* "The Prostate Cell: Structure and Function" (G. P. Murphy, A. A. Sandbert, and J. P. Karr, eds.), Part B, pp. 131–163. Liss, New York.

Sondik, E. (1988). Incidence, survival and mortality trends in prostate cancer. *In* "A Multidisciplinary Analysis of Controversies in the Management of Prostate Cancer" (D. S. Coffey, M. I. Resnick, F. A. Dorr, and J. P. Karr, eds.), pp. 1–16. Plenum, New York.

Stanbridge, E. J. (1990). Human tumor suppressor genes. *Annu. Rev. Genet.* **24**, 615–657.

Steinberg, G. D., and Isaacs, J. T. (1993). Pharmacological approaches to the management of metastatic prostatic cancer. *In* "Cancer Chemotherapy" (J. A. Hickman, and T. R. Tritton, eds.), pp. 322–343. Blackwell Scientific Publications, London.

Stiens, R., and Helpap (1981). Regressive changes in the prostate after castration. A study using histology, morphometries and autoradiography with special reference to apoptosis. *Pathol. Res. Pract.* **172**, 73–87.

Story, M. T., Livingston, B., Baeten, L., *et al.* (1989). Cultured human prostate-derived fibroblasts produce a factor that stimulate their growth with properties indistinguishable from basic fibroblast growth factor. *Prostate* **15**, 355–365.

Sumiya, H., Masai, M., Akimoto, S., Yatani, R., and Schimazaki, J. (1990). Histochemical examination of the expression of ras p21 protein and R-1881-binding protein in human prostatic cancers. *Eur. J. Cancer* **26**, 786–789.

Thastrup, O., Cullen, P. J., Drøbak, B. K., Hanley, M. R., and Dawson, A. P. (1990). Thapsigargin, a tumor promoter, discharges intracellular Ca^{2+} stores by specific inhibition of the endoplasmic reticulum Ca^{2+}-ATPase. *Proc. Natl. Acad. Sci. U.S.A.* **87**, 2466–2470.

Thigpen, A. E., Davis, D. L., Milatovich, A., Mendonca, B. D., Imperato-McGinley, J., Griffin, J., Franke, U., Wilson, J. D., and Russel, D. W. (1992). Molecular genetics of steroid 5α-reductase 2 deficiency. *J. Clin. Invest.* **90**, 799–809.

Troll, W., Klassan, A., and Janoff, A. (1970). Tumorigenesis in mouse skin. Inhibition by synthetic inhibitors of proteases. *Science* **169**, 1211–1213.

Umbas, R., Schalken, J. A., Aalders, T. W., Carter, B. S., Karthaus, H. F. M., Schaafsman, H. E., Debruyne, F. M. J., and Isaacs, W. B. (1992). Expression of the cellular adhesion molecule E-caderin is reduced or absent in high-grade prostatic cancer. *Cancer Res.* **52**, 5104–5109.

Unbeless, J., Dano, J. K., Kellerman, G. M., and Reich, E. (1974). Fibrinolysis associated with oncogeneic transformation. *J. Biol. Chem.* **249**, 4295–4305.

Verhagen, A. P. M., Aadlers, T. W., Ramaekers, F. C. S., Debruyne, F. M. J., and Schalken, J. A. (1988). Differential expression of keratins in the basal and luminal compartments of rat prostatic epithelium during degeneration and regeneration. *Prostate* **13**, 25–38.

Verhagen, A. P. M., Ramaekers, C. S. F., Aadlers, T. W., Schaafsma, H. E., Debruyne, F. M. J., and Schalken, J. A. (1992). Colocalization of basal and luminal cell-type cytokeratins in human prostate cancer. *Cancer Res.* **52**, 6182–6187.

Viola, M., Fromowitz, F., Oravez, S., Deb, S., Finkel, G., Lundy, L., Hard, P., Thor, A., and Schlom, J. (1986). Expression of ras oncogene p21 in prostate cancer. *N. Engl. J. Med.* **314**, 133–137.

Walsh, P. C., and Jewett, H. J. (1980). Radical surgery for prostatic cancer. *Cancer (Philadelphia)* **45**, 1906–1911.

Ward, J. M., Reznik, G., Stinson, G. F., Lattuada, C. P., Longfellow, D. A., and Cameron, T. P. (1980). Histogenesis and morphology of naturally occurring prostatic carcinoma in the ACI/Seg Hap BR rat. *Lab. Invest.* **43**, 517–522.

Walensky, L. D., Coffey, D. S., Chen, T-H, Wu, T-C., and Pasternack, G. R. (1993). A novel M_2 32,000 nuclear phosphoprotein is selectively expressed in cells competent for self-renewal. *Cancer Res.* **52**, 4720–4726.

Weinberg, R. A. (1989). Oncology, antioncogene, and the basis of multistep carcinogenesis. *Cancer Res.* **49**, 3713–3721.

Wenderoth, U. K., George, F. W., and Wilson, J. D. (1983). The effect of a 5α-reductase inhibitor on androgen-mediated growth of the dog prostate. *Endocrinology (Baltimore)* **113**, 569–573.

Wigler, M., and Weinstein, I. B. (1976). Tumor promoters induce plasminogen activator. *Nature (London)* **259,** 1232–1234.

Wilson, J. D. (1984). The endocrine control of sexual differentiation. *Harvey Lect.* **79,** 145–172.

Wyllie, A. H., Kerr, J. F. R., and Currie, A. R. (1980). Cell death: The significance of apoptosis. *Int. Rev. Cytol.* **68,** 251–306.

Yan, G., Fukabori,Y., Nikolaropoulos, S., Wang, F., and McKeehan, W. L. (1992). Heparin-binding keratinocyte growth factor is a candidate stromal to epithelial cell andromedin. *Mol. Endocrinol.* **6,** 2123–2128.

Young, J. L., Percey, C. L., and Asire, A. J. (1981). Surveillance, epidemiology and end results: Incidence and mortality data, 1973–1977. *Natl. Cancer Inst. Monogr.* **57,** DHHS Publ. No. (NIH) 81-2330.

Zuckerman, S. (1936). The endocrine control of the prostate. *Proc. R. Soc. Med.* **29,** 1557–1567.

Index

A